Automatic Transmissions & Transaxles

by

Chris Johanson
ASE Certified Master Technician

James E. Duffy
Automotive Writer

Publisher
The Goodheart-Willcox Company, Inc.
Tinley Park, Illinois

Cover photo courtesy of ZF Transmission Group, Saarbrucken, Germany

The Goodheart-Willcox Company, Inc. Brand Disclaimer: Brand names, company names, and illustrations for products and services included in this text are provided for educational purposes only, and do not represent or imply endorsement or recommendation by the author or the publisher.

Goodheart-Willcox Publisher Safety Notice: The reader is expressly advised to carefully read, understand, and apply all safety precautions and warnings described in this book or that might also be indicated in undertaking the activities and exercises described herein to minimize risk of personal injury or injury to others. Common sense and good judgment should also be exercised and applied to help avoid all potential hazards. The reader should always refer to the appropriate manufacturer's technical information, directions, and recommendations; then proceed with care to follow specific equipment operating instructions. The reader should understand these notices and cautions are not exhaustive.

The publisher makes no warranty or representation whatsoever, either expressed or implied, including but not limited to equipment, procedures, and applications described or referred to herein, their quality, performance, merchantability, or fitness for a particular purpose. The publisher assumes no responsibility for any changes, errors, or omissions in this book. The publisher specifically disclaims any liability whatsoever, including any direct, indirect, incidental, consequential, special, or exemplary damages resulting, in whole or in part, from the reader's use or reliance upon the information, instructions, procedures, warnings, cautions, applications or other matter contained in this book. The publisher assumes no responsibility for the activities of the reader.

Library of Congress Cataloging-in-Publication Data

Johanson, Chris.
    Automatic transmissions & transaxles / by Chris Johanson, James E. Duffy.
      p. cm.
    Includes index.
    ISBN 1-59070-426-6
     1. Automobiles—Transmission devices, Automatic—Maintenance and repair. 2. Automobiles—Transaxles—Maintenance and repair. I. Title: Automatic transmissions and transaxles II. Duffy, James E. III. Title.

TL263.J65.2005
629.2'446'0288—dc22

2004059024

2

Introduction

At one time, fixing an automatic transmission involved taking it apart, cleaning and inspecting the internal components, and replacing worn parts. While this took skill and attention to detail, it was a relatively simple procedure. Since there were only a few types of automatic transmissions, technicians could become proficient in a short period of time. Many automatic transmission technicians learned by experience, making mistakes and learning from them.

Today, it is almost impossible to get started on a career in automatic transmission and transaxle repair without formal training and a great deal of study. The modern technician must deal with about 75 different transmissions and transaxles. The high cost of parts and labor prevent the technician from using trial and error to diagnose a troublesome transmission. Today's technicians must develop the same troubleshooting and repair skills needed in the past, but must also learn to diagnose complex electrical and electronic problems.

Automatic Transmissions & Transaxles provides the latest information on the design, construction, diagnosis, service, and repair of the automatic transmissions and transaxles used in late-model vehicles. This text has been carefully designed so all pertinent components are identified and operating principles are fully explained before troubleshooting, service, and repair are discussed.

Each chapter of this text begins with a list of objectives that provide focus for the chapter. A list of important technical terms is also presented at the beginning of each chapter. Be sure to look for these terms as you read through the chapters. They will be printed in ***bold italic type*** when first used in the body of the chapter and will be defined when introduced. Figure references are printed in **bold type** for ease of identification. Warnings and cautions are provided when the risk of injury or property damage is present. Notes are used to highlight information that supplements the material presented in the body of the text.

ASE certification is becoming increasingly important. In a few years, it may be difficult to get a job as an automotive technician if you are not certified. *Automatic Transmissions & Transaxles* is a valuable resource for those preparing for ASE Certification Test A2, Automatic Transmission/Transaxle. The content is correlated to the ASE/NATEF Task List. ASE-type questions at the end of each chapter help students prepare for the questions encountered on the ASE tests. In addition, Chapter 20 is devoted to information about preparing for and taking ASE tests.

A *Workbook for Automatic Transmissions & Transaxles* is available to complement this textbook. The workbook contains additional test questions, as well as jobs to help you develop your hands-on service techniques. The chapters in the workbook are correlated to those in the textbook.

By choosing *Automatic Transmissions & Transaxles*, you have made a good start in preparing for a career in the field of automotive service and repair.

Chris Johanson
James E. Duffy

How to Use the Color Key

Colors are used throughout *Automatic Transmissions & Transaxles* to help illustrate different transmission and transaxle components. Color is also used to help clarify hydraulic flow diagrams. The following key shows what each color represents.

Transmission and Transaxle Components

Torque converter, gears, shafts, and related components

Holding members (bands, clutches, etc.)

Hydraulic control components (valve body, oil pump, hydraulic valves, servos, accumulators, etc.)

Electronic control components (sensors, solenoids, etc.)

Case, housings, and oil pans

General Illustrations

Primary component or feature

Secondary components or feature

General components

General components

Hydraulic Flow Diagrams

The color keys used for the hydraulic flow diagrams found in this textbook are indicated on the individual diagrams. The keys will vary from one diagram to another, depending on the transmission or transaxle manufacturer. Therefore, be sure to check the color key on each diagram before studying the illustration.

Table of Contents

Chapter 1

Introduction to Automatic Transmissions and Transaxles

After studying this chapter, you will be able to:
- ❏ Explain the basic purpose of a transmission.
- ❏ Identify the differences between manual and automatic transmissions.
- ❏ Describe the differences between automatic transmissions and automatic transaxles.
- ❏ Identify major automatic transmission and transaxle components.
- ❏ Explain the basic operation of an automatic transmission or transaxle.
- ❏ Trace the development of modern automatic transmissions and transaxles.

Technical Terms

Gear ratio	Planetary gears	Gaskets
Reduction gear	Holding members	Seals
Direct drive	Case	Manual linkage
Overdrive	Oil pans	Throttle linkage
CV axles	Bushings	Transmission fluid
Continuously variable transaxles (CVTs)	Ball bearings	Transmission fluid cooler
Fluid couplings	Hydraulic pump	Transmission fluid filter
Torque converters	Pressure regulator	Infinitely variable transmission (IVT)
Input shafts	Hydraulic control system	Toroidal transmission
Output shafts	Electronic control system	

Introduction

Automatic transmissions and transaxles have been used for more than 60 years. They have been consistently modified and improved, evolving from early inefficient designs to the smooth-shifting, efficient units of today. Most modern transmissions and transaxles are controlled by an onboard computer and provide almost the same fuel economy as manual models.

To service late-model automatic transmissions or transaxles, the technician must possess considerable knowledge and skill. This chapter will introduce you to the fundamentals of automatic transmissions and transaxles. The basic principles covered here will be expanded upon in later chapters.

The Purpose of Transmissions

All transmissions, whether manual or automatic, have the same basic purposes:

❑ To transmit power from the engine to the drive wheels when necessary.

❑ To disconnect the running engine from the drive wheels during gear changes and when the vehicle is not moving.

❑ To reverse the direction of power flow when the vehicle must be backed up.

❑ To multiply engine torque as needed.

In simplest terms, a transmission modifies engine torque and speed to match the vehicle's needs. For example, moving a vehicle from a stop requires a great deal of engine torque, or turning force. At low speeds, however, an engine produces relatively little torque. The transmission must multiply engine torque to get the vehicle moving. It does this by reducing speed to increase torque.

The relationship of the speed of the transmission's input shaft to the speed of its output shaft is called the *gear ratio*. The transmission uses a set of at least two gears that cause the output shaft speed to be much lower than the input shaft speed. This set of gears is called a *reduction gear*. **Figure 1-1** shows a simple reduction gear.

At higher speeds, the vehicle does not require as much engine torque to keep it moving. The engine would be turning very fast if the transmission output speed remained slower than the input speed. High engine speed will cause poor fuel economy and rapid engine wear. Therefore, the transmission must be shifted into successively higher gears as vehicle speed increases. Shifting into higher gears changes gear ratios, so the speed of the output shaft approaches and eventually equals or exceeds the speed of the input shaft.

Most modern transmissions have at least four forward gears, with the highest gear being either direct drive or overdrive. A *direct drive* gear causes the input and output

shafts to turn at the same speed. The *overdrive* gear causes the output shaft to turn faster than the input shaft, **Figure 1-2.** Overdrive allows the engine to turn at a relatively slow speed, increasing fuel economy and reducing engine wear.

Both manual and automatic transmissions can accomplish any of the previously mentioned jobs. There are many similarities between manual and automatic transmissions. All transmissions have a way of keeping the engine from stalling when the vehicle is stopped, all use gears and shafts to obtain different ratios, and all have a way to reverse the direction of vehicle travel. However, with a manual transmission, the gear selection decision must be made by the vehicle's operator. The driver slides the transmission gears in and out of engagement using a gearshift lever. The gears are meshed in different combinations to achieve the desired gear ratios. The driver must also operate a manual clutch to connect and disconnect the engine from the transmission when stopping or changing gears.

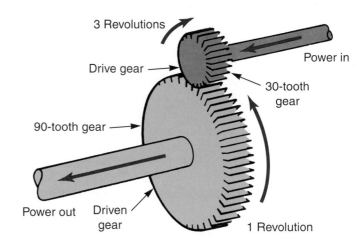

Figure 1-1. *Gear reduction produces the torque needed to move the vehicle from a stop. In this illustration, a 30-tooth gear is turning a 90-tooth gear. The 30-tooth gear must make three revolutions to turn the 90-tooth gear once. This multiplies torque three times but cuts speed to one-third.*

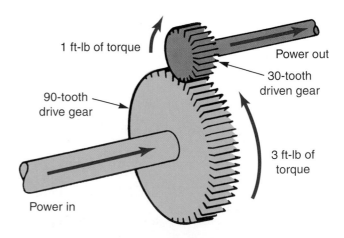

Figure 1-2. *In this example, overdrive is accomplished by using the 90-tooth gear to turn the 30-tooth gear. This increases speed three times but cuts torque to one-third.*

With automatic transmissions, on the other hand, gear selection decisions are made by an automatic control system. Instead of a manual clutch to connect and disconnect the engine from the transmission, automatic transmissions use fluid couplings or torque converters to transfer power from the engine to the transmission.

Automatic transmissions use planetary gearsets, which do not slide in and out of engagement. In operation, one of the gears in the gearset is locked in place. The remaining unlocked gears are driven by engine power and comprise the input and output. Different gear ratios are achieved by different combinations of locked and unlocked gears. The gears are operated by holding members called clutches and bands. The clutches and bands are controlled by a hydraulic control system. Late-model automatic transmissions have hydraulic systems controlled by on-board computers. Vehicles with automatic transmissions are easier to drive than those with manual transmissions. They are also more durable for heavy-duty operation, such as trailer towing. The major differences between manual and automatic transmissions/transaxles are shown in **Figure 1-3.**

The ideal transmission will transmit engine power with no slipping. *Slipping* can be defined as failure to transmit all engine power to the other drive train components. In other words, a slipping transmission will lose both speed and torque between its input and output shafts.

Early automatic transmissions were so inefficient and slipped so much that they were called "slush boxes." Modern automatics are efficient and smooth. Except for first and reverse gears, modern automatic transmissions permit no slippage. They transmit as much engine power as manual transmissions. It is now possible for an automatic transmission to be more efficient than a manual. With an automatic transmission, there is no need to release the accelerator pedal during shifts, and then reaccelerate to maintain vehicle speed.

Transmissions and Transaxles

Until about 20 years ago, nearly all vehicles had a rear-wheel drive arrangement that used a transmission to transfer power to the rest of the driveline. The rear-wheel drive transmission transmits power in a straight line, from the front of the vehicle to the back. The differential and the final drive assembly are contained in a separate housing at the rear axle.

Today, most automobiles use front-wheel drive systems equipped with transaxles. Transmissions and transaxles perform the same function. The major differences include the arrangement of the parts and the fact that the differential and the final drive assembly (sometimes called the ring-and-pinion assembly) are an integral part of the transaxle. Transaxles have two output shafts, one for each wheel. These shafts are attached to the **CV axles.** Engine power is transmitted sideways through a chain or gears at some point in the transaxle.

The advantages of transaxles include reduced weight and increased fuel economy. On trucks, however, weight and fuel economy are less of a factor than durability. The rear-wheel drive train is usually used on these vehicles. Therefore, it is important that both transmissions and transaxles be understood completely.

Four-wheel drive vehicles have a transfer case attached to the rear of the transmission. The transfer case sends power to the front wheels. **Figure 1-4** illustrates the layouts of modern rear-wheel drive vehicles with automatic transmissions and front-wheel drive vehicles with automatic transaxles.

Automatic Transmission and Transaxle Development

The modern automatic transmission was not the result of a single invention. Some components used in automatic transmissions were developed long before the automobile itself. Planetary gear principles were known during the time of the Roman Empire and eventually appeared on the Ford Model T. Fluid couplings were used to drive machinery in 19th century mills. The first automotive fluid couplings were used on English cars in the 1920s and on Chrysler vehicles in the mid 1930s.

The 1938 Oldsmobile is widely considered the first car to have an automatic transmission. These early Oldsmobile Hydra-Matics had planetary gears operated by

Transmission Function	Manual Transmission	Automatic Transmission
Engaging and disengaging engine drive wheels.	Driver operated clutch.	Fluid coupling or torque converter.
Reversing directions.	Sliding gears.	Planetary gears.
Changing gear ratios to match vehicle speed.	Sliding gears operated by driver.	Planetary gears operated by a hydraulic control system.

Figure 1-3. *Although their basic function is the same, there are differences between automatic transmissions and manual transmissions.*

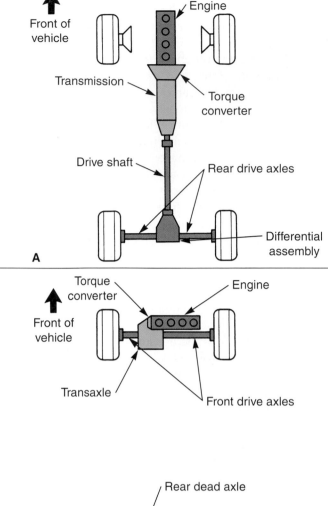

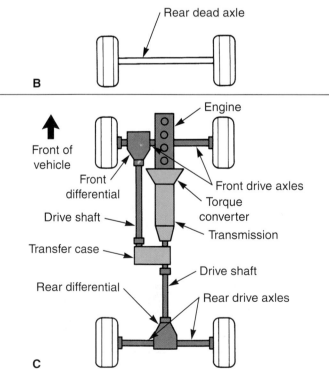

Figure 1-4. *Various automotive drive trains. A—Rear-wheel drive vehicle with a longitudinal engine and a transmission. B—Front-wheel drive vehicle with a transverse engine and a transaxle. C—Four-wheel drive vehicle with a transmission and a transfer case.*

a hydraulic control system. However, they used a manual clutch in place of a fluid coupling. The clutch was used to engage the transmission when the vehicle was first started, after which the hydraulic system made all the shifts. **Figure 1-5** shows this early design. In 1939 (1940 models), Oldsmobile introduced an updated version of this transmission, replacing the clutch with a fluid coupling. Cadillac began using this transmission in 1940. This was the first fully automatic hydraulic transmission. See **Figure 1-6.**

The first transmission to use a torque converter instead of a fluid coupling was the 1948 Buick Dynaflow, **Figure 1-7.** This transmission relied entirely on a complex torque converter with multiple stators and turbines for torque multiplication. Although the torque converter provided a much more efficient transfer of power than a fluid coupling, the Dynaflow had no gear changes, and the planetary gears were always in direct drive. The planetary gears were also used for manual low and reverse. The Dynaflow was smooth but extremely inefficient.

The first transmission that resembled modern transmissions was designed by Ford and Borg-Warner and offered in 1950. It used a torque converter and a hydraulic control system that automatically changed gears. The first models were two-speed types, while later designs had three speeds. A similar Borg-Warner transmission appeared on Studebakers. The Studebaker transmission, as well as one offered by Packard, contained the first version of the lockup torque converter, which eliminated slippage in certain gears. See **Figure 1-8.** The Studebaker automatic was the first to use a one-way clutch to obtain different gear ratios.

At first, all automatic transmissions had cast iron cases. Aluminum bell housings and tailshaft housings were used on some Ford and Chrysler transmissions during the 1950s. The first transmission to use an all-aluminum case was the Chevrolet Turboglide installed in the 1958 models, **Figure 1-9.** The Chrysler Torqueflite followed in 1960. All automatics were being designed with aluminum cases by 1965. Around this time, simplified gear trains were introduced, some with only two forward gears. A late-1960s transmission might have fewer than half the parts of a comparable model produced ten years earlier.

The Chrysler Corporation reintroduced the lockup torque converter in 1977. It was controlled by hydraulic pressure, and the lockup clutch applied only in third gear. Later lockup torque converters were operated electrically or electronically. Lockup clutches were applied in all gears but first and reverse.

In the early 1960s, a few vehicles, such as the Corvair, Volkswagen, and Porsche, used rear engines and were equipped with transaxles. A few imported front-wheel drive cars had transaxles. However, the front-wheel drive automatic transaxle was unknown in the United States until the introduction of the 1966 Oldsmobile Toronado. The Toronado's engine was mounted longitudinally (facing forward), and its transaxle used a conventional cast iron differential bolted to the transmission case. In 1976,

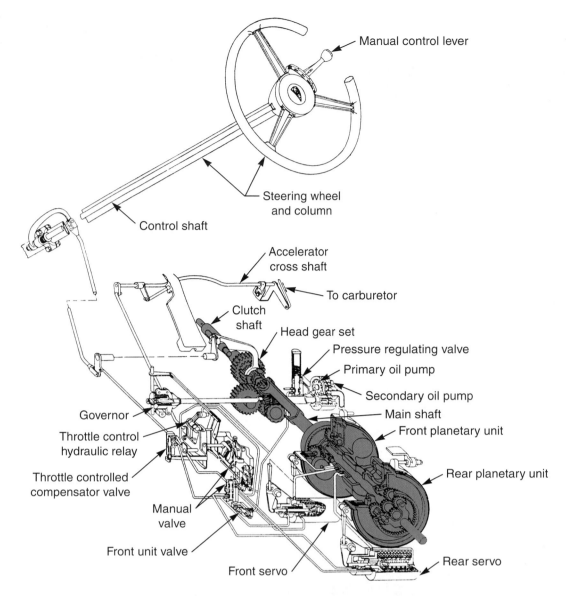

Figure 1-5. *Diagram of an early Oldsmobile automatic transmission. A clutch was used to engage the transmission when the vehicle was started from a stop.*

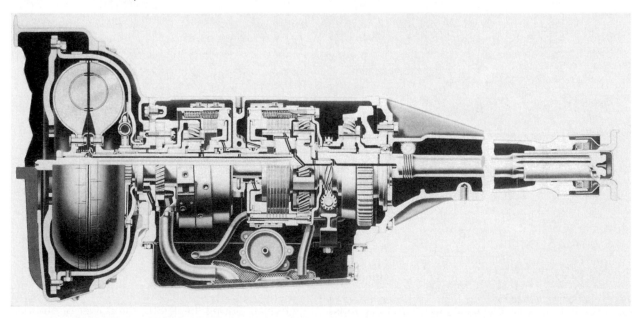

Figure 1-6. *Cross-section of a Cadillac Hydra-Matic transmission. This transmission used a fluid coupling instead of a clutch.*

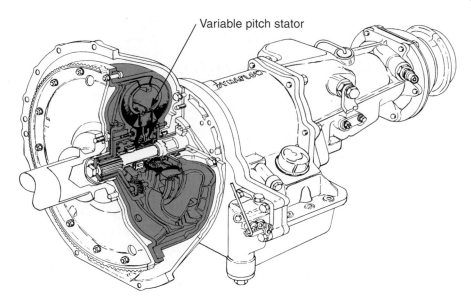

Variable pitch stator

Figure 1-7. *Dynaflow transmission. Note the variable-pitch stator. Opening the throttle changed the angle of the stator blades for increased torque during rapid acceleration.*

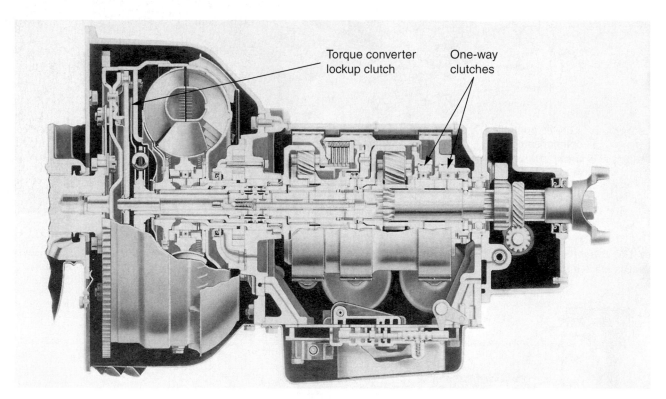

Torque converter lockup clutch   One-way clutches

Figure 1-8. *Cross-section of a Studebaker automatic transmission.*

Honda and Nissan introduced the first transverse (side facing) engine and transaxle combinations sold in the U.S. By 1980, all domestic manufacturers were producing front-wheel drive cars with transaxles. At present, almost all passenger cars use front-wheel drive and a transaxle. Many modern transaxles use a belt and pulley mechanism to change gear ratios. These transaxles are called ***continuously variable transaxles***, or ***CVTs***.

The first attempt to use electricity to control the transmission was made in 1963, with the introduction of an electric passing gear solenoid on the Pontiac Tempest, **Figure 1-10.**

Electric passing gears were used on other transmissions during the 1960s, and some electrically operated lockup torque converters were used during the 1970s. When the first computerized engine controls were introduced, the transmission was often equipped with pressure switches to tell the computer when the transmission was in high gear. During the 1980s, the power of on-board computers increased to the point that they could apply the converter clutch. Later they were able to control transmission shifting and internal pressures. Early electronic transmissions were simply updated hydraulic models, with only a few functions being performed by the computer and the output solenoids. Modern transmissions and transaxles are fully controlled by the computer.

Automatic Transmission and Transaxle Components

Automatic transmissions are made of many separate components and systems. See **Figure 1-11** and **1-12.** Some of the most important automatic transmission and transaxle components are discussed in the following sections. These components will be covered in more detail in later chapters.

Fluid Coupling/Torque Converter

Fluid couplings and *torque converters* are fluid-filled units installed between the engine's crankshaft and the transmission. They consist of two sets of blades. One set of blades is driven by the engine, and the other set of blades is connected to the transmission's input shaft. The blade set connected to the engine is called the *impeller,* and the blade set connected to the input shaft is called the *turbine.* See **Figure 1-13.** A hydraulic pump in the transmission forces fluid into the converter. Inside the converter, the fluid is spun by the impeller blades. As the fluid is thrown from the impeller blades, it strikes the turbine blades. See **Figure 1-14.** Power is transmitted from the impeller to the turbine through the fluid. When the vehicle is stopped, the fluid from the impeller continues to strike the turbine, but the fluid allows enough slippage between the impeller and the turbine to prevent engine stalling.

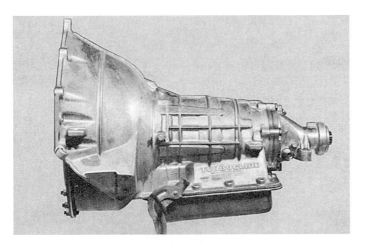

Figure 1-9. *1958 Chevrolet Turboglide transmission. This transmission used an all-aluminum case.*

Figure 1-10. *Passing gear solenoid used on a 1963 Pontiac Tempest.*

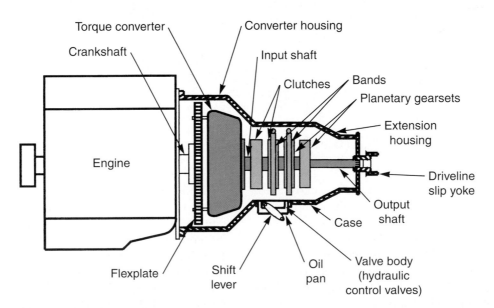

Figure 1-11. *This diagram shows the major components of an automatic transmission. Note how the components fit in relation to each other.*

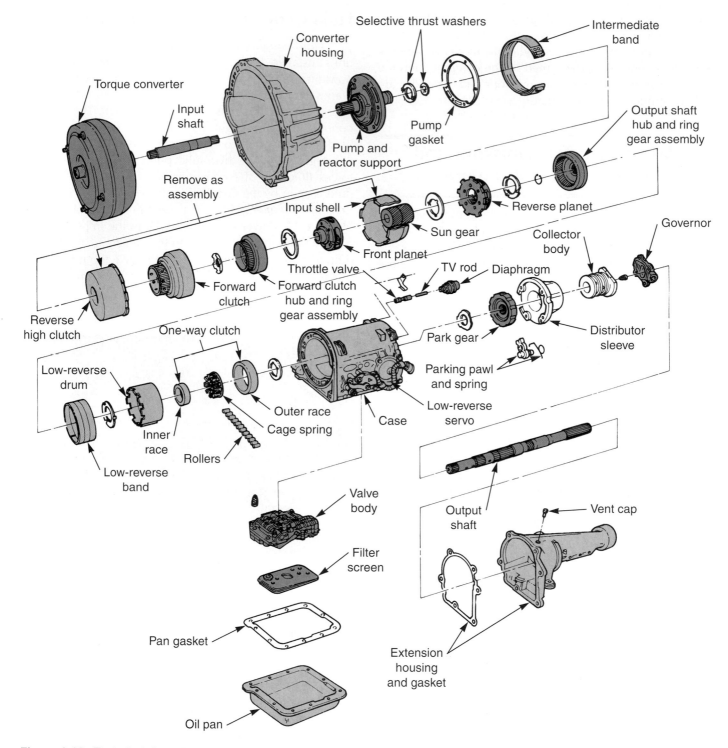

Figure 1-12. *Exploded view of a typical automatic transmission. (Ford)*

The fluid coupling was widely used on early automatic transmissions. Fluid couplings, however, slip excessively and are very inefficient at transmitting power at low speeds. Therefore, fluid couplings have been replaced by torque converters. In addition to an impeller and a turbine, a torque converter uses a device called a *stator*. The stator redirects the fluid to reduce slipping, **Figure 1-15.** All transmissions made since the 1960s use torque converters.

Modern transmissions are equipped with *lockup torque converters.* These torque converters are equipped with an internal clutch called a *converter lockup clutch*. The converter lockup clutch locks the transmission input shaft to the converter cover. The clutch is applied to lock the turbine to the cover. Since the input shaft is attached to the turbine, slippage is eliminated. The lockup clutch is disengaged at low speeds to prevent engine stalling.

Input and Output Shafts

All transmissions have **input shafts** and **output shafts**. For strength, the shafts are made of heat-treated steel. The shafts are splined to attach to other parts of the transmission for power transfer. Every transmission must have separate input and output shafts, **Figure 1-16** and **1-17.** Most

automatic transaxles have at least one hollow shaft. The solid shaft can turn inside the hollow shaft to permit power transfer to each side of the vehicle. See **Figure 1-18.**

Planetary Gears

Planetary gears are used in all automatic transmissions and transaxles. The term planetary comes from the resemblance of the gear assembly to the solar system. The basic planetary gear consists of a central **sun gear** surrounded by **planet gears** that are housed in a **planet carrier.** A **ring gear** with internal teeth surrounds the sun and planet gears. **Figure 1-19** shows the main parts of a planetary gear assembly. The advantage of the planetary gear is that the gears remain in mesh at all times. This prevents gear clash when shifting. Different gear ratios can be obtained by holding or driving different parts of the planet gear assembly. Simple planet gears can be combined into more complex units, such as the one shown in **Figure 1-20.** Complex planetary gear sets are used to obtain many gear ratios.

Holding Members

Holding members are the units that hold or drive the various parts of the planetary gear assembly to drive the vehicle. Holding members consist of friction material, which is similar to that used on manual clutches or brake shoes, bonded to a metal backing. **Clutches** and **bands** are the most common holding members. Clutches are a series of flat, ring-shaped plates.

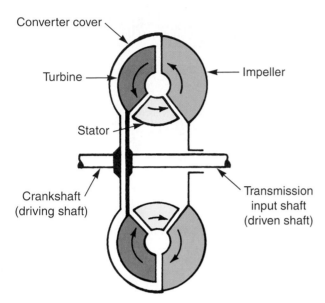

Figure 1-13. *This sectional view of a torque converter shows the three main parts: the impeller, the turbine, and the stator. All modern converters have these three parts arranged in this way. Modern converters also contain an internal clutch to eliminate slippage.*

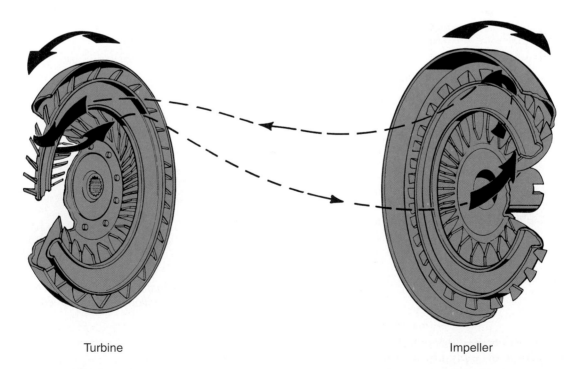

Turbine   Impeller

Figure 1-14. *Engine power flows from the pump, or impeller, to the turbine. The impeller causes the fluid to rotate. Then, the fluid causes the turbine to rotate. Following the arrows, you will notice that the fluid returning from the turbine strikes the impeller in the opposite direction of impeller rotation. (General Motors)*

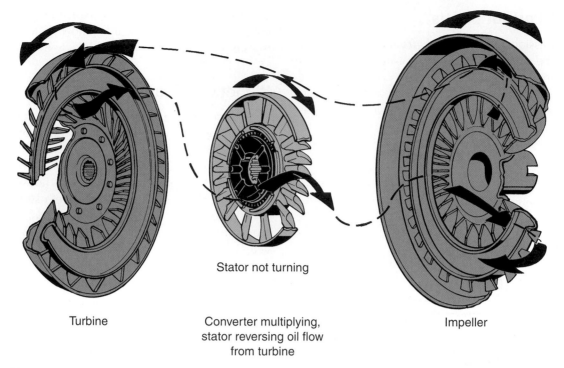

Turbine   Converter multiplying,   Impeller
                                stator reversing oil flow
                                    from turbine

Stator not turning

Figure 1-15. *The stator reverses the flow of the fluid leaving the turbine. The fluid then strikes the impeller in the same direction as impeller rotation. This multiplies engine power. (General Motors)*

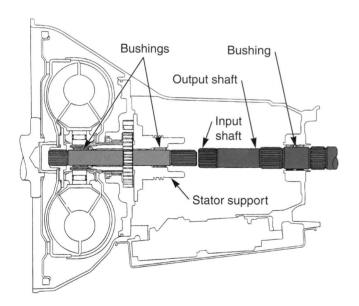

Figure 1-16. *The input and output shafts are installed at the centerline of the transmission, but they are not connected. The input shaft is splined to the converter turbine. The output shaft is splined to the final drive unit. Connecting the input and output shafts is the job of the other transmission components. (Ford)*

Figure 1-17. *Typical output shaft.*

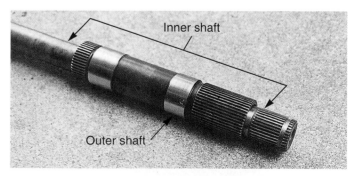

Figure 1-18. *Some shafts are hollow to permit other shafts to rotate inside them. Hollow shafts are commonly used on transaxles.*

They are applied (pressed together) by a hydraulically operated ***clutch piston,*** or ***apply piston.*** The clutch plates are alternately splined, and applying them holds two parts together. One set of plates is lined with friction material and the other set is bare steel. Clutches are commonly used to deliver engine power to the planetary gears, and hold parts of the gear assembly to the case. The entire assembly of clutch plates, the apply piston, and related

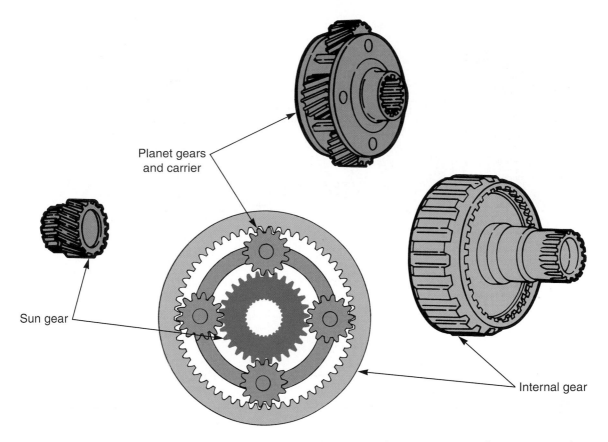

Planet gears
and carrier

Sun gear

Internal gear

Figure 1-19. *This figure illustrates the three main components of a simple planetary gearset: the sun gear, planet gears and planet carrier, and the internal, or ring, gear. (General Motors)*

Figure 1-20. *Simple planetary gears can be combined into multiple, or compound, gearsets to obtain many different gears. This photograph shows a gearset containing two sun gears, two planet carriers, and a ring gear. Many other combinations are possible.*

components is called a ***clutch pack***. A typical clutch pack is shown in **Figure 1-21.**

Bands wrap around ***drums*** (cylindrical transmission parts) to hold them stationary. The band is tightened against the drum by a hydraulic piston called a ***servo***, which is operated by hydraulic pressure. **Figure 1-22** shows a band, servo, and related linkage.

Another type of holding member is the ***one-way clutch***, **Figure 1-23.** A one-way clutch is a mechanical device that allows the central hub to turn in one direction, but causes it to lock up when it tries to turn in the opposite

direction. One-way clutches are always used in combination with other holding members to hold or drive parts of the planetary gearsets.

Case and Housings

The ***case*** is the main support for the other transmission or transaxle parts. It also contains passages to deliver pressurized fluid between various parts of the transmission. Modern cases are made of aluminum that is cast into the proper shape. After casting, the case is machined where necessary to form mating surfaces for the other components. Oil passages are then drilled in the case where necessary. Bushings may be pressed into the case at wear points. **Figure 1-24** shows two common transmission and transaxle cases. On most modern vehicles, the ***housings*** around the torque converter and the output (or tail) shaft are cast as an integral part of the case. When used, separate bell housings (housing around the torque converter) and tailshaft housings are bolted to the case.

Oil Pans

All transmissions or transaxles contain one or more ***oil pans***. The main purpose of the oil pan is as a storage place for extra transmission fluid. Airflow over the oil pan helps remove heat from the fluid. Some oil pans contain a

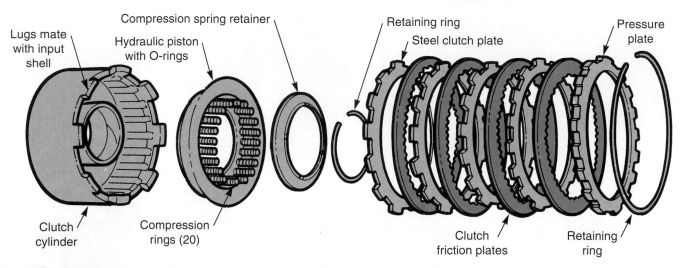

Lugs mate with input shell

Compression spring retainer

Hydraulic piston with O-rings

Retaining ring

Steel clutch plate

Pressure plate

Clutch cylinder

Compression rings (20)

Clutch friction plates

Retaining ring

Figure 1-21. *This exploded view of a multiple disc clutch shows all the components that make up a common clutch pack. Clutch packs can be used to transmit power or to lock parts of the gear train to the case. (Ford)*

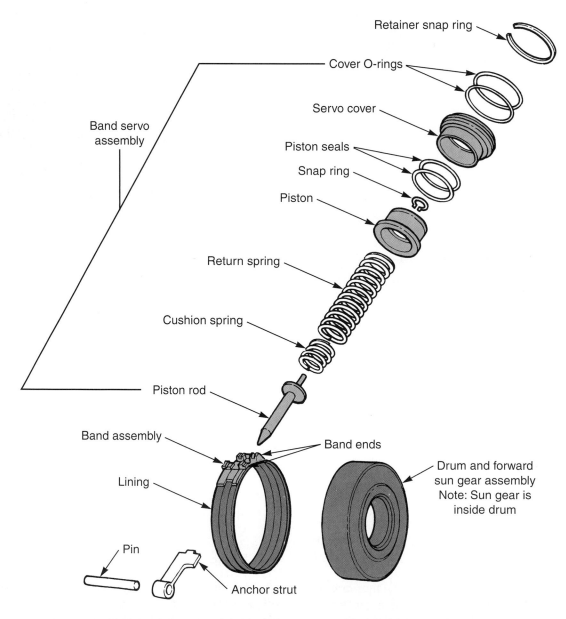

Retainer snap ring

Cover O-rings

Band servo assembly

Servo cover

Piston seals

Snap ring

Piston

Return spring

Cushion spring

Piston rod

Band assembly

Band ends

Drum and forward sun gear assembly Note: Sun gear is inside drum

Lining

Pin

Anchor strut

Figure 1-22. *The band and servo shown here are used to grab a drum and stop it from turning. Bands are always used to lock a part of the gear train to the case. (Subaru)*

magnet to trap the metal particles produced as the transmission components wear. Oil pans are usually made of sheet metal. A few pans are made of cast aluminum.

Bushings and Bearings

Bushings and bearings allow parts to move against each other with minimal friction. *Bushings,* **Figure 1-25,** provide a sliding contact with the moving part and require good lubrication. They are installed where rotating part passes through a stationary part or two rotating parts are in contact with each other.

Ball bearings or *roller bearings* provide a rolling contact for reduced friction. Bearings are usually used where there is a heavy load, such as the output shaft, or an internal part that is subjected to high pressure. *Thrust bearings* are used in other places where parts are rotating in relation to each other under heavy pressures. A *thrust washer* separates moving parts, but it is made of a single piece of flat metal. Some thrust washers are available in different thicknesses and are used to adjust transmission shaft back-and-forth movement, or *endplay.* Typical roller bearings, thrust bearings, and washers are shown in **Figure 1-26.**

Hydraulic Pump and Pressure Regulator

The *hydraulic pump* provides all the hydraulic pressure used in the automatic transmission or transaxle. An extension at the rear of the torque converter drives the pump on transmissions. A separate shaft attached to the converter is often used to drive transaxle oil pumps. Whenever the engine is running, the converter is turning and causing the pump to turn. Therefore, whenever the

Figure 1-23. *A one-way clutch assembly. Note how each roller fits into a ramp on the inside of the outer gear. This design permits the internal roller to rotate in one direction but locks it when it tries to turn the other way.*

Figure 1-24. *Common transmission and transaxle case designs are shown here. Power enters the transmission case at the front and exits at the rear. Power enters the transaxle case at the engine connection and exits in two directions. A few transaxles are designed so power exits in only one direction.*

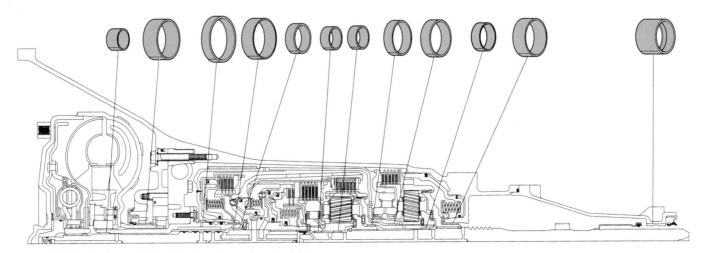

Figure 1-25. *The average transmission or transaxle contains many bushings to support moving parts. Bushings may be installed between the case and a moving shaft, or between two turning parts. (General Motors)*

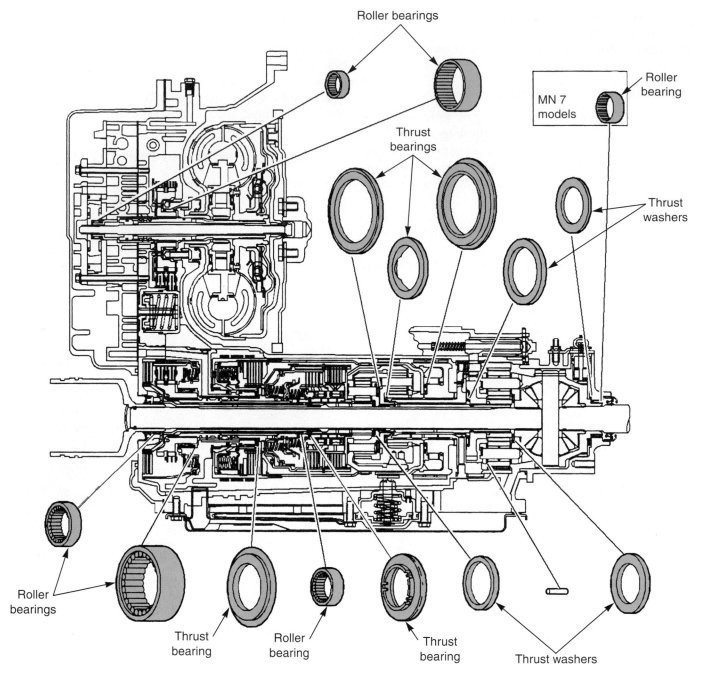

Figure 1-26. *Modern transmissions and transaxles contain many bearings. Some bearings are round types that serve the same purpose as bushings. Others are flat rings that keep moving parts from rubbing against each other. (General Motors)*

engine is running, the pump is producing pressure. There are several types of hydraulic pumps. These will be discussed in a later chapter.

The hydraulic pump draws transmission fluid from the bottom or side oil pan, which is usually called the *sump*. The pump pressurizes the fluid for use by the other hydraulic system components. See **Figure 1-27**. Modern transaxles with two oil pans may have more than one pump. The system shown in **Figure 1-28** contains three pumps. A three-gear scavenger pump removes the fluid from the bottom pan and pumps it into the side pan. The primary and secondary pumps pick up the oil from the side pan and pump it to the other hydraulic system

components. On very old vehicles, a pump was installed on the output shaft and it produced pressure only when the vehicle was moving. Output shaft pumps are no longer used.

The ***pressure regulator*** is installed in the outlet line from the pump and controls overall transmission pressures, usually called line pressures.

If the pump output becomes too great, the pressure regulator valve opens, dumping oil back into the oil pan. Once the pressure returns to normal, the valve closes. A pressure regulator is shown in **Figure 1-29**. The pressure regulator may be installed in the pump housing or in the valve body.

Hydraulic Control System

The **hydraulic control system** is a set of hydraulic parts and passages that performs the following functions:

❏ Applies and releases the holding members to obtain the needed gear ratios at any vehicle speed and throttle position.

❏ Controls system pressures for proper shift feel and long holding member life as loading and acceleration vary.

❏ Keeps transmission fluid flowing to the torque converter, transmission cooler, and lubricating system.

The major component of the hydraulic control system is the valve body, which contains the control valves and related springs. The valve body also contains the manual valve, which is connected by linkage to the shift lever in the passenger compartment. Moving the manual valve directs pressure to other parts of the hydraulic system. The valve body may also contain accumulators, check balls, and spacer plates. A typical valve body is shown in **Figure 1-30.** Other parts of the hydraulic system include governor valves, throttle linkage, band servos, clutch apply pistons, and the passages that connect them. Some of these passages and parts may be located in the transmission case.

Most modern transmissions and transaxles use an **electronic control system** to operate the hydraulic components. An on-board computer processes information from input sensors. It then uses this information to operate

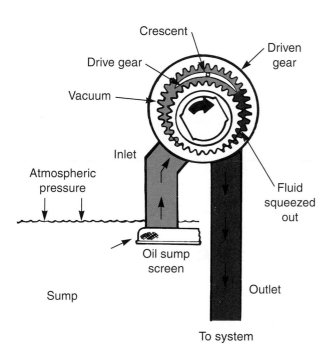

Figure 1-27. *Simplified hydraulic pump. Fluid is drawn in as the gears move apart. It is then carried around by the gear teeth and pressurized as the teeth come together. (Ford)*

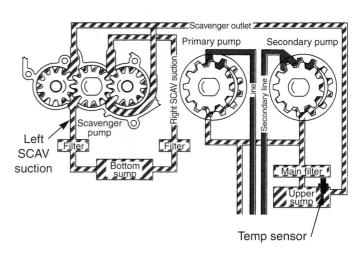

Figure 1-28. *Some transmissions and transaxles use more than one pump. The transaxle oil diagram here shows three pumps: a primary pump, a secondary pump, and a scavenger pump. (General Motors)*

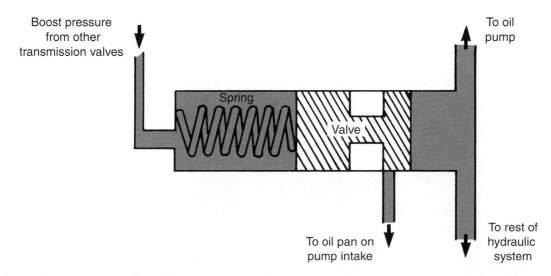

Figure 1-29. *Simplified pressure regulator. If system pressure becomes too great, the valve moves against spring pressure and oil is routed back to the oil pan.*

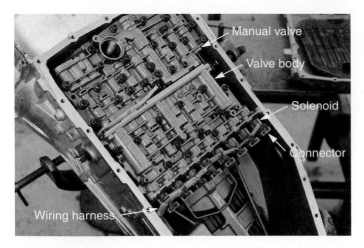

Figure 1-30. *Valve bodies contain the hydraulic valves, check balls, springs, restrictors, screens, and other components needed to control the holding members. Note that this valve body uses electrical solenoids to control transmission operation.*

solenoids and other output devices installed in the transmission or transaxle to control pressure flow through the hydraulic system. See **Figure 1-31. Figure 1-32** illustrates the input and output devices used with a typical electronically controlled transmission. The construction and operation of hydraulic and electronic control components will be discussed in detail in later chapters.

Gaskets and Seals

Gaskets and *seals* are used to keep fluid from leaking out of the transmission or transaxle and to keep pressure from leaking internally. Gaskets are used where major components are joined together. For example, a gasket is used to seal the oil pans to the case. Gaskets are also used in the valve body and where the front pump and extension housing are attached to the case. See **Figure 1-33.**

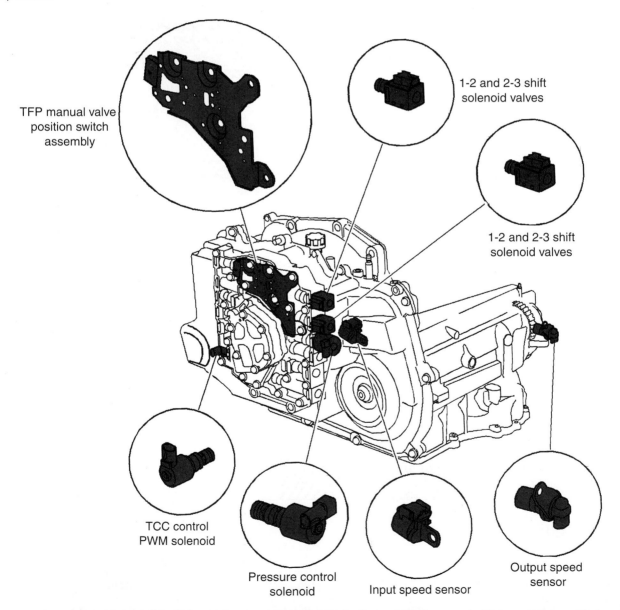

Figure 1-31. *Sensors and solenoids used on a modern electronically controlled transaxle. The sensors provide input signals to the computer, which in turn sends output signals to the solenoids. (General Motors)*

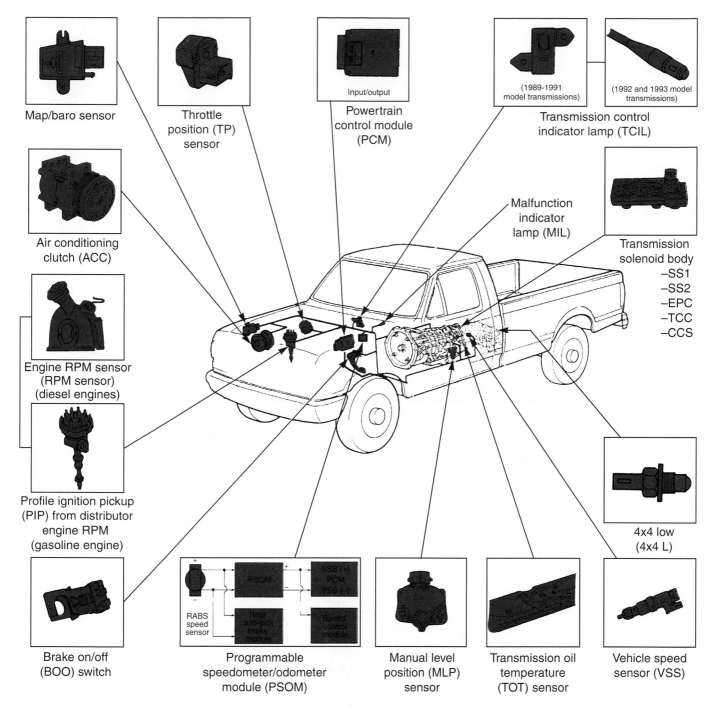

Map/baro sensor

Throttle position (TP) sensor

Powertrain control module (PCM)
Input/output

Transmission control indicator lamp (TCIL)
(1989-1991 model transmissions)
(1992 and 1993 model transmissions)

Air conditioning clutch (ACC)

Malfunction indicator lamp (MIL)

Transmission solenoid body
−SS1
−SS2
−EPC
−TCC
−CCS

Engine RPM sensor (RPM sensor) (diesel engines)

Profile ignition pickup (PIP) from distributor engine RPM (gasoline engine)

4x4 low (4x4 L)

Brake on/off (BOO) switch

PSOM
RABS speed sensor
Rear anti-lock brake module
VSS (+) PCM VSS (−)
Speed control module

Programmable speedometer/odometer module (PSOM)

Manual level position (MLP) sensor

Transmission oil temperature (TOT) sensor

Vehicle speed sensor (VSS)

Figure 1-32. *The computer that controls a modern transmission or transaxle is connected to many input sensors and output devices. In some vehicles, a single computer may be used to control the transmission or transaxle, the engine, and other systems. (Ford)*

Seals are used at moving parts, such as the torque converter, drive shaft, and various internal rotating parts. They are also used as sliding pressure seals at the band servos and clutch apply pistons. **Figure 1-34** shows a servo seal. Some seals are used to seal stationary parts. These seals are usually called *O-rings*.

A *seal ring* is a special type of seal that prevents leaks in pressure passages between parts that rotate in relation to each other, **Figure 1-35.** For example, oil pressure from the valve body may be directed through a stationary support to

a rotating clutch drum. Seal rings keep the pressure from being lost.

Manual Linkage and Throttle Linkage

To provide drive input, two types of linkage are used. The *manual linkage* connects the shift lever to the manual valve inside the transmission or transaxle. The linkage can be a cable or a series of rods and levers. Cables are commonly used on modern vehicles. **Figure 1-36** shows a

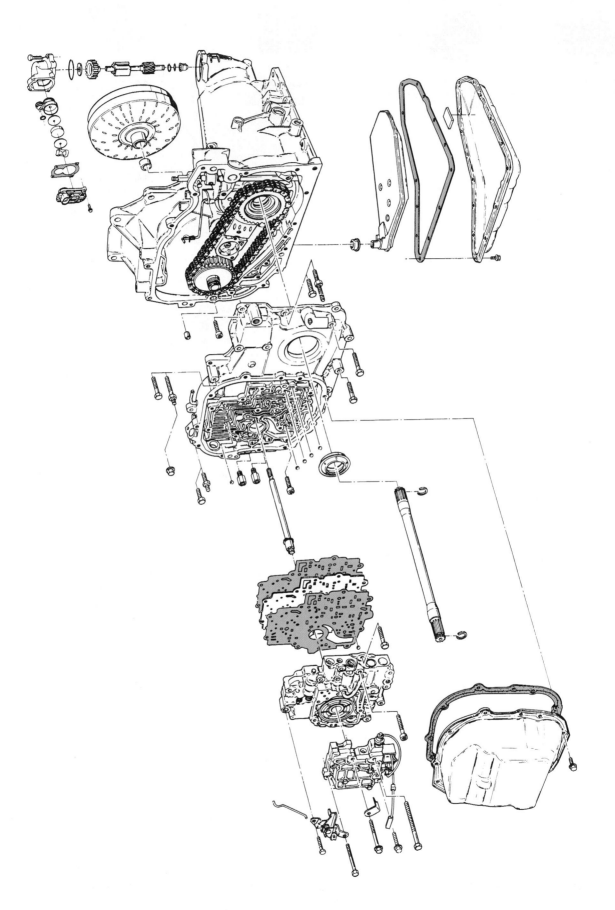

Figure 1-33. *Gaskets are used throughout every transmission and transaxle. Some gaskets seal internal pressure passages. Others keep fluid from leaking out of the unit. (General Motors)*

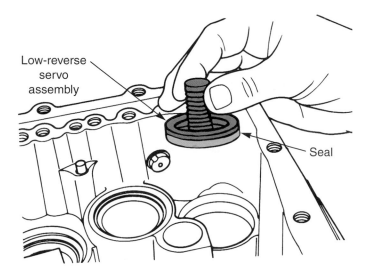

Figure 1-34. *Lip seals are used to seal band servos, such as the one shown here, or to seal clutch pack pistons. Lip seals are moved outward by pressure and provide a tighter seal with less resistance than would be possible with O-rings. (DaimlerChrysler)*

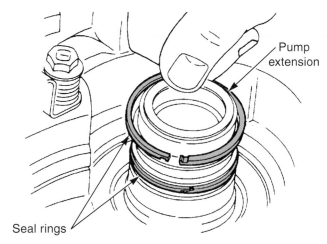

Figure 1-35. *Metal or Teflon® seal rings are used to seal pressure passages. Seal rings usually provide a seal between rotating and stationary parts. (General Motors)*

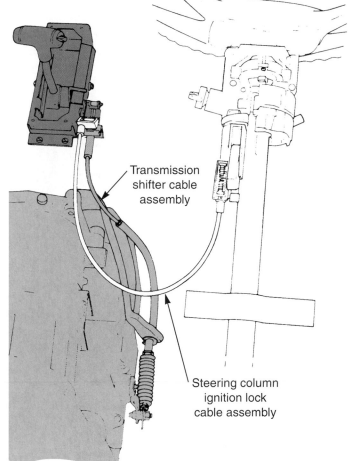

Figure 1-36. *The shifter linkage is used to transfer the driver's shift selection to the transmission or transaxle. It can be a series of rods and links, or, as shown here, a cable arrangement. (Ford)*

commonly used manual linkage arrangement. The shifter is mounted on the steering column or on a center console. In the Park position, the manual linkage operates a park lock inside the transmission or transaxle. The **park lock** is a lever that engages a toothed wheel on the output shaft, **Figure 1-37.** When the park lock is engaged, the vehicle cannot roll.

Throttle linkage connects the engine's throttle plate to the transmission's throttle valve. On some transmissions, throttle linkage is used to apply extra pressure for a forced downshift (or passing gear). On other transmissions and transaxles, the throttle linkage controls all shift speeds. On some older cars, a *vacuum modulator* controlled shift speeds. Vacuum modulators are still used on a few vehicles. Most modern throttle linkage is cable operated, **Figure 1-38.** On

newer vehicles, the computer controls shift speeds and throttle linkage has been eliminated.

Transmission Fluid

Transmission fluid is a combination of petroleum oils and various additives. It is pressurized by the pump and used to operate the hydraulic system. Since the transmission fluid splashes on the holding members, it contains additives that help the holding members to grip when they are applied. Transmission fluid also lubricates gears, bearings, and moving parts. Some transmission parts are lubricated by fluid under pressure. Other components depend on splash lubrication. Transmission fluid also carries away heat and helps seal in pressure.

There are several types of transmission fluid used in modern vehicles. Proper fluid is important to transmission and transaxle operation. At one time, almost all automatic transmissions and transaxles used the same type of fluid. Today, however, many domestic and imported vehicles use special fluids. Never add transmission fluid to any

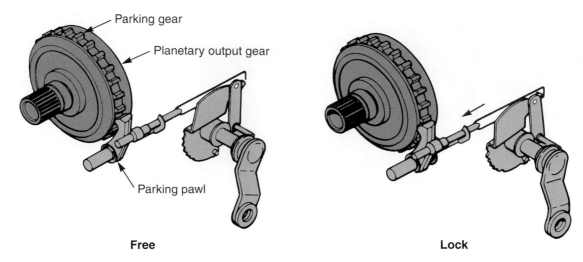

Free   **Lock**

Figure 1-37. *The parking gear linkage is attached to the transmission/transaxle shifter linkage. Moving the linkage causes the parking pawl to engage the parking gear on the output shaft. (Subaru)*

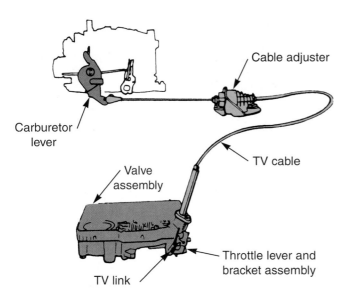

Figure 1-38. *On many older vehicles, throttle linkage was used to control shift speeds and transmission pressures. The linkage was connected between the carburetor or throttle body and a throttle valve in the transmission. Increased throttle opening caused the throttle valve to increase pressures in the transmission and to delay the upshift. (General Motors)*

automatic transmission or transaxle without first finding out what type of fluid should be used. Some manufacturers allow the use of a common fluid, such as Dexron III, if a special friction modifier is added to the fluid.

Transmission Fluid Cooler

Transmission and transaxle operation causes the fluid to get very hot. This heat must be removed to keep the fluid from breaking down and to keep the holding members from becoming so hot that they begin to slip and burn.

To accomplish this, the transmission fluid is pumped through metal lines to a **transmission fluid cooler** in the

vehicle's radiator. In the cooler, the fluid gives up its heat to the engine coolant and then returns to the transmission. On most transmissions, the fluid goes directly from the torque converter to the cooler. This is because the torque converter produces most of the heat generated by the transmission, especially at low speeds. **Figure 1-39** shows the path of transmission fluid through the cooler and back to the transmission. A few vehicles use a separate ***direct-air cooler***. Direct-air coolers are also available as add-on units. A direct-air cooler is a single tube bent into one or more U-shaped forms. These forms are surrounded by fins. The direct air cooler is installed ahead of the radiator. Air passing through the cooler removes heat by direct contact.

Transmission Fluid Filter

All transmissions and transaxles produce some metal shavings and particles of friction material as they wear. Additionally, transmission fluid breaks down from heat and age. The aging fluid develops solid deposits, which circulate in the fluid. The metal shavings, friction material particles, and solid deposits must be removed from the fluid as it circulates through the hydraulic system. Failure to remove these impurities can result in sticking transmission control valves and worn parts.

The ***transmission fluid filter*** catches and removes these impurities as the fluid passes through it. Some filters are fine mesh screens. Most filters, however, are made of felt or filtration paper enclosed in a metal or plastic housing. The filter is always installed on the suction side of the pump so it can remove impurities before they reach the pump and other hydraulic system components. The filter is located in the bottom of the transmission or transaxle so it is always covered by transmission fluid, **Figure 1-40.** A few transaxles have two filters. Filters should be changed as part of periodic transmission or transaxle maintenance.

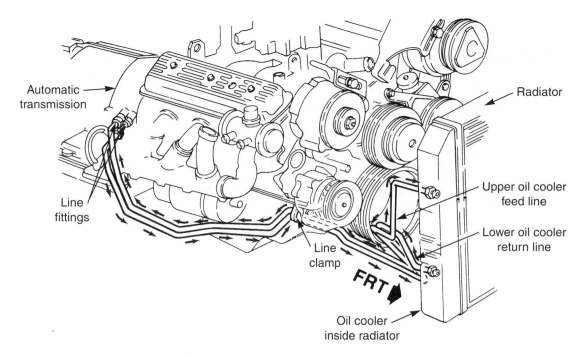

Figure 1-39. *To cool the transmission fluid, an oil cooler is often installed in the vehicle's radiator. Transmission oil pressure forces fluid through the cooler, where it gives up heat to the engine coolant. The cooled fluid then returns to the transmission. Hot fluid usually comes directly from the torque converter, where most transmission heat is generated. (General Motors)*

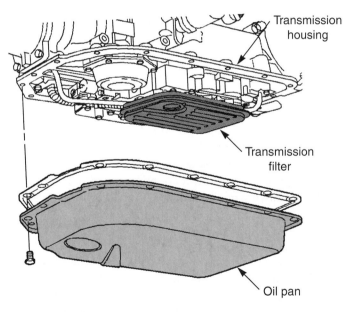

Figure 1-40. *The transmission filter is located at the lowest point in the transmission, usually in the lower oil pan. This allows it to pick up transmission fluid, even when the fluid level is slightly low. (General Motors)*

Automatic Transmission Operation

The following is a brief discussion of how an automatic transmission operates. The principles discussed here will be explained in more detail in later chapters.

When the engine is running and the transmission is in Park or Neutral, the pump produces pressure to keep the torque converter filled. No holding members are applied, and no power reaches the planetary gears. The converter impeller turns the turbine and input shaft, but the power stops at the planetary gears. In Park, the parking gear holds the output shaft stationary.

When the transmission is placed in Drive, oil flows through the manual valve to one or more holding members. The holding members apply, causing the planetary gears to connect the input and output shafts. Engine power goes through the converter impeller, through the turbine and input shaft, and into the planetary gears. It exits the planetary gears and tries to turn the output shaft. If the brakes are applied, the turbine, input shaft, and gears do not move. Fluid from the impeller striking the turbine creates friction and heat, which is removed by the cooler in the radiator. Once the vehicle starts moving, the planetary gears reduce input shaft speed and increase torque to get the vehicle moving.

As vehicle speed increases, the hydraulic control system moves various valves to change which holding members are applied. This changes the rotation of the planet gears and shifts the transmission into a higher gear. Power continues to flow through the torque converter, input shaft, planet gears, and output shaft.

As vehicle speed continues to increase, other valves move to obtain higher gears until the transmission is at its highest gear ratio. At some point, the control system applies the lockup clutch for increased fuel economy. With the lockup clutch applied, the impeller and turbine turn at the same speed.

When the vehicle is brought to a stop, the control system again moves various valves to lower the gear ratios to match vehicle speed and engine load. The control system releases the converter lockup clutch as the vehicle approaches the completely stopped position.

In Reverse, the manual valve sends pressure to the proper holding members. They apply and hold the planetary gears to place the transmission in Reverse. Power goes through the impeller, turbine, and input shaft to the planetary gears. The planetary gears reverse the direction of rotation before delivering power to the output shaft.

Future Transmissions

Transmission technology is always advancing, and some entirely new concepts are being developed. The following section discusses transmissions that may appear in future vehicles.

Toroidal or Infinitely Variable Transmission (IVT)

Figure 1-41 shows an *infinitely variable transmission (IVT)*, also known as a *toroidal transmission*, that is currently being used in Asia. This type of transmission may eventually be installed in vehicles made for American and European roads. The IVT consists of a set of movable discs that rotate between drive and driven races. Varying the angle of the discs varies the speed ratio between the drive and driven races. Since sending power through the discs reverses the direction of rotation, a second set of movable discs are installed behind the driven race. The driven race powers a final output race through the discs. This again reverses power flow so that power exits the transmission in the same direction as it entered. The combination of ratios

Figure 1-41. *Infinitely variable transmissions (IVTs) like the one shown here are currently being used in Asia. In this IVT, a planetary gearset provides reverse. (NSK)*

between the two sets of discs results in the final transmission ratio.

The angles of the movable discs are varied by servos that are operated by the transmission's hydraulic system. An onboard computer monitors and adjusts the hydraulic system outputs based on inputs from engine and road speed sensors.

Electronically Shifted Manual Transmission

Many modern manual transmissions operate without a clutch pedal. Electronically operated devices control clutch operation and gear selection. The driver moves a shifter, or "paddle," that sends signals to an onboard computer. The computer controls solenoids or motors to shift gears and also applies and releases the clutch. On some vehicles with this type of transmission, the clutch pedal is used only to start moving the vehicle from rest. Other vehicles with electronically shifted manual transmissions have no clutch pedal at all.

The transmission controls now in use can be connected to the engine and anti-lock brake computers. The computers could provide fully automatic shifting. Some manufacturers are working on automatic versions of manual transmissions. A few of these designs are being used in large trucks, usually with a torque converter replacing the manual clutch.

Summary

Manual and automatic transmissions have the same basic purposes. They must connect and disconnect the engine and drive wheels, multiply engine power as dictated by vehicle speed and load, and provide a way to reverse the direction of power flow.

Manual and automatic transmissions have many similarities. Major differences are that automatic transmissions use of a torque converter instead of a clutch, planetary gears instead of sliding gears, and a hydraulic control system that makes shifting decisions for the driver. Modern automatic transmissions and transaxles are almost as efficient as manual models. Transmissions and transaxles differ only in the layout of parts and the fact that the transaxle contains the final drive assembly and the differential.

The modern automatic transmission has been gradually refined over the years. The first automatic transmission was introduced more than 60 years ago. Over the years, automatic transmissions with torque converters, simplified gear trains, and aluminum cases were developed and placed in service. Lockup torque converters, transaxles, and electronic control systems have been introduced and gradually perfected during the last 20 years.

Fluid couplings and torque converters are fluid-filled units that contain a set of impeller blades driven by the engine and a set of turbine blades connected to the input shaft. All modern transmissions use a torque converter instead of a fluid coupling. Modern torque converters use a lockup clutch to eliminate slippage.

All transmissions and transaxles have separate input and output shafts and planetary gears. A basic planetary gear consists of a central sun gear surrounded by planet gears that are housed in a planet carrier. A ring gear with internal teeth surrounds the sun and planet gears. Simple planet gear sets can be arranged into complex gear units to obtain different gear ratios.

The parts of the planetary gear assembly are held or driven by holding members. Clutches and bands are common holding members that are operated by hydraulic pressure. One-way clutches are mechanically operated holding members.

The transmission or transaxle case is the support for the other parts of the transmission. It contains oil pressure passages. The oil pan is a reservoir for extra fluid. The hydraulic pump provides all the pressure used in the automatic transmission or transaxle. The pressure regulator is installed in the outlet line to the pump and controls overall transmission pressures.

The hydraulic control system provides pressure to the holding members and determines when shifts should occur. It also keeps transmission fluid flowing to the torque converter, transmission cooler, and lubricating system. On most modern transmissions and transaxles, an electronic control system operates the hydraulic components.

Manual linkage connects the shift lever to the manual valve inside the transmission or transaxle. Throttle linkage connects the engine's throttle plate to the transmission's throttle valve and controls shift speeds.

Transmission fluid is made of oils and various additives. The fluid is pressurized, and it operates the holding members. It also lubricates moving parts, carries away heat, and helps seal internal parts. The proper fluid is important to transmission and transaxle operation.

To cool the transmission fluid, it is pumped from the torque converter to a cooler in the radiator. A few vehicles have direct-air coolers.

To remove debris from the transmission fluid, a filter is installed on the pump intake. Filters can be screens or some other type of filtering material in a plastic housing. Filters should be changed periodically.

Automatic transmission operation varies from one gear to another. In Neutral and Park, no holding members are applied and the power flow stops at the planetary gears. In Drive or Reverse, holding members apply and cause the planetary gears to connect the input and output shafts. Engine power goes through the impeller, through the turbine and input shaft, and into the planetary gears. The hydraulic control system varies gear ratios by applying and releasing different holding members.

Review Questions—Chapter 1

Please do not write in this text. Write your answers on a separate sheet of paper.

1. One of the primary jobs of any transmission is to disconnect the engine from the _____ _____ when the vehicle is not moving.

2. Which of the following gears should be selected to obtain the best acceleration from stop?
   (A) Reduction gear.
   (B) Direct drive.
   (C) Overdrive.
   (D) Passing gear.

3. If the transmission output speed is more than the input speed, the transmission is in a(n) _____ gear.

4. Which of the following gears should be selected to obtain the best fuel economy?
   (A) Reduction gear.
   (B) Direct drive.
   (C) Overdrive.
   (D) Passing gear.

5. Briefly explain why placing an automatic transmission in drive with the brakes applied does not kill the engine.

6. Planetary gears are so named because they resemble the _____ system.

7. Clutches and bands are examples of _____ _____.

8. The _____ _____ serves as a storage place for extra transmission fluid.

9. _____ are used to adjust transmission shaft endplay.
   (A) Bushings
   (B) Shim packs
   (C) Thrust washers
   (D) None of the above.

10. The hydraulic pump is driven by the _____ _____ whenever the engine is running.

11. The governor, oil passages, and servos are part of the _____ control system.

12. An on-board computer operates the _____ control system.

13. The _____ _____ is a cable or a series or rods and levers that connect the shift lever to the manual valve.

14. Transmission fluid _____.
    (A) carries away heat produced in the transmission
    (B) helps the holding members grip when they are applied
    (C) helps seal in pressure
    (D) All of the above.

15. The transmission fluid filter is always located on the _____ side of the hydraulic pump.

ASE-Type Questions—Chapter 1

1. In any transmission, gears are used to do all the following *except:*
   (A) provide direct drive.
   (B) disconnect the engine and drive wheels.
   (C) reverse the vehicle.
   (D) provide overdrive.

2. Technician A says early transmissions were inefficient. Technician B says a slipping transmission is an efficient transmission. Who is right?
   (A) A only.
   (B) B only.
   (C) Both A and B.
   (D) Neither A nor B.

3. Which of the following materials is most commonly used to make modern transmission and transaxle cases?
   (A) Cast Iron.
   (B) Wrought Iron.
   (C) Magnesium.
   (D) Aluminum.

4. The planetary ring gear has _____ teeth.
   (A) internal
   (B) external
   (C) internal and external
   (D) no

5. Which of the following is *not* part of a clutch pack?
   (A) Apply piston.
   (B) Steel plates.
   (C) Servo.
   (D) Splined hub.

6. Technician A says bands are applied by servos. Technician B says bands wrap around drums. Who is right?
   (A) A only.
   (B) B only.
   (C) Both A and B.
   (D) Neither A nor B.

7. The modern transmission/transaxle case is made of _____.
   (A) cast iron
   (B) steel
   (C) plastic
   (D) aluminum

8. The hydraulic control system contains all the following components *except:*
   (A) governor.
   (B) parking lock.
   (C) accumulators.
   (D) valve body.

9. Technician A says gaskets seal moving parts. Technician B says that O-rings seal stationary parts. Who is right?
   (A) A only.
   (B) B only.
   (C) Both A and B.
   (D) Neither A nor B.

10. Available transmission oil cooler configurations include all of the following *except:*
    (A) in the radiator—factory installed.
    (B) in the radiator—aftermarket.
    (C) in front of the radiator—factory installed.
    (D) in front of the radiator—aftermarket.

Chapter 2

Shop Safety and Environmental Protection

After studying this chapter, you will be able to:
- ❏ Identify the major causes of accidents in the workplace.
- ❏ Explain why accidents must be avoided.
- ❏ Describe ways to maintain a safe work area.
- ❏ List safe work procedures.
- ❏ Identify types of environmental damage caused by improper auto shop practices.
- ❏ Help prevent environmental damage by properly disposing of automotive wastes.

Technical Terms

Material safety data
  sheet (MSDS)

Safety glasses

Face shield

Safety shoes

Steel-toed shoes

Core value

Refrigerant

Environmental Protection
  Agency (EPA)

Introduction

Making repairs in a safe manner protects the technician, the vehicle, and the shop. This chapter examines various unsafe conditions and work practices, and explains how to correct or avoid them. It also covers the proper handling and disposal of waste products generated in the automobile repair shop.

Carelessness: The Usual Cause of Accidents

Accidents often result in painful injuries that keep you from working or enjoying your free time. Some accidents can be fatal. Even slight injuries are annoying, and they may impair your ability to work and play. Although an accident may not cause personal injury, it can result in property damage and may cost you your job. Damage to vehicles or shop equipment can be expensive and time consuming to fix.

Accidents frequently occur when the technician becomes careless. Unfortunately, even experienced technicians become rushed and careless, taking shortcuts instead of following proper work procedures. Therefore, falls, injuries to hands and feet, fires, explosions, electric shocks, and even poisonings occur in auto repair shops. Carelessness in the shop can also lead to long-term bodily harm from prolonged exposure to harmful liquids, vapors, and dust. Lung damage, skin disorders, and even cancer can result from contact with these toxic substances. For these reasons, the technician must keep safety in mind at all times. Do not let carelessness make you a casualty.

Types of Accidents

There are many ways an accident can occur in the automotive shop. Nevertheless, most accidents are caused by one of two factors:
- ❑ Failure to maintain a safe workplace.
- ❑ Failure to perform service procedures properly.

Examples of failure to maintain a safe workplace include failing to properly maintain tools and equipment; allowing old parts, containers, or other trash to accumulate around the shop; and ignoring water or oil spills.

Examples of improper service procedures include using the wrong methods to perform repairs, using defective or otherwise inappropriate tools, not wearing proper protective equipment, and not paying close attention while performing the job.

The best way to prevent accidents is to maintain a neat workplace, to use safe methods and common sense

when making repairs, and to wear protective equipment when needed.

Maintaining a Safe Workplace

Return all tools and equipment to their proper storage places. See **Figure 2-1**. This saves time in the long run and reduces the chance of accidents and theft. Never leave equipment or tools out where others can trip over them.

Keep workbenches and work areas clean. A clean workbench reduces the chance of tools or parts falling

Figure 2-1. *Keep your tools clean and arranged in an easy-to-find manner. (Snap-on Tool Corp.)*

from the bench to the floor, where they could be lost or damaged. A falling tool or part can land on your foot, causing injury. A clean work area reduces the possibility that critical parts will be lost or that a fire will start in oily debris. See **Figure 2-2.**

Clean up spills before they are tracked around the shop. People are often injured when they slip on floors coated with oil, antifreeze, or water. Gasoline spills can be extremely dangerous, since even the smallest spark can ignite the vapors, causing an explosion and fire.

Know what types of chemicals are stored on the shop premises. Chemicals commonly used in the automotive shop include transmission fluid, motor oil, antifreeze, hot tank solutions, and parts cleaners. Chemical manufacturers provide a *Material Safety Data Sheet* (**MSDS**) for every chemical they produce. These sheets list all the known dangers of the chemical, as well as the first aid procedures to follow in the event of skin, eye, or respiratory system contact. There should be a MSDS on file for every chemical used in the shop. Read the appropriate MSDS before working with any unfamiliar chemical. See **Figure 2-3.**

Make sure your work area is well lighted. Poor lighting makes it hard to see what you are doing, leading to accidental contact with moving parts or hot surfaces.

Overhead lights should be bright and centrally located. Portable lights, or droplights, should be in proper operating condition. Droplight cages should be in place. Always use a rough-service lightbulb in incandescent droplights, **Figure 2-4.** These bulbs are more rugged than normal lightbulbs and are less likely to shatter if they break. Do not use a high-wattage bulb in a droplight. Lightbulbs get very hot and can melt the light socket or cause burns.

Never overload electrical outlets or extension cords by operating several electrical devices from a single outlet. See **Figure 2-5.** Do not operate high-current electrical devices through extension cords.

Inspect electrical cords and compressed air lines frequently to ensure that they are in good condition. Also, check for improper air hose connections. Do not close vehicle doors on electric cords or air hoses. Never run electrical cords through water puddles or use them outside when it is raining.

Make sure all shop equipment, such as grinders and drill presses, is equipped with the safety guards provided by the manufacturer, **Figure 2-6.** These guards should only be removed for service operations, such as changing grinding wheels. Never operate equipment without the proper guards in place.

Figure 2-2. *Keep workbenches clean and orderly.*

ACME Chemical Company
Material Safety Data Sheet
Product Name: Acetylene

24-hour Emergency Phone: Chemtrec 1-800-424-9300   Outside United States 1-905-501-0802

Trade Name/Syn: Acetylene   **NFPA Ratings**
Chemical Name/Syn: Acetylene, Ethyne, Acetylen, Ethine   Health: 0
CAS Number: 74-86-2   Flammability: 4
Formula: C_2H_2 Reactivity: 0

Hazards Identification

Simple Asphyxiant. This product does not contain oxygen and may cause asphyxia if released in a confined area. Maintain oxygen levels above 19.5%. May cause anesthetic effect. Highly flammable under pressure. Spontaneous combustion in air at pressures above 15 psig. Acetylene liquid is shock sensitive.

Effects of Exposure-Toxicity-Route of Entry

Toxic by inhalation. May cause irritation of the eyes and skin. May cause an anesthetic effect. At high concentrations, excludes an adequate oxygen supply to the lungs. Inhalation of high vapor concentrations causes rapid breathing, diminished mental alertness, impaired muscle coordination, faulty judgment, depression of sensations, emotional instability, and fatigue. Continued exposure may cause nausea, vomiting, prostration, loss of consciousness, eventually leading to convulsions, coma, and death.

Hazardous Decomposition Product

Carbon, hydrogen, carbon monoxide may be produced from burning.

Hazardous Polymerization

Can occur if acetylene is exposed to 250°F (121°C) at high pressures or at low pressures in the presence of a catalyst. Polymerization can lead to heat release, possibly causing ignition and decomposition.

Stability

Unstable--shock sensitive in its liquid form. Do not expose cylinder to shock or heat, do not allow free gas to exceed 15 psig.

Fire and Explosion Hazard

Pure acetylene can explode by decomposition above 15 psig; therefore the UEL is 100% if the ignition source is of sufficient intensity. Spontaneously combustible in air at pressures above 15 psi (207 kPa). Requires very low ignition source. Does not readily dissipate, has density similar to air. Gas may travel to source of ignition and flash back, possibly with explosive force.

Conditions to Avoid

Contact with open flame and hot surfaces, physical shock. Contact with copper, mercury, silver, brasses containing >66% copper and brazing materials containing silver or copper.

Accidental Release Measures

Evacuate all personnel from affected areas. Use appropriate protective equipment. Shut off all ignition sources. Stop leak by closing valve. Keep cylinders cool.

Ventilation, Respiratory, and Protective Equipment

General room ventilation and local exhaust to prevent accumulation and to maintain oxygen levels above 19.5%. Mechanical ventilation should be designed in accordance with electrical codes. Positive pressure air line with full face mask or SCBA. Safety goggles or glasses, PVC or rubber gloves in laboratory; and as required for cutting or welding, safety shoes.

Figure 2-3. *This material safety data sheet (MSDS) is for acetylene.*

Figure 2-4. *Always use a rough-service bulb in a droplight. Keep the cage closed.*

Figure 2-5. *An overloaded electrical outlet poses a risk of fire.*

When servicing any piece of equipment, be sure it is turned off and unplugged. Read the equipment service literature before beginning any repair.

Closely monitor tool and equipment condition and make repairs when necessary. This includes such varied tasks as replacing damaged leads on test equipment, checking and adding oil to hydraulic jacks, and regrinding the tips on screwdrivers and chisels.

Never leave open containers of antifreeze in the shop or outside. Ethylene glycol antifreeze will poison any animal (or person) that drinks it and will create an extremely slippery floor if spilled.

Know where the shop fire extinguishers are located and know how to operate them. Make sure you know what type of fire extinguisher is used on each type of fire. See **Figure 2-7.** Periodically check that the fire extinguishers are in working order and have them checked periodically by qualified personnel.

Performing Work Procedures Properly

The following work procedures may add a little time to the job, but they will make your job and your life easier.

Never wear open jackets, scarves, or shirts with long, loose sleeves. They can be caught in moving parts. If your hair is long, keep it away from moving parts by securing it under a hat or tying it back.

Remove rings and other jewelry when working in the shop. These items can also be caught in moving parts. Metal rings and jewelry can cause a short circuit if the metal parts form a path between a positive terminal and ground.

The automotive technician spends a good deal of time working under the vehicle. Therefore, the chance of eye injury is great. Eye protection should always be worn when working in a situation that could result in dirt, metal, or liquids being thrown into your face. **Figure 2-8** shows some typical eye and face protection. The *safety glasses*

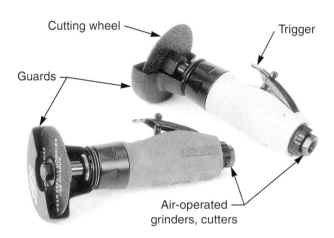

Figure 2-6. *An air-operated cutter set. Note the use of safety guards. These tools operate at speeds up to 20,000 RPM. Wear safety glasses when using power tools. (Snap-on Tool Corp.)*

protect your eyes from flying metal and dirt, while a **face shield** provides protection when working with solvents and other harmful liquids.

In any shop, there is always the danger of falling objects. Some parts are very heavy and can cause severe foot damage. It is also very common for the technician to be injured by getting his foot caught under a lift arm or tire when a vehicle is lowered. **Safety shoes** have a steel insert in the toe. This insert absorbs any damage from falling objects or pinching hazards. These shoes are sometimes called **steel-toed shoes.**

To protect your hands from hot, sharp, or burred surfaces, heavy gloves can be worn. To protect your hands from harmful liquids, such as cleaning solvents, rubber or latex gloves can be used.

Some clutch discs contain asbestos—a powerful cancer-causing substance. To avoid breathing dust inside the bell housing or clutch assembly, wear a respirator when performing clutch service, **Figure 2-9.** Never use compressed air to blow dust from the clutch assembly.

Fire Extinguishers and Fire Classifications

Fires	Type	Use		Operation
Class A Fires Ordinary Combustibles (Materials such as wood, paper, textiles.) *Requires... cooling-quenching.*	**Soda-acid** Bicarbonate of soda solution and sulfuric acid	Okay for use on A Not for use on B C D		Direct stream at base of flame.
Class B Fires Flammable Liquids (Liquids such as grease, gasoline, oils, and paints.) *Requires...blanketing or smothering.*	**Pressurized Water** Water under pressure	Okay for use on A Not for use on B C D		Direct stream at base of flame.
Class C Fires Electrical Equipment (Motors, switches, etc.). *Requires... a nonconducting agent.*	**Carbon Dioxide (CO_2)** Carbon dioxide (CO_2) gas under pressure	Okay for use on B C Not for use on A D		Direct discharge as close to fire as possible, first at edge of flames and gradually forward and upward.
	Foam Solution of aluminum sulfate and bicarbonate of soda	Okay for use on A B Not for use on C D		Direct stream into the burning material or liquid. Allow foam to fall lightly on fire.
Class D Fires Combustible Metals (Flammable metals such as magnesium and lithium.) *Requires...blanketing or smothering.*	**Dry Chemical**	Multi-purpose type Okay for A B C Not okay for D	Ordinary BC type Okay for B C Not okay for A D	Direct stream at base of flames. Use rapid left-to-right motion toward flames.
	Dry Chemical *Granular-type material*	Okay for use on D Not for use on A B C		Smother flames by scooping granular material from bucket onto burning metal.

Figure 2-7. *A chart showing fire and fire extinguisher classifications. The proper extinguisher must be used to effectively put out fires. Using the wrong extinguisher may lead to electrocution.*

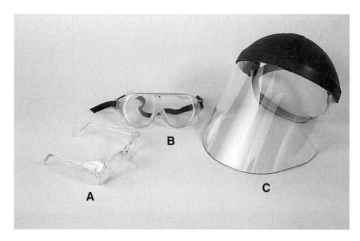

Figure 2-8. *Eye protection should be worn while working in the shop. A—Safety glasses. B—Safety goggles. C—Face shield.*

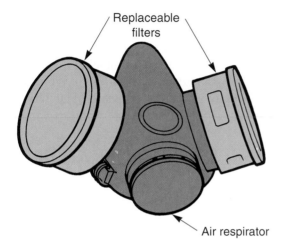

Replaceable filters

Air respirator

Figure 2-9. *An approved respirator will protect the respiratory system from harmful products, such as brake dust.*

Remove the dust with an approved vacuum-collection system and clean clutch housings or components with a liquid cleaner to prevent creating dust.

Study work procedures before beginning any job that is unfamiliar. Never assume that a procedure you have used in the past will work on all vehicles.

Always work carefully. Speed is not as important as doing the job right and avoiding injury. Avoid co-workers who refuse to work carefully or who tend to engage in horseplay.

Use the right tool for the job. Using the wrong tool is asking for an accident or at least a broken tool. Never use a 12-point hand socket with an impact wrench. Special impact sockets should always be used with an impact wrench. Do not use low-quality tools or damaged tools.

Learn how new equipment and tools work before attempting to use them. This is especially true for air-operated tools, such as impact wrenches and air chisels, and large electrical devices, such as drill presses. These tools are very powerful and can cause serious injury if improperly used. A good way to learn about new equipment is to read the manufacturer's instructions.

When working on electrical systems, avoid creating a short circuit with a jumper wire or metal tool. Not only will this damage the vehicle's components or wiring, but it may also develop enough heat to cause a severe burn or to start a fire.

Always lift with your legs, not your back. Make sure you are strong enough to lift the object to be moved. If an object is too heavy to lift by yourself, get help.

Do not smoke in the shop. You may accidentally ignite an unnoticed gasoline leak. A burning cigarette may also ignite oily rags or paper cartons.

Never attempt to raise a vehicle with a damaged jack or a jack that does not have the proper capacity rating. After raising a vehicle with a jack, always support it with jack stands. Never use boards or cement blocks to support a vehicle. When using a lift, be sure to place the lifting pads under the frame or in a location that can support the vehicle's weight. See **Figure 2-10.**

When using a transmission jack, always securely attach the transmission, transaxle, or transfer case to the jack with a safety chain before lowering the unit from the vehicle. If the part is not secured to the jack, it can fall off if the jack is moved suddenly or if the part hangs up on an underbody component.

Once the unit is secured, lower the jack slowly, making sure all connections to the vehicle have been removed. If a connection is still attached, raise the jack and remove the attachment. Once the part has cleared the vehicle body, lower it completely before moving it from the jack.

Closely watch the engine as the transmission jack is lowered. If the engine mounts are broken, the engine can fall out of the vehicle once the transmission or transaxle has been removed.

Do not run any engine in an enclosed area without proper ventilation. Running engines emit carbon monoxide, a deadly gas that builds up quickly. Because carbon monoxide is colorless and odorless, it cannot be detected without special equipment.

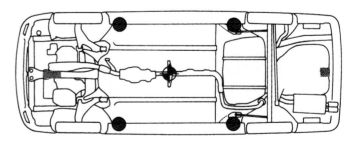

█████ Floor jack locations   ⊕ Approximate center of gravity

● Frame contact hoist, twin post hoist or scissors jack (emergency) locations

Figure 2-10. *When lifting a vehicle, make sure the lift pads are positioned at the recommended lift points.*

When working on or near a running engine, stay away from rotating parts. Never reach between moving engine parts for any reason. Do not leave a running vehicle unattended. The vehicle may slip into gear or overheat while you are away. Whenever you must work on a running vehicle, apply the parking brake to prevent the vehicle from moving if it accidentally slips into gear.

When road testing a vehicle, be alert and obey all traffic laws. Do not become so absorbed in diagnosing a problem that you forget to watch the road. Be alert for the actions of other drivers.

Preventing Environmental Damage

Automotive repair shops generate wastes that can cause considerable environmental damage. Technicians are often guilty of carelessly disposing of solid and liquid wastes, and damaging the atmosphere by improper repair procedures.

Typical solid wastes produced in automotive repair shops include scrap parts, tires, and cardboard boxes. Liquid wastes include transmission fluid, motor oil, antifreeze, brake fluid, and cleaning solvents.

Improper repair procedures include allowing refrigerant gases to escape into the atmosphere and adjusting or modifying engines so they emit excessive amounts of pollutants.

Allowing these practices to continue lowers the quality of our lives by contaminating the air we breathe, the water we drink, the soil we use to grow food. The economic burden of dealing with waste will grow ever larger. If we ignore the effects of irresponsible waste production and dumping, our descendants will have to deal with the problem.

Preventing Solid Waste Contamination

Never allow solid wastes to build up around the shop. It is unsightly, dangerous, and expensive. Recycle used parts and scrap materials whenever possible. It makes good economic sense to recycle, since almost every rebuildable part has a return value, usually called a *core value.* Check with your local parts supplier to determine which parts can be sent back for rebuilding.

Do not throw away parts boxes, old tires, and salable scrap metals unless you are sure that they cannot be recycled. The value of paper, scrap tires, and scrap metals depends on current market conditions. Recyclers are often listed in the telephone book. They can give you advice on what to do with recyclable materials. If solid wastes cannot be recycled, they should be disposed of responsibly, not by illegal dumping or burning.

Preventing Liquid Waste Contamination

One of the most common ways that automotive technicians damage the environment is by pouring used transmission fluid, antifreeze, motor oil, brake fluid, or gear oil on the ground. This immediately contaminates the soil. In addition, these liquids sink farther into the ground every time that it rains, eventually contaminating the water table, which could be the source of your drinking water. Do not pour liquid wastes into drains. Municipal waste treatment plants cannot handle antifreeze, brake fluid, or petroleum products, such as transmission fluid. Most automotive fluids also contain poisonous additives and heavy metals absorbed from the vehicle during use. In most areas, such dumping is illegal.

In many areas, local waste management companies accept used oil and antifreeze for recycling. Several 55-gallon drums should be kept in the shop to store liquid wastes. See **Figure 2-11.** The used oil and antifreeze is then re-refined and reused. Some used oil is burned by power plants to produce electricity, eliminating the used oil and reducing the dependence on imported crude oil. A recently developed process converts old motor oil into diesel fuel.

Preventing Damage to the Atmosphere

Modifications that damage or defeat the purpose of a vehicle's emission control systems can cause considerable damage to the atmosphere. Adjusting carburetors for a richer mixture, changing the manufacturer's timing settings, and disconnecting emission controls will all

Figure 2-11. *Fifty-five-gallon drums should be used to store liquid wastes. Never pour wastes onto the ground, into the sewer system, etc. (Pennzoil Co.)*

increase emissions. Some seemingly harmless actions, such as installing a cooler thermostat (thermostat designed to maintain a lower engine operating temperature) or a non-stock air cleaner, can also cause a rise in vehicle emissions. Not only are these actions illegal, but they rarely increase power and mileage as much as expected.

Never discharge air conditioner **refrigerant** into the atmosphere. Studies have shown that certain refrigerants, such as R-12, cause extensive damage to the ozone layer, leading to increased ultraviolet ray damage. Even "safe" refrigerants, such as R-22 and R-134a, contribute to ozone layer loss. This is another area covered by federal and, in some cases, state law.

Additionally, it makes good economic sense to recover and reuse refrigerants. R-12 refrigerant costs over seven times more than it did just a few years ago. If you do not have a machine that can recover and recycle refrigerant, such as the one in **Figure 2-12,** refer the customer to a shop that is equipped to do the job properly.

The Environmental Protection Agency

Improper disposal of vehicle wastes, as well as modifying, disabling, removing, or otherwise tampering with engine emission controls, is a federal crime, with severe penalties. Vehicle waste disposal and emissions laws are enforced by the **Environmental Protection Agency** (**EPA**). The EPA investigates suspected violations and often

Figure 2-12. *A refrigerant recovery and recycling machine must be used when servicing an air conditioning system. (RTI Technologies, Inc.)*

conducts "sting" operations to catch violators. Some states have additional laws protecting the environment.

Additional information about waste disposal and vehicle emissions can be obtained from the Environmental Protection Agency. The EPA has ten regional offices and six field offices. For the address of the nearest EPA office, write, call, or e-mail:

Automotive/Emissions Division
United States Environmental Protection Agency
401 M Street S. W.
Washington, DC 29460
(202) 260-7647

Summary

The most common cause of accidents is carelessness. Many technicians become rushed and forget to do the job safely. An accident may result in personal injury, long-term bodily harm, or damage to equipment or property.

Many accidents are caused when technicians fail to correct dangerous conditions in the work area, such as oil spills or tripping hazards. Other accidents are caused when technicians take shortcuts instead of following proper repair procedures.

The best ways to prevent accidents are to maintain a neat workplace, to use proper methods of repair, and to wear protective equipment when needed. It is up to the technician to study the job beforehand, work safely, and take the steps necessary to prevent accidents. Always use common sense when working on vehicles, and avoid people who do not.

Careless production and disposal of wastes cause extensive environmental damage. The two main ways in which an automotive shop can cause environmental damage are carelessly disposing of wastes and repairing vehicles in such a way that they pollute the atmosphere. Environmental rules should always be followed to prevent damage to the air, water, or soil. In many cases, federal and state law requires proper disposal of wastes and proper vehicle repairs.

Review Questions—Chapter 2

Please do not write in this text. Write your answers on a separate sheet of paper.

1. The most common cause of accidents is _____.

2. Accidents often happen when technicians try to take _____ instead of following proper repair procedures.

3. Manufacturers provide Material Safety Data Sheets (MSDS) for every _____ that they make.

4. A tool that draws a lot of current should have its own
_____.
   (A) bench
   (B) electrical outlet
   (C) operator
   (D) All of the above.

5. List three precautions to take when working with electrical cords and compressed air lines.

6. Guards on shop equipment should never be removed, except for _____.

7. Explain why it is a good idea to wear steel-toed shoes whenever you are in the shop.

8. Always lift heavy objects with your _____, not your _____.

9. Almost every rebuildable part has a return value, usually called a _____ _____.

10. Used oil and antifreeze are examples of _____ wastes.

ASE-Type Questions—Chapter 2

1. Technician A says that the one way to prevent accidents is to avoid people who are careless. Technician B says that one way to prevent accidents is to keep an orderly workplace. Who is right?
   (A) A only.
   (B) B only.
   (C) Both A and B.
   (D) Neither A nor B.

2. If all tools and equipment are returned to their proper storage places, which of the following will occur?
   (A) They will be hard to find the next time.
   (B) They will be easier to steal.
   (C) They cannot be tripped over.
   (D) All of the above.

3. The manufacturer provides a Material Safety Data Sheet for all dangerous _____.
   (A) procedures
   (B) tools
   (C) chemicals
   (D) working conditions

4. Which of the following is likely to happen if a standard lightbulb is used in place of a rough-service bulb in a droplight?
   (A) Broken bulb.
   (B) Burns.
   (C) Either A or B.
   (D) None of the above.

5. Technician A says rings and jewelry should not be worn in the shop because they can be caught in moving parts. Technician B says that metal jewelry should not be worn in the shop because it can cause a short circuit between a positive terminal and ground. Who is right?
   (A) A only.
   (B) B only.
   (C) Both A and B.
   (D) Neither A nor B.

6. When should you use a 12-point socket with an impact wrench?
   (A) When an impact socket is not available.
   (B) When the bolt to be removed is not very tight.
   (C) When the air supply to the impact wrench is weak.
   (D) Never.

7. Technician A says that after raising a vehicle with a floor jack, the vehicle can be supported by boards or cement blocks. Technician B says that when using a transmission jack, the transmission should be attached to the jack with a safetychain. Who is right?
   (A) A only.
   (B) B only.
   (C) Both A and B.
   (D) Neither A nor B.

8. An automotive technician can cause environmental damage by _____.
   (A) carelessly disposing of wastes
   (B) repairing vehicles so that they pollute the atmosphere
   (C) discharging refrigerants into the atmosphere
   (D) All of the above.

9. Spilled ethylene glycol antifreeze can cause all of the following *except:*
   (A) slipping.
   (B) poisoning.
   (C) fires.
   (D) environmental damage.

10. Technician A says that changing the manufacturer's timing settings may cause a rise in vehicle emissions. Technician B says that installing a non-stock air cleaner will have no effect on vehicle emissions. Who is right?
   (A) A only.
   (B) B only.
   (C) Both A and B.
   (D) Neither A nor B.

Chapter 3

Special Service Tools and Service Information

After studying this chapter, you will be able to:
- ❑ Identify special tools used in automatic transmission and transaxle service.
- ❑ Demonstrate the use of common measuring tools.
- ❑ Select the proper tool for the job at hand.
- ❑ Discuss the correct and incorrect ways to use certain types of tools.
- ❑ Describe the various sources of service information available to the automobile technician.

Technical Terms

Special service tools
Spanner wrenches
Punches
Drifts
Snap ring pliers
Picks
Press
Drivers
Pullers
Spring compressor
Seal protector
Tubing cutter
Tubing flaring tools

Bending tools
Transmission alignment
   tools
Screw extractors
Reamers
Holding fixtures
Transmission jacks
Oil cooler and line
   flusher
Converter flusher
Parts washers
Gauges
Micrometer
Hole gauge

Telescoping gauge
Calipers
Feeler gauge
Straightedges
Dial indicator
Torque wrench
Jumper wire
Nonpowered test light
Self-powered test light
Scan tool
Multimeters
Schematic

Introduction

In addition to the common hand tools used in repair shops, **special service tools,** or **specialty tools,** are needed to service parts of the automatic transmission or transaxle. These are tools exclusively designed to service a specific component or a specific class of components. This contrasts with most hand tools, which are designed to be used on many parts for many purposes. *Spanner wrenches, snap ring pliers, drivers,* and *pullers* are all examples of special service tools.

As mentioned, some special service tools are designed to service a specific component and must often be obtained through the tool or vehicle manufacturer. These tools are identified in vehicle service manuals. **Figure 3-1** shows some of the special service tools used to service automatic transmissions. **Figure 3-2** shows some of the tools that can be used for automatic transaxles. The part number appearing with each tool is a coded description of that tool. This number is specified when ordering.

Service manuals also contain information on when to use a tool and how to use it properly. Most service manuals contain the address for the tool manufacturer and information on how to order. Prices of special service tools can be obtained from the tool manufacturer. A few specialty tools can be made, **Figure 3-3.**

Specialty Wrenches and Sockets

Wrenches are tools used to grip and turn fasteners or other parts. Some wrenches have jaws; others have sockets. Special hand wrenches may have openings to fit a fastener with an unusual head shape. Fasteners with special heads are often used when space problems prohibit common tools from being used. Sometimes, the wrench handle is specially shaped to fit in areas where clearances are tight. **Figure 3-4** shows a wrench designed to fit a fastener having both an unusual head shape and a hard-to-access location.

Spanner wrenches are special service tools used to tighten large parts or to hold parts in place while other tools are used. Most spanner wrenches are one-piece units. Some spanner wrenches, like the one in **Figure 3-5A,** have lugs that fit into holes in the part. Others, like the wrench shown in **Figure 3-5B,** have provisions for bolting to the parts they hold.

Special *sockets* are available that have shapes or depths to fit fasteners that are different from standard or to fit where clearance is a problem. See **Figure 3-6.** These sockets are designed to fit standard handles and drive sizes, such as 3/8" or 1/2" ratchets and extensions. They are used the same way as any other socket.

Punches and Drifts

Punches and **drifts** are removal and installation tools designed to be struck with a hammer. They are used to remove pins, plugs, and other pressed-in parts from bores. The punch or drift is placed against the part to be removed. The top of the tool is struck with the hammer. A punch and a drift are included among tools shown in **Figure 3-7.** Note that drifts can be made from discarded steel or brass rods.

Snap Ring Pliers and Picks

Snap ring pliers are special service tools used to remove and install *snap rings,* or *retainer rings.* As shown in **Figure 3-8,** snap ring pliers have special jaws that fit and grasp the snap ring. A collection of typical snap ring pliers is shown in **Figure 3-9.**

 Note: Eye protection should always be worn when working with snap rings. A flexed ring can shoot into your face with considerable force.

Picks are used for prying on parts and for jobs requiring a small pointed tool. (Refer back to **Figure 3-7.**) These tools can be used to get under, pry up, and remove snap rings when there is no other way of removing them.

Presses

In many cases, a part is held in a mating part with an **interference fit.** This means that the part is slightly larger than the space into which it is fit. A shaft that is fit into a smaller hole is an example. The parts fit together tightly, and they cannot be removed easily. A **press fit** is one type of interference fit. In this type of fit, considerable pressure is applied to the mating pieces to get them to fit together.

Often, gears and bearings have an interference fit with the shafts on which they ride. These parts can be installed on or removed from a shaft with a **press.** A press is a piece of equipment consisting essentially of a frame and a ram. It is used for applying pressure on an assembly.

The most common press found in repair shops today is the *hydraulic press,* **Figure 3-10.** This press has a hydraulic ram. Pressure is developed by a hand-operated piston or, on larger presses, by an electric pump.

Special *adapters* are often used to make the pressing operation easier. The adapter shown in **Figure 3-11** is used for removing bearings. The tool fits under the bearing. It provides a firm surface for the bearing to rest on. Adapters are usually adjustable to permit use with many sizes of gears and bearings.

Universal remover A **J-7004-1**	Dial indicator set E **J-8001**	Handle E **J-8002**
Holding fixture and base E **J-8763-02**	Oil pump body and cover alignment band E **J-21368**	Rear seal installer E **J-21426**
Pump oil seal installer E **J-25016**	Piston compressor E **J-22269-01**	Bushing remover A **J-23062-14**
Clutch spring compressor E **J-23327**	Clutch spring compressor adapter E **J-25018-A**	Clutch spring compressor press E **J-23456**
Universal remover E **J-23907**	Oil pump remover and end play checking fixture E **J-24773-A**	End play checking fixture adapter E **J-25022**
End play checking fixture adapter E **J-34725**	Bushing remover A **J-25019-4**	Bushing installer A **J-25019-9**
Bushing installer A **J-25019-12**	Bushing remover A **J-25019-14**	Bushing remover A **J-25019-16**
Bushing and universal remover set * **J-29369-1** * **J-29369-2**	Bushing remover A **J-24036**	Servo cover compressor E **J-29714**
Output shaft support fixture E **J-29837**	Inner overrun clutch seal protector E **J-29882**	Inner forward clutch seal protector E **J-29883**
2-4 band apply pin tools E **J-33037**	Snap ring pliers A **J-34627**	Bushing set A **J-34196**
Dial indicator stand and guide pin set E **J-25025-B**		

E – Essential tool   A – Available tool

Figure 3-1. *Special service tools make transmission service and repair much easier. The tools shown above are for one type of transmission. (General Motors)*

Transmission support fixture base	Handle	Handle	Torque converter pressurization kit
E **J3289-20**	E **J7079-2**	E **J8092**	A **J21369-D**
Forward clutch spring compressor	Compressor screw and frame		Universal remover
E **J23327-1**	E **J23456**		E **J23907**
Dial indicator stand and guide pin set	#30 Torx bit or equivalent		#40 Torx bit or equivalent
E **J25025-A**	A **J25359-4**		A **J25359-5**
Output shaft aligning and loading tool	Adapter plug		Torque converter end-play fixture
E **J26958-16**	E **J26958-10**		E **J29830**
Converter seal installer	Anaroid modulator checking tool		"C" ring remover/installer - output shaft
E **J28540**	A **J36619**		E **J34757**
Transmission support fixture	Bushing installer		Pump bearing - installer and remover
E **J28664-B**	E **J25019-6**		E **J28698**
Axle seal installer	Bushing and universal remover set		Turbine shaft seal installers and sizer (1 seal)
E **J29130**	E **J26941** A **J29369**		E **J29569** E **J29829**
Bearing installer - drive sprocket support	Clutch assembly/final drive remover and installer		1-2 and reverse bands apply pin gauge
E **J28677**	E **J33381**		E **J33382**
Input shaft end play tool	Input clutch piston seal protector		Third clutch piston seal protector
E **J33386**	E **J34091**		E **J34092**
Thermo element height gauge	Output shaft loading tool adapter		Left side axle seal installer
E **J34094-A**	E **J34095**		E **J34115**
Input seal installer	Driven sprocket support bearing installer		Driven sprocket support bearing remover
E **J34741**	E **J34126**		E **J34129**

E–Essential tool   A–Available tool

Figure 3-2. *These special service tools are used to repair automatic transaxles on some front-wheel drive vehicles. (General Motors)*

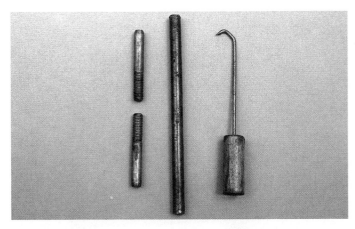

Figure 3-3. *Tools can sometimes be made from scrap parts. These tools have been made from discarded metal shafts. The seal hook was made to remove inner clutch drum hub seals.*

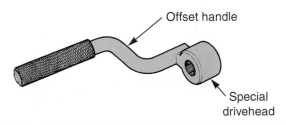

Figure 3-4. *A hand wrench for servicing a transmission assembly. Notice the offset handle on this special wrench. The head is shaped to fit an unusual fastener.*

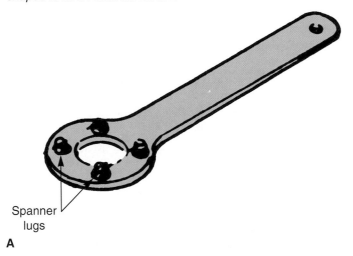

A

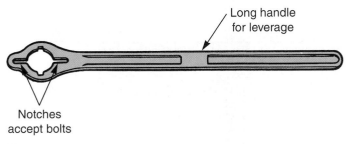

B

Figure 3-5. *Spanner wrenches are used on certain transmission and transaxle parts. A—This spanner wrench has special lugs that fit into holes in the drive pinion flange. B—This type of spanner wrench is bolted to the part on which it is used.*

Figure 3-6. *This special socket is made for a drive pinion lock nut. A socket such as this is used when a standard socket will not fit because of clearance problems.*

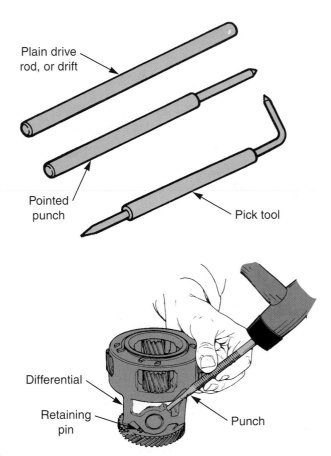

Figure 3-7. *An assortment of special service tools. In this illustration, the pinion shaft retaining pin is being removed with a hammer and a punch. (General Motors)*

Another type of press is the mechanical *arbor press.* It is hand operated and is used for light-duty jobs. It performs the same function as a hydraulic press, but its ram-pressure capability is much less.

Drivers

There are several different types of **drivers** used in automatic transmission and transaxle service. One type is used to install and remove bearings and seals in housings. This driver is used with a hydraulic press or an arbor press. If a press is not available, a hammer can sometimes be

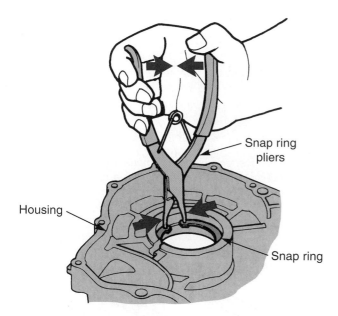

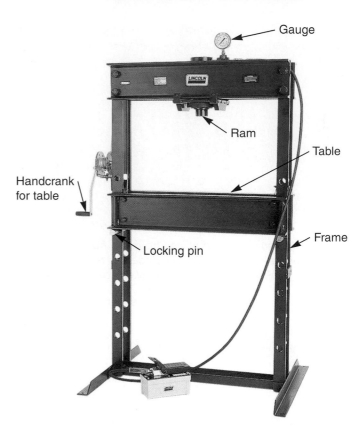

Figure 3-8. *Snap ring pliers are designed for removing snap rings. The illustration shows a snap ring being installed in a transmission housing. (Honda)*

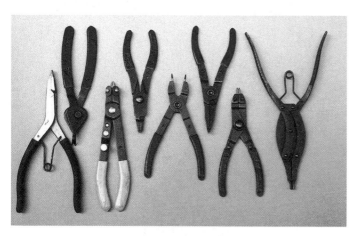

Figure 3-9. *The technician should have many sizes and types of snap ring pliers to remove and install wide variety of snap rings encountered when servicing transmissions and transaxles.*

Figure 3-10. *The hydraulic press is used to remove many pressed-on parts, such as axle bearings and automatic transmission bushings. (Lincoln Automotive)*

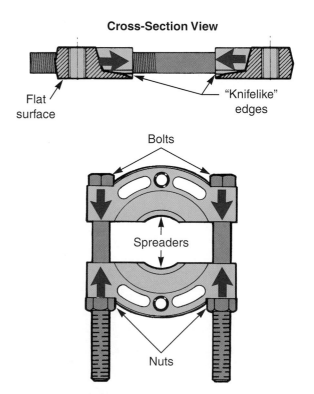

Figure 3-11. *Bearing adapters are used to make bearing removal jobs easier and speed the process. The adapter is installed under the bearing.*

used with the driver, **Figure 3-12.** When using either method, always wear eye protection. Also, always ensure that the part is properly aligned in the bore.

Before using a driver, always make sure that it is the proper size for a particular job. It should fit the seal or bearing to be installed or removed properly. The shape of the driver is critical. If the driver does not match, the part will be damaged. See **Figure 3-13.**

Figure 3-14 shows drivers being used to install and remove *bushings,* which are a type of bearing. The outside diameter of the driving surface should be slightly smaller than the bore diameter. This allows the driver to push the bushing into or out of a bore without contacting the bore itself. Drivers for bearings that are to be installed to specified depths have a shoulder that is larger than the housing bore, as shown in **Figure 3-14A.** The bushing is at the proper depth when the shoulder bottoms on the housing. **Figure 3-14B** shows a driver used for removing a bushing. This driver does not need or have such a shoulder.

Pullers

Pullers are often used to remove parts that are pressed together. A puller is a device that exerts pressure on such parts to separate them. There are many kinds of pullers.

Figure 3-15 shows some of the pullers that you might use to remove drive train parts. A *wheel puller,* **Figure 3-15A,** is a tool that uses screw threads to exert pressure. It is used to remove gears, pulleys, or bearings from a central shaft. The puller bolts attach the screw and plate assembly to the part to be removed. The central screw is then tightened to pull the part from the shaft.

The operation of another kind of puller is similar to that of the wheel puller, **Figure 3-15B.** This kind of puller is used to remove seals or other parts that are lightly pressed in. The tool is fastened onto the part by the threaded connection. Tightening the central screw, then, causes the part to be pulled out. One disadvantage of this type of puller is that threading the tool into the pressed-in part usually ruins the part.

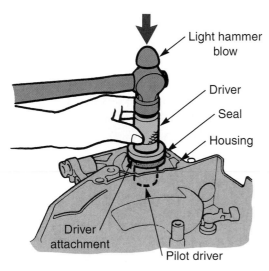

Figure 3-12. *A hammer is used on a driver to install the oil seal in a torque converter housing. Note the pilot driver is used to ensure that the seal is driven in straight. It aligns the driver and attachment. (Honda)*

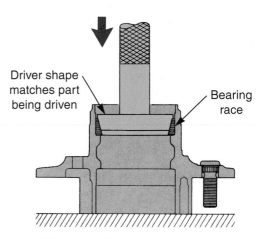

Figure 3-13. *When installing a bearing race, check that the driver closely matches the inner shape of the race or damage may result. (DaimlerChrysler)*

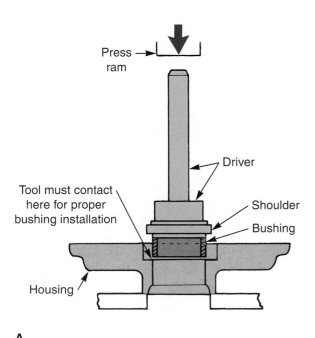

A

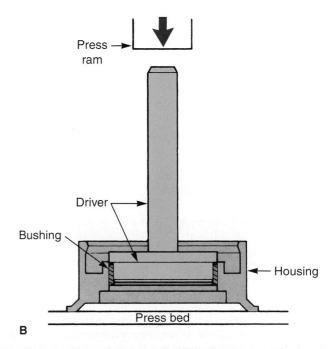

B

Figure 3-14. *Make sure the driver diameter matches the part being installed. A—Driver being used with hydraulic press for installation of a pump bearing, or bushing. B—Driver being used with hydraulic press for removal of the forward clutch hub bushing. (Ford)*

A *jaw puller,* a third kind of puller, is shown in **Figure 3-15C.** This *bearing puller,* as it is also called, is used to remove bearings, gears, or other parts that are pressed onto shafts or into housings. It consists of a central block, with a center screw and jaws that pivot from the block. The puller jaws are placed over the bearing to be removed. The jaws are securely tightened in place with bolts or a thumbscrew that draws the jaws in toward the center. The center screw is then tightened to pull the bearing from the shaft.

Another tool used to remove parts is the *slide hammer,* or *impact puller,* **Figure 3-15D.** Its pulling force comes from impact, instead of screw force. The puller is installed on the part to be removed. This is sometimes done by threading the guide rod into an existing threaded hole on the part. Other times, an adapter, which is attached to the threaded end of the guide rod, is used. Then, the slide, which is weighted, is pulled sharply against the shoulder of the guide stop. The resulting impact is transmitted to the part, and the part is removed from position.

To remove oil seals, a special adapter is sometimes used with the slide hammer. This adapter consists of a sheet metal screw that is brazed or welded onto a nut. The internal threads of the nut match the external threads of the slide hammer guide rod. To remove a seal, a small hole is drilled in the metal portion of the seal. The adapter screw is then threaded tightly into the hole. The guide rod is threaded into the adapter nut. After the adapter is attached to the guide rod, the slide portion of the slide hammer can be moved sharply against the shoulder of the guide stop to remove the seal. If the seal is difficult to remove or the hole becomes stripped, another hole can be drilled on the opposite side of the seal, and the removal process can be repeated.

Special precautions should be taken when using pullers:

❑ *Always* wear eye protection. Failure to do so could result in irreversible eye damage.

❑ *Always* install the puller properly. An improperly installed puller could cause injury or damage.

❑ *Never* apply force through the rolling elements of a ball or roller bearing. The bearing will most likely be ruined, and it may fly apart, causing additional damage and injury.

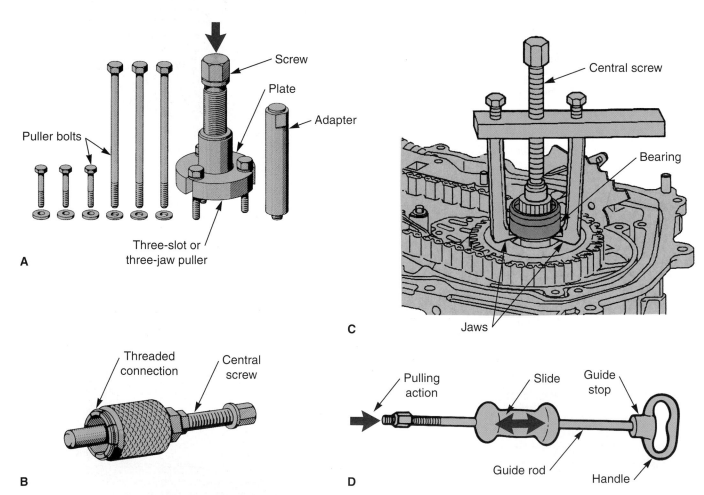

Figure 3-15. *Pullers are used to remove many transmission and transaxle components. A—This kind of puller is used to remove pressed-on parts that have threaded holes to accept puller bolts. B—This puller threads into the part to be removed. The part is usually ruined in the pulling process. C—The jaw puller is used to remove many antifriction bearings. It must be carefully installed. D— The slide hammer can be used to remove many kinds of bearings and seals. This tool will accept any number of adapters. (Mercedes-Benz, Honda, DaimlerChrysler)*

Spring Compressors

A *spring compressor* is used to remove and install coil springs. In operation, the tool compresses the spring. This shortens the spring and takes pressure off the parts against which it is seated, allowing the spring and associated parts to be removed.

Spring compressors are widely used in automatic transmission overhaul. For example, a spring compressor is used to compress the clutch release spring of the transmission. This will take the pressure off the spring retainer and snap ring for the clutch apply piston. In doing so, the parts can be removed without damage. See **Figures 3-16** and **3-17.**

⚠ **Warning: Use extreme care when removing and installing springs. A loaded spring can cause serious injury if it is released suddenly. Be sure you are wearing eye protection when working with springs.**

Seal Protectors

A *seal protector* is a metal or plastic sleeve. It is used to prevent seal damage when a seal is installed over a shaft with a sharp area, such as a *keyway,* **Figure 3-18.** Once the seal is in place, the protector can be removed.

Tubing Tools

It is sometimes necessary to repair or replace tubing that connects the automatic transmission to the transmission *oil cooler* in the vehicle radiator. To do this properly, some special tubing tools are needed. There are three general types of tubing tools: cutters, flaring tools, and bending tools.

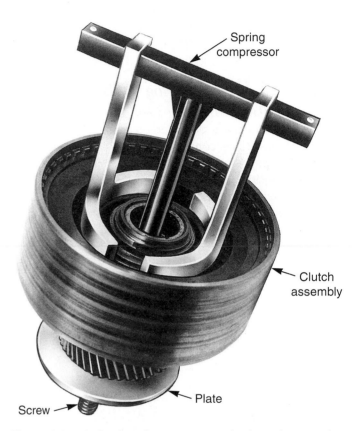

Figure 3-17. *A clutch spring compressor is shown in operation. Once the spring retainer is depressed, the retaining snap ring can be removed. Then, the compressor is carefully released and the clutch disassembled. (Owatonna Tool)*

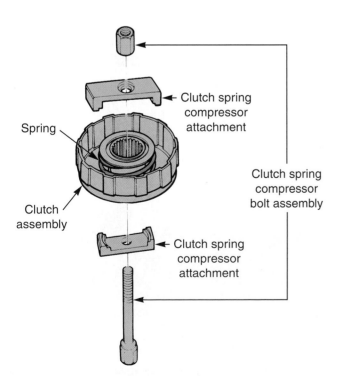

Figure 3-16. *The clutch spring compressor is a vital tool for repairing automatic transmissions. Automatic transmissions have clutch assemblies that require its use.*

Figure 3-18. *Using seal protectors prevents damage when seals are installed over shafts. (Owatonna Tool)*

Tubing Cutters

A *tubing cutter* is shown in **Figure 3-19A.** The tool essentially consists of a cutting wheel and opposing rollers that guide the tool and provide a surface against which to cut.

In cutting the tubing, the tubing cutter is gradually tightened, and the tool is rotated completely around the tubing. The cutting wheel should be tightened in small steps to prevent deforming the tubing. After the tubing is cut, it should be *deburred* to remove any sharp projections. Many tubing cutters have a *reamer blade* for this purpose.

Tubing Flaring Tools

Tubing flaring tools are used to *flare* the tubing, which means to make an enlarged lip at the end of the tubing. Flaring tools consist of a ram and a block that contains different-size holes to match various tubing sizes. See **Figure 3-19B.**

To flare the end of a length of tubing, place it in the proper hole with the recommended amount of tubing exposed. Tighten the tubing firmly in the block. Then, tighten the ram into the tubing until the flare is formed.

Bending Tools

If tubing is bent in a more or less sharp angle, it will usually kink at the inside of the bend. To prevent this, special **bending tools** should be used. There are several kinds of bending tools available, all of which accomplish the job handily.

Alignment Tools

Transmission alignment tools are used to align some internal transmission parts during assembly. These parts are usually moving or are supports for moving parts. Transmission alignment tools will vary.

Flywheel Turning and Locking Tools

Often, the engine flywheel must be turned or held stationary so bolts can be reached during transmission or transaxle removal or installation. *Flywheel turners* move the flywheel by engaging two or more teeth on the flywheel ring gear. The handle is long enough to provide good mechanical advantage, which enables the flywheel to be turned easily. See **Figure 3-20.**

A *flywheel holder* is used to keep the flywheel from turning. One end of the flywheel holder has teeth that engage the teeth on the flywheel ring gear. The other end of the tool is bolted to the engine block or to another stationary part of the drive train.

Organizing Trays

An *organizing tray,* **Figure 3-21,** can be useful when you are working on components with many small, similar

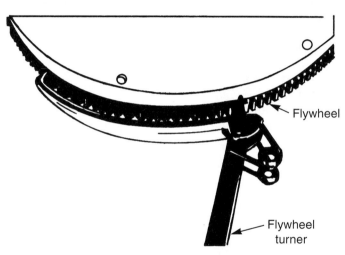

Flywheel

Flywheel turner

Figure 3-20. *Flywheel turners are useful when the flywheel must be turned by small amounts, such as when converter or pressure plate bolts are being installed.*

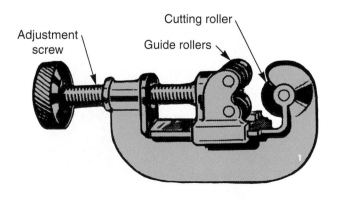

Adjustment screw

Cutting roller

Guide rollers

A   Tubing cutter

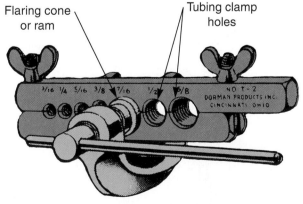

Flaring cone or ram

Tubing clamp holes

B   Tubing flaring tool

Figure 3-19. *Tubing tools are a necessity when transmission cooler lines must be repaired or new lines must be run. A—The tubing cutter is used to make a clean cut in tubing. B—Tubing can be flared easily with a flaring tool. (Dorman Products)*

parts. The tray will provide a way to organize parts so they can be reassembled in the same position as they were originally. Organizing trays are handy when working on automatic transmission valve bodies. They can also help prevent the accidental switching of identical parts, which must be returned to their original locations. (Though such parts look identical, they will have worn differently; thus, they should not be switched.)

Assembly Lube

Many manufacturers recommend using assembly lube instead of petroleum jelly to hold parts in place during reassembly. This lube can also be used to lubricate seals for installation. See **Figure 3-22.**

Drain Pans

If oil is allowed to drip on the floor during repairs, it will create a safety hazard. Further, if tracked onto carpets and other floor treatments, oil can cause damage. **Drain pans** are handy for catching oil that drips from assemblies as they are drained or disassembled, **Figure 3-23.**

Screw Extractors

Screw extractors are used to remove broken screws or other broken fasteners from parts. Screw extractors are hardened shafts designed to grip a broken fastener from the inside. To use an extractor, first, drill a hole in the broken fastener. (This hole must be the proper size for the extractor to be used, and it must be exactly in the center of the shaft.) Then, lightly tap the extractor into the hole and turn it with a wrench. The fastener will unscrew in the process. This is shown in **Figure 3-24.**

Other tools are available for extracting fasteners from parts. These are designed to grip the fastener from the outside. They can be used only if the broken fastener extends from the part. A special *stud puller* is an example. Sometimes, vise grip pliers can be used for this purpose.

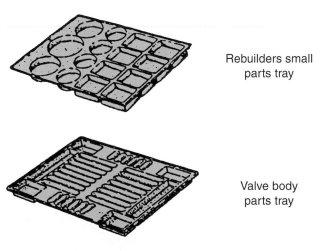

Rebuilders small
parts tray

Valve body
parts tray

Figure 3-21. *Parts trays are useful when disassembling components with many small parts, such as transmissions, transaxles, and rear axle assemblies. (Hayden)*

Figure 3-22. *As its name implies, assembly lube is used to hold parts in place during transmission and transaxle assembly. It is also used to lubricate seals to reduce the possibility of damage during installation.*

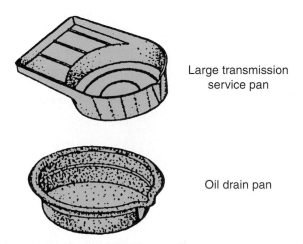

Large transmission
service pan

Oil drain pan

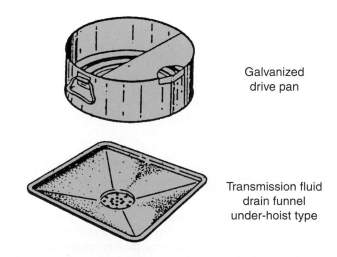

Galvanized
drive pan

Transmission fluid
drain funnel
under-hoist type

Figure 3-23. *The drain pans shown here are useful for catching oil. This helps to prevent messes and accidents. (Hayden)*

Reamers

Some transmission repairs call for reamers. *Reamers* are cutting tools used to salvage damaged transmission parts by enlarging a hole cut in the metal of the part. If a bearing surface or transmission valve bore is damaged, the reamer is used to enlarge the hole and remove the damage. Then, an oversized part is installed in the newly enlarged hole. The reamer is often part of a repair kit that includes the new oversized part. **Figure 3-25** shows a special reamer used to repair a valve body bore in a certain type of valve body. Note that it has two cutting sections. Each section is a different diameter. The two cutting diameters allow the reamer to ream two different diameter openings in the valve bore in one operation. After the reamer enlarges the bore, new oversized valves can be installed. Reamers should be turned by hand to ensure that the new bore surface is as smooth and accurate as possible.

Lifting and Holding Equipment

Transmissions and transaxles are large, heavy, and awkward. To assist in removing and servicing these components, special equipment, such as holding fixtures and transmission jacks, is needed.

Holding Fixtures

Holding fixtures, like the one shown in **Figure 3-26,** are handy for holding transmissions and transaxles. Some holding fixtures are designed to be used with a specific manufacturer's components, while others can be used with any transmission or transaxle.

Transmission Jacks

Transmissions and transaxles are heavy assemblies and can cause serious injury if they fall. Lifting them into place during installation can also pose problems. To prevent this, **transmission jacks** should be used when removing and installing transmissions and transaxles. See **Figures 3-27** and **3-28.**

Cleaning Equipment

Transmissions and transaxles collect dirt from the road. The dirt mixes with leaking oil. Heavy grease deposits result. Internal parts usually require cleaning because of the heavy oils used. Further, where there is internal damage, heavy debris and metal deposits will be present. For these reasons, special cleaning equipment is needed.

Line and Converter Flushers

In many cases, by the time an automatic transmission or transaxle requires overhaul, the transmission fluid has become contaminated with debris. The fluid has been

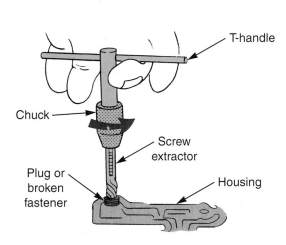

Figure 3-24. *A screw extractor is used for removing broken fasteners from parts. A hole must be drilled in the fastener before the kind of extractor pictured can be used. (General Motors)*

Figure 3-25. *This reamer is used to enlarge a valve body bore for use with a new oversized valve. The two cutting diameters allow the reamer to cut necessary large and small valve bores in one operation. (Sonnax)*

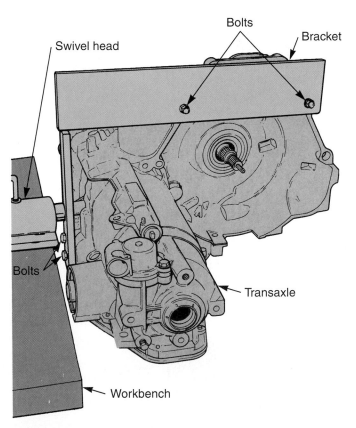

Figure 3-26. *A holding fixture simplifies the job of overhauling a transaxle. (General Motors)*

badly overheated, which caused it to break down and nonmetallic materials to deteriorate. Varnish and carbon have built up throughout the transmission as a result of the fluid. A defective transmission oil cooler can cause the fluid to become contaminated with water and antifreeze. The transmission parts must be thoroughly cleaned. Special equipment is needed to clean and flush the oil cooler and lines, as well as the torque converter.

The *oil cooler and line flusher,* **Figure 3-29,** uses solvent to remove contaminants from the cooler lines. Once the proper connections are made, solvent is forced through the lines, cleaning and flushing them in the

process. A final blast of compressed air removes the solvent, and the cooler and lines are ready for reuse. Some technicians prefer to use cooler flush in aerosol cans, **Figure 3-30.**

The *converter flusher,* **Figure 3-31,** is used to remove debris from inside the automatic transmission torque converter. It is a special device that moves a pulsating flow of solvent in and out of the converter. In addition, the converter flusher rotates the converter to further agitate the solvent inside the converter. This agitation and flow through the converter removes varnish and debris. The flushing action is usually accomplished in a few minutes.

Parts Washers

After you have removed all the parts from a transmission or transaxle assembly, everything should be cleaned. Problems can be hard to see when a part is covered with oil, grease, or carbon deposits. Closer part inspection can

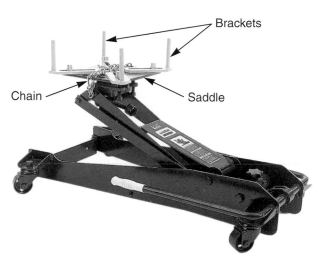

Figure 3-27. *A transmission jack, like the under-the-vehicle model shown, makes transmission removal much easier. (Lincoln Automotive)*

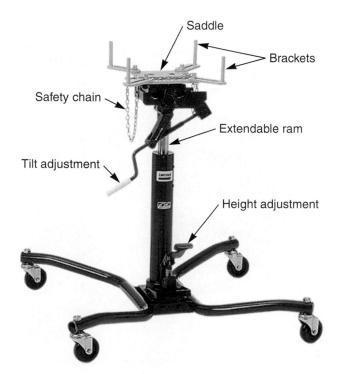

Figure 3-28. *This transmission jack is designed to be used when the vehicle is on a hoist. (Lincoln Automotive)*

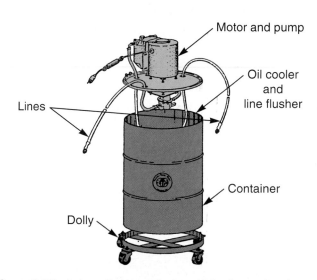

Figure 3-29. *Automatic transmission cooler lines often become filled with sludge, antifreeze, or metal. They must be cleaned when a transmission is overhauled. Cooler lines can be flushed easily with this pressurized equipment. (Hayden)*

Figure 3-30. *Aerosol cooler flush. The pressure in the can forces a cleaning agent through the cooler.*

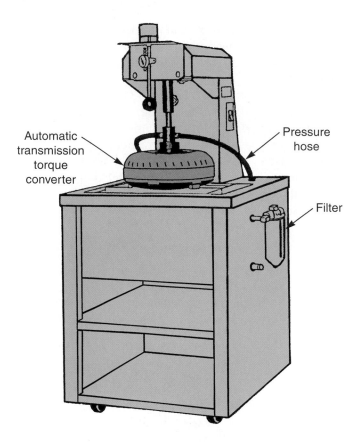

be done during and after cleaning. ***Parts washers*** are often used for cleaning. Different types are available, depending on the part construction and type of material. See **Figure 3-32.**

One type of parts washer is the *cold-soak tank.* It can be thought of as a sink that runs cleaning solvent instead of water. The solvent aids in dirt and grease removal. Most parts washers have a pump to recirculate the solvent. Parts are left to soak in the solvent. When necessary, brushes and scrapers are used for cleaning. Since the solvent is much harsher than water, you should always protect your hands and face when using it.

An ***immersion cleaner,*** or ***hot tank,*** is a more powerful type of parts washer. The part to be cleaned is placed in the tank, which is usually filled with strong corrosive chemicals. Cleaning is done automatically.

Disposal of Vehicle Maintenance Products

Certain products involved in vehicle maintenance, such as rust removers, parts cleaners, and degreasers, may contain hazardous materials. These products may contain toxic chemicals or strong acid or alkaline solutions. They must be disposed of properly.

Petroleum products used for vehicle operation, such as automatic transmission fluid and gear oil, are also materials that require special handling. If you generate hazardous waste, you might be subject to regulations that govern the disposal of these waste products. In particular,

Figure 3-31. *To avoid the expense of replacement, many shops clean sealed torque converters with a converter flusher. (Hayden)*

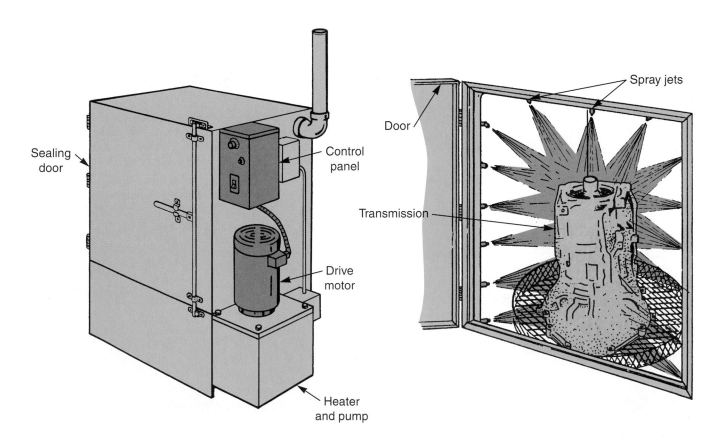

Figure 3-32. *This parts washer will clean the interior and exterior of large parts with spray action. (Hayden)*

any business that maintains or repairs vehicles, heavy equipment, or farm equipment is subject to the requirements of the Resource Conservation and Recovery Act. Check with your state's hazardous waste management agency or Regional EPA (Environmental Protection Agency) office for more information.

Measuring Tools

Many transmission and transaxle components require special measuring tools in the reassembly process and, afterwards, to ensure that they have been put together correctly. *Gauges* are used for this purpose. A gauge is an instrument used to measure something. In automatic transmission and transaxle service, *pressure gauges, size gauges,* and *angle gauges* are often used. Another gauge commonly used measures torque. This gauge is an integral part of a *torque wrench.*

Pressure Gauges

Pressure gauges measure pressure of a fluid. Most automotive gauges use the **gauge pressure** scale. This means that normal *atmospheric pressure* (14.7 psi [101 kPa] at sea level) is chosen as a zero reference pressure. These automotive gauges, then, measure pressure relative to that of the surrounding atmosphere. The zero reference is denoted 0 psig, where the *g* stands for *gauge;* it is equivalent to 14.7 psi (101 kPa). In many cases, this distinction for gauge pressure is unimportant, and just *psi* is used.

Pressure gauges are used in the troubleshooting of automatic transmissions and transaxles. They allow the technician to precisely measure internal oil pressure. All automatic transmissions have at least one pressure connection, called a *tap,* or *plug,* where the gauge can be attached. The gauge is installed at the tap, and the engine is started. Transmission pressures can then be checked for each of the transmission operating positions. The actual pressures are checked against the specifications given in the manufacturer's service manual. Any variation from the specified pressures means there is a problem in the transmission hydraulic system. See **Figure 3-33.**

A **vacuum gauge** is also shown installed in **Figure 3-33.** This type of pressure gauge is used to measure a **vacuum,** which is pressure below normal atmospheric pressure. Thus, vacuum gauges measure pressures *below* 0 psig (0 kPag). Vacuum pressure is often given in terms of *inches (millimeters) of mercury (in. Hg vacuum or mm Hg vacuum).* On the vacuum pressure scale, 0" Hg (0 mm Hg) is equivalent to 0 psig. A vacuum pressure of 2" Hg (52 mm Hg) is approximately equal to -1 psig, and 10" Hg (259 mm Hg) equals about -5 psig.

Vacuum gauges are useful for measuring engine manifold vacuum when diagnosing an automatic transmission that has a *vacuum modulator.* The vacuum gauge is attached to the intake manifold, and the engine is started.

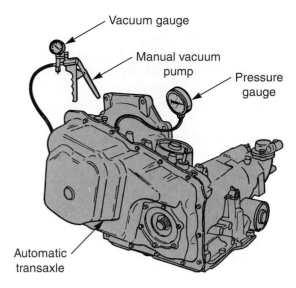

Figure 3-33. *Pressure gauge used here to check internal transmission pressures. This is an aid to locating internal fluid leaks or other problems. Manual vacuum pump is used to simulate vacuum condition. Vacuum pressure is indicated on vacuum gauge. (General Motors)*

Lower-than-normal vacuum indicates an engine problem that could affect transmission operation. A vacuum gauge can also be connected to the modulator line at the modulator. If the vacuum is much less at the modulator than at the intake manifold, the modulator line is plugged or is leaking.

Size Gauges

Size gauges are linear measurement devices. They are used for measuring part dimensions and clearances—the space between two parts. An example of one type of size gauge is the *micrometer,* **Figure 3-34.** This type of gauge has a graduated scale from which very accurate measurements can be read. *Feeler gauges,* which consist of metal strips made to exact thicknesses, are another example of size gauges. Feeler gauges are placed between two mating parts to measure the opening, or gap, between them.

Size gauges are often used prior to assembly to determine what clearances of mating parts will be after they are assembled. This eliminates the problem of assembling and reassembling parts to make adjustments.

Outside Micrometers

A **micrometer,** or **mike,** is a measuring device used to make very accurate measurements. An **outside micrometer** is pictured in **Figure 3-34.** This micrometer has a frame, a sleeve, a thimble, a spindle, and an anvil. The opening between the anvil and spindle can be made larger or smaller by turning the thimble. The size of the opening can be read on the sleeve and thimble. Micrometers are available in inch-based and metric-based versions.

To use an outside micrometer, place the part to be measured between the anvil and spindle. See **Figure 3-34.**

Turn the thimble to close the mike until it tightens lightly against the part. The part should be able to move between the anvil and spindle with a light drag. Do not overtighten the micrometer. This will distort the frame and cause inaccurate readings.

The basic steps of reading a micrometer are similar for both inch-based and metric-based micrometers. The following is an explanation of reading the inch-based micrometer, **Figure 3-35.** With the part in place, read the greatest numbered division visible on the sleeve. These numbers are tenths of an inch (0.1″). Then, read the fractional increment visible on the sleeve. Each division is 25 thousandths of an inch (0.025″). Add these readings together and write them down. Next, read the number on the thimble that matches the centerline on the sleeve. Each division is 1 one-thousandth of an inch (0.001″). To obtain the final reading, add the sleeve and thimble readings together.

Note that a fourth decimal place is added to the final reading if there is an auxiliary scale, known as a *vernier,* on the sleeve of the mike. To find the fourth decimal place, which will be in ten-thousandths of an inch, determine which vernier line aligns exactly with one of the lines on the thimble.

The following is an explanation of reading the metric-based micrometer, **Figure 3-36.** With the part in place, read the greatest division visible above the centerline of the sleeve. Each division is 1 millimeter. Add 0.5 millimeter if a following division mark is visible below the centerline. (On some micrometers, 1 mm divisions fall below the centerline, and 0.5 mm divisions fall above.) Write the number down. Add to this the number on the thimble that aligns with the centerline on the sleeve. Each division is 1 one-hundredth of a millimeter (0.01 mm).

Hole Gauges and Telescoping Gauges

These tools are used to make accurate measurements of the inside diameter of bushings and other internal surfaces. The **hole gauge** is used to measure small-diameter holes or openings. The gauge is inserted into the hole and adjusted to fit, **Figure 3-37A.** Then, it is removed and measured with an outside micrometer. The **telescoping gauge** is generally used to measure larger openings. The gauge is placed in the hole, properly tightened, and locked. See **Figure 3-37B.** It is then withdrawn from the hole and measured with a micrometer. The micrometer reading is the size of the opening.

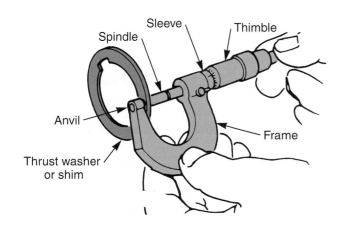

Part number	Thickness
90441–PLS–000	4.00 mm (0.157 in.)
90442–PL5–000	4.05 mm (0.159 in.)
90443–PL5–000	4.10 mm (0.161 in.)
90444–PL5–000	4.15 mm (0.163 in.)
90445–PL5–000	4.20 mm (0.165 in.)
90446–PL5–000	4.25 mm (0.167 in.)
90447–PL5–000	4.30 mm (0.169 in.)
90448–PL5–000	4.35 mm (0.171 in.)
90449–PL5–000	4.40 mm (0.173 in)

Figure 3-34. *The micrometer is used on machined parts to check that parts fall within tolerance. It is shown being used to check a thrust washer for proper thickness. (Honda)*

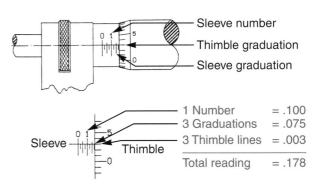

1 Number	= .100
3 Graduations	= .075
3 Thimble lines	= .003
Total reading	= .178

Figure 3-35. *Study basic steps for reading a micrometer graduated in thousandths of an inch. Read number, then sleeve graduations, then thimble. Add these three values to obtain reading. Total reading above is 0.178 inches. (Starrett)*

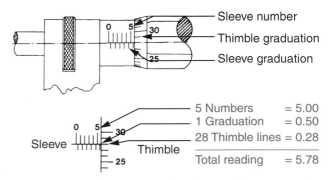

5 Numbers	= 5.00
1 Graduation	= 0.50
28 Thimble lines	= 0.28
Total reading	= 5.78

Figure 3-36. *Metric micrometer is read like an inch-based micrometer. However, note metric values for sleeve and thimble. Total reading above is 5.78 millimeters. (Starrett)*

Inside Micrometers

An **inside micrometer, Figure 3-38,** is used for measuring larger inside diameters or other part openings. The instrument is placed within the bore or other opening and adjusted to fit. It is read in the same manner as an outside micrometer.

Depth Micrometers

A **depth micrometer** is similar to the outside micrometer, but it is used to measure the depth of openings in machined surfaces. **Figure 3-39** shows a depth micrometer being used to check the depth of a bearing bore in a transmission case.

Calipers

Instruments called **calipers** make up a whole class of tools used for external or internal measurements. **Outside** and **inside calipers, Figure 3-40,** essentially consist of a pair of movable legs. In measuring, the legs are adjusted to fit the dimension in question. The caliper can then be laid over a rule, and the span can be measured. Calipers of this type are less accurate than micrometers.

The **vernier caliper** is a type of sliding caliper. It is an accurate tool, and is often used for measuring inside and outside diameters. This type of caliper is a ruled instrument having a set of jaws—one fixed and one movable. The movable jaw has a sliding scale, or vernier, that is part of

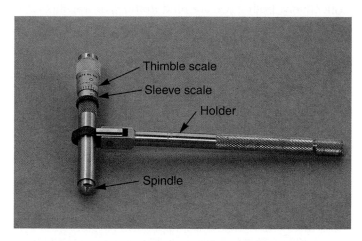

Figure 3-38. *Inside micrometers, such as the one shown here, are used to measure larger holes and openings. (Starrett)*

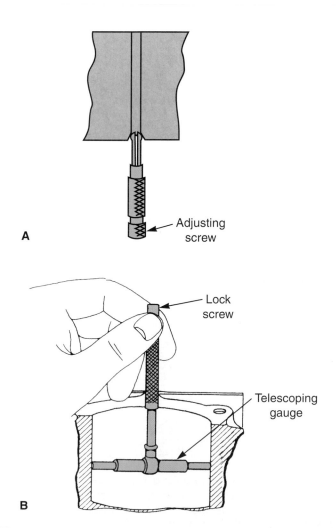

Figure 3-37. *Tools used to measure diameters of small holes and openings. A—Hole gauge. B—Telescoping gauge. (Vaco)*

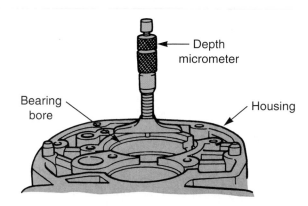

Figure 3-39. *The depth micrometer measures depth of bores and depressions in machined parts. The graduated scale (not shown) of the depth micrometer is read in the same manner as the outside micrometer. (Ford)*

Figure 3-40. *Outside and inside calipers used for making rough measurements on part openings. (Starrett)*

it. A component is placed within the jaws of the caliper, and the movable jaw is adjusted to fit the components. The sliding scale moves with the movable jaw, as they are a unit. **Figure 3-41** shows a typical use of a vernier caliper.

Vernier calipers come in inch-based and metric-based versions. Readings can be obtained to 0.001″ and 0.02 mm, respectively. Both types are read in more or less the same way. The measurement marked by 0 on the vernier scale is read off the fixed scale. To this, the amount on the vernier is added. This amount will be read from the vernier line that coincides with a line on the fixed scale. Only one vernier line will line up exactly with a line from the fixed scale.

A *dial caliper* is another type of sliding caliper. With this type of measuring tool, the vernier is replaced by a dial gauge. Otherwise, the two instruments look and operate the same. The dial gauge reads directly to 0.001″ or 0.02 mm. It makes precision measurements easier to read.

Feeler Gauges

A **feeler gauge** is used to measure spaces, or gaps, between two surfaces. A *flat feeler gauge* is made of a thin strip of steel or brass. A *wire feeler gauge* is made of a wire of specified diameter. Feeler gauges are usually sold in sets. A set will contain a number of gauges of different diameters or thicknesses. See **Figure 3-42.**

A feeler gauge is used by inserting it in the gap to be measured. Different thicknesses are tried. The correct gauge will drag slightly when pulled between the parts. The clearance, then, will be given by the size written on the gauge.

Straightedges are used to check for warping of part surfaces. These hardened-steel bars are manufactured so that one edge is very straight. They serve as the standard against which to compare. The tool is placed directly on the part, and the part surface is checked in relation to the straightedge. If a feeler gauge over a certain size can be inserted between the part and straightedge, the part is excessively warped.

Dial Indicators

A **dial indicator** is a gauge that is used on moving parts to measure small amounts of movement. These instruments are frequently used to check gear *backlash,* or clearance between teeth, shaft *end play,* cam lobe lift, and similar kinds of part movements. The gauge, which has a dial face, is operated by a plunger. Typical scale graduations are in thousandths of an inch or hundredths of a millimeter, but finer divisions are available. The gauge can be *zeroed* (adjusted to zero) by turning the dial face assembly.

The dial indicator is mounted on or near the component to be checked. The contact point, or plunger, is positioned so it contacts the movable portion of the component with a slight pressure. The gauge assembly should be secured so that only the plunger can move. Once the gauge is firmly in place, it should be zeroed. Again, this is done by turning the dial face. The movable portion of the component is then moved back and forth or rotated, and a movement amount will be indicated on the dial face, **Figure 3-43.**

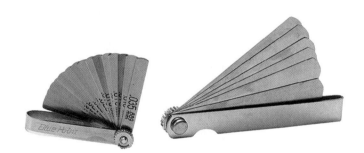

Figure 3-42. *Feeler gauges are useful for checking small clearances. (Snap-on Tool Corp.)*

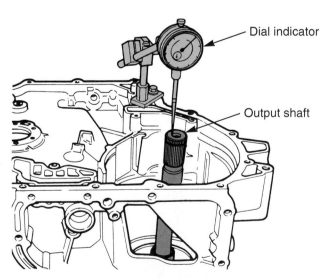

Dial indicator

Output shaft

Figure 3-43. *This dial indicator is being used to check output shaft end play. If the dial indicator is installed and adjusted properly, end play can be checked to a thousandth of an inch or better. (Ford)*

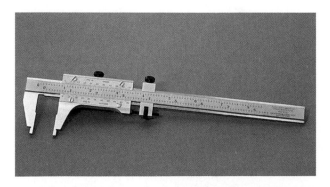

Figure 3-41. *A vernier caliper being used to check the thickness of a clutch disc. (Honda)*

Torque Wrenches

A **torque wrench** is used to apply a *measured* turning force, or torque, to a threaded fastener. The amount of torque is indicated on the tool while the force is applied. Torque wrench scales usually read in *foot-pounds (ft lb)* or in metric units of *Newton-meters (N·m)*.

In **Figure 3-44,** a torque wrench is being used to verify the amount of preload (mild pressure) on an installed bearing. In this procedure, enough torque is applied to the tightened fastener to cause it to move. Reading the scale at this point will indicate the amount of torque on the bolt, which equates to the amount of preload on the bearing.

Every threaded fastener has a torque value. Some parts of the vehicle require very precise torquing procedures. These procedures and the torque values are given in the vehicle service manual. If the specifications are not available, standard torque references, **Figure 3-45,** must be used to determine the proper torque.

Figure 3-44. *This torque wrench is being used to tighten a fastener to the proper torque value. (General Motors)*

Bolt Torque

Bolt size	Grade 5		Grade 8	
	N·m	ft lb (in lb)	N·m	ft lb (in lb)
1/4-20	11	(95)	14	(125)
1/4-28	11	(95)	17	(150)
5/16-18	23	(200)	31	(270)
5/16-24	27	20	34	25
3/8-16	41	30	54	40
3/8-24	48	35	61	45
7/16-14	68	50	88	65
7/16-20	75	55	95	70
1/2-13	102	75	136	100
1/2-20	115	85	149	110
9/16-12	142	105	183	135
9/16-18	156	115	203	150
5/8-11	203	150	264	195
5/8-18	217	160	285	210
3/4-16	237	175	305	225

A

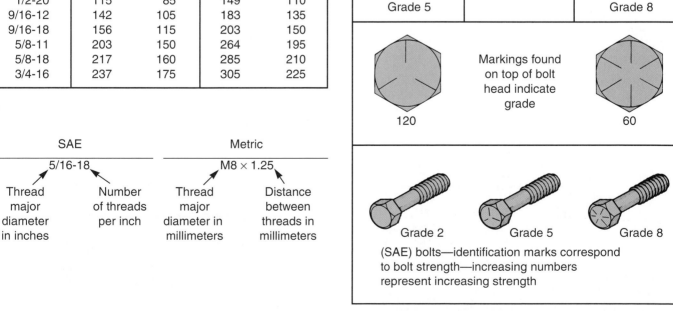

Metric bolts—identification class numbers correspond to bolt strength—increasing numbers represent increasing strength

C

SAE	Metric
5/16-18	M8 × 1.25
Thread major diameter in inches	Number of threads per inch	Thread major diameter in millimeters	Distance between threads in millimeters

B

SAE Classification		
Grade 5		Grade 8
120	Markings found on top of bolt head indicate grade	60
Grade 2	Grade 5	Grade 8

(SAE) bolts—identification marks correspond to bolt strength—increasing numbers represent increasing strength

D

Figure 3-45. *Torque references should be consulted when torque requirements of a particular bolt is unknown. A—This table shows the maximum torque values for various sizes and grades of threads. B—This reference explains how to interpret SAE (inch) and metric thread notations. C—Strength of metric bolts may be determined by identification class number on head of bolt. D—Strength of SAE bolts may be determined by identification markings embossed on bolt head. (DaimlerChrysler)*

Electrical and Electronic Test Equipment

To service modern electronically controlled transmissions and transaxles, you will need a variety of electrical and electronic test equipment.

Jumper Wires

A *jumper wire,* or *jumper,* is a short piece of insulated wire with an alligator clip on each end. It is commonly used to bypass components and to apply voltage to a component or section of a circuit. It may be used to determine whether current is flowing through a switch, relay, solenoid, electrical connection, wire, or other electrical component. Jumper wires should be equipped with a fuse or circuit breaker to reduce the chance of damaging a circuit.

Figure 3-46 shows how a jumper wire is used to check a switch. Prior to bypassing the switch with the jumper wire, the bulb would not light when the switch was in the "on" position (contacts should have been made, or closed). If attaching the jumper around the switch turns the light on, then the switch is bad. If it does not, the problem is elsewhere in the circuit.

Test Lights

A *nonpowered test light,* **Figure 3-47,** is used to check a circuit for power. This type of test light essentially consists of a probe with a light and a lead with an alligator clip. The light is powered off the circuit. To use a nonpowered test light, connect the one lead to ground. Then touch the probe to a desired test point in the circuit to check for voltage at that point. If there is voltage, the light will glow. If the light does not glow, the circuit is open between the battery and the test point. If there is voltage at the next test point nearer the power source, the open lies somewhere between the two test points.

A *self-powered test light* is used to check for circuit *continuity,* or whether a circuit is complete. This device resembles the nonpowered test light, but it has an internal battery. To use this type of test light, the normal source of power (car battery or feed wire) must be disconnected. The test leads should be connected across the circuit component or components in question. The circuit under test must be isolated, or disconnected from any circuits in parallel. If the light glows when the leads are connected, the component or circuit has continuity. If it does not glow, there is an open somewhere between the two test leads.

> **Caution: Electronic components can be destroyed by careless use of jumper wires or test lights. Do not use a jumper wire or test light to check any electronic devices unless it is specifically allowed by the manufacturers' service instructions. When testing any nonelectronic device with a jumper wire or test light, be sure the device is disconnected from all electronic components.**

Scan Tools

The *scan tool* is used to obtain trouble codes and other information from the vehicle's on-board computer. Scan tools are small computers that can communicate with the vehicle's computer. They consist of a tester housing about the size of a VCR tape cassette. See **Figure 3-48.** The housing has a small screen that displays information. When started, the scan tool screen displays menus that are used to access most of the information. Menus are lists of available test procedures and diagnostic operations. Below the screen is a keypad or a series of pushbuttons that allows the technician to select various menu screens and make tests. The screen displays trouble codes and diagnostic test results as the technician requests them.

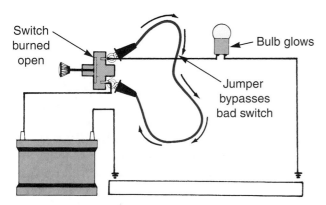

Figure 3-46. *A jumper wire can be used for diagnosis. In this example, the wire is connected across the terminals of a switch that may be defective. If the light now comes on, the bypassed switch is the problem.*

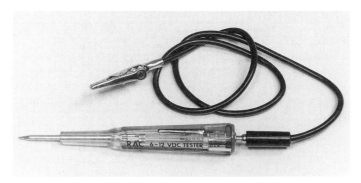

Figure 3-47. *A nonpowered test light. To use this kind of test light, connect the clip to a good ground and then probe the positive side of the circuit. If the light comes on, electrical power is available to the point being probed. The circuit must be energized (have voltage available) to make this test.*

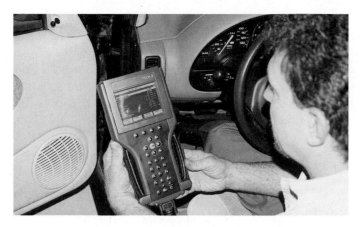

Figure 3-48. *A scan tool is used to perform various computer control system troubleshooting functions. Scan tools can be used to retrieve trouble codes and obtain readings from sensors and output devices. Some scan tools can be used to reprogram the vehicle computer. The scan tool shown here is used to troubleshoot OBD II systems.*

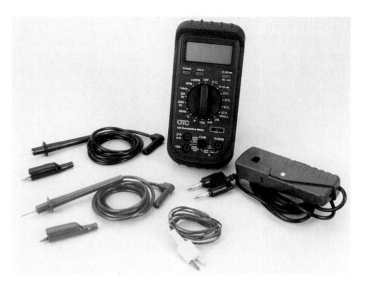

Figure 3-49. *Multimeters can be used to check electrical values. The electrical leads shown to the left in the photograph allow the multimeter to check voltage and resistance. These leads can also be used to check low-amperage circuits. The clamp-on pickup on the right allows the multimeter to be used to check high-amperage circuits without damaging the meter. (OTC)*

The scan tool is attached to the vehicle's diagnostic connector through a cable and communicates with the vehicle computer. The simplest scan tools will display trouble codes. A more sophisticated scan tool will display trouble codes and check the operation of the input sensors and output devices. Some scan tools are able to reprogram the vehicle's computer with updated information.

There are two main types of scan tools. The *dedicated scan tool* will interface (communicate) with the computer systems of one manufacturer. These tools will check all the computer operated systems of that manufacturer. The *multi-system scan tool* will interface with the computer systems of many manufacturers. This is sometimes called a generic scan tool. It may be able to diagnose all parts of every system or only selected portions.

To use any scan tool, make sure the ignition switch is turned off and plug the tool's connector into the diagnostic connector. Connect the tool to a power source, if necessary. Then follow the instructions on the scan tool screen to retrieve trouble codes or perform other diagnostic routines.

Multimeters

Modern **multimeters,** such as the one in **Figure 3-49,** can read all major electrical values (voltage, resistance, and amperage) and may be able to read voltage waveforms and provide other information. All modern multimeters are digital types that display the reading as a number. Older multimeters were analog types, which used a needle that moved against a calibrated scale. The modern multimeter will contain the individual meters discussed in the following sections.

Note: Electrical values, such as voltage and resistance, will be explained in Chapter 7.

Voltmeter

The voltmeter section of the multimeter is connected to read the voltage from the positive and negative sides of a circuit. To read the voltage at any connection, connect the leads to the connection and to ground. Then select the proper voltage range (if necessary) and observe the reading.

The voltmeter can also be connected to read the voltage across a connection as current flows through it. If the connection has high resistance, current will try to flow through the meter, creating a voltage reading. Voltage higher than the specified figure means that the connection must be cleaned or replaced.

Ohmmeter

Ohmmeters are used to check for continuity and exact resistance values. These are always measured in ohms. To make an ohmmeter check, turn on the multimeter and set it to ohms. Most modern digital ohmmeters will select the correct range automatically. Then attach the leads to the correct wires or terminals as explained in the service manual. When checking wires or relay contacts for continuity, the resistance should be at or near zero. Other parts, such as solenoid windings, temperature sensors, or throttle position sensors, should have a specific amount of resistance. If the reading is zero or infinity, the part is defective. Resistance of temperature sensors should change with changes in temperature. Resistance of throttle position sensors should change when the throttle is opened.

An analog ohmmeter often does a better job of checking the operation of throttle position sensors than a digital ohmmeter, **Figure 3-50.** Set the meter scale to the position that will read the normal resistance range of the throttle position sensor being tested (generally 0–5 ohms). Then connect the leads to the correct wires as identified in the service literature. Move the throttle while observing the ohmmeter needle. It should rise smoothly. If the needle jumps or moves erratically, the throttle position sensor is defective.

Ammeter

The multimeter will also have an ammeter setting capability. Most are capable of handling amperage of up to ten amps. When measuring greater amperage, many modern ammeters are equipped with an inductive pickup. This pickup is clamped over the current-carrying wire. The pickup reads the magnetic field created by current flowing through the wire and converts it into an amperage reading.

 Note: Many electronic components can be severely damaged by careless use of multimeters. Always check the manufacturer's literature before testing any electronic part.

Waveform Meters

The waveform created by the operation of some electrical and electronic devices can be observed to obtain various types of diagnostic information. Some multimeters can display waveforms. Another device for displaying waveforms is a lab scope, which some shops may have.

A typical waveform is shown in **Figure 3-51**. This waveform shows a solenoid being energized (turned on) and then de-energized (turned off). Waveform meters may be part of a multimeter, or they may be separate units. The

advantage of waveforms is that they can indicate a problem more accurately than just voltage or resistance readings. For instance, the solenoid waveform shown in **Figure 3-52** is not shaped exactly like the standard waveform in **Figure 3-51.** Note the slope during the solenoid "on time" signal. This slope could indicate a problem in the solenoid windings or connections.

Service Information

To properly service any automatic transmission or transaxle, the technician must have many sources of service information. Automatic transmissions and transaxles are complex assemblies and their designs vary greatly from manufacturer to manufacturer. Often the same transmission has been updated between two model years. The sources of service information described in the following sections can simplify the technician's job.

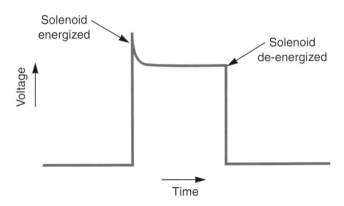

Figure 3-51. *This illustration is a general representation of a waveform that occurs during normal operation of a transmission solenoid. The voltage is at zero until the solenoid is energized. The voltage rises sharply when the solenoid is energized. It then drops slightly once the solenoid plunger is pulled in, and then stays at this level until the solenoid is de-energized.*

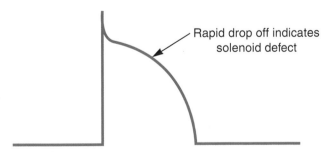

Figure 3-52. *Compare this waveform with the waveform shown in Figure 3-51. The voltage rises when the solenoid is energized, but then drops off before the solenoid is de-energized. The drop may indicate that the solenoid winding is shorted or that it is drawing excessive current. It may also indicate a problem with the computer solenoid driver. This type of problem would not be caught if the voltage were simply checked with a voltmeter.*

Figure 3-50. *An analog (needle type) multimeter, such as the one shown here, is often useful when checking throttle position sensors for defects. If the meter needle does not rise smoothly when the throttle is opened and closed, there may be a bad contact inside the sensor.*

Note: Much of the service literature described below is available on CD-ROM. CDs can be inserted in a computer equipped with a CD drive. The information can then be accessed and read on the computer screen.

Manufacturers' Service Manuals

Manufacturers' *service manuals* contain the information needed to repair transmissions and transaxles. They contain information specific to one vehicle make or model. Manufacturers' service manuals are divided into diagnosis and overhaul manuals.

Diagnostic Manuals

Many vehicle manufacturers publish separate *diagnostic manuals*. These manuals contain only the information needed to diagnose problems. The diagnostic manual consists of troubleshooting procedures and discussion of common problems. Diagnostic manuals also contain scan tool data and other specifications. The repair of components is covered in the overhaul manual.

Overhaul Manuals

Overhaul manuals contain comprehensive information needed to disassemble, inspect, and reassemble vehicle components. This information is confined to disassembly and reassembly only, and does not cover transmission and transaxle diagnosis. A typical overhaul manual is shown in **Figure 3-53.**

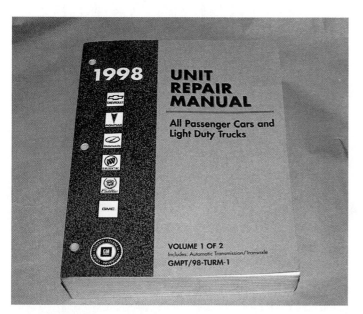

Figure 3-53. *A typical transmission/transaxle overhaul manual. This manual contains information on how to disassemble and reassemble the transmission or transaxle, and how to service internal parts. It contains clearance, fastener torque, and other repair information, but does not include diagnostic information.*

General Service Manuals

General service manuals are more general than manufacturers' manuals and contain service information for many types of vehicles. Motor, Chilton, and Mitchell are typical general service manuals. Like manufacturers' manuals, these manuals may be available in diagnosis, overhaul, and other variations. Many general manual publishers have separate transmission and transaxle manuals.

Schematics

It is almost impossible to repair modern vehicles without occasionally referring to a *schematic.* A schematic is a graphic representation of a hydraulic, pneumatic, or electrical system. Tracing the flow of power along the path shown in the schematic allows the technician to tell which components should be energized under what conditions. Techniques for reading an electrical schematic are given in Chapter 7. A common oil-flow schematic is shown in **Figure 3-54.** Reading oil-flow schematics is covered in Chapter 9.

Troubleshooting Charts

Troubleshooting charts contain the logical steps used to determine the cause of a problem. A typical troubleshooting chart is shown in **Figure 3-55.** To use a troubleshooting chart, begin at the top and follow the instructions. In most cases, the chart will call for a yes or no answer at certain points. Depending on your answer, you will be directed to further diagnostic steps. If you follow the chart closely, you will reach the proper diagnostic conclusion.

Telephone Hotlines

If all other sources of information are exhausted, many technicians call a manufacturer's or part supplier's *telephone hotline.* Some general service manual manufacturers provide technical support services over a technical hotline. Calling these hotlines will connect you with a technical-support person. Hotline personnel often have information gathered from actual repair and diagnosis situations. This is a way for the technician to obtain real life information that would otherwise not be available. Hotline personnel will also have access to the latest update information from manufacturers' engineering departments.

Some vehicle manufacturer's hotlines are available only to the technicians who work for that manufacturer's dealerships. Other hotlines are available by subscription. These hotlines can be accessed after a yearly fee is paid. Some parts manufacturers' hotlines are available to anyone. These are primarily aimed at the technician who has questions about the manufacturers' parts.

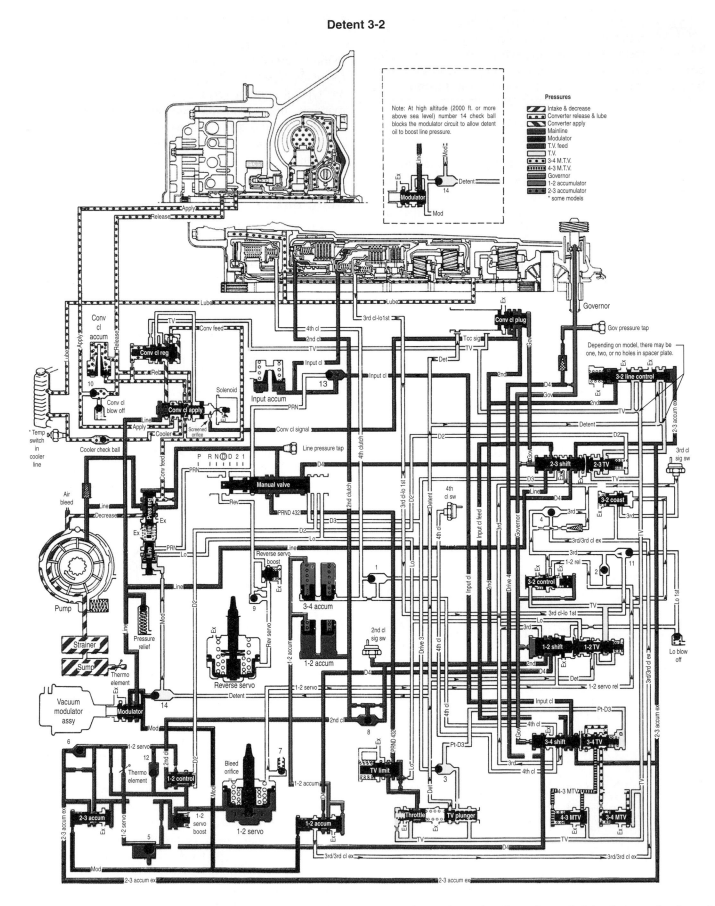

Figure 3-54. *A transmission/transaxle hydraulic pressure schematic, or oil flow schematic, allows the technician to trace the flow of oil pressure through the transmission or transaxle. Tracing the oil flow circuits allows the technician to determine which hydraulic component could be the cause of a problem. (General Motors)*

Flowchart to Diagnose Concerns on an Electronically Controlled Transmission

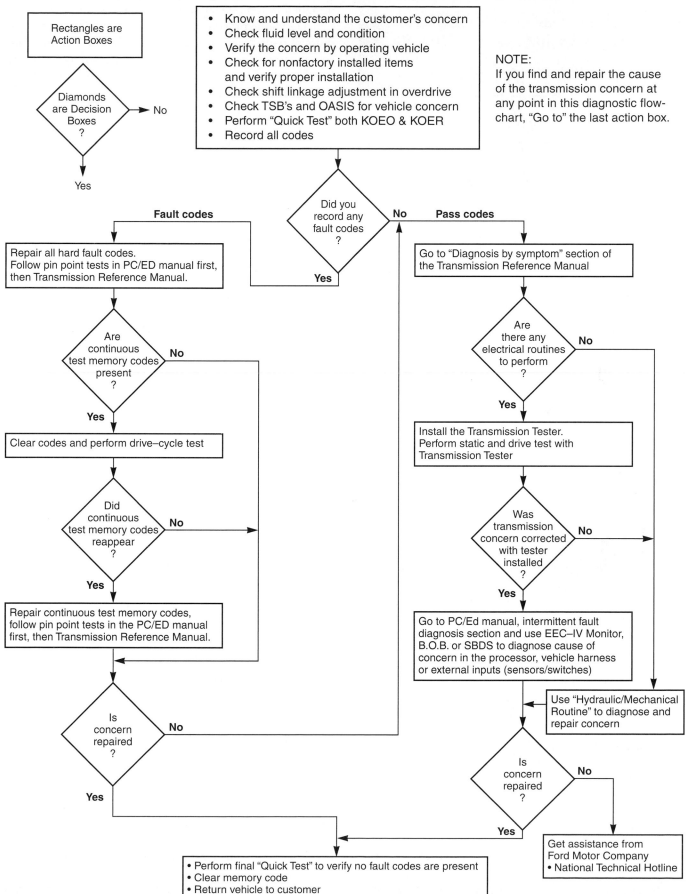

Figure 3-55. *Troubleshooting flowcharts, such as the one shown here, specify the steps a technician should take to most efficiently diagnose a problem. Many technicians find that the flowchart is easier to follow than the same information in text form. (Ford)*

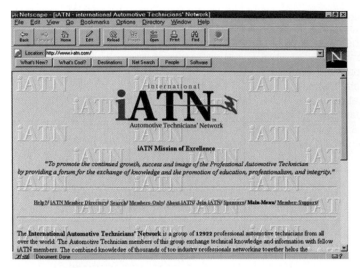

Figure 3-56. *The Internet can be a good source of technical information. Some Internet sites allow technicians to ask questions and share information about difficult problems.*

Another use for technical hotlines is computer reprogramming. If a vehicle computer needs updated software, the software can sometimes be obtained from the manufacturer over the phone line. How this is done will be explained in a later chapter.

Internet Resources

The Internet has become a valuable source of automotive repair information. Many manufacturers have websites (locations on the Internet) that contain technical information. Organizations such as the International Automotive Technicians Network (iATN) provide a way for technicians from around the world to help each other by way of e-mail. See **Figure 3-56.**

Summary

Special service tools are often needed to repair transmission and transaxle parts. These tools are made specifically for troubleshooting or repairing a particular component or a group of similar components.

There are many special service tools, most of which can be purchased from independent parts suppliers. Other tools must be purchased from special tool manufacturers, listed in vehicle service manuals.

Special wrenches are designed to turn fasteners that have odd head shapes or that are positioned so that they cannot be reached with the more common hand wrenches.

Punches and drifts are special service tools for removing small pins or shafts. They are designed to be used with a hammer. Picks and snap ring pliers are designed to remove snap rings or other small parts.

Hydraulic presses are useful for removing or installing press-fit parts. Various adapters must be used during disassembly and reassembly of bearings to securely hold them in place. Presses should always be used very carefully.

Drivers are used to install bearings or seals in housings. They must be used carefully so that the part is installed correctly. Some drivers are used in combination with a hydraulic press, while others are designed to be struck with a hammer.

Pullers are used to remove pressed together parts, such as bushings, bearings, or seals. These pullers operate by applying pressure through screw threads or by the impact of a weight sliding on a shaft.

Spring compressors are used to compress coil springs, which removes pressure from snap rings so that they can be easily removed. Seal protectors are used to prevent seal damage when installing seals over shafts.

Special tools are required to repair or make tubing connections. Cutting tools cut the tubing in the most efficient manner. Flaring tools are used to flare the tubing ends. Bending tools are used to bend tubing without kinking it.

Alignment tools are used to keep parts in alignment. Flywheel locking and turning tools are used to turn or lock the flywheel, to gain access to fasteners or to lock them thoroughly.

Organizing trays are used to keep small parts in proper order. They are useful when working on drive train components containing many small parts. Drain pans will prevent a potentially dangerous mess by catching oil that would otherwise end up on the shop floor.

Screw extractors are used to remove broken fasteners from parts. Some extractors are designed to grip the fastener from the outside. Others are used internally by drilling a hole in the broken fastener.

Holding fixtures are useful for holding large heavy transmissions and transaxles. Some holding fixtures can be used on only one kind of transmission, while others are designed to fit many types of transmissions and transaxles.

Transmission jacks make transmission and transaxle removal and replacement much easier. Some jacks are used under the vehicle, while others are used when the vehicle is raised on a hoist. To prevent injury and accidents, they should always be used if available.

Drive train components collect large amounts of dirt, sludge, and metal particles, and they must be carefully cleaned. Line flushers, converter flushers, and parts cleaners are necessary to thoroughly clean drive train parts. They should be used according to the manufacturer's directions.

There are many gauges available. Size gauges are used to measure a size or clearance. Angle gauges measure angles. Pressure gauges are used to diagnose automatic transmissions. They measure internal transmission oil pressures. Vacuum gauges measure intake manifold vacuum to diagnose vacuum modulator problems.

Micrometers are able to make very precise size measurements. To accurately use a micrometer, you must proceed very carefully. The micrometer must be carefully tightened on the part. The micrometer must also be carefully read. Micrometer reading varies for inch-based and metric-based micrometers.

Hole and telescoping gauges are used to make very accurate measurements of inside diameters. A depth micrometer is used to measure the depth of openings in machined surfaces. Calipers have movable legs or jaws to gauge the size of openings.

Feeler gauges are made in different thicknesses. They are used to gauge openings, or gaps. They come in flat and wire varieties. Straightedges are flat bars used to gauge the straightness, or flatness, of parts against which they are placed. Straightedges are often used in combination with feeler gauges.

Dial indicators are used for moving drive train parts to measure small amounts of movement. They are easy to use but must be carefully installed, adjusted, and read.

The torque wrench is used to apply torque to a part and to measure this torque at the same time. Torque wrenches are usually used to tighten fasteners. They are sometimes used to measure bearing preload. There are many varieties of torque wrenches. Always consult the manufacturer's tightening specifications before tightening a part.

Various sources of service information can be used to obtain tightening values and sequences, repair procedures, and other specifications. The appropriate service information must be used to properly repair an automatic transmission or transaxle. Common sources of service information include manuals, troubleshooting charts, telephone hotlines, and the Internet.

Review Questions—Chapter 3

Please do not write in this text. Place your answers on a separate sheet of paper.

1. _____ tools are designed to be used on many parts for many purposes.

2. _____ _____ are special tools used to tighten large parts or to hold a part while another tool is used.

3. A press fit is an example of a(n) _____ fit.
    (A) clearance
    (B) shrink
    (C) interference
    (D) running

4. When using a driver to install and remove bushings, the outside diameter of the driving surface should be slightly _____ than the bore diameter.

5. A(n) _____ _____ is used to prevent damage when a seal is installed over a shaft with a sharp edge.

6. _____ can be used to remove broken screws or other broken fasteners from parts.
    (A) Stud pullers
    (B) Screw extractors
    (C) Vise grip pliers
    (D) All of the above.

7. An immersion cleaner is _____.
    (A) generally filled with strong, corrosive chemicals
    (B) is used to remove debris from inside a torque converter
    (C) forces solvent through the transmission cooler lines
    (D) None of the above.

8. When using an outside micrometer, the part to be measured is placed between the _____ and the _____.

9. _____ calipers can be used to take measurements that are accurate to 0.001″ or 0.02 mm.

10. A jumper wire is a short piece of _____ wire.

11. In addition to the parts used in a nonpowered test light, a powered test light contains a _____.

12. Which of the following could be destroyed by careless use of jumper wires or test lights?
    (A) Bulbs.
    (B) Mechanical relays.
    (C) Electronic relays.
    (D) Solenoids.

Match each of the following scan tool parts with its function.

13. Screen _____
14. Keypad _____
15. Cable _____

    (A) Connects scan tool to vehicle computer.
    (B) Displays information.
    (C) Selects menus.

16. Which type of scan tool can be used on the vehicles of one manufacturer only?

17. All modern multimeters can read at least three major electrical values. Name them.

18. What does it mean when an actual waveform does not match the standard waveform?

Match each of the following types of service literature with its description.

19. Manufacturer's overhaul manual _____

20. General service manual _____

21. Troubleshooting chart _____

22. Schematic _____

23. Troubleshooting manual _____

(A) Repair procedures for many kinds of vehicles.

(B) Written instructions on how to diagnose a problem.

(C) List of parts prices and availability.

(D) A graphic representation of a vehicle system.

(E) A graphic representation of diagnostic procedures.

(F) Information on how to rebuild parts on one type of vehicle.

24. The technician can reach persons with service information over a telephone _____.

25. Manufacturers websites that contain technical information can be found on the _____.

ASE-Type Questions—Chapter 3

1. All of these statements about spanner wrenches are true *except:*
    (A) they are sometimes used to tighten large fasteners.
    (B) they are sometimes used to hold parts as they are tightened by another wrench.
    (C) they are designed to fit standard 3/8-in. or 1/2-in. ratchets.
    (D) some are bolted to the part that they hold.

2. Technician A says that a punch should be placed against the part to be removed before striking the punch with a hammer. Technician B says that both punches and chisels can be struck with a hammer. Who is right?
    (A) A only.
    (B) B only.
    (C) Both A and B.
    (D) Neither A nor B.

3. Technician A says that snap ring pliers can be used for both removing and installing snap rings. Technician B says that picks should never be used to remove snap rings or retainers. Who is right?
    (A) A only.
    (B) B only.
    (C) Both A and B.
    (D) Neither A nor B.

4. All of these can be installed by the use of drivers *except:*
    (A) seals
    (B) O-rings
    (C) bearings
    (D) bushings

5. Technician A says that some pullers remove parts by the use of screw force or impact. Technician B says that some pullers remove parts by the use of hydraulic pressure. Who is right?
    (A) A only.
    (B) B only.
    (C) Both A and B.
    (D) Neither A nor B.

6. In drive train service, the spring compressor is usually used in automatic transmission service to remove:
    (A) diaphragm springs
    (B) clutch release springs
    (C) leaf springs
    (D) torsion springs

7. Technician A says that special tools should be used to cut, flare, and bend tubing. Technician B says that using a tubing bender will cause the tubing to kink. Who is right?
    (A) A only.
    (B) B only.
    (C) Both A and B.
    (D) Neither A nor B.

8. Technician A says that flywheel turners are bolted to the engine or another stationary part of the drive train. Technician B says that flywheel turners engage the teeth on the flywheel ring gear. Who is right?
    (A) A only.
    (B) B only.
    (C) Both A and B.
    (D) Neither A nor B.

9. Cold-soak tanks and hot tanks are types of _____.
    (A) parts washers
    (B) converter flushers
    (C) oil cooler flushers
    (D) drain pans

10. A vacuum gauge is a type of pressure gauge that is most commonly used to measure _____.
    (A) automatic transmission pressures
    (B) intake manifold vacuum
    (C) bolt torque
    (D) coil spring release tension

11. All of these are types of size gauges *except:*

    (A) micrometers
    (B) calipers
    (C) torque wrenches
    (D) dial indicators

12. A flywheel is being checked for excessive warping. Technician A says that a feeler gauge and straightedge can be used to make this check. Technician B says that a dial indicator can be used to make this check. Who is right?

    (A) A only.
    (B) B only.
    (C) Both A and B.
    (D) Neither A nor B.

13. Technician A says that some threaded fasteners do not have torque values. Technician B says that an experienced technician does not need to use a torque wrench. Who is right?

    (A) A only.
    (B) B only.
    (C) Both A and B.
    (D) Neither A nor B.

14. An ohmmeter is used to measure _____.

    (A) voltage
    (B) current
    (C) resistance
    (D) temperature

15. Technician A says a general service manual contains information on many types of vehicles. Technician B says an overhaul contains information about diagnosis, disassembly, and reassembly. Who is right?

    (A) A only.
    (B) B only.
    (C) Both A and B.
    (D) Neither A nor B.

Cutaway of an electronically controlled, five-speed automatic transmission. Gears, chains, and bearings are the building blocks of all transmissions and transaxles. (Ford)

Chapter 4

Gears, Chains, and Bearings

After studying this chapter, you will be able to:
- ❑ Identify certain basic parts of a gear.
- ❑ Discuss gear reduction and overdrive.
- ❑ Name and describe types of gears found in automatic transmissions and transaxles.
- ❑ Calculate gear ratio.
- ❑ Describe the construction and explain the operation of a planetary gearset.
- ❑ Discuss common gear problems.
- ❑ Describe the construction and explain the operation of a chain drive.
- ❑ Summarize general methods of chain drive lubrication and inspection.
- ❑ Discuss aspects of the different types of bearings.
- ❑ Summarize fundamental methods for servicing bearings.

Technical Terms

Gear	Working depth	Scoring	Oil clearance
Spur gears	Clearance	Welding	Thrust bearings
Helical gears	Gear backlash	Chain	Thrust washers
End thrust	Gear ratio	Sprockets	End clearance
Hypoid gears	Gear reduction	Roller chain	Retainer plates
Planetary gearsets	Direct drive	Silent chain	Collars
Faces	Overdrive	Chain tensioner	Spline
Face width	Wear	Bearings	Splined shaft
Root	Failure	Friction bearing	Keyways
Whole depth	Pitting	Antifriction bearing	Key
Pitch circle	Spalling	Radial loads	
Circular thickness	Peening	Axial loads	

Introduction

All transmissions and transaxles, no matter how complex, are made up of simple parts. Parts such as *gears, chains,* and *bearings* are the building blocks of any transmission or transaxle. These parts can be arranged in many ways to create durable and efficient devices that can transmit and modify engine power as needed.

Engine power is transmitted by the action of gears. Engine torque is multiplied and rpm is reduced through gears. Vehicle direction is reversed through gears. Increasingly, these actions are accomplished by chains and sprockets. Simple gears and chains are arranged into some complex transmission or transaxle mechanisms, but their basic design does not vary.

All shafts and other rotating parts of the transmission or transaxle revolve on bearings. Without bearings, the moving parts of the transmission would encounter too much friction, resulting in wasted engine power and premature wearing of parts. Automatic transmissions and transaxles use many kinds of bearings, but all bearings serve the same purpose: to reduce friction and improve efficiency.

In this chapter, you will learn how gears, chains, and bearings function. You must understand how they operate before you begin to study more complex automatic transmission and transaxle parts. In later chapters, you will see how these simple parts are combined in transmissions and transaxles.

The inspection and service procedures covered in this chapter should always be performed when transmissions or transaxles are overhauled. Learning to perform these tasks *now* will allow you to concentrate on overall service procedures as they are presented in later chapters.

Gear Drives

A simple **gear** is a toothed wheel. In operation, gears engage other gears or mechanical parts for the purpose of transmitting power. Such a system is referred to as a **gear drive.** The gear that transmits the power is called the *drive gear;* the gear that receives the power is called the *driven gear.* Gears provide *positive* (non-slipping) power transmission. Gears that are engaged are said to be *in mesh.* Gears can be used to increase or decrease turning speeds. They are often used to increase torque. Another use is maintaining proper *timing* between two rotating parts, such as an engine crankshaft and camshaft.

A gear that is used in a transmission or transaxle is usually made of high-strength steel. In its manufacture, a gear *blank*—a steel disk without any features—is produced from relatively soft steel. The gear teeth are cut into the blank on a special machine. The most common type of

gear-making machine is called a *gear shaper.* Once the teeth are cut, the parts of the gear that will contact other moving parts are ground and polished to a smooth finish. The finished gear is hardened by a heat-treating process. The process makes a durable gear and allows for a precision fit because the gears can be machined to closer tolerances.

Gears that are used in light service are often made of brass, aluminum, fiber material, or plastic. These materials are relatively cheap, and they permit quieter operation. Some examples are speedometer drive gears, engine distributor gears, and the internal gears found in many electric motors.

Types of Gears

There are a number of different types of gears available. Many of these are commonly used on automotive transmissions and transaxles. Types of gears include *spur gears, helical gears, hypoid gears,* and *planetary gearsets.*

Spur Gears

Spur gears are a type of gear used to connect parallel shafts. The teeth on spur gears are always cut straight across, or *axially.* In other words, they run parallel to the axis of rotation. Spur gears are found most often in manual transmissions, where gears must be manually moved in and out of engagement. The straight cut of the teeth allows a smooth engagement and disengagement. Spur gears are inexpensive to manufacture. A disadvantage to these gears is that they become noisy if the clearance between mating gears becomes too large. Typical spur gears are shown in **Figure 4-1.**

Helical Gears

Helical gears are similar to spur gears. They, too, are used to connect parallel shafts. However, the teeth of helical gears are cut at an angle, called a *helix angle,* across the gear surface. A helical gear is shown in **Figure 4-2.** Helical gears operate more smoothly and quietly than spur

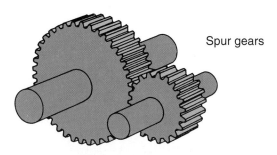

Spur gears

Figure 4-1. *The simplest gear found in automotive drive trains is the spur gear. Spur gear teeth are cut axially, or parallel to the gear axis of rotation. These gears are strong, but they may become noisy if the clearance between the teeth of the mating gears becomes too great. (Deere & Co.)*

gears. The engagement of individual teeth is a gradual sliding motion, and several teeth are engaged at once. Helical gears are often used for the forward gears of manual transmissions, where high gear speeds make smooth, quiet operation all the more important. The sliding action of the teeth causes more friction than encountered by spur gear teeth. As a result, helical gears must be kept well lubricated.

A thrusting action occurs along the shaft whenever power is being transmitted through a helical gear. This action, which is due to the helix angle, is called **end thrust**. See **Figure 4-3.** Two meshing helical gears will try to push each other in opposite directions. If end thrust is too large, the gears will disengage. *Thrust washers* or *thrust bearings* are often used to control end thrust. These components are detailed later in this chapter.

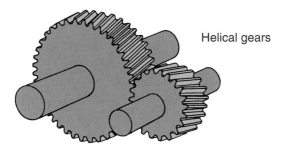

Figure 4-2. *Helical gear teeth are cut at an angle. Helical gears are used in many places, including transmissions and final drives. Helical gears provide smoother and quieter power transmission than do spur gears because the meshing of teeth is more gradual. (Deere & Co.)*

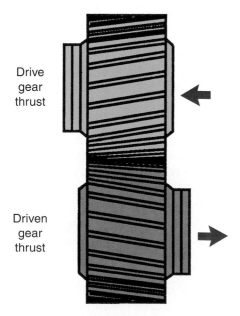

Figure 4-3. *The action of helical gears tends to push the gears away from each other. This puts pressure on the gears and on the shaft on which they ride. The pressure must be controlled by other parts in the drive train, or gear movement will be excessive.*

Hypoid Gears

Hypoid gears are shown in **Figure 4-4.** These gears are used in applications where power flow must be diverted by some angle. While hypoid gears are not used in automatic transmissions, they are commonly found in the final drives of automatic transaxles used with longitudinal (forward facing) engines.

Planetary Gearsets

Planetary gearsets are gear assemblies used in automatic transmissions and in manual transmissions with *overdrive units.* They provide *gear reduction* and *direct drive.* They also provide *overdrive.*

A planetary gearset, as shown in **Figure 4-5,** consists of three types of gears: a **sun gear** in the center; **planet gears,** or **planet pinions,** surrounding and meshed with the sun gear; and a **ring,** or **annulus, gear** meshed with the planet gears. The planet gears are attached to a **planet,** or **planet-pinion, carrier,** yet they are free to rotate.

Hypoid gears

Figure 4-4. *The hypoid gear is a variation of the bevel gear. The hypoid system is used when the gear shaft axes are at different levels. Hypoid gears are used extensively in rear-wheel drive vehicles. The drive gear can be placed lower in the rear axle housing to allow the drive shaft hump to be smaller. (Deere & Co.)*

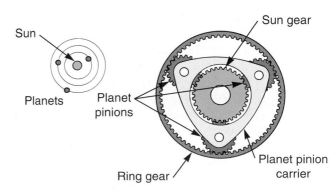

Figure 4-5. *A planetary gearset somewhat resembles our solar system. Planetary gears are in constant mesh. Gears rotate or are held stationary in different combinations to achieve different gear ratios. A single gearset can be used to provide reduction, direct drive, overdrive, and reverse. (Deere & Co.)*

The unique feature of a planetary gearset is that all gears are in constant mesh. In driving power through the gearset, one of the three gear types—sun, ring, or planet—will be held stationary; the other two will rotate, serving as input and output. Different combinations of input, output, and stationary are used to achieve different effects—for example, an increase or decrease of speed. A system will be in neutral when all three gears are free.

Gear Nomenclature

There are terms to describe almost every possible dimension of a gear. Design drafters and machinists involved in gear manufacturing are very familiar with these terms. Automotive technicians do not need to know every term. This is because servicing transmission or transaxle gears usually means replacing the *damaged* gear with an exact duplicate of the original. However, to aid in obtaining the proper replacement gear, the technician should know a few common gear terms. Some of these are given in **Figure 4-6.**

The contact surfaces of gear teeth are called *faces.* The width of the tooth measured parallel to the gear axis is the *face width.* The base of the gear tooth is called the *root.* The distance from the *outside diameter* of the gear to the *root diameter* is the *whole depth.* Stated another way, the whole depth is the distance from the top of the tooth to the root. The face width and whole depth should be measured and the number of teeth should be counted when obtaining replacement gears.

An imaginary circle around the gear located at about midway between the outside diameter and the root diameter is the *pitch circle.* Pitch circles of two mating gears are tangent to one another. *Circular thickness* is the thickness of a tooth measured between its faces and along the pitch

circle. The *working depth* is the overlap measured between outside diameters of the two gears. Finally, there must always be some space between the root of one tooth and the top of the mating tooth. This space is referred to as *clearance.*

Gear Backlash

Gear backlash is a term the technician will commonly use. Backlash, in gears, is the amount by which the space between neighboring gear teeth exceeds the circular thickness of a mating gear tooth. The difference is figured at the pitch circles. In general, the greater the amount, the more play there will be between two meshing gears. The degree of backlash must be correct for quiet gear operation and long gear life. On some gear trains, the backlash can be corrected by adjusting the positions of the gears. On others, the gears must be replaced to obtain the proper backlash. Some gears, by design, have fewer backlash problems than others. Helical gears, for example, are better in this respect than spur gears.

Gear Ratio

You are probably familiar with the term *gear ratio.* It is the speed relationship between two gears. The gear ratio is determined by the difference in the number of teeth between two gears. Mathematically, it is the ratio of the number of teeth on the driven gear to the number of teeth on the drive gear. This is the same as saying it is the number of teeth on the driven gear *divided by* the number of teeth on the drive gear.

In **Figure 4-7**, the driven gear has 90 teeth and the gear that drives it, or the drive gear, has 30 teeth. The ratio between the two gears is found to be 90/30 (90 divided by 30), which equals 3/1 and is expressed as 3:1 (read as "3 to 1").

The gear ratio relates the speed of the drive gear to the speed of the driven gear. A 3:1 gear ratio is an example of what is called *gear reduction.* A 3:1 gear reduction means that it takes three revolutions of a drive gear to turn the driven gear through one complete revolution. Refer to **Figure 4-7A.** A set of gears with a ratio of 3:1 is considered a *reduction gearset.* This is because the speed of the output shaft will be less than the speed of the input.

While a reduction gearset reduces speed, it multiplies torque. For example, with a 3:1 reduction gearset, 1 ft lb (1.35 N·m) of torque on the drive gear generates 3 ft lb (4.05 N·m) on the driven gear. Refer to **Figure 4-7B.**

To sum up these mathematical relationships:

gear ratio = number of teeth on driven gear/number of teeth on drive gear

= rpm of drive gear/rpm of driven gear

= ft lb on driven gear/ft lb on drive gear

A vehicle requires different gear ratios for different operating conditions. For instance, internal combustion engines develop minimal torque at low rpm. To overcome this problem, reduction gearsets are used in the

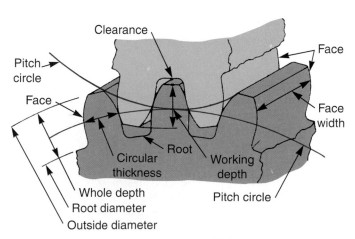

Figure 4-6. *Basic gear terms should be learned to help understand gear operation. Gear faces are the contact surfaces of gear teeth. The faces carry the load of the gears. Depth and width of the teeth are critical for proper gear operation. Backlash is the clearance at the pitch circle between faces of two mating teeth.*

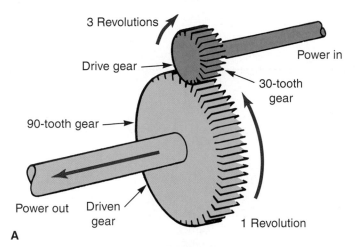

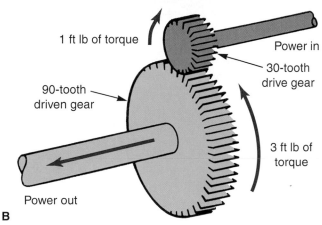

Figure 4-7. *A reduction gearset arrangement, which reduces speed while increasing torque, is shown. A—The output shaft completes one revolution for every three of the input shaft. B—A torque of 1 ft lb (1.35 N·m) on the input shaft translates to 3 ft lb (4.05 N·m) on the output shaft.*

transmission. A typical reduction in low gear is about 3:1. For example, a 900-rpm engine speed would be reduced at the transmission output to 300 rpm. At the same time speed is being reduced, torque is being multiplied. If 50 ft lb (67.5 N·m) of torque enters the gears, 150 ft lb (202.5 N·m) leaves. The reduction gearset, then, serves as a torque-multiplying device to get the vehicle moving at low speeds.

As vehicle speed increases, less torque is needed to keep the vehicle moving, due to its *inertia* (resistance to change in state of motion). Also, the engine is turning faster and is producing more power. The vehicle can then be shifted into a gear with less reduction, such as 2:1. As an example, at the transmission output, an engine speed of 1500 rpm will be reduced to 750 rpm, and an engine output of 75 ft lb (101.25 N·m) will be increased to 150 ft lb (202.5 N·m).

When the vehicle is moving at highway speeds, gear reduction is no longer needed or desired. The vehicle is shifted into high gear, which allows the engine to run at reduced speed while maintaining vehicle speed. In this gear, which is known as direct drive, gears are not actually used to transmit power. *Direct drive* has a gear ratio of 1:1. With this gear ratio, there is no change in either speed or torque between the engine output and the transmission output.

For greater fuel economy at higher speeds, **overdrive** is often employed. Most *overdrive gearsets* have a ratio of around 1:1.5 (sometimes written as 0.66:1). This means that an engine speed of 1000 rpm will exit the transmission at 1500 rpm. With this type of ratio, there is a speed increase through the gearset. This allows the engine to run more slowly than a higher gear ratio would permit at the same speed of travel. Since the engine is running more slowly, it will use less fuel and last longer. Some of the newest 6- and 7-speed transmissions have two overdrive gears. Overdrive ratios on these transmissions may be as high as .5:1.

Reverse Gear

Simple gears can be arranged to reverse direction. In the normal action of two mating gears, the drive gear and driven gear rotate in opposite directions. A third gear will restore the original direction of rotation, which has been reversed by the second gear. A manual transmission, for instance, employs a **countershaft gear assembly** between input and output shafts. The assembly consists of a shaft and several gears, which rotate as a unit. See **Figure 4-8.**

To change directions, a vehicle has a *reverse gear,* which is part of the transmission. **Figure 4-8** shows reverse gear engaged. Note the use of the **reverse idler gear,** or simply, the **idler gear,** turning the output shaft in the same direction as the countershaft gear assembly, which is opposite to the direction of the input shaft. The reverse idler gear does not affect the gear ratio.

Planetary Gear Ratios

Planetary gear ratios are harder to calculate than sliding gear ratios. The gear components interlock and rotate around a common center. This creates movements that cannot be calculated by simple division. When several planetary components are combined into a compound gearset, gear ratio calculation is further complicated.

Some planetary gear ratios can be calculated by counting the number of teeth on the driving and driven gears to obtain a ratio and then adding one (1.0) to the calculated figure. Adding one compensates for the rotation of the gears about a common center. However there are many exceptions to this rule depending on the following:

❑ Exact size and relationship of the meshing gears.
❑ Planet carrier size in relation to gear size.
❑ Whether the gears are combined into compound gearsets.
❑ Whether the gear is a reduction gear or an overdrive gear.

Rather than master the complex mathematical computations, most technicians simply accept the transmission or transaxle manufacturers' published gear ratios.

A method of roughly calculating the planetary gear ratio is to assemble the planetary gear components while

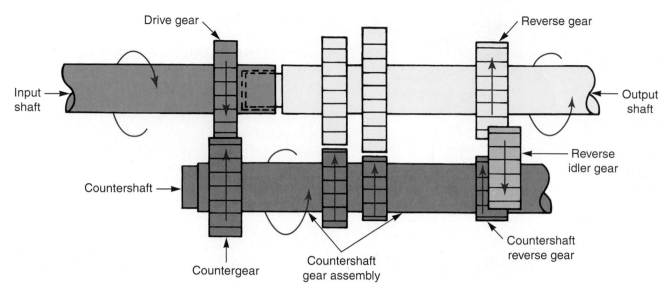

Figure 4-8. *Simplified drawing of a transmission with reverse gears engaged. Countergear causes countershaft gear assembly to rotate opposite to direction of input shaft. Countershaft reverse gear meshes with reverse idler gear, which meshes with reverse gear of output shaft. Reverse gears rotate in the same direction because of the reverse idler gear. Output shaft rotation is opposite of input shaft.*

the transmission is apart. After assembly, make a chalk mark on the input and output shafts. Then simulate the holding members—turn and hold the gearset parts that would be operated to obtain the gear you are checking. One or two persons can turn the input shaft and the planetary gearset parts that would be locked to the shaft. Another person can hold the gear parts that would be stationary in this gear. While turning the input shaft, observe how many revolutions (and fractions of a revolution) it must be turned to obtain one revolution of the output shaft. This is the approximate gear ratio.

Gear Inspection

Gears should be inspected whenever there is evidence of gear problems or when they are removed from the vehicle for any reason. Some of the problems to look for are covered in these next few paragraphs.

Wear is erosion of a surface caused by relative motion between two parts. Gears are subject to wear, and given time, a gear will wear out from use. In most automotive transmissions and transaxles, the various gears are made to be durable. These gears will operate satisfactorily for many miles. Some gears, such as those of the rear axle assembly, will typically last for the life of the vehicle.

Gear wear can be responsible for gear *failure*—a total breakdown in structure so the gear can no longer fulfill its purpose. A gear with normal wear from high mileage will appear worn but in good condition. A gear removed from a well-maintained vehicle with very high mileage may show considerable wear. In either case, the faces will show an even wear pattern and will appear smooth and shiny. They should be light gray in color.

At some point, gear wear may become so extreme that the teeth are too thin to support the loads placed on them. Such gears usually develop excessive backlash, which increases the shock loads on the teeth when they are first engaged. When this happens, a gear tooth may develop a crack and break. The broken tooth falls to the bottom of the gear housing; however, the gear may operate for a while longer. This increases the shock loads on the remaining teeth, causing other teeth to break. Sooner or later, the gear will fail completely.

Early gear failure is usually caused by a transmission or transaxle defect, lack of lubrication, or abuse. The most common cause of gear tooth failure on a low mileage vehicle is abuse. Placing heavy loads on a vehicle not designed to carry them can overstress the teeth and cause gear failure. Loading the gears too rapidly will cause shock loads, which can overload the teeth and break the gear. On a manual transmission vehicle, this may happen when the clutch is engaged abruptly with the engine running at high speed. On an automatic transmission vehicle, placing the car in gear at high engine speeds can break gear teeth.

Figure 4-9 shows some common types of gear *damage.* Left uncorrected, these problems of gear damage will lead to gear failure. A common gear problem is *pitting.* Pitting is evidenced by small holes, or pits, in the gear teeth. The pits reduce the gear contact area. If the cause of the problem is not corrected, the pits can become larger and more plentiful. Pitting is usually caused by abrasive particles, corrosion, or a lack of lubricant. These can be attributed to dirt, metal debris, or water in the lubricant.

Severe pitting can lead to *spalling,* a condition in which sections of the gear teeth flake or chip off. Spalling, like pitting, can also be caused by lack of lubricant or severe gear loads.

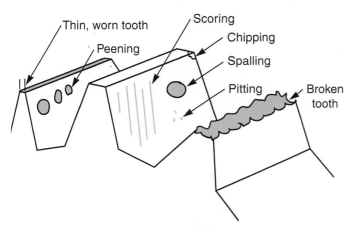

Figure 4-9. *Excessive gear wear and some other problems leading to complete gear failure. Note that, in reality, gear teeth are likely to show just one kind of damage.*

Another cause of gear failure is **peening,** which appears as indentations in the gear faces. Peening is usually caused by severe loads on the gear teeth or by large particles of foreign matter in the lubricant.

Scoring is another cause of gear failure. It appears as scratches on gear teeth faces and is caused by improper machining or lack of lubrication. If the scoring is not corrected, it will lead to spalling or a condition called **welding.** Welding occurs when mating gear teeth overheat, melt together, and then pull apart as the gear turns. This will quickly destroy the gear.

Sometimes lack of lubrication will cause the gears to overheat. Overheating will undo the heat treating applied in manufacturing to harden the gears. The gears become relatively soft, and they soon wear out. Overheated gear teeth usually turn a different color than the undamaged part of the gear. This color can vary anywhere from yellow to deep blue.

Chain Drives

A **chain drive** consists of a series of links, or a **chain,** and two or more gears called **sprockets.** The chain serves to transmit motion between the sprockets. The links form a chain, or *flex belt.* Chain drives have a direct mechanical connection, similar to that of two meshing gears. The sprocket teeth match the openings in the chain links and mesh with them to transmit power. One sprocket transmits power. It is called the *drive sprocket.* The other one receives the power and is called the *driven sprocket.*

Chain drives transmit power in a straight line between the sprockets. This means of power transmission is compact. For this reason, chain drives are often used to connect torque converters to the transmission mechanical system in front-wheel drive vehicles where space is limited. Chain drives are found not only in transaxles, but in four-wheel drive transfer cases. These will be explained in later chapters. **Figure 4-10** is an example of a chain drive assembly.

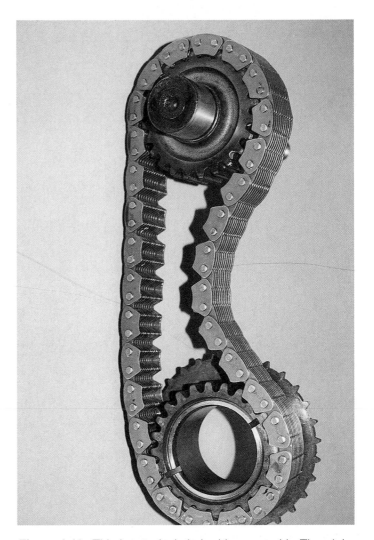

Figure 4-10. *This is a typical chain drive assembly. The major parts of the chain drive are the chain and drive sprockets. Every system has at least two sprockets.*

Types of Chains

There are two types of chains used on modern vehicles. The most common is the **roller chain, Figure 4-11.** A roller chain employs rollers that rotate on *drive pins.* The rollers contact the sprocket teeth. The spaces between the rollers match the size of the teeth. The rollers provide a rolling contact, which reduces friction between the chain and the sprocket.

Roller chain can be thought of as alternating *pin links* and *roller links.* Refer to **Figure 4-11.** Roller links are coupled together with pin links. An enlarged view of a pin link is shown in **Figure 4-12.** Many chains have a master pin link, which is held together with a clip. The clip can be removed to separate the chain for easy replacement. Chain links vary in design, but all serve the same function.

The **silent chain, Figure 4-13,** is made up of a series of flat metal links, which provide quiet operation. These links have teeth that engage matching teeth in the sprockets. The links are placed side by side. This creates a solid contact for engaging the sprocket. The links are held

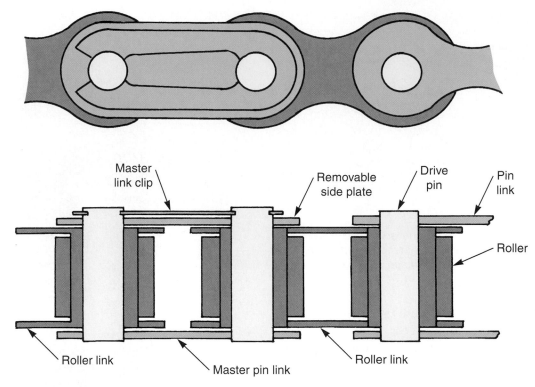

Figure 4-11. *Roller chains are made up of pin links and roller links. The rollers mesh with the sprocket teeth, providing positive drive.*

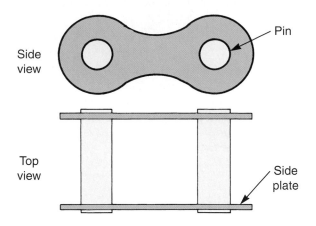

Figure 4-12. *A pin link consists of two side plates riveted to two pins.*

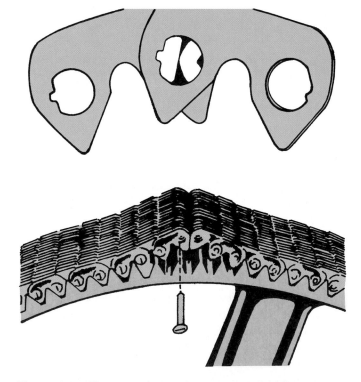

Figure 4-13. *The construction of the silent chain differs from that of other chains. On a silent chain, the links are a series of toothed leaves that extend the entire width of the chain. The links engage the sprocket. They are held in place by pins. The pins keep the links in alignment. This is just the opposite of the roller chain.*

in alignment by pins that pass completely through the chain. This is opposite from the roller chain, wherein the pins (or rollers) engage the sprockets, and the links perform the task of alignment. The construction of the silent chain allows some internal movement of the chain parts. This reduces vibration and noise and partially compensates for wear and sprocket misalignment.

Chain Tensioners

Every chain will eventually wear. This can cause the chain to lose its tension. To prevent this, many chain drives have a ***chain tensioner*** to take up the slack. Most chain tensioners are spring-loaded devices similar to that shown in **Figure 4-14.** They maintain constant pressure on the return, or *nondrive,* side of the chain.

One drawback of a *spring-loaded chain tensioner* is that as the chain wears, the tensioner exerts less tension on it. The spring extends further to take up the slack, and, eventually, the chain is too worn for the tensioner to have any effect. In most cases, the chain will become noisy before it fails completely.

Another type of tensioner is the *hydraulic chain tensioner.* This type uses system oil pressure to keep constant tension on the chain, even as parts wear. It is not as common as the spring-loaded tensioner.

Chain Drive Ratio

To find chain drive ratio, divide the number of teeth on the driven sprocket by the number of teeth on the drive sprocket. **Figure 4-15** shows how to find the ratio for a final drive that employs a chain drive.

In another example, a driven sprocket has 30 teeth and the drive sprocket has 10 teeth. Dividing 30 by 10 gives 3. The gear ratio in this example is 3:1. This is a reduction ratio because for every 3 revolutions of the input, the output completes only 1 revolution. In other words, the speed of the output is less than the speed of the input.

If the driven sprocket had 10 teeth, and the drive sprocket had 30 teeth, the gear ratio would be 1:3. This would be an overdrive ratio because the speed of the output would be greater than the speed of the input.

If the drive and driven sprockets have an equal number of teeth, the ratio is 1:1. The input and output turn together. There is no multiplication of torque or increase of speed. Note that the length of the chain has no effect on the ratio.

Chain Lubrication

Even the most efficient chains require lubrication. One advantage of chain drives is that they are usually enclosed. This way, they are kept lubricated, as well as protected from dirt and moisture.

There must always be some provision for chain lubrication. The chains are usually designed so they pass through a pool of oil, which collects in the bottom of the chain case. In other cases, oil is sprayed on the chain through a calibrated oil hole.

Chain Inspection

A chain drive should be inspected whenever there is evidence of chain problems or when the chain is removed from the vehicle for any reason. Chain and sprocket wear usually shows up as a loose chain. A loose chain is hard on sprocket teeth. If the chain is loose, there is a very good chance the sprocket is damaged; therefore, always be sure you check both the chain and the sprocket.

Sometimes a chain will become so loose that it jumps off the sprockets. Usually the chain will become noisy before this happens. A worn loose chain that is still in place will have too much slack—there will be too much play in the chain. This will be more noticeable on the side that is not under tension. If the chain is removed from the sprockets, lay it sideways as shown in **Figure 4-16.** Any noticeable sagging means the chain is worn out and should be replaced.

The sprockets should be checked for excessive wear or cracks. The tooth faces of roller chain sprockets may appear as hooked teeth. In extreme cases, even these hooks will be worn away. Sprocket problems often resemble gear problems. The tooth faces of silent chain sprockets will often show a series of indentations where the chain makes contact.

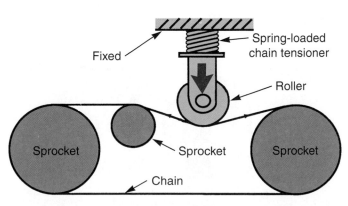

Figure 4-14. *Like any mechanical part, chains and sprockets will wear. Wear makes the chain loose. A loose chain will be noisy and may jump off the sprockets. To keep the chain from becoming too loose, a chain tensioner is often used. Most tensioners are spring loaded. A spring is compressed against the chain, and this puts tension on the chain, taking up the slack.*

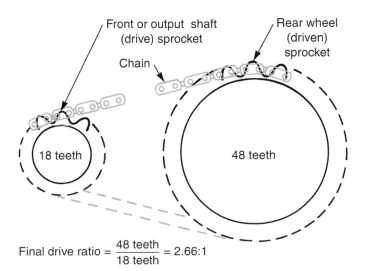

$$\text{Final drive ratio} = \frac{48 \text{ teeth}}{18 \text{ teeth}} = 2.66{:}1$$

Figure 4-15. *Chain drive ratios are figured in the same way that gear ratios are. The number of teeth on the driven sprocket is divided by the number of teeth on the drive sprocket. The resulting figure is the ratio. The length of the chain has no effect on the ratio.*

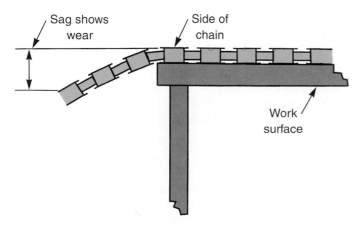

Figure 4-16. *Chains that are in place should be checked for looseness. A removed chain may be checked on a workbench by laying the chain on its side and projecting it off the edge. A chain that sags even slightly should be replaced.*

A worn or damaged sprocket should be replaced. The chain may be reused, in this case, depending on its condition. If a chain must be replaced, the sprockets should be replaced at the same time. Otherwise, the old sprocket may damage the new chain.

A problem encountered in chain drives is misalignment of sprockets. When chains are used, there is little room for misalignment, which is one of the disadvantages of chain drives. Misalignment will cause excessive chain and sprocket wear and noise. In extreme cases, the chain will jump out of engagement with the sprockets.

For proper alignment, sprocket shafts must be parallel. Check distances between shafts on both front and back sides of the sprockets; they should be equal. Also, for proper alignment, the two sprockets must be in line with each other, or fall in the same plane. This can be checked by placing a straight edge across the flat surfaces of the sprockets.

Similar to the chain drive in general layout is the belt and pulley drive used on some modern transaxles. This system uses a metal belt and two pulleys to transfer power. Changing the diameter of the pulleys changes the gear ratio. This type of drive is discussed in more detail in Chapter 10.

Bearings

When two objects rub together, resistance to movement arises between them. This resistance is friction. Bearings are used to reduce friction. This section will discuss how bearings accomplish this task, as well as types of bearings and inspection and service procedures, but first, a few more words about friction.

Friction in Bearings

Friction is a part of everyday life. Often, it is useful. Friction between your shoes and the ground, an example of static friction, allows you to walk without falling down. Friction holds assembled parts together. It prevents threaded parts from unthreading and provides the holding power for press-fit parts. Friction is what makes a brake system work.

There are times when friction can be a handicap. Friction between a part that is supposed to turn freely and a stationary part will turn power into heat and wear. Friction can be a severe problem in the transmission or transaxle, where the rotating parts are moving at high speeds. In addition, drive train action often puts side loads on a shaft, subjecting one side of a stationary part to more friction.

Bearings are devices used to reduce friction between rotating and stationary parts. Further, they guide and support rotating parts, to prevent damage from misalignment or excessive clearances. The relationship between a rotating shaft, a stationary housing, and a bearing is shown in **Figure 4-17.**

Types of Bearings

There are two major classes of bearings—*friction bearings* and *antifriction bearings.* **Figure 4-18** illustrates the two types. The **friction bearing** is pressed into place and does not move. A rotating shaft slides around on the friction bearing surface. The **antifriction bearing** uses an assembly of balls or rollers contained within a housing, where they are aligned and free to roll. The bearing is pressed into place; however, the one side (inner or outer) is free to move with the rotating member because of the rolling elements. This changes *sliding,* or *kinetic, friction* between the mating surfaces to *rolling friction.* A detailed discussion about each type of bearing follows.

Friction Bearings

Friction bearings are used for supporting rotating shafts. A rotating shaft slides around the supporting surface of the bearing, which is stationary. These bearings have a one-piece or two-piece construction. *Sleeves* and *bushings* are one-piece friction bearings. Bushings are small sleeves. Two-piece bearings are called *insert bearings.*

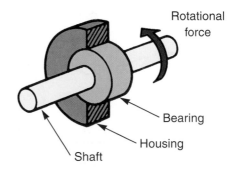

Figure 4-17. *Note the relationship between rotating shaft, stationary housing, and bearing. The bearing ensures the shaft is properly aligned with the housing. It also serves as a buffer. It incurs the wear, rather than the shaft or housing, making servicing easier and cheaper. (Federal Mogul)*

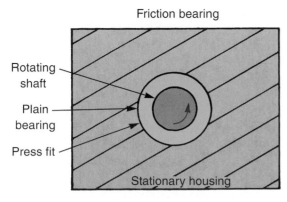

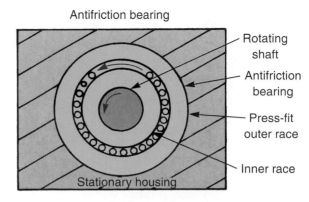

Figure 4-18. *The two main types of bearings are shown. Note that sliding friction is seen in the friction bearing, while rolling friction is seen in the antifriction bearing.*

Friction Bearing Material

Friction bearings are composed of many materials. Most bearings used in transmissions are made of a steel shell with a coating of a soft metal, such as copper, brass, lead, or aluminum. Sometimes *babbit metal,* which is an alloy of tin or lead, copper, and antimony, is used for the coating.

Some bearings used in transmissions are solid brass or bronze. One thing is true about any friction bearing, no matter what its composition—it is always designed to wear before the shaft metal. This is done so the bearing, which is relatively cheap, will wear out instead of the expensive shaft.

Friction Bearing Lubrication

A rotating shaft can contribute to friction through two types of loads—*radial* and *axial.* **Radial loads** are perpendicular to the axis of rotation. **Axial,** or **thrust, loads** are parallel to the axis of the shaft. To reduce friction from these loads, the bearing must be lubricated.

For proper bearing lubrication, the shaft and bearing must be separated entirely by a film of lubricant, **Figure 4-19.** When enough lubricant is present, the sliding action takes place in the lubricant between the shaft and the bearing; the shaft *rides* on the film.

In order to accommodate the bearing lubricant, a clearance is needed between the shaft and bearing. An **oil clearance, Figure 4-20,** must allow the lubricant to enter and circulate properly. If the clearance is too tight, the lubricant cannot form an adequate surface between the mating parts. If it is too loose, the lubricant will leak out too quickly to maintain the surface separation. The shaft and bearing are said to have a **running fit** when the oil clearance is sufficient to enable parts to turn freely and receive proper lubrication.

The lubricant itself may be a type of oil, supplied by pressure, splash, or immersion. It may be a heavy grease, applied periodically. Friction bearings used in automatic transmissions are often supplied with oil from the transmission oil pump. Bearings used in standard transmissions

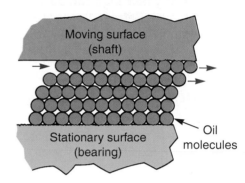

Figure 4-19. *A film of oil between two surfaces can act as a bearing. Oil molecules serve as tiny ball bearings. The oil film is what makes the use of friction bearings feasible. (Federal Mogul)*

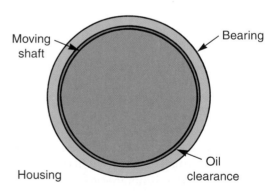

Figure 4-20. *The oil clearance between a bearing and a shaft is usually a few thousandths of an inch. This clearance must be loose enough to allow the oil to reach the bearing and shaft surfaces. However, it must not be so loose the oil leaks out too quickly. (Federal Mogul)*

are lubricated by splash from the turning gears. Plain pilot bearings in the clutch assembly are greased only when the clutch is serviced.

Most friction bearings will have slots and grooves for lubricant. The grooves usually run the entire length of the bearing surface. They allow the lubricant to be picked up and evenly distributed around a moving shaft. **Figure 4-21** shows a number of friction bearings. Notice the grooves and slots in these bearings.

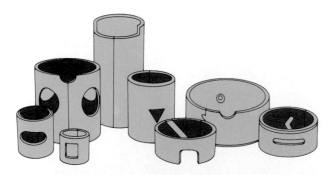

Figure 4-21. *An assortment of friction bearings, showing grooves and slots for lubrication. (Federal Mogul)*

Note: Many bearings can be inspected without removal. Refer to the appropriate service information to determine whether a particular bearing should be removed prior to inspection.

Friction Bearing Removal

Bushings and sleeves, friction bearings of one-piece construction, are press fit into position. There are a couple of ways to go about removing them. The best way is to use a puller made for the particular bearing. The puller is secured to the bearing. Applying pressure will remove the bearing from the housing.

A different method of bearing removal can be used when the bearing housing is a *blind hole*—that is, open on one side, blocked off on the other. An example of this is the clutch pilot bearing. Pack or fill the area behind the bearing with heavy grease. Insert a shaft that is of the same diameter as the bearing opening into the bearing. Put on eye protection and strike the exposed end of the shaft with a hammer. The hammer force will be transferred into the grease. The region behind the bore becomes pressurized. The pressure will be applied to the bearing, and the bearing will be forced from the bore.

A *cape chisel* or a *round-nose cape chisel* is an ideal tool for removing small bushings where both ends are accessible. A *pin punch* is another tool that can be used. Where applicable, these tools can be used to remove sleeves and insert bearings as well. Always use care to avoid damaging the bearing housing.

Friction Bearing Inspection

Whenever a transmission or transaxle is disassembled, all bearings should be closely inspected. Clean the bearings thoroughly before inspection. Use a solvent that will not damage or coat the bearings.

During inspection, check for wear and damage. A normally worn bearing will have smooth contact surfaces. Bearings that are to be reused should be checked for proper size. Often, a bearing will look good but will be worn excessively. Bearing clearances often measure only up to a few thousandths of an inch, or several hundredths of a millimeter.

One quick way to check a friction bearing is to place it onto the shaft on which it rides. If the fit feels loose, the bearing, and possibly the shaft, is worn. This method depends on your knowing what a loose bearing fit feels like. It is hard to tell without experience.

If you are not sure, check actual dimension of bearing inside diameter using a caliper or other such instrument. If the bearing has excessive clearance in one area only, it has been subjected to heavy loads on one side. Check that the shaft is not bent or misaligned. Any bearing with excessive clearance should be replaced.

In many cases, when the bearing is replaced, the shaft might also have to be replaced. Do not assume that replacing the bearing will solve the problem. Proper operation of the assembly depends on both the bearing and the shaft. Always check the shaft where it mates with the bearing. If there is no obvious damage, use a micrometer or sliding caliper to check for proper diameter and taper. Check the shaft in several places.

Another problem to look for is scoring of the bearing. Scoring will appear as scratches or ridges partially or totally inscribed about the bearing surface. Scoring is usually caused by lack of lubrication or by abrasive particles in the lubricant. If the bearing is badly scored, the shaft is probably also damaged.

A bearing that has been in use for some time may have pitting or even large craters in its surface. Pitting starts with dirt or water in the lubricant. Over time, large pits, or craters can develop.

Sometimes, a missing ground strap between the engine and car body can cause craters. The engine electrical components will ground through the drive train. This electrical action will cause metal transfer, leading to pitting.

Sometimes, a friction bearing will be operated without lubrication or under other abusive conditions. This can lead to severe bearing damage. Bearings may overheat, spin in their housing, or weld themselves to the shaft. A sign of a bearing that has been subjected to overheating is a change in the material's color. Indications of spun or welded bearings will usually be obvious. In many cases, the shaft or housing must also be replaced or repaired.

Friction Bearing Installation

Bearing drivers are made to closely fit into a bearing. The bearing will bottom against a slightly larger area on the driver. Pressure on the driver is transmitted to the bearing, and the bearing is forced into its housing.

Bearings should be installed with the proper-size driver, preferably, on a hydraulic press. If a press is not available, some bearings can be hammered into place, using a driver to avoid part damage. Strike the driver with a hammer. Do *not* strike the bearing with the hammer. Also, attempt hammering only if the proper-size driver is available.

The bearing should be installed carefully. If the bearing has any oil supply holes, line them up with the oil holes in the housing for the bearing. Make sure the bearing enters the housing bore squarely and does not become cocked. A poorly installed bearing will quickly destroy itself and the shaft it supports.

After the bearing is installed, additional service may be required. Some bearings will require *reaming* and *honing*, to widen and smooth the bearing inside diameter to fit the mating shaft. Some bearings must be *staked.* This involves taking a punch to the bearing surface, which produces a raised dimple on the side of the bearing against the housing. This action locks the bearing in place and keeps it from spinning or slipping out of position. After reaming, honing, or staking, the bearing should be thoroughly cleaned.

Replacement bearings should be checked to ensure they have the proper clearance with the shaft. Before reassembly, that is, before inserting the shaft into the bearing, lightly lubricate the bearing surface. Use the same kind of lubricant that the bearing will have during normal operation. This will ensure adequate lubrication the first time a load is placed on the bearing. If the bearing is only lubricated during service (a clutch pilot bearing, for example), be sure to apply the proper amount of grease before reassembly.

Antifriction Bearings

Antifriction bearings contain rolling elements that operate within a housing made up of one or two pieces of metal. The rolling elements can be balls or rollers. In general, the pieces of the housing are called **races.** For conical antifriction bearings, the races are sometimes called *cones.* For long life, all of these parts are made of heat-treated, high-strength steel.

A typical antifriction bearing, with inner race, outer race, and rolling element, is shown in **Figure 4-22.** Note that antifriction bearings do not always have both an inner and outer race to contain the rolling elements. Sometimes, one of these is omitted. Then the elements are in direct contact with the mating surface.

In addition to rolling elements and races, many antifriction bearings contain a *cage* to keep the rolling elements in position. Also, some bearings are prelubricated, and these have seals to keep the lubricant in and to keep dirt and moisture out. The placement of seals is shown in **Figure 4-23.**

Antifriction bearings may be pressed or slipped into position—onto a shaft, for example, or into a stationary housing. Frequently, the bearing is pressed onto the shaft, and then the shaft and bearing are slipped into place in the stationary housing. In some cases, such as rear axle pinion gear bearings, the bearing is pressed into the stationary housing.

The advantage of the antifriction bearing is that it uses a rolling motion, rather than a sliding motion like the friction bearing. The rolling motion produces less friction. The antifriction bearing is usually used where rotating parts are highly stressed or where it is difficult to supply adequate amounts of lubricant. The antifriction bearing is more

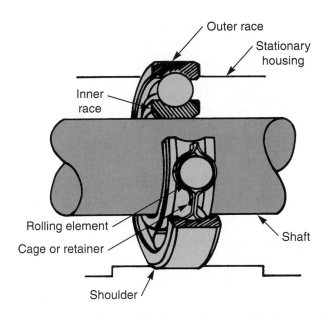

Figure 4-22. *A typical antifriction bearing. Note the three major parts—inner race, outer race, and rolling element.*

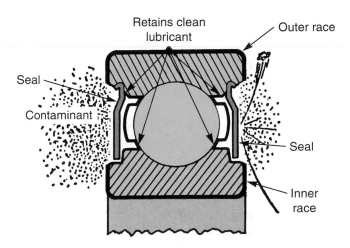

Figure 4-23. *Some bearings have a seal, as shown in this partial sectional view of a ball bearing. The seal serves to keep bearing lubricant in and dirt and moisture out. Bearings of a transmission, transaxle, or rear axle assembly will not have a seal.*

efficient than the friction bearing, but it is more expensive. Sometimes, however, size, clearance problems, or the back-and-forth movement of the shaft prevents the use of an antifriction bearing.

Types of Antifriction Bearings

The two major types of antifriction bearings are *ball bearings* and *roller bearings,* **Figure 4-24.** Ball bearings are used for clutch *throwout bearings,* which are used for clutch release. They are also used for pilot bearings and in some transmissions and transaxles. Roller bearings of various types are used as pilot bearings and also in transmissions and rear axle assemblies. *Tapered roller bearings, needle bearings,* and *thrust bearings* are all variations of the roller bearing.

Ball Bearings

Although they vary in size, ball bearings used in automotive drive trains are all of the same basic design; however, methods of lubrication vary. Clutch throwout bearings and some *rear axle bearings* are greased by the manufacturer and are not intended to be regreased. Ball bearings used in other parts of the drive train are lubricated by oil splashed from other moving parts.

Roller Bearings

As mentioned, there are several types of roller bearings. **Straight roller bearings** are used in manual and automatic transmissions. This type of bearing usually comes as a one-piece unit. See **Figure 4-25.**

A **tapered roller bearing** is shown in **Figure 4-26.** Rollers of this bearing assembly are tapered. The outside diameter of the inner race (cone) and the inside diameter of the outer race (cup) are both tapered, also, to fit the rollers. The assembled bearing, however, including its inside diameter, is cylindrically shaped. Tapered roller bearings are useful for heavy loads. They are used in rear axles, front-wheel drive axles, and transaxle final drives.

Needle bearings, Figure 4-27, are another type of roller bearing. Needle bearings perform the same function as other roller bearings, but their diameter is much smaller in relation to their length. Needle bearings have tiny rollers that resemble needles, hence, the name *needle bearing.*

Thrust bearings, Figure 4-28, are flat, disclike bearings that resemble washers. Thrust bearings are made up of needle rollers. The rollers are arranged *radially.* This means they radiate out along imaginary lines projected from a center point. This kind of bearing is sometimes called a **Torrington bearing.**

Bearing Load Factors

Antifriction bearings can be subjected to radial loads and thrust loads. **Figure 4-29** shows these. Applied radial loads are directed perpendicular to a shaft's rotational axis.

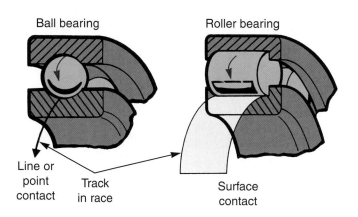

Figure 4-24. *Two types of antifriction bearing—ball and roller— are shown here. (Federal Mogul)*

Figure 4-25. *An example of a straight roller bearing. Notice that the bearing shown here has only an outer race. The axle shaft is functioning as the inner race. The rolling elements fit right up against the shaft.*

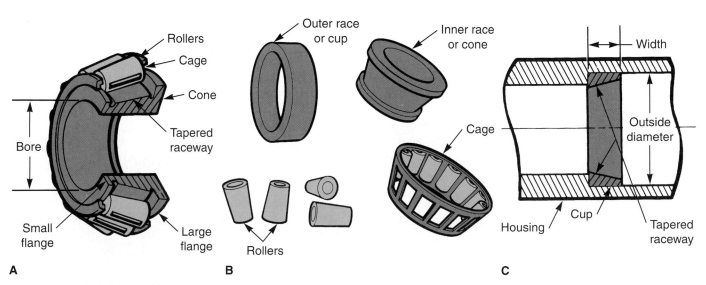

Figure 4-26. *Tapered roller bearing. A—This sectional view of an assembled tapered roller bearing shows the bearing with the outer race removed. B—This shows various parts of a tapered roller bearing. C—This sectional view shows the outer race, or cup, installed in a housing.*

Thrust loading is directed parallel to the axis. Thrust loading is caused by the back and forth movement of the shaft. The design of helical and bevel gears inherently creates thrust loads, which must be compensated for in bearing design. In some cases, the bearings are *preloaded* to reduce the deflection of parts and minimize thrust variation. Most bearing loads are *combination loads,* which are the result of both thrust and radial loads.

Ball and roller bearings are affected differently by loads. The ball bearing absorbs both thrust loads and radial loads on the ball surface. The straight roller bearing absorbs radial loads on the rolling surface, and minor thrust loads are absorbed on the sides of the rollers. The tapered roller bearing is designed to absorb both radial loads and thrust loads on the roller surfaces.

Applied loads will not be perfectly distributed about the bearing. Several factors can cause loading to be greater on one side or another. Gravity is a more obvious factor. The manufacturing tolerances of gears and shafts—the fact that they are not perfectly round or symmetrical—is another. Helical gears, too, by their design, will cause loading to be greater on one side. These conditions will cause the bearing elements to be alternately loaded and

unloaded as the shaft rotates. **Figure 4-30** depicts this action. Loading intensity is illustrated by the shading of the bearings—the darkest being the point of greatest intensity. In conclusion, the bearing must be manufactured to withstand not only heavy loads, but also to withstand the effects of *changing* loads.

Bearing Service Precautions

Upcoming paragraphs cover servicing aspects of antifriction bearings. This includes antifriction bearing removal, inspection, and installation. Before you study these aspects, there are some special precautions you must be aware of and take whenever servicing antifriction bearings. They are as follows:

❑ *Always* wear eye protection when removing or installing a bearing.

❑ Do *not* apply excessive force to a bearing or shaft during removal or installation.

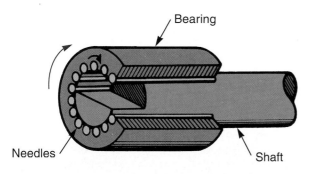

Figure 4-27. *The rollers of a typical needle bearing are longer and thinner than those used with other roller bearings. Note that the bearing shown has only a single race.*

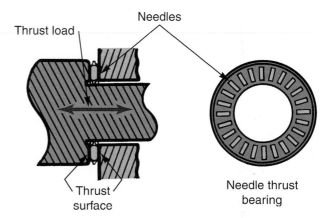

Figure 4-28. *Note a thrust bearing using needle rollers. The needles are arranged radially; their axial centerlines radiate outward from a center point. (Federal Mogul)*

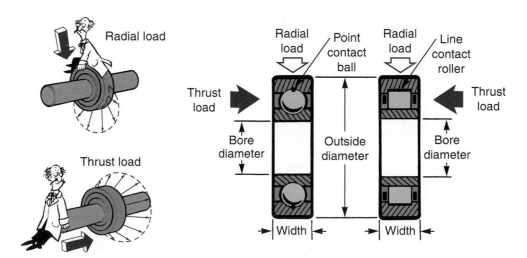

Figure 4-29. *Two types of loads placed on a bearing are radial and thrust. The two loads acting upon a bearing at the same time may be referred to singly as a combination load. Most drive train bearings are subjected to combination loads. (Federal Mogul)*

❑ Do *not* apply any force through the rolling elements of an antifriction bearing. Only apply force to the press-fit side of a bearing. The bearing could be damaged or could burst apart and cause injury.

❑ *Never* spin a bearing with compressed air. The rolling elements could be thrown out with explosive force and severely hurt you or someone else. Mechanics have been killed by using compressed air to spin bearings.

Antifriction Bearing Removal

Some bearings are held to shafts by snap rings or by large nuts that thread onto the shaft. In these cases, bearing removal is simple. The snap ring or nut is removed, and the bearing is slid from the shaft. In other cases, the bearing is pressed on, and special methods must be used to remove it.

The best way to remove a pressed-on antifriction bearing from a shaft that has not been removed from its installation is to use a bearing puller. Some pullers are made for a particular bearing; however, others are universal designs that can be used on almost any bearing. These pullers have adjustable jaws that can be tightened around the bearing. Using a puller, the bearing can be forced from its shaft.

An ideal way to take a bearing off a shaft that is removed from a vehicle is to use a hydraulic press. A support is placed under the bearing, and pressure is applied to the shaft to press the shaft from the bearing. Remember! Pressure should only be applied to the *press-fit* race to avoid applying force through the rolling elements, which could damage the balls, rollers, or races.

Refer to **Figure 4-31** for examples of proper and improper bearing removal.

Some bearings are press fit into a housing instead of around a shaft. A brass drift can be used to drive the race from the housing. Look for a notch cut into the shoulder of the outer race and carefully apply the drift to it to remove the bearing.

Sometimes, a *bearing retainer,* which is a force-fit metal ring or plate, must be removed before removing a bearing. This would apply to some rear axle bearings. The

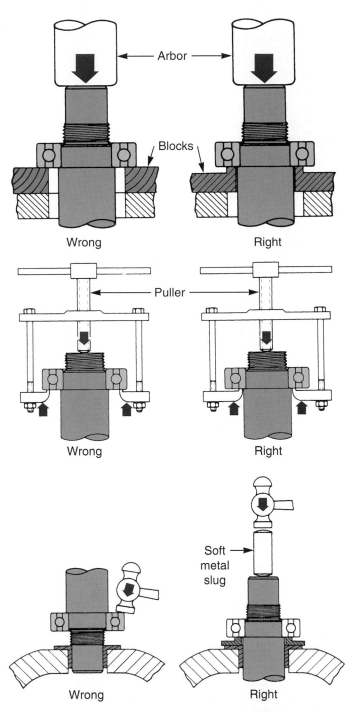

Figure 4-31. *Different methods of bearing removal. Do you see the difference between right and wrong ways of removal in each drawing? (FAG Bearings)*

Figure 4-30. *Complicating the bearing loading process is the fact that bearing loads vary about the bearing, according to gravity and shaft tolerances. Bearing rollers are alternately loaded and unloaded as the shaft rotates. (Federal Mogul)*

retainer is removed with a chisel or torch—carefully, to avoid damaging the shaft. In some cases, it is possible to cut the bearing from the shaft with a cutting torch. The torch must be used with caution to prevent overheating or otherwise damaging the shaft. If there is any way to avoid using a torch, do so.

 Warning: Wear eye protection when cutting off a retainer plate.

Antifriction Bearing Inspection

Bearings should be thoroughly cleaned before inspection. Often, what appears to be a bearing defect is dirt that collected on the bearing during disassembly. At other times, actual bearing defects are hidden by dirt or sludge. If you are cleaning a bearing with compressed air, always hold both races so the bearing does not spin. Remember! Never spin bearings with compressed air.

After cleaning, rotate the bearing by hand as shown in **Figure 4-32.** Note any places where you feel roughness or binding. If you find any problems, replace the bearing.

Visually inspect the bearing for signs of wear and damage, including conditions of scoring, overheating, and corrosion. Bearings may show signs of extreme wear or damage due to problems of overloading and lack of lubrication. Spalling and *brinneling,* **Figure 4-33,** are indications of these problems. Check for a bent or otherwise damaged cage. Cage damage is usually caused by careless installation.

Identifying Replacement Bearings

It is very difficult to match bearings by a visual inspection. It is, therefore, important to make sure the

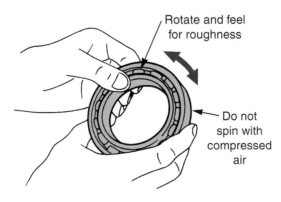

Figure 4-32. *Hold clean bearing like this during inspection and rotating outer race by hand. Do not use compressed air to inspect. An unloaded bearing spun at high speeds can come apart with explosive force! (Nissan)*

replacement bearing you get is the bearing you need. If the old bearing has a part number, **Figure 4-34,** compare it with the number of the replacement bearing. If the old bearing does not have a number, or if the number cannot be read, consult an appropriate parts catalog to ensure that the bearing is correct for the make and model of the vehicle you are working on. In some cases, you may need to obtain the serial number of the vehicle or of a particular component (transmission, rear axle, etc.).

Storing Bearings

Cleanliness is vital to the successful operation of antifriction bearings. Bearings should be left in their original wrappings until just before installation. Be careful not to drop any bearings; since the hardened steel is brittle, it may chip or develop a shock crack.

Antifriction Bearing Installation

The best way to press a bearing on is with a bearing driver and a hydraulic press. A driver of the proper diameter is selected to fit the bearing. Pressure on the driver from the hydraulic press is transferred to the bearing, and the bearing is forced onto the shaft.

If a hydraulic press is not available or cannot be used, some antifriction bearings can be driven into place by tapping on the race that is to be press fit. Remember not to hammer on the free race when installing a bearing. This will cause damage to the rolling elements and races and can cause injury.

When press fitting a bearing, watch closely to ensure that it is not being cocked or pressed in too far. If the shaft has a shoulder, the bearing should firmly bottom on the shoulder. Also, make sure dirt or other debris is removed from the bearing before installation. Refer to **Figure 4-35** for examples of proper and improper bearing installation.

If a shaft is too large, a bearing can be heated and expanded somewhat. This might enable the bearing to be dropped into place over the shaft. If you decide to use this method, make sure you do not overheat the bearing. Doing so will destroy the properties of the metal and ruin the bearing.

Just as a bearing can be expanded by heating, it can also be shrunk by cooling. Therefore, if a bore is too small for a bearing, you can try chilling the bearing in a freezer. This may shrink it just enough to allow installation into the bore.

After the bearing is installed, check it for proper operation and alignment. Some bearings, like pinion and front axle bearings, must be preloaded, or adjusted so they are under mild pressure. This is done to prevent bearing looseness. When bearings must be preloaded, follow the manufacturer's procedures exactly. Bearing preloading is discussed again in *Manual Transmission* and *Rear Axle Assembly* chapters herein.

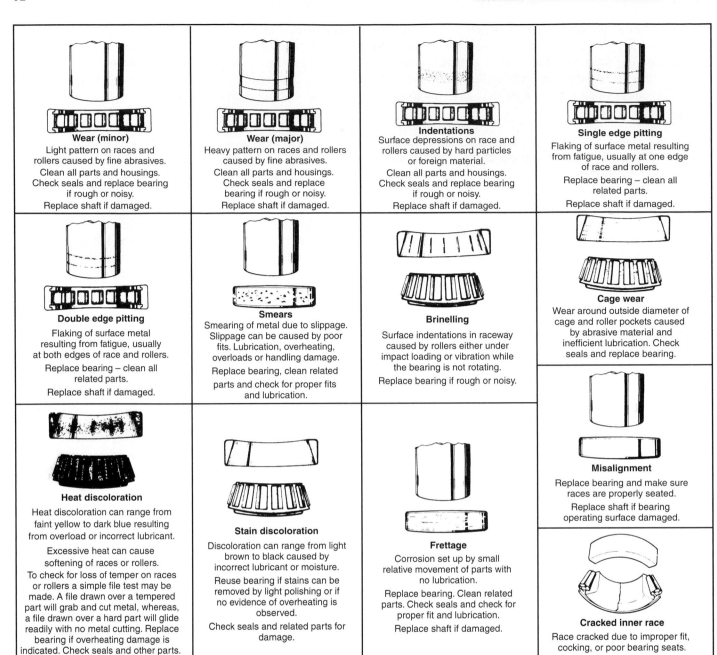

Wear (minor)
Light pattern on races and rollers caused by fine abrasives. Clean all parts and housings. Check seals and replace bearing if rough or noisy. Replace shaft if damaged.

Wear (major)
Heavy pattern on races and rollers caused by fine abrasives. Clean all parts and housings. Check seals and replace bearing if rough or noisy. Replace shaft if damaged.

Indentations
Surface depressions on race and rollers caused by hard particles or foreign material. Clean all parts and housings. Check seals and replace bearing if rough or noisy. Replace shaft if damaged.

Single edge pitting
Flaking of surface metal resulting from fatigue, usually at one edge of race and rollers. Replace bearing – clean all related parts. Replace shaft if damaged.

Double edge pitting
Flaking of surface metal resulting from fatigue, usually at both edges of race and rollers. Replace bearing – clean all related parts. Replace shaft if damaged.

Smears
Smearing of metal due to slippage. Slippage can be caused by poor fits. Lubrication, overheating, overloads or handling damage. Replace bearing, clean related parts and check for proper fits and lubrication.

Brinelling
Surface indentations in raceway caused by rollers either under impact loading or vibration while the bearing is not rotating. Replace bearing if rough or noisy.

Cage wear
Wear around outside diameter of cage and roller pockets caused by abrasive material and inefficient lubrication. Check seals and replace bearing.

Heat discoloration
Heat discoloration can range from faint yellow to dark blue resulting from overload or incorrect lubricant. Excessive heat can cause softening of races or rollers. To check for loss of temper on races or rollers a simple file test may be made. A file drawn over a tempered part will grab and cut metal, whereas, a file drawn over a hard part will glide readily with no metal cutting. Replace bearing if overheating damage is indicated. Check seals and other parts.

Stain discoloration
Discoloration can range from light brown to black caused by incorrect lubricant or moisture. Reuse bearing if stains can be removed by light polishing or if no evidence of overheating is observed. Check seals and related parts for damage.

Frettage
Corrosion set up by small relative movement of parts with no lubrication. Replace bearing. Clean related parts. Check seals and check for proper fit and lubrication. Replace shaft if damaged.

Misalignment
Replace bearing and make sure races are properly seated. Replace shaft if bearing operating surface damaged.

Cracked inner race
Race cracked due to improper fit, cocking, or poor bearing seats.

Figure 4-33. *Refer to the above to learn about common antifriction bearing defects and their causes. (General Motors)*

Related Components

Many components are used with gears and bearings in the drive train. Some of these are explained below.

Thrust Washers

Thrust washers are used between the end of a rotating shaft and a stationary housing or between two rotating shafts. They prevent excessive axial, or back-and-forth, shaft movement by limiting the amount of *end clearance,* or the distance between the two parts. This type of shaft movement is often called *end play.*

Figure 4-36 is a typical thrust washer. Like friction bearings, thrust washers are made of a metal that is purposely softer than the parts against which they operate. Most thrust washers have a key or tab extension. This is to lock the washer to one component, so all wear occurs on the side contacting the other component. The grooves in the thrust washer allow lubricant to flow between the contact surfaces of the washer and rotating part. Note that Torrington bearings are often used as thrust washers.

Most thrust washers are replaced when end clearance becomes too large. Thrust washers for some drive train units are available in different thicknesses. This way, improper end play in the unit can be corrected by selecting a thicker or thinner washer.

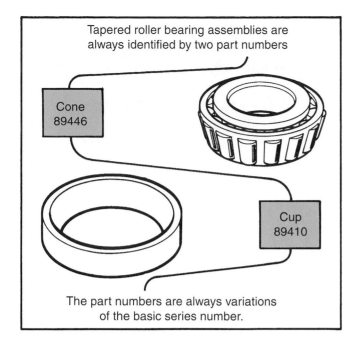

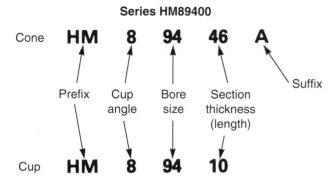

Figure 4-34. *Always try to identify the part number of the original bearing before ordering replacements. An example is shown here. If the number is missing, you will need to gather information about the vehicle to obtain a replacement part. (Federal Mogul)*

Retainer Plates

Retainer plates are used to hold gears and bearings in position. They are attached to a stationary housing with bolts, machine screws, or snap rings. Most retainer plates have a center hole for a shaft to pass through. On a few transaxles, the retainer plate is a solid casting and seals in lubricant.

Collars

Collars are another type of retainer. These are steel rings used to hold gears or bearings on shafts, as shown in **Figure 4-37.** Collars are usually used where space is too limited to allow a different kind of retainer. They are usually a press fit.

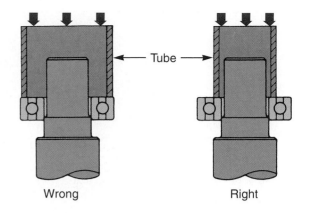

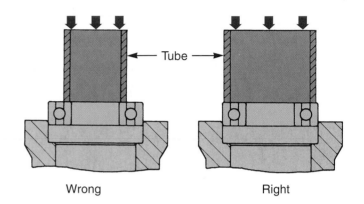

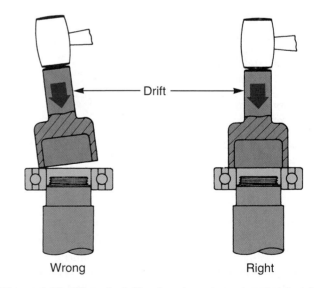

Figure 4-35. *When installing bearings, keep in mind that force should never be applied through the rolling elements. Also, when pressing a bearing onto a shaft or into a housing, always make sure the bearing is going on or in straight. Analyze these diagrams showing different methods of installing bearings. Do you see the difference between right and wrong ways of installation in each drawing? (FAG Bearings)*

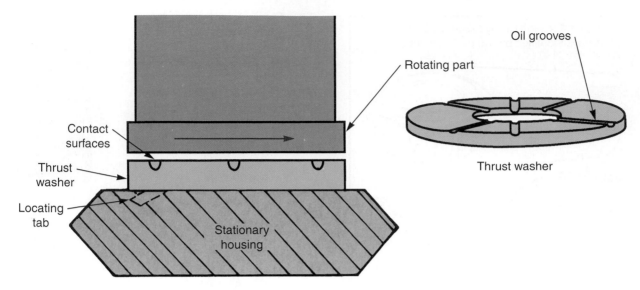

Figure 4-36. *Thrust washers are installed between two parts where there is relative rotational motion between the two. Thrust washers keep back-and-forth movement of shafts to a minimum. This is a concern when using helical gears, which tend to cause this sort of movement. Thrust washers also reduce thrust caused by friction and wear.*

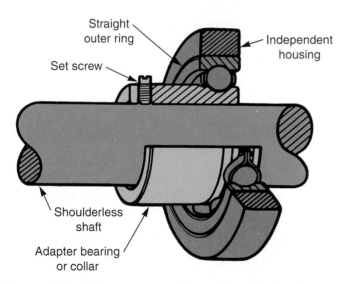

Figure 4-37. *Collars are generally used to retain bearings on shafts, as shown here. They are also sometimes used to retain gears. (Federal Mogul)*

Collars are most often removed with a chisel, which is used to split the part open. There are other ways to remove a collar, too, but they usually expand and distort the collar. A collar, in any case, should never be reused. If it is not destroyed in the removal process, it will most likely be distorted and not fit properly.

Splines

A *spline* is a type of slot that is cut into a shaft or other part. The slot runs *longitudinally,* or lengthwise to the axis

of rotation. A spline may be straight, running parallel to the axis of the shaft or part, or it may be a spiral, running at an angle. A *splined shaft* will have many equally spaced, parallel splines. Splines can be internal or external. Internal splines of one component are designed to mate with external splines of another component. The two components, then, rotate as a unit.

Splines serve the important purpose of allowing two parts to move back and forth in relation to each other, while allowing power to be transferred through them. They are used on the *sliding* gears of manual transmissions and on slip yokes of drive shaft assemblies. They are used on clutch friction discs to drive manual transmission input shafts and on torque converter turbines to drive automatic transmission input shafts. They are used on other drive train parts as well. External splines are shown on the end of an input shaft in **Figure 4-38A.** This is a typical splined shaft. **Figure 4-38A** also shows internal splines on a typical planetary gearset. **Figure 4-38B** shows the two parts assembled.

Keys and Keyways

To keep a shaft and machine part, such as a gear or pulley, from rotating in relation to each other, keys and keyways are often used. Slots, or *keyways,* are cut in the two mating parts, and a *key* is inserted in the slots. Keys and keyways come in many variations. The most common for automotive use are the *straight key* and the *Woodruff key,* **Figure 4-39.** Sometimes, a single external spline is cut into a part and used as a key. A matching slot is cut into the mating part. The spline and slot are locked in place, and the two parts rotate as a unit.

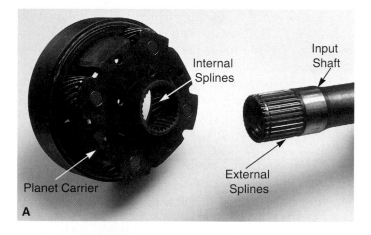

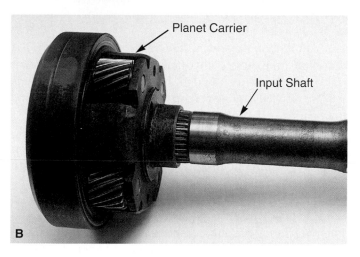

Figure 4-38. *Splines prevent rotation between two parts. A—Note how the internal splines on the planetary gearset match the external splines on the input shaft. B—When the input shaft slides into the planetary gearset during assembly, the fit between the internal and external splines causes the two parts to rotate as a unit.*

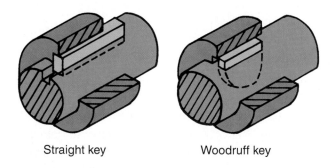

Figure 4-39. *Keys are used to prevent rotation between a machine part and a shaft.*

Summary

Vehicle drive trains are made up of many simple parts such as gears, chains, and bearings. Engine speed can be increased or decreased at the output of a gear or chain drive. A vehicle's direction can also be reversed by these parts. Friction can be reduced using bearings.

Simple gears are toothed wheels that engage one another to transmitting power. Speed and torque are increased (or decreased) through drive train gears. To improve fuel economy at higher vehicle speeds, overdrive gears are used.

Several classes of gears are used in vehicle drive trains. Spur gears are used to connect parallel shafts. The teeth on spur gears are straight and parallel with the shaft. Spur gears are sometimes used in manual transmissions. Spur gears are the simplest kind of gear. They can be noisy if the clearance is too large.

Helical gears have teeth that are cut at an angle. Helical gears operate more smoothly and quietly than spur gears. The angled cut of helical teeth puts a sideways thrust on the gears and shafts. This is often called end thrust.

Herringbone gears overcome some of the thrust and noise problems of helical and spur gears. Herringbone gears are sometimes called double-helical gears. These gears are seldom used in drive trains.

Bevel gears connect shafts that run at an angle. The angle of shafts used with drive train bevel gears is usually 90°. Similar to the bevel gear is the spiral bevel. Bevel gear teeth are straight, and spiral bevel gear teeth are curved.

Hypoid gears are similar to bevel gears, but the shaft centerlines do not intersect as they do with bevel gears. Otherwise, hypoid gears resemble spiral bevel gears.

Worm gears connect shafts that run at a 90° angle. Worm gears are useful when a large speed reduction is needed. They are seldom used on vehicle drive trains.

Rack-and-pinion gears consist of a flat bar with teeth cut into it (rack) and a small gear (pinion). Rack-and-pinion gears are usually used on the vehicle steering system.

A planetary gearset consists of a sun gear in the center, planet gears surrounding and meshed with the sun gear, and a ring gear meshed with the planet gears. The planet gears are attached to a planet carrier but are free to revolve within it. All planetary gearset gears are constantly in mesh with each other. To change gear ratios, different parts of the gearset are driven or held stationary. Planetary gearsets are used in overdrives and automatic transmissions.

Gear ratio is the speed relationship between gears. It can be summed up as the ratio between the number of teeth in any two meshing gears. A vehicle requires different gear ratios for different operating conditions. A vehicle must also have a reverse gear. Two or more gears can be arranged into a gear train that will provide an output shaft rotation that is opposite to engine rotation.

Given time, gears will wear out. Normal gear wear will occur after many miles. Low mileage gear failures are usually caused by abuse, overloading, or poor maintenance. Common tooth failure conditions are pitting, spalling, peening, scoring, and welding. Gear teeth may also overheat and lose their strength. Overheated gears wear out quickly.

Many gears are a slip fit with their shafts. Slip fit gears are held in position by other parts of the drive train or by snap rings. If the gear must turn with its shaft, it is held by a key or by splines. Sometimes the gear and shaft are machined from a single casting. Some gears are pressed onto the shaft they ride on. Removing these gears requires a gear puller or a hydraulic press.

Chain drives are often used where space is limited or to transmit power between a large distance. Every chain drive assembly consists of a chain and at least two sprockets. Some systems use more than two sprockets. Other transaxles use a metal drive belt and pulleys to transfer power.

There are two major classes of drive chains. The roller chain consists of a series of pins, usually encased in rollers, which are held in position by links. The pins or rollers engage the sprocket teeth. The silent chain consists of a series of leaves that are toothed to engage the sprocket teeth. The leaves are held in position by alignment pins.

Drive chain wear is compensated for by automatic tensioners that place a calibrated load on the chain. These tensioners are usually spring loaded.

Drive chains can be checked for wear on or off the vehicle. A loose chain should always be replaced with a new one. If the sprockets show any tooth wear, they should be replaced. The chain and sprockets are usually replaced at the same time.

Bearings are used to reduce friction between moving and stationary parts. Bearings also guide and support rotating parts. The two major classes of bearings are friction and antifriction. Friction bearings are often pressed into a housing. They have no moving parts.

A rotating shaft slides around the surface of a friction bearing. Friction is reduced by a film of lubricant between the shaft and bearing. The lubricant may be supplied by pressure or splash. Some bearings are permanently lubricated by grease. Most friction bearings have slots to aid in oil distribution.

Friction bearings should always be checked for wear and proper size during any kind of service. Worn or damaged bearings should be replaced. In many cases, the shaft that rides on the bearing should also be replaced. Replacement bearings should be installed with a driver. Oil holes should be aligned with the matching holes in the housing before installation.

Antifriction bearings contain balls or rollers that rotate between races. The balls and rollers change sliding friction into rolling friction.

Ball bearings are used in clutches, transmissions, and transaxles. Roller bearings are used in transmissions and rear axles. Tapered roller, needle, and thrust bearings are all varieties of roller bearing.

Bearing loads usually change with shaft rotation, so that the bearing is alternately loaded and unloaded. This makes the use of high quality bearings critical.

The best way to remove an antifriction bearing is with a special puller or a hydraulic press. *Always* wear eye protection when removing a press-fit bearing. *Never* spin an antifriction bearing with compressed air.

Carefully clean and inspect antifriction bearings. A bearing with any kind of defect must be replaced. Always make sure you obtain the proper replacement bearing. New bearings should be left in their original wrappings until needed.

Antifriction bearings should be installed with the proper bearing driver. A hydraulic press is preferred. Never put driving pressure through the rolling elements, as this will damage the bearing and may cause injury. After installing the bearing, check that it rolls freely, with no binding.

Thrust washers are used to prevent excessive back-and-forth shaft movement. Retainer plates are devices used to hold gears and bearings in position. Collars are used to hold gears or bearings on shafts. Splines allow two rotating parts to move backward and forward in relation to each other while power is transferred through them. The internal splines on one component are designed to mate with external splines on another component. Keys and keyways keep two parts rotating with each other. The key is placed in two slots cut in the mating parts.

Review Questions—Chapter 4

Please do not write in this text. Write your answers on a separate sheet of paper.

1. *True or false?* Some gear drives are capable of providing an increase in speed relative to the drive input.

2. Which of the following statements regarding gear manufacture is *not* true?
   (A) Gear blanks are made from a relatively hard steel.
   (B) The faces of a gear are polished to a smooth finish.
   (C) Gears are hardened in a heat-treating process.
   (D) Gear teeth are cut by a machine called a gear shaper.

3. List and explain six types of gears.

4. The greater the gear _____, the more play there will be between two meshing gears.

5. A drive sprocket has 10 teeth, and the driven sprocket has 40 teeth. What is the drive ratio?

6. Describe what happens when gear teeth wear thin.

7. A failure condition of gears evidenced by flaking and chipping off of metal is known as _____.

8. Peening in gears may be caused by severe _____ or by large foreign particles in the _____.

9. A failure condition of gears evidenced by scratches on the tooth faces is known as _____.

10. _____ drives are found in transaxles and in four-wheel drive transfer cases.

11. How do roller chains and silent chains differ?

12. A(n) _____ can be used to take up slack in a loose chain of a chain drive.

13. What should you look for when inspecting a chain?

14. Sleeves, bushings, and bearing inserts are all types of _____ bearings.

15. What does color change in a bearing indicate?

16. In some applications, a friction bearing must be _____ to keep it from spinning or slipping out of position.

17. Antifriction bearings contain _____ elements that operate within a housing.
    (A) sliding
    (B) rolling
    (C) stationary
    (D) None of the above.

18. _____ bearings contain needle rollers that are arranged radially.

19. Antifriction bearings should remain in their original _____ until just before installation.

20. Back-and-forth, or axial, shaft movement is often referred to as _____ _____.

ASE-Type Questions—Chapter 4

1. All of these are achieved through the action of gears *except:*
    (A) friction reduction.
    (B) torque multiplication.
    (C) power transmission.
    (D) direction reversal.

2. All of these would result if bearings were not used in the drive train *except:*
    (A) excessive friction between moving parts.
    (B) wasted engine power.
    (C) increased turning speeds.
    (D) premature wearing of parts.

3. All of these are parts of a planetary gearset *except:*
    (A) sun gears.
    (B) spur gears.
    (C) planet gears.
    (D) ring gears.

4. All of these are common gear problems *except:*
    (A) pitting.
    (B) scoring.
    (C) splining.
    (D) spalling.

5. Chain drives transmit power in a straight line between:
    (A) pulleys.
    (B) keyways.
    (C) CV joints.
    (D) sprockets.

6. Technician A says chain tensioners are designed to take up slack in chains. Technician B says chain tensioners maintain pressure on the drive side of the chain. Who is right?
    (A) A only.
    (B) B only.
    (C) Both A and B.
    (D) Neither A nor B.

7. When finding chain drive ratio, Technician A says the number of teeth on the driven sprocket should be added to the number of teeth on the drive sprocket. Technician B says the number of teeth on the driven sprocket should be divided by the number of teeth on the drive sprocket. Who is right?
    (A) A only.
    (B) B only.
    (C) Both A and B.
    (D) Neither A nor B.

8. Technician A says it is not necessary to replace sprockets when a chain is replaced. Technician B says it is not necessary to replace a chain when its sprockets are replaced. Who is right?
    (A) A only.
    (B) B only.
    (C) Both A and B.
    (D) Neither A nor B.

9. All of these functions are served by bearings *except:*
    (A) reducing friction between parts.
    (B) guiding and supporting rotating parts.
    (C) preventing damage from misalignment.
    (D) transmitting power.

10. All of these are used as a coating for friction bearings *except:*
    (A) brass.
    (B) copper.
    (C) chrome.
    (D) lead.

11. Technician A says if the oil clearance is too tight, the oil cannot form an adequate surface between mating parts. Technician B says a loose oil clearance will allow oil to leak out too quickly to maintain surface separation. Who is right?
    (A) A only.
    (B) B only.
    (C) Both A and B.
    (D) Neither A nor B.

12. To increase service life, all parts of an antifriction bearing are made of:
    (A) copper.
    (B) low-carbon steel.
    (C) high-strength steel.
    (D) babbitt metal.

13. All of these are roller bearing variations *except:*
    (A) needle bearings.
    (B) thrust bearings.
    (C) ball bearings.
    (D) Torrington bearings.

14. Technician A says it is acceptable to apply a light force through the rolling elements of an antifriction bearing. Technician B says it is acceptable to spin the rolling elements of an antifriction bearing with compressed air. Who is right?
    (A) A only.
    (B) B only.
    (C) Both A and B.
    (D) Neither A nor B.

15. Technician A says an antifriction bearing can be heated to facilitate installation. Technician B says an antifriction bearing can be cooled to facilitate installation. Who is right?
    (A) A only.
    (B) B only.
    (C) Both A and B.
    (D) Neither A nor B.

16. Thrust washers are used:
    (A) to hold chain links.
    (B) around bevel gears.
    (C) to connect two sprockets.
    (D) between the end of a rotating shaft and its housing.

17. Technician A says thrust washers have grooves that allow oil to flow between the washers and the rotating parts. Technician B says thrust washers are generally replaced when end clearance becomes too large. Who is right?
    (A) A only.
    (B) B only.
    (C) Both A and B.
    (D) Neither A nor B.

18. All of these are used to attach a retainer plate to a stationary housing *except:*
    (A) cotter pins.
    (B) machine screws.
    (C) bolts.
    (D) snap rings.

19. All of these statements about collars are true *except:*
    (A) they are steel rings.
    (B) they are a type of retainer.
    (C) they should never be reused.
    (D) they keep a shaft from rotating.

20. Which of these is one of the most common keys for automotive use?
    (A) Spur key.
    (B) Wood key.
    (C) Woodruff key.
    (D) Torrington key.

Chapter 5

Sealing Materials, Fasteners, and Lubricants

After studying this chapter, you will be able to:
- ❑ Identify the types of seals found in automatic transmissions and transaxles.
- ❑ Explain proper methods of removing and installing seals.
- ❑ Explain the importance of proper gasket installation.
- ❑ Describe how sealants are used in automatic transmission and transaxle repair.
- ❑ Summarize the different types of transmission fluid.

Technical Terms

Seals
Lip seals
Dust shield
Ring seals
O-ring
Static seals
Dynamic seals
Seal rings
Gaskets
External gaskets
Internal gaskets
Form-in-place gaskets
Sealants or sealers
Hardening sealers
Nonhardening sealers
Form-in-place gaskets
RTV (room temperature vulcanizing) sealer

Anaerobic sealer
Thread-locking compounds
Thread sealers
Adhesives
Epoxies
Fasteners
Threaded fasteners
Nonthreaded fastener
Bolt
Nut
Cap screw
Screws
Machine screws
Major diameter
Minor diameter
Bolt size
Nut size
Head size

Bolt length
Thread length
Thread pitch
Thread pitch gauge
Right-hand threads
Left-hand threads
Bolt grade
Tensile strength
Grade markings
Torque specifications
Tightening sequence
Snap ring
Automatic transmission fluid (ATF)

Introduction

Automatic transmission and transaxle components must have adequate *lubricant* to operate properly. This means there must be protection against leaks, which could reduce lubricant supply. Components must also be protected from damaging dirt and moisture. This chapter describes the ways lubricants are retained in transmission or transaxle components and the ways dirt and moisture are kept out. In addition, it identifies different types of *fasteners* and major classes of transmission and transaxle lubricants.

This chapter will prepare you to perform some basic procedures required to successfully overhaul transmissions and transaxles. Later chapters will refer to descriptions and procedures covered here.

Seals

Seals are devices that keep fluids, such as lubricant, within a component or a desired space. They also keep contaminants, such as air, dirt, and water, from getting into such places. Seals may be used for one or both of these purposes. In some applications, they are used to separate two different fluids. Seals are usually used with rotating shafts or with shafts that slide back and forth axially. Special seals are sometimes used to seal stationary parts. Some seals allow controlled leakage of fluid to provide some fluid for lubrication of moving parts. These are called *nonpositive seals.*

Types of Seals

Several different types of seals are found throughout the transmission or transaxle. Some are used to seal rotating shafts. Stationary seals are used to seal many nonrotating parts. The type used depends on the specific application. Common seals include *lip seals, ring seals, seal rings,* and *boots.*

Lip Seals

Lip seals are used on rotating shafts. A typical lip seal is shown in **Figure 5-1**. The casing is a rigid support for the other seal components. It is often made from stamped steel. The sealing element is designed to contact the rotating shaft. Most sealing elements are made from oil-resistant, synthetic rubber. The contact area is made in the form of a lip. A steel coil spring, often called a *garter spring* or a *garter,* is wrapped around the sealing element. It places pressure on the sealing element to hold it tightly against the shaft. Some lip seals do not have a spring.

Figure 5-2 shows typical sealing action of a lip seal. Spring pressure and *internal pressure*—fluid pressure in the seal cavity—presses the sealing element tightly against the shaft. An automatic transmission *front pump seal* typifies this condition. If there were not pressure behind the seal,

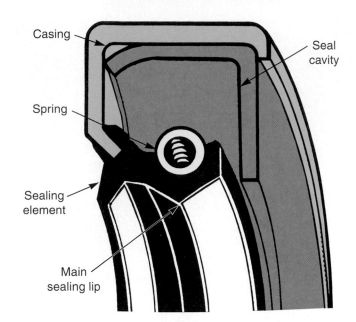

Figure 5-1. *Components of a lip seal. The casing acts to hold seal in position when installed. It is usually press fit into a stationary housing. The sealing element has the lip that contacts the shaft. The main sealing lip is placed nearer, or toward, fluid to be retained. Optional spring puts pressure on sealing element to assure contact with the shaft at all times. Note that the shape of the components may vary. (Ford)*

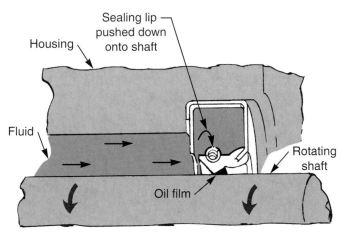

Figure 5-2. *Note how a lip seal works. Pressure helps the spring hold the lip against the shaft. A film of oil between the lip and shaft helps to seal and reduces wear. Slight leakage at the lip is usually acceptable. (Caterpillar)*

as would be the case for the transmission *rear seal,* the garter spring alone would keep the sealing element in contact with the shaft.

A thin oil film at the sealing lip reduces friction and provides additional sealing, **Figure 5-2**. If the seal is properly sized and installed, the oil film will not leak past the lip. Some lip seals have a felt *wiper* on the outside of the seal to catch small amounts of lubricant that get past the lip.

Different conditions require that different kinds of lip seals be used. Some of the variations are shown in **Figure 5-3**.

Figure 5-3A shows some of the ways a garter spring seal can be made. Notice the various lip constructions and placements of the garter. The design of the lip and garter spring depends on the sealing needs of certain shafts and certain applications. Variations in pressure, shaft end play, and exposure to outside elements affect the design of the seal.

Figure 5-3B shows a variation of the lip seal. This type uses a *finger spring,* which is a cone-shaped series of *fingers* that place pressure on the sealing lip. The function of the finger spring is similar to that of the garter spring; however, the finger spring places less pressure on the shaft and lip. This reduces seal and shaft wear; however, some sacrifice is made in sealing ability. This type of seal is used in lower-pressure systems or for sealing components lubricated with grease.

Figure 5-3C illustrates lip seals that have no springs. These seals are used when the main requirement is to hold back contaminants. They may also be used in places where *internal* fluid pressure will force the sealing element against the shaft. These seals would not be appropriate for applications where they would be subjected to *external* pressure. Most seals under external pressure require a spring to ensure a good seal.

Figure 5-3D shows seals that have more than one lip. These seals are used where external pressures are high or where seals are exposed to harsh conditions. The outer lip stops oil that has gotten past the inner lip, or main sealing lip, and it protects the inner lip from damage. *Double lip seals* are often used on parts that are exposed to the outside elements. Examples include some transmission rear, or output shaft, seals.

A variation of the double lip seal has lips pointing away from each other, as shown in **Figure 5-3E.** These seals are used in some transaxles, where it is necessary to keep transmission fluid and differential gear oil separated.

Some seals are protected by a **dust shield, Figure 5-4.** Dust shields are used in places where the seal would be exposed to large amounts of dirt and water, such as the front wheel bearing assemblies on front-wheel drive vehicles and the transmission and pinion oil seals of rear-wheel drive vehicles. A massive assault from these elements would quickly destroy the seal, allowing lubricant to leak out. The dust shield acts as a deflector, or baffle, to keep most of the dirt and water away from the seal. The seal can then handle the small amount of material that does get through to it. Dust shields are very effective in helping to reduce seal failure.

Ring Seals

Ring seals are used to seal stationary parts or hydraulic pistons that slide in their bores. The ring seal fits into a shallow groove cut into one or both parts. Ring seals can be round, half-round, or square in cross section. See **Figure 5-5.** In automatic transmissions and transaxles, the round seal, or **O-ring,** is most common. Most O-rings have a small diameter through the cross section, as compared to

Components

SAE-ASTM Approved nomenclature

Garter spring seals

A

Finger spring seals

B

No spring seals

C

1. Sealing element
2. Outer case
3. Inner case
4. Bonded case
5. Molded case
6. Garter spring
7. Finger spring
8. Spacer
9. Clinch

Multiple lip seals

D

E

Figure 5-3. *Study different kinds of lip seals used in drive trains. A—Design of sealing lip and garter spring depends on usage. B—This variation of lip seal employs a finger spring. C—Some lip seals require no springs. Fluid pressure may force sealing element against shaft. D—Some seals may have two or more lips. These seals are often used to seal exposed drive train parts. E—Lips point away from each other on this double lip seal. These seals are used where it is necessary to keep two fluids from mixing together. (Federal Mogul)*

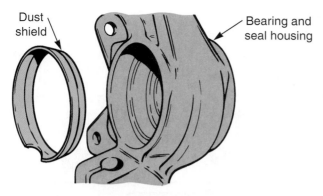

Figure 5-4. *A dust shield keeps dirt and water away from sealing elements. This reduces seal failures in parts of drive train exposed to excessive dirt and water. (Ford)*

between the seal ring and bore allow leakage. The rings are installed in grooves machined in the shaft. They are commonly used in automatic transmissions.

Metal seal rings resemble small piston rings. Some metal rings are locking types, with small latch tabs to hold the ends of the ring together. Others are the straight, or butt, type. This type depends on the size of the surrounding bore for a tight fit. See **Figure 5-7.**

Teflon seal rings are one-piece rings made of Teflon and plastic. Teflon seal rings resemble square-cut ring seals (square rings), although Teflon seals are usually made in bright colors. They are often used in original installations because they allow lower-cost assembly than metal rings.

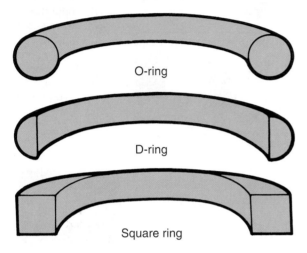

Figure 5-5. *Types of ring seals. O-rings and D-rings are almost always made of synthetic rubber. Square rings may be made of rubber, metal, or Teflon. These rings are all designed for specific applications. (Ford)*

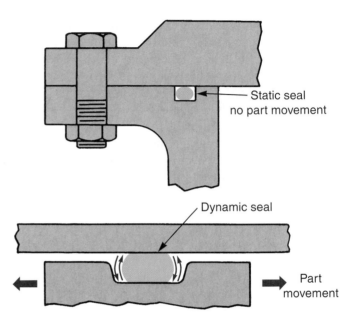

Figure 5-6. *Static and dynamic seals. Static seals are used to seal parts that will not move in relation to each other. Dynamic seals seal where there is relative motion between two parts. (Deere and Co.)*

the *annulus,* or ring. O-rings are used with both stationary and moving parts, while *half-round rings* (D-rings) and *square rings* are usually used with moving parts.

Stationary ring seals fit into nonmoving components and are found in many parts of automatic transmissions and transaxles. They are often called **static seals.** They are intended to form a perfect seal. The ring seals used on hydraulic pistons found inside automatic transmissions—on the clutch apply piston, for example—are called **dynamic seals.** They flex slightly to allow a small amount of fluid leakage, just as piston rings in an engine are expected to allow some *blowby* of compressed gases. Slight fluid leakage keeps the seals lubricated. See **Figure 5-6.**

Seal Rings

Seal rings (not to be confused with ring seals) are metal or Teflon® rings. They seal oil pressure directed to or through passageways in rotating shafts or drums. Seal rings are nonpositive seals. They are designed to allow slight leakage for lubrication of moving parts. Unlike dynamic seals, nonpositive seals do not flex. Slight clearances

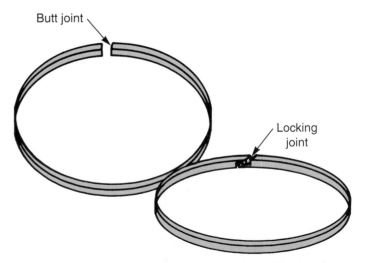

Figure 5-7. *Seal rings with locking and butt joints are shown here. Be careful how you handle these as they are very brittle.*

Both types of seal rings seal in hydraulic fluid to maintain pressure. They do this by forming a mechanical seal between the moving parts. The outside diameter of the ring seats against the inner surface, or bore, of a rotating part. The side of the ring, moved by pressure, seats against one side of the shaft groove. This prevents practically all fluid leakage. See **Figure 5-8.**

The seal ring should rotate with the bore. If not, the bore surface will wear, allowing pressure loss. (The side of the ring will rotate against the shaft groove, and the groove will incur slight wear.) Note that the rings used to seal *servo pistons* and *accumulators,* which will be discussed in Chapter 9, do not rotate.

Metal seal rings can be removed without being damaged, although most manufacturers recommend replacing the rings when a transmission is overhauled. Teflon seal rings cannot be removed without being destroyed. In many cases, the original Teflon rings can be replaced with metal rings.

Seal Service

A leaking seal should be replaced as soon as it is discovered. When a seal leaks *out,* a component will eventually run out of lubricant and be destroyed. The leaking lubricant often causes messy, slippery floors, and it may leak on and damage other vehicle components. A seal that leaks *in* will allow lubricant to become contaminated, which may result in damage to components.

Seal Leaks

Almost all seal problems show up first as leaks. Although seal leaks can be caused by several problems, the most common is a worn or damaged seal. All seals will wear out, given time, and it is not unusual for a seal in a vehicle with high mileage to leak. This is usually caused by wear and hardening of the sealing element. Sometimes, the leak occurs because a lip seal garter spring becomes weak. **Figure 5-9** points out areas of a lip seal that commonly wear. Ring seals usually become hard and

shrink. When this happens, they no longer conform to the grooves they are sealing. A new ring seal generally solves the problem.

Sometimes, the shaft or housing for the seal wears out, causing a leak. When replacing a lip seal having many miles of service, always check the shaft for signs of wear. The shaft tends to wear where the sealing lip contacts it. See **Figure 5-10.** If this area is not worn, a standard replacement seal will stop the leak. Many times, however, the shaft will be so worn that the standard replacement will not correct the problem. In most of these instances, the shaft must be replaced.

In some cases, however, the problem can be corrected by using a replacement seal that will contact the shaft to either side of the contact area of the original seal (worn area). On more expensive parts, the shaft can sometimes be repaired by pressing a sleeve over the worn shaft area. A seal is then fitted to the sleeve. The sleeve adds to the diameter of the shaft, so a larger lip diameter will be required to fit the sleeve.

When checking a shaft for signs of wear, check for excessive play in the shaft. If there is too much *lateral,* or side-to-side, movement, the seal cannot form a good

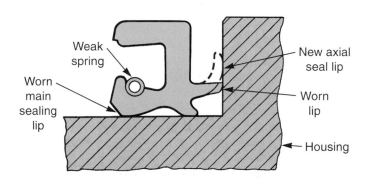

Figure 5-9. *Depicted are conditions of wear and damage that can affect performance of a lip seal. Any one of these problems can cause a leak. (Ford)*

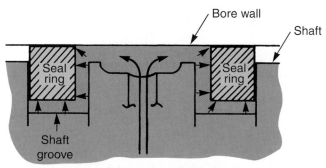

Figure 5-8. *This is a partial view of a shaft in a bore. Seal rings are installed in the shaft grooves. Pressurized hydraulic fluid keeps seal rings tight against the side of their grooves.*

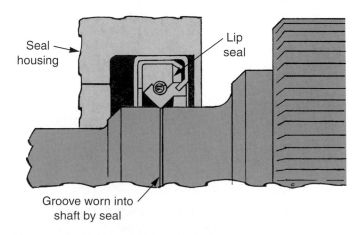

Figure 5-10. *A groove worn in a shaft can cause a leak. A new seal alone may not solve this problem. In many instances, you will find that the shaft must be replaced or a sleeve must be installed. (Caterpillar)*

sealing surface. The shaft or shaft bearings should be replaced. If a seal seems to have failed at low service mileage, check for shaft or housing problems. Also, check for excessive dirt or metal particles, which will cause premature wear.

One of the most common causes of seal leakage following a repair is using the wrong seal. The wrong size lip seal will leak almost immediately. An improperly sized ring seal may seal for a few minutes or hours, but it will eventually leak. A seal that is not the right size should never be used. See **Figure 5-11.**

The other most common cause of seal leakage after a repair job is improper installation, evidenced by a cocked or distorted seal. See **Figure 5-12.** O-rings that are improperly installed will leak immediately. A lip seal that is not perfectly aligned with the shaft will not make good contact with the shaft. The seal will begin leaking, wear unevenly, and possibly damage the shaft.

Shaft or housing damage attributed to other causes can damage a seal and cause it to leak. Examples of bore and shaft damage are shown in **Figure 5-13.** Scratches and dents from hammering out an old seal, grooves, sharp keyway edges, corrosion, and improper machining can all cause seal leakage. A loose or bent shaft will cause seal distortion and leaks, even with a new seal. All these problems should be looked for and eliminated before installing a new seal.

Seal Removal

The following precautions should be observed during any seal removal operation:
- ❑ *Always* wear eye protection when you are using a seal puller or driving the seal from its housing.
- ❑ Do *not* damage the seal bore or the shaft.
- ❑ After removing the seal, check all related parts for damage that could result in a seal leak.

Seals should be removed carefully. Most static O-ring seals are compressed between two stationary parts. Some

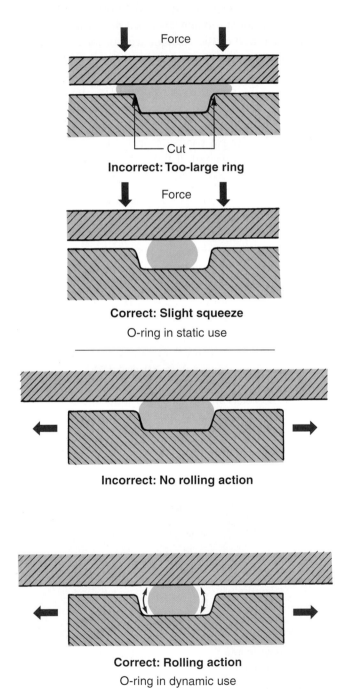

Figure 5-11. *Proper sizing of the O-ring is important. Note what happens when proper and improper sizes of static and dynamic seals are used. (Deere and Co.)*

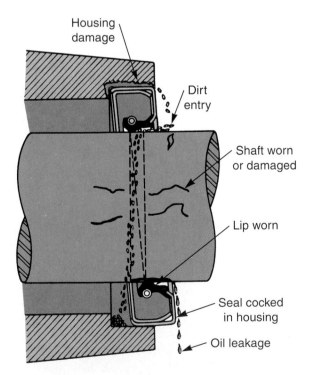

Figure 5-12. *A cocked lip seal can cause many problems. The most obvious problem is leaking, but a cocked seal can also cause shaft and housing damage. The housing will usually be damaged when the seal is installed. The shaft will wear because of the uneven pressure placed on it by the lip. (Deere and Co.)*

O-rings are held by a retainer plate. After the plate or component is disassembled, the O-ring can be lifted or pried out of its groove. Dynamic O-rings are removed by removing the piston from its cylinder to expose the O-ring. The O-ring is then carefully pried from its groove.

Most lip seals are pressed into a bore that is part of a transmission or transaxle component. A special tool, such as that shown in **Figure 5-14,** should be used to pry out the old seal. Alternate pressure on the seal from side to side. This will help the seal come out easily and minimize damage to the bore.

Another way to remove the seal is to use a slide hammer with special puller jaws. The jaws are pushed through the seal and then expanded. In lieu of this, drill a small hole on the side of the seal in the metal casing. Thread a screw attached to a slide hammer into the hole. Then use the slide hammer to pull the seal from the bore. To remove the seal evenly and prevent it from becoming wedged, drill two holes in the metal casing, one directly across from the other. The slide hammer can then be alternated from one side of the seal to the other during removal.

If you can reach the seal from behind, you can drive it out of its bore with a seal driver or a hammer and a soft metal drift. Be careful not to damage the housing!

Once the old seal has been removed, check the housing, shaft, and all related parts. If leakage was a problem, these parts may provide clues as to the cause. Correct any problems before installing a new seal.

If a standard replacement seal is to be used, check its part number against that of the old seal to ensure they match. If they do not, you may have the wrong replacement seal or it could be that a new number has since been assigned to the particular seal. Otherwise, it is possible that the old seal was the wrong one. If there was a problem with the original seal, this could well have been the cause.

Seal Installation

Two general points should be made in regard to seal installation prior to discussing specifics. First, a seal should not be removed from its wrappings until it is ready to be installed. This will help keep the seal clean. Also, the seal contact surfaces should be coated with lubricant prior to assembly, **Figure 5-15.** For this, you should use the same type of lubricant used in the part or assembly.

If a ring seal is to be installed, lubricate the groove. Carefully install the ring in the groove. If an O-ring is being used between two stationary parts (a static seal), make sure it fits properly in the groove before reassembling the component. If the O-ring is the dynamic type, check it for proper size around its piston *and* in its bore.

If installing a lip seal, lubricate the shaft, as well as the sealing lip. Again, use the correct lubricant. If the seal has a metal casing and it is not precoated by the manufacturer, coat the casing lightly with a *nonhardening sealer,* **Figure 5-16.**

If the seal will be installed into a housing bore with the shaft already installed, make sure the shaft is free from burrs or scratches that could damage the seal. If necessary,

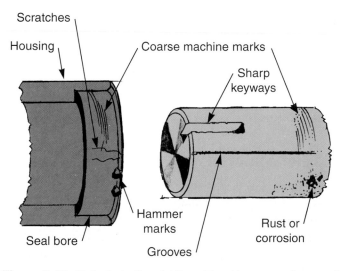

Figure 5-13. *Defects in the shaft and housing can ruin a seal during installation. Sharp edges on shafts can cut the seal as it is slipped into place. Housing defects can damage the seal casing. (Caterpillar)*

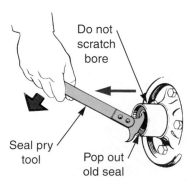

Figure 5-14. *The best way to remove a seal is with a special seal puller. If a puller is not available, seals can be removed by several other methods. Always be careful not to damage the seal housing during removal. (Lisle)*

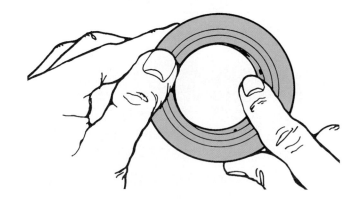

Figure 5-15. *Always lubricate the sealing lip of a new seal before installation. Prelubrication will make the seal easier to install and will provide an oil film at the lip during break-in. (Ford)*

polish the shaft with *crocus cloth,* which is an abrasive cloth that has very fine particles of iron oxide. In addition, to protect a seal from drilled holes, splines, keyways, square shaft ends, and other sharp spots, a temporary collar may be formed over the spot with a piece of smooth, heavy paper; otherwise, a plastic sleeve may be placed over the spot.

Slide the seal over the shaft until it reaches the housing, making sure the seal is oriented in the proper direction. Using a driver that closely matches the outside diameter of the seal, drive the seal squarely into the housing until it is properly seated in the bore. After installation, make sure that the garter spring (if used) did not come loose and that there was no damage from the driving action. **Figure 5-17** shows a seal driver being used to install a seal in a bore.

In an emergency, a large socket or pipe that closely matches the outside diameter of the seal can be used as a driver. It is also possible to install a seal with a hammer, if done carefully. The seal should be lightly tapped in.

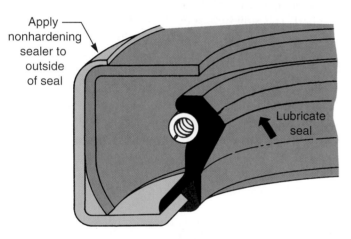

Figure 5-16. *If the seal has a metal casing, apply nonhardening sealer to the casing before installation. This step is not necessary if the casing has a flexible coating. (Deere and Co.)*

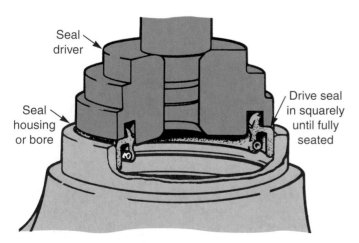

Figure 5-17. *An improperly installed seal will be ruined immediately. Watch the seal carefully as it is being installed to ensure that it goes in squarely with the bore. (Ford)*

Alternate hammer taps from side to side until the seal is seated. Failing to do it this way will cause the seal to become cocked and bent.

Gaskets

Gaskets are a type of static seal used on nonmoving parts. Gaskets used on drive trains range from the simple gasket used to seal a differential inspection cover to the complex gaskets that seal automatic transmission valve bodies. Gaskets are never used to seal moving parts.

Gaskets used in automatic transmissions and transaxles are normally cut from a single piece of gasket material. In contrast, some *engine* gaskets are made of multiple layers of materials. Gaskets are used to prevent the leakage of lubricant between two parts that are bolted together. There are both *external* and *internal* gaskets. *External gaskets* keep fluid from leaking out of a component. They also keep dirt or moisture from entering. *Internal gaskets* prevent loss of pressure or crossover of fluids within a component.

Gasket material is flexible. This allows gaskets to conform to imperfections in mating surfaces. **Figure 5-18** is an exaggerated illustration of this. When the parts are tightened, the gasket compresses and deforms, filling in the low spots, scratches, and other imperfections in the mating surfaces. This results in a leakproof seal.

Gasket Materials

The majority of gaskets used in drive trains are made of oil-resistant paper. These are sometimes called *chloroprene gaskets.* This material is often mixed with a substance such as rubber to add flexibility and strength.

Where mating parts will be subject to expansion and contraction, gaskets made from a synthetic rubber called *neoprene* are often used. Natural rubber cannot be used because petroleum-based lubricants, like transmission fluid, will make it swell and fail in service.

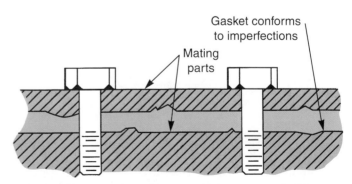

Figure 5-18. *Gaskets are made of relatively soft materials. This flexibility enables the gasket to conform to imperfections in the gasket mating surfaces.*

Some gaskets are made of metal, such as copper or aluminum. These gaskets have a ridge running around their perimeter. When mating parts are tightened down, the ridge is compressed, providing the seal. Metal gaskets are used in only a few places and are often replaced with other types of gaskets when the vehicle is serviced.

Thick gaskets of cork or felt are usually used in providing seals for assembly covers made of sheet metal. The thickness of the gasket helps compensate for distortion of the sheet metal cover, **Figure 5-19.** These thick gaskets are often coated with Teflon to reduce sticking when the gasket is removed.

Some gaskets are made from a silicone sealer. These are called **form-in-place gaskets.** The silicone normally comes in a tube. It is applied directly to one of the mating part surfaces. The parts are assembled while the silicone gel is still wet to the touch. More will be said about form-in-place gaskets later in this chapter.

Gasket Problems

Gaskets are sometimes a source of problems. More often, they are blamed for problems really caused by other components. If a gasket *appears* to be leaking, make sure that it is not a cracked part that is leaking or a nearby component leaking onto the gasket area.

Some gasket leaks are caused by loose bolts. Tightening the bolts properly may stop the leak. Sometimes, however, pressure from within the component will force out a piece of gasket. This is called a **blown gasket.** Blown gaskets are usually caused by loose fasteners or warped mating parts that do not tightly hold the gasket. Also, *overtightening* attaching bolts can crush the gasket, resulting in a weak spot that will blow out much more easily.

Gasket Service

Gasket service involves disassembling parts, removing the old gasket, cleaning mating surfaces, and installing a new gasket. Some basic rules for removing and replacing gaskets are:
- ❑ *Verify* the cause of the problem before disassembly. Often, what appears to be a gasket problem is a defective part or an improper part installation.
- ❑ *Avoid* causing part damage during disassembly or gasket removal. Although the gasket material is designed to compensate for scratches, dents, and nicks, do not push your luck. Aluminum parts are especially easy to damage.
- ❑ *Thoroughly* clean the gasket surfaces. If the old gasket must be scraped off, use a dull scraper on the mating surfaces to avoid damaging them. This is especially important for automatic transmission valve bodies, where the slightest scratch can cause an internal leak and pressure loss.

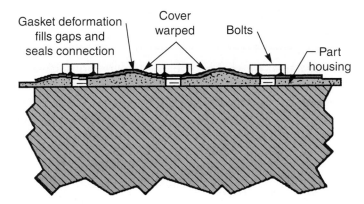

Figure 5-19. *Thick gaskets can sometimes compensate for warped sheet metal cover plates or pans.*

- ❑ *Thoroughly* wash the parts with solvent and blow them dry with air pressure. Avoid using rags to dry internal transmission or transaxle parts. Rag lint can damage bearings or clog passages and valves.
- ❑ Check the new gasket shape. It should conform to the mating part surfaces and the old gasket. Make sure that the gasket contains all the required holes and that all gasket holes match holes in the mating parts.
- ❑ Use sealer *only* if needed, especially where automatic transmissions and transaxles are concerned. If the manufacturer recommends using sealer, place a thin coat on the gasket.
- ❑ Hand-tighten all bolts before using a wrench. Make sure all bolts will start in their respective holes.
- ❑ Tighten the bolts in sequence. If a *tightening sequence* is not given, start from the middle and work toward the edges. Tighten the bolts in a crisscross pattern to prevent distortion of the mating surfaces.
- ❑ Tighten all bolts in small steps. After the final tightening sequence, recheck all the bolts.
- ❑ Do *not overtighten* the bolts. This is especially important when installing pans or other sheet metal parts.

Making a Gasket

Sometimes, a replacement gasket is not available, and you must make a gasket from a sheet of gasket material. Simple gaskets can be made easily by either of two methods.

One method is to lay out a sheet of gasket material on the component that requires a gasket or lay the component over the gasket material. Use a pencil to trace the outlines of the sealing surface on the gasket material. Use scissors to cut out the gasket material to the desired pattern.

The second method is to lay the sheet of gasket material over the component and use a brass or ball peen hammer to lightly tap the gasket against the sharp edges of the sealing surface. Tap out one or two bolt holes first. Then, thread bolts into the holes to hold the gasket material. Go back to tapping around the sealing surface until you obtain the proper gasket shape. Trim any rough edges with scissors.

Sealants

Sealants, or *sealers,* are used for various reasons. They can improve the sealing ability of a gasket. They can take the place of a gasket. They can also be used to hold a gasket in place during reassembly. Sometimes, sealers are used to improve the sealing ability of dynamic seals and hose connections.

Types of Sealers

There are several types of sealers. They have different properties and are designed for different purposes. Always read the manufacturer's label and a service manual before selecting a sealer. Study the use and characteristics of the sealer types shown in **Figure 5-20.**

Hardening Sealers

Hardening sealers are used on parts that will remain assembled for long periods, possibly the life of the vehicle. They are used on some threaded fittings and fasteners, for example. Sometimes, they are used for filling uneven sealing surfaces. Hardening sealers are usually resistant to heat and chemicals.

Nonhardening Sealers and Shellac

Nonhardening sealers are used on parts that will be taken apart occasionally, such as transmission pans. They are also used on some *threaded fasteners* and fittings and on hose connections. They are resistant to most chemicals and to moderate heat.

Shellac is a gummy and sticky substance that remains pliable. It is frequently used on gaskets of fibrous materials, both for sealing and for holding gaskets in place during assembly. Shellac can be classified as a nonhardening sealer.

Form-in-Place Gaskets

Form-in-place gaskets are sometimes used instead of conventional gaskets. They are applied, in fluid form, directly to parts to be sealed. The gasket is formed when the parts are tightened together and the substance *cures* (dries, or sets up). Form-in-place gaskets are commonly made of either *RTV sealer* or *anaerobic sealer.*

RTV Sealer

RTV (room temperature vulcanizing) sealer is silicone sealer. It cures to a rubberlike consistency to form the gasket. RTV sealer is usually used on thin, flexible flanges.

RTV sealer normally comes in collapsible tubes. The substance is usable for some time after a tube is opened. After the expiration date, which is generally printed on the package, the sealer may not cure properly; in which case, it would not make a good seal. Always check the expiration date before using the sealer. Use a new tube if in doubt.

To apply the RTV sealer, **Figure 5-21,** lay a 1/8" (3 mm) diameter continuous bead around the edge of the part. Keep the bead to the inside of the mounting holes—preferably, encircle them. Do not permit breaks in the bead, as a leak will result. Assemble and tighten components while the RTV sealer is still wet to the touch. You should have about 10 minutes to get the parts assembled before the sealer begins to dry. If necessary, RTV sealer may be removed with a rag before it cures.

Type	Temperature range	Use	Resistant to	Characteristics
shellac	−65 to 350 F (−54 to 177 C)	general assembly: gaskets of paper, felt, cardboard, rubber, and metal	gasoline, kerosene, grease, water, oil, and antifreeze mixtures	dries slowly sets pliable alcohol soluble
hardening gasket sealant	−65 to 400 F (−54 to 205 C)	permanent assemblies: fittings, threaded connections, and for filling uneven surfaces	water, kerosene, steam, oil, grease, gasoline, alkali, salt solutions, mild acids, and antifreeze mixture	dries quickly sets hard alcohol soluble
nonhardening gasket sealant	−65 to 400 F (−54 to 205 C)	semipermanent assemblies: cover plates, flanges, threaded assemblies, hose connections, and metal-to-metal assemblies	water, kerosene, steam, oil, grease, gasoline, alkali, salt solutions, mild acids, and antifreeze solutions	dries slowly nonhardening alcohol soluble

Figure 5-20. *This table shows some types of sealers and their uses. No one sealer can be used for every sealing job. (Fel-Pro)*

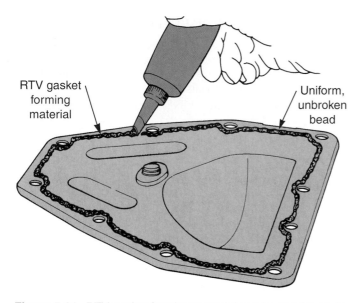

Figure 5-21. *RTV gasket forming material can be used to make gaskets on many parts. (DaimlerChrysler)*

Anaerobic Sealer

Anaerobic sealer cures to a plasticlike substance in the absence of air. It will remain fluid as long as it is exposed to the atmosphere. Anaerobic sealer is used with tightly fitting, thick parts, which air cannot penetrate. It is used between two smooth, true surfaces, not on thin, flexible flanges, which may be penetrated by air. Note that anaerobic sealers will not harden in their containers because the air space in the containers provides enough air to prevent hardening.

Anaerobic sealer should be applied sparingly. Lay about a 1/16" (2 mm) diameter continuous bead on one surface only. It is recommended that the sealer encircle each mounting hole. The parts should be assembled and tightened within 15 minutes of applying the sealer.

Before any kind of sealer is applied, a few preliminary steps must be taken. All loose material and lubricant must be removed from the part contacting surfaces, which are usually called the mating surfaces. To do this, scrape or wire brush the mating surfaces. Use a shop rag and solvent to clean any oil or grease from these surfaces. The mating surfaces must be clean and dry before applying sealer. Check that mating surfaces are reasonably flat in areas where the sealer is to be applied. Finally, before applying any kind of sealer, always refer to the manufacturer's instructions to be sure you use the correct type.

Thread-Locking Compounds and Thread Sealers

Thread-locking compounds are used to prevent threaded fasteners, such as *bolts, screws,* and *nuts,* from loosening. These compounds are a type of anaerobic sealer. When they are applied to a threaded fastener and the fastener is tightened, the lack of air causes them to harden. Thread-locking compounds make the removal of a fastener very difficult. They should be used only when the vehicle manufacturer recommends them.

Before applying thread-locking compound, threads should be cleaned and sprayed with a special primer. This will reduce the setting time of the compound. Apply the compound as shown in **Figure 5-22.** Then, install the fastener.

It will require considerable turning effort to loosen a fastener installed with thread-locking compound. The fastener will also be more difficult to remove once the "lock" is broken. This is because the thread-locking compound remains in the threads and tends to jam the fastener. If a fastener will not loosen, light heat may soften the compound and allow the lock to be broken.

Thread sealers are used to seal any threaded fasteners that extend into the oil-lubricated interior of a component. Oil will seep out through the threads if a sealer is not used. Thread sealers should be applied only after the threads have been thoroughly cleaned.

Adhesives

Adhesives are used to hold parts together. They are *not* used as sealants. Adhesives are sometimes used to hold gaskets in place or hold small parts together during assembly. Before applying adhesives, make sure the surfaces to be bonded are clean. Some adhesives are applied in a two-step process; a *setting agent* is applied first; then the adhesive is applied. Other adhesives are two-part mixtures applied in a one-step process. These are sometimes called *epoxies.* Epoxies are mixed just before application.

Fasteners

Fasteners are devices used to hold, or fasten, parts together. Fasteners can be divided into threaded and

Figure 5-22. *Thread-locking compounds are used when a threaded fastener must not loosen. Thread-locking compounds remain liquid in the container and can be easily applied to the threads of the fastener. They harden and form a solid bond when the fastener is tightened. (Deere and Co.)*

nonthreaded varieties. Nuts, bolts, and screws are all types of *threaded fasteners.* Snap rings are a type of *nonthreaded fastener.*

Threaded Fasteners

The most common types of threaded fasteners used on automatic transmissions and transaxles are bolts and nuts. A *bolt* is a threaded rod with a *head* on one end. The head is usually six-sided (*hex head*), although some special bolts have square or round heads. A *nut* has inside threads and, commonly, a six-sided outer shape.

Bolts can be threaded into nuts or parts with threaded holes. A bolt that threads into a part instead of a nut is called a *cap screw.* The threads of the bolt and nut or threaded part provide a holding force to keep parts assembled.

Although *screws* resemble bolts, they are generally smaller than bolts. Screws are generally designed to thread directly into a nut or threaded hole. Screw heads are generally designed to accept screwdrivers, although a few screws have hex heads and accept wrenches or sockets. *Machine screws* are commonly used on automatic transmissions and transaxles. They are usually smaller than cap screws and are flat across the bottom.

Fastener Classification

Fasteners are made in many different sizes, strengths, and thread patterns. Fasteners are classified accordingly.

Screw Size Terminology

There are a variety of terms used to refer to the size of threaded fasteners. The automotive technician must know what these terms mean. Some of the most commonly used terms relating to fasteners are:

- ❏ *Major diameter.* The larger diameter pertaining to screw threads. Applies to both internal and external threads.
- ❏ *Minor diameter.* The smaller diameter pertaining to screw threads. Applies to both internal and external screws.
- ❏ *Bolt size.* Refers to the major diameter of the bolt threads.
- ❏ *Nut size.* Refers to the minor diameter of the nut threads.
- ❏ *Head size.* Size of hex head measured across the flats. This is the wrench size.
- ❏ *Bolt length.* Length measured from the bottom of the bolt head to the end of the threaded section.
- ❏ *Thread length.* Measure of external thread length.
- ❏ *Thread pitch.* The distance measured across crests of adjacent threads. Thread notes on drawings of metric fasteners give pitch in millimeters. Conventional (nonmetric) fasteners denote the number of threads per inch, which is the reciprocal of pitch. (Conventional threads conform to thread standards in the United States, Canada, and England.)

Thread Types

The fasteners used in automatic transmissions and transaxles have one of three basic types of threads:

- ❏ **Coarse threads.** Called UNC, which stands for Unified National Coarse.
- ❏ **Fine threads.** Called UNF, which stands for Unified National Fine.
- ❏ **Metric threads.** Called SI, which stands for International System of Units (English translation).

Never attempt to interchange thread types, as the threads will be damaged. Metric threads can be mistaken for conventional threads if they are not inspected carefully. A *thread pitch gauge,* either conventional or metric, should be used to check any threads you are unsure of. See **Figure 5-23.**

In addition to these thread variations, both *right-* and *left-hand threads* are used on modern vehicles. To tighten a bolt or screw with **right-hand threads,** it must be turned clockwise. Right-hand threads are more common than left-hand threads. A fastener with **left-hand threads** must be turned counterclockwise to tighten it. Sometimes, the letter *L* is stamped on fasteners with left-hand threads. Left-hand threads are reserved for special situations.

Bolt Grade

Bolt grade refers to the amount of *pulling,* or *stretching,* force a fastener can withstand before it stretches or, in some cases, breaks. (Ductile materials will stretch before breaking; brittle materials will break almost immediately when a tensile force, or longitudinal pulling force, is applied.) Bolt grade, then, is an indication of *tensile strength,* which is measured in pounds per square inch. Tensile strengths vary among bolts of the same size and thread pattern. Bolts are made of different metals. Some of these metals are stronger than others.

Bolt head markings are used to identify the tensile strength of a bolt. These markings are called *grade markings.* Conventional bolts are marked with lines, or slash marks—the more lines, the stronger the bolt. Grade 8 (6 lines) is the strongest. A metric bolt is marked with a

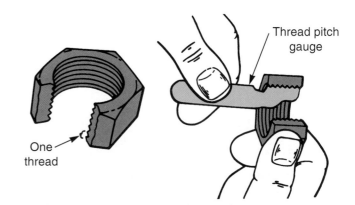

Figure 5-23. *A thread pitch gauge can be used to check thread pitch or number of threads per inch.*

numbering system—the larger the number, the stronger the bolt. Metric bolts range from 4.6 to 10.9. See **Figure 5-24.**

Never replace a bolt with one of a lower grade. A lower grade bolt is weaker, and it can break, causing the failure of a major part. This can lead to a vehicle breakdown or a potentially dangerous situation.

Fastener Torquing and Tightening Sequence

Torque specifications are tightening values for threaded fasteners. These specifications are determined by the vehicle manufacturer. Torque specifications are critical for automatic transmission and transaxle components, as well as other vehicle parts.

It is very important to tighten all fasteners properly. An overtightened bolt will stretch and possibly break. The threads can also be damaged, or *stripped.* Gaskets can be smashed or broken. An undertightened bolt can work loose and fall out. Part movement can also shear the loose bolt or break a gasket, causing leakage.

Vehicle manufacturers also provide a *tightening sequence,* or *pattern,* for threaded fasteners. The sequence ensures that parts are fastened evenly. Uneven tightening can cause part warping or breakage and gasket leaks. Tightening is usually done gradually and follows a crisscross pattern, starting in the middle of the part and working outward in steps. This results in an even pressure across the entire mating surface of the parts.

You should follow these rules to properly torque threaded fasteners:
- ❏ Pull steadily on the handle of the torque wrench. Jerking the torque wrench will cause excessive pointer deflection and invalid readings.
- ❏ Clean and oil the fastener threads before tightening.
- ❏ *Avoid* using *swivel, extension,* and *crowfoot sockets,* if possible. They will cause invalid readings.

- ❏ If using the *flex-bar,* or *beam-type,* torque wrench, look straight at the scale. Viewing from an angle produces a false reading. Never let the *pointer shaft* contact any other part of the vehicle.
- ❏ Obtain and use the manufacturer's recommended tightening sequence and torque specifications.
- ❏ Torque the fasteners in steps. Never run one fastener up to full torque before all bolts have been partially torqued.
- ❏ Recheck torque once after all fasteners have been fully torqued and once after the component has been run up to normal operating temperatures and cooled off.

Removing Broken Fasteners

Occasionally, a fastener will break while it is in the part. Various methods of removing broken fasteners are shown in **Figure 5-25.** The method used depends on whether the broken fastener extends from the part or not.

If a fastener shaft extends from a part, several methods can be used to remove it. A slot can be cut into the fastener shaft, and the fastener can be turned out using a screwdriver. A pipe or a metal rod can be welded to a broken fastener shaft. The rod can then be turned to unscrew the fastener. Vise grip pliers and stud pullers can be used to tightly grasp the fastener shaft for removal.

If the fastener is broken inside the part, a screw extractor can be used. Screw extractors were described briefly in Chapter 3. Occasionally, the broken fastener can be removed by striking one edge with a chisel. This may get the fastener turning so it can be removed.

Snap Rings

A *snap ring* is a split ring used for holding parts on shafts or inside bores. The snap ring is generally used when clearance will not permit the use of a threaded fastener. The snap ring fits tightly in or around a groove that is machined to accept it.

The two major kinds of snap rings are *internal snap rings* and *external snap rings.* **Figure 5-26** shows the two types. Snap rings can be removed and installed with snap ring pliers, as shown. If a snap ring is bent or distorted during removal, it should be replaced.

Automatic Transmission Fluids

Automatic transmissions and transaxles require a special type of lubricating oil, or fluid. The viscosity of *automatic transmission fluid (ATF),* as it is called, is lower than that of manual transmission gear oils. Lighter oil is necessary to operate the automatic transmission hydraulic system and provide efficient power transfer through the torque converter. To provide proper lubrication of the planetary gears and bearings, these fluids contain special EP

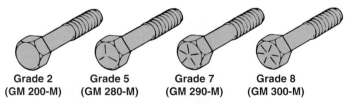

Grade 2   Grade 5   Grade 7   Grade 8
(GM 200-M)   (GM 280-M)   (GM 290-M)   (GM 300-M)

Customary (inch) bolts – Identification marks correspond to bolt strength – Increasing numbers represent increasing strength.

Metric bolts – Identification class numbers correspond to bolt strength – Increasing numbers represent increasing strength

Figure 5-24. *It is very important to use a fastener that is strong enough. Always replace a fastener with one that is the same grade or stronger. Bolt grades can be determined by markings on the head of the bolt.*

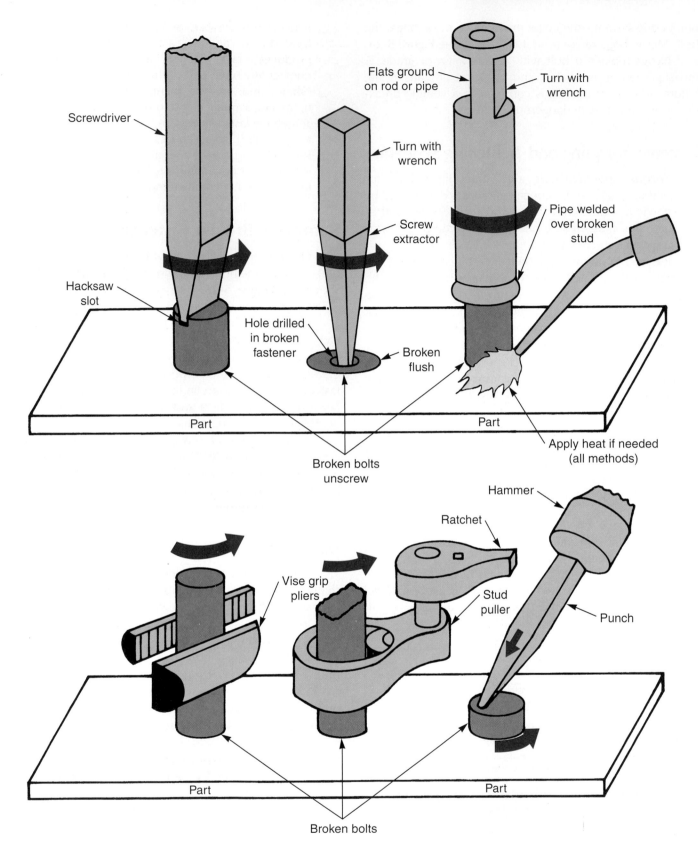

Figure 5-25. *A broken fastener can be removed in many ways. Common methods are shown here. Broken fasteners are usually easier to remove if part of the fastener is sticking out of the part.*

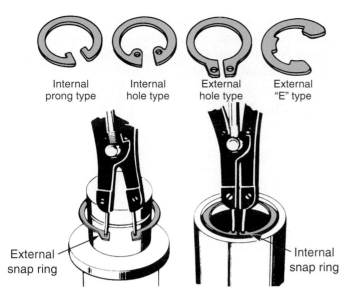

Internal prong type   Internal hole type   External hole type   External "E" type

External snap ring

Internal snap ring

Figure 5-26. *There are two kinds of snap rings—internal and external. Snap rings are often used in transmissions and transaxles.*

(extreme-pressure) additives. EP additives are composed of long-chain molecules (molecules that are chemically tied into long strings, or chains). The long-chain molecules resist being squeezed from between moving parts and, therefore, provide a constant oil film between these parts.

Automatic transmission fluids can be divided into different classes, according to the additives they contain. ATFs contain additives that help them withstand heat, increase their lubricating ability, and reduce foaming. The major difference in the various ATF classes is in the use of additives that affect the coefficient of friction of the fluid. The coefficient of friction describes how much friction the oil allows when it is compressed between two surfaces. While there should be as little friction as possible in gears and bearings, holding members require some friction to operate.

Since *holding members* are always soaked in transmission fluid, the fluid's coefficient of friction determines how well they can hold. A particular fluid's coefficient of friction is varied according to the types of transmissions in which it will be used. Transmission fluid requirements vary according to the type of holding members used. The pressure applied on the holding members is another factor.

Transmission Fluid Types

In the past, there were only two types of transmission fluid. Today, however, there are many types of fluid. Until 1972, whale oil was used in all transmission fluids. Whale oil was replaced with petroleum-based oils after whale hunting was banned by the United States and most other countries. Additives have been developed to match the friction and lubrication qualities of whale oil. However, nothing quite matches the whale oil qualities, and fluid must be changed more often on all vehicles made between 1970 and 1990. Many modern vehicles use synthetic or semisynthetic transmission fluids, which may not need changing unless the vehicle is in severe service. In any vehicle, using the correct fluid is very important.

Failure to use the proper fluid can cause performance problems and may shorten transmission or transaxle life. The following transmission fluids are used in modern vehicles. In some cases, they can be used to replace the original fluids in older transmissions.

Current Fluids

The following transmission fluids are used in late-model transmissions and transaxles. New fluid specifications are always being developed, and the technician must consult the latest service literature to be sure the proper fluid is being used.

Note: Transmission fluid color varies according to the type of fluid and manufacturer. If the color of the replacement fluid does not match the color of the original fluid, make sure the correct fluid is being added. It is acceptable to mix colors as long as the new fluid is the correct type.

Dexron Type II, IIE, and III

Dexron fluids have been used in many vehicles, including all General Motors cars and trucks made after 1965, Chrysler products throughout the 1980s, and many common imported vehicles. The original Dexron has been updated and renamed Dexron II, Dexron IIE, and Dexron III. Some transmissions and transaxles using Dexron III can go 100,000 miles (160,000 km) between transmission fluid changes. However, if the vehicle is used for heavy-duty or severe service, the transmission fluid may need to be changed more often.

Most transmissions or transaxles can be topped off or refilled with Dexron III in place of the older Dexron types. However, older types of Dexron should not be used in a transmission or transaxle designed for Dexron III. A few manufacturers, such as Nissan, recommend using original Dexron instead of Dexron II or Dexron III.

Mercon, Mercon V

Mercon has been used in Ford Motor Company transmissions and transaxles since 1988. Mercon V was released in 1997. Some oil companies market a combination Dexron/Mercon transmission fluid. In most cases, this fluid will work well in Ford Motor Company vehicles. Some transaxles, however, may develop shift problems unless dedicated Mercon is used. Dexron/Mercon can also be used in General Motors vehicles.

Chrysler 7176/ATF+3/ATF+4

Chrysler 7176, also called Mopar ATF+3 or ATF+4, has been used in most DaimlerChrysler cars, SUVs, and trucks for about the last ten years. 7176 fluid can also be used on some early Chrysler transmissions and transaxles when there is a lockup torque converter shudder problem (discussed in later chapters). A letter often follows the number 7176, such as 7176B. Increasing letters indicate improvements in the 7176 additive mix. Therefore, you should not use an earlier letter series fluid in a transaxle requiring a later series. Aftermarket 7176 transmission fluid is usually identified as ATF+3 or ATF+4. ATF+3 has replaced all earlier 7176 fluids and can be used in any Chrysler vehicle. ATF+3 fluid can also be used to replace Diamond ATF in the transaxles of some Mitsubishi vehicles. ATF+4 is used on some Chrysler vehicles made after the 1999 model year.

Honda ATF

Special Honda fluid should be used when refilling or topping off Honda and Acura transaxles. In some areas, this fluid is only available from Honda dealers. Honda ATF is also used in some Isuzus. Honda ATF should not be used in other transmissions.

Types C-3 and C-4

C-3 and C-4 fluids are used in heavy-duty transmissions in large trucks and tractors. They may also be used in off-road equipment. C-3 and C-4 have extra EP additives. C-4 is an improved version of C-3. Units using C-3 may generally be upgraded to C-4. Some transmissions and hydraulic systems can be switched from C-3 or C-4 to Dexron III. Always check with the transmission maker before using C-4 in a C-3 system or replacing C-3 or C-4 with Dexron.

Other Current Fluids

In addition to the fluids discussed above, special fluids currently being used in other vehicles include:
❏ Toyota "T-IV" for Lexus and some Toyota cars.
❏ BMW CA 2634 for BMWs with 5-speed automatics.
❏ Mercedes-Benz fluid for Mercedes-Benz vehicles with electronic transmissions.
❏ Saturn transaxle fluid 21005966+ for Saturn transaxles.
❏ Diamond ATF for some Mitsubishis and Mitsubishi-built DaimlerChrysler automobiles.
❏ VW/Audi #60020000 for Volkswagon and Audi vehicles.

These fluids may be difficult to get at any place other than the vehicle dealer. Some fluids can be replaced with Dexron or Mercon if friction-modifying additives (discussed below) are added.

Older Fluids

Fluids designed for use in older vehicles are still available in many areas. However, they will eventually be discontinued as the transmissions and transaxles that require them become fewer. The following fluids should be used only in the transmissions they were originally intended for.

Type F

Type F fluid was used in many older Ford vehicles. Some older imports, mostly those using Borg-Warner or Aisin-Warner transmissions and transaxles, require Type F. Type F fluid should not be used in any Ford transmission made after 1980. The Type F fluid in some transmissions can be replaced with Mercon, depending on the year of the vehicle and the transmission type. Some imported vehicles continued to use Type F until the late 1980s. Type F was the recommended power steering fluid for many Ford products until recently. It is still widely available.

Types CJ and H

Types CJ and H were used in some Ford Motor Company vehicles in the 1970s and 80s. However, they are no longer sold. Transmissions using types CJ or H can be refilled with Dexron or Mercon. Check the latest Ford recommendations to determine which type should be used in a particular transmission or transaxle.

Type A

Type A fluid can be purchased in many supermarkets and discount stores. However, it is an obsolete fluid and should never be used in any modern transmission or transaxle. It is specified for use in certain very old transmissions. However, the Type A being produced now varies widely in quality, and most manufacturers recommend that it be replaced with a current fluid when replacement is necessary.

Fluid Additives

There are two main types of transmission fluid additives: friction modifiers and problem-solving additives. Each type is discussed below.

Friction Modifiers

Some of the fluids discussed above may be hard to obtain. Many oil companies and parts suppliers make friction modifiers that can be added to commonly available

Dexron III/Mercon fluid. These modifiers are supplied in small bottles, **Figure 5-27**. Adding the modifier will give Dexron III/Mercon the same friction characteristics as the specialized fluid. Typically, these additives are used to convert Dexron to Chrysler C-3, Honda, Toyota "T," Mitsubishi, and other vehicle-specific fluids. Most of these additives are added to the fluid when the transmission or transaxle filter and fluid are changed. Some manufacturers do not approve the use of friction-modifying additives and recommend using the correct fluid only.

Friction-modifying additives should be used only when called for. Use in transmissions that do not require them may result in harsh shifts or other problems.

Problem-Solving Additives

Some additives are sold as transmission or transaxle "tune-ups" or "performance improvers." These additives usually contain solvents to loosen sludge, combined with chemicals to increase fluid viscosity. Other transmission additives are marketed as cures for leaks. Most of these additives contain alcohol or other chemicals that cause the transmission's rubber seals to swell. These problem-solving additives seldom improve the operation of a defective transmission or transaxle, although they may occasionally seal a slight leak.

There is no need for additives if the transmission or transaxle is operating properly and the right fluid is being used. The technician should be cautious about using additives to solve a problem. A single application of an additive usually has no harmful effect on transmission or transaxle operation. However, more than one application may lead to problems by causing the seals to swell excessively or modifying the thickness or friction properties of the fluid. If the additive loosens too much sludge, the released debris may clog the filter.

Checking Transmission Fluid Level

On most vehicles, transmission fluid is checked using a dipstick. The level is always checked with the engine running and the transmission in park or neutral, as required. The transmission fluid should be warm—the dipstick should be too hot to touch comfortably when the proper temperature is reached. If the fluid level is low, add fluid in small amounts through the dipstick tube until the level is correct. A few vehicles use a filler plug instead of a dipstick. On these vehicles, the fluid should just touch the bottom of the plug opening with the transmission warmed up. Do not overfill an automatic transmission, or the planetary gears will whip the fluid into a foam.

 Caution: If you are in doubt about the type of automatic transmission fluid a particular vehicle uses, do not guess! Different fluid types are not compatible with each other, nor are they interchangeable. The proper type of fluid is generally printed on the transmission dipstick. If not, check the owner's manual or the service manual for the proper fluid.

Summary

Seals are used to keep lubricant in and contaminants out of components. Lip seals are used on rotating or sliding shafts. Ring seals are used to seal stationary parts or hydraulic pistons. There are many kinds of lip and ring seals. Some seals are protected by dust shields in places where the seal would be exposed to dirt and water.

Most seal defects are first noticed when the seal begins leaking. A leaking seal should be replaced as soon as possible. There are several methods of removing, inspecting, and replacing seals. When a seal is replaced, all related parts, such as shafts and housings, should be inspected and replaced if defective. In most cases, the replacement seal must exactly match the old one.

Gaskets are used to seal stationary part surfaces. Gaskets are made of relatively soft materials so they can conform to imperfections in the mating parts they seal. Many different types and thicknesses of gasket materials are used in automatic transmissions and transaxles.

Many gasket leaks are caused by warped or damaged mating surfaces or loose fasteners. Gaskets are replaced by disassembling the parts, removing the old gasket, cleaning and inspecting the mating surfaces carefully, and installing a new gasket.

Sealants, or sealers, can be used to improve the sealing ability of a gasket, take the place of a gasket, or hold a gasket in place during reassembly. Sometimes, sealers are used to improve the sealing ability of dynamic seals and hose connections.

Form-in-place gaskets are sometimes used instead of conventional gaskets. They are applied, in fluid form, directly to parts to be sealed. RTV (room temperature vulcanizing) sealer is silicone sealer. It cures to a rubber-like consistency to form the gasket.

Figure 5-27. *Friction modifiers give Dexron III/Mercon the same friction characteristics as specialized fluids.*

Adhesives are used to hold parts together. They are not used as sealants. Adhesives are sometimes used to hold gaskets in place or hold small parts together during assembly.

Fasteners are used to hold parts together. Fasteners can be divided into threaded and nonthreaded varieties.

Automatic transmission fluids are thinner than gear oils. Two major kinds of automatic transmission fluids are Dexron III and Mercon. Other types are available. The proper fluid should always be used to avoid transmission damage.

The two main types of transmission fluid additives are friction modifiers and problem-solving additives. Friction modifiers can be added to Dexron III/Mercon to give them the same friction characteristics as specialized fluids. Problem-solving additives are often sold as transmission or transaxle "tune-ups" or "performance improvers." These additives contain solvents to loosen sludge, combined with chemicals to increase fluid viscosity. Other additives are marketed as cures for leaks. These additives generally contain alcohol or other chemicals that cause the transmission's rubber seals to swell.

Transmission fluid should be checked with the engine running and the transmission in park or neutral, as required. The transmission fluid should be warm. If the fluid level is low, add fluid through the dipstick tube or filler plug opening until the level is correct.

Review Questions—Chapter 5

Please do not write in this text. Write your answers on a separate sheet of paper.

1. A(n) _____ is a device that keeps a desirable element, such as a lubricant, from getting out of a component.

2. _____ are commonly used to seal rotating shafts.

3. What is the purpose of a dust shield?

4. What are O-rings used for?

5. What, besides the seal itself, should you inspect when replacing a lip seal with many miles of service?

6. What three precautions should you observe during seal removal?

7. During installation, the metal casing of a lip seal that is not precoated by the manufacturer should be coated with _____ sealer.

8. Gaskets provide a kind of _____ seal between non-moving parts.

9. Cite basic rules you should observe when removing and replacing gaskets.

10. What is RTV sealer?

ASE-Type Questions—Chapter 5

1. Technician A says that seals and gaskets are used to keep lubricants inside of components. Technician B says that seals and gaskets are used to keep dirt and moisture out of components. Who is right?
   (A) A only.
   (B) B only.
   (C) Both A and B.
   (D) Neither A nor B.

2. All the following statements about seals are true *except:*
   (A) they can be used to separate two different fluids.
   (B) they are only used to seal rotating or sliding shafts.
   (C) boots are a type of seal.
   (D) some seals are made of metal.

3. To ensure that a sealing lip makes good contact with a rotating shaft, lip seals often use:
   (A) garter springs.
   (B) leaf springs.
   (C) return springs.
   (D) torsion springs.

4. Technician A says that a lip seal with a finger spring is commonly used to seal components lubricated with grease. Technician B says that a finger spring exerts less pressure on the shaft and lip than a garter spring. Who is right?
   (A) A only.
   (B) B only.
   (C) Both A and B.
   (D) Neither A nor B.

5. Technician A says that lip seals without springs are commonly used to hold back contaminants. Technician B says that lip seals without springs are used in applications where external pressure will force the sealing element against the shaft. Who is right?
   (A) A only.
   (B) B only.
   (C) Both A and B.
   (D) Neither A nor B.

6. The most common ring seal used in drive trains is the:
   (A) square seal.
   (B) half-round seal.
   (C) triangular seal.
   (D) round seal.

7. Technician A says that seal rings are nonpositive seals. Technician B says that seal rings can be made from metal or Teflon. Who is right?
   (A) A only.
   (B) B only.
   (C) Both A and B.
   (D) Neither A nor B.

8. Almost all seal problems show up first as:
   (A) noises.
   (B) lubricant leaks.
   (C) shaft wear.
   (D) part overheating.

9. Which of these would be the least likely cause of low-mileage seal failure?
   (A) Hardening of the sealing element.
   (B) Wrong size seal.
   (C) Improper seal installation.
   (D) Damaged or bent shaft.

10. Technician A says that excessive lateral movement of a shaft may cause a seal to leak. Technician B says that an improperly sized seal will leak immediately. Who is right?
    (A) A only.
    (B) B only.
    (C) Both A and B.
    (D) Neither A nor B.

11. Technician A says that some lip seals can be removed by carefully prying them from their bore. Technician B says that some static O-ring seals can be removed by using a special puller with a slide hammer. Who is right?
    (A) A only.
    (B) B only.
    (C) Both A and B.
    (D) Neither A nor B.

12. Before seal installation, a shaft can be polished with:
    (A) polishing compound.
    (B) steel wool.
    (C) crocus cloth.
    (D) fine sandpaper.

13. All of these can be used to install a seal in a housing bore except:
    (A) a hammer.
    (B) pliers.
    (C) a socket.
    (D) a pipe.

14. Gaskets are never used to seal:
    (A) aluminum parts.
    (B) oil-filled parts.
    (C) stationary parts.
    (D) moving parts.

15. All of these are used as gasket materials except:
    (A) oil-resistant paper.
    (B) aluminum.
    (C) cork.
    (D) ceramic.

16. An oil pan gasket is leaking. Technician A says that bolts could have been overtightened, damaging the gasket material. Technician B says that the bolts may be loose. Who is right?
    (A) A only.
    (B) B only.
    (C) Both A and B.
    (D) Neither A nor B.

17. Technician A says that when servicing an apparent gasket leak, the first step is to clean the gasket surfaces. Technician B says that the first step when servicing an apparent gasket leak is to make sure that the gasket is the cause of the leak. Who is right?
    (A) A only.
    (B) B only.
    (C) Both A and B.
    (D) Neither A nor B.

18. Which of these can be made relatively easily if an exact replacement is not available?
    (A) Gaskets.
    (B) Lip seals.
    (C) O-rings.
    (D) Ring seals.

19. Hardening sealers can be used in all of these applications except:
    (A) threaded fittings.
    (B) uneven surfaces.
    (C) permanent assemblies.
    (D) hose connections.

20. Technician A says that RTV sealer is commonly used to make form-in-place gaskets. Technician B says that form-in-place gaskets must never be used in place of conventional gaskets. Who is right?
    (A) A only.
    (B) B only.
    (C) Both A and B.
    (D) Neither A nor B.

21. Anaerobic sealers will harden in the absence of:
    (A) moisture.
    (B) air.
    (C) oil.
    (D) gasket compression.

22. Technician A says that epoxies are mixed just before being applied. Technician B says that thread-locking compounds should be used on all threaded fasteners. Who is right?
    (A) A only.
    (B) B only.
    (C) Both A and B.
    (D) Neither A nor B.

23. All of these are types of threaded fasteners *except:*
    (A) nuts.
    (B) cap screws.
    (C) snap rings.
    (D) machine screws.

24. All of these are used to describe threaded fasteners *except:*
    (A) major diameter.
    (B) thread pitch.
    (C) head length.
    (D) thread length.

25. Technician A says that most threaded fasteners have left-hand threads. Technician B says that the threads of metric and conventional bolts will interchange even if the head sizes are different. Who is right?
    (A) A only.
    (B) B only.
    (C) Both A and B.
    (D) Neither A nor B.

26. Technician A says that tightening bolts in the proper sequence is very important when reassembling any drive train part. Technician B says that tightening bolts to the correct torque is very important when reassembling any drive train part. Who is right?
    (A) A only.
    (B) B only.
    (C) Both A and B.
    (D) Neither A nor B.

27. Technician A says that a bent snap ring should be replaced. Technician B says that when a threaded fastener is broken inside of a part, the part must be replaced. Who is right?
    (A) A only.
    (B) B only.
    (C) Both A and B.
    (D) Neither A nor B.

28. All the following relate to lubricant viscosity *except:*
    (A) weight.
    (B) thickness.
    (C) resistance to foaming.
    (D) resistance to flow.

Chapter **6**

Hydraulics and Pneumatics

After studying this chapter, you will be able to:
- ❏ Explain why a hydraulic system can transfer motion and power.
- ❏ Discuss different hydraulic components used in modern automatic transmissions and transaxles.
- ❏ Discuss basic pneumatic systems and compare them to hydraulic systems.

Technical Terms

Hydraulics

Hydrodynamics

Hydrostatics

Hydraulic pressure

Apply piston

Output piston

Pascal's law

Hydraulic leverage

Hydraulic pumps

Nonpositive displacement pumps

Positive displacement pumps

Gear pump

Internal gear pump

Crescent pump

Rotor pump

Sliding-vane pump

Vane pump

Hydraulic valves

Check valve

Pressure relief valve

Control valves

Spool valve

Valve body

Valve bores

Pressure regulator valves

Throttle valves

Actuators

Directional control valves

Pneumatics

Open system

Closed system

Atmospheric pressure

Compressed-air system

Vacuum

Vacuum modulator valve

Introduction

Hydraulics and pneumatics are widely used in automatic transmissions and transaxles. A *hydraulic system* is used to develop and control hydraulic pressure and flow. Automatic transmissions and transaxles use a hydraulic system to transfer power from the engine to the drive shaft. A *pneumatic system* uses the pressure differential between the outside atmosphere and engine vacuum to control portions of the transmission's hydraulic system.

This chapter will identify the basic principles and major components of hydraulic and pneumatic systems. The knowledge gained by studying this chapter will enable you to more fully understand the components discussed in detail in later chapters.

Hydraulics

In general, **hydraulics** is the science and technology of liquids at rest and in motion. **Power hydraulics** deals specifically with the *transmission of power* through liquids. This branch of hydraulics looks for practical ways to produce and use hydraulic power, which is what a **hydraulic system** does.

The major advantage that hydraulic systems have over other types of power transmission is that they are simple and provide a compact way to multiply power. Hydraulic power systems operate more quietly than other forms of power transmission. The components in these systems last longer because they are continuously lubricated by hydraulic fluid. Further, hydraulic systems are controlled rather easily, and many components are controlled automatically.

Types of Hydraulic Systems

A hydraulic system may be classified as *hydrodynamic* or *hydrostatic*. The type of system it is, described below, depends on the type of hydraulic power used.

Hydrodynamic Systems

Hydrodynamics is the study of *dynamic,* or moving, liquids. In a **hydrodynamic system,** power is transmitted when a moving fluid strikes or acts upon a device composed of blades or vanes. The energy in the fluid is transmitted to the blades or vanes, causing them to move. In this type of system, the moving fluid delivers the force that results in a transfer of power. A water wheel, **Figure 6-1,** is an example of a hydrodynamic system. *Fluid couplings* and torque converters are examples of hydrodynamic devices. These devices are complete in themselves, or are *self-contained*; they are usually replaced as a unit.

Hydrostatic Systems

Hydrostatics is the study of *static* liquids, or liquids at rest. This branch of hydraulics studies the nature of fluids

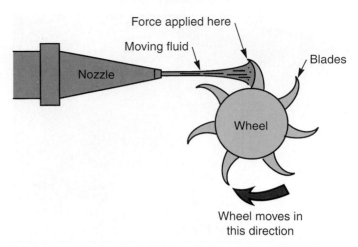

Figure 6-1. *Study this simple hydrodynamic system. Moving fluid strikes blades attached to a wheel. The force of the fluid striking the blades causes the wheel to revolve. This principle is used to operate fluid couplings and torque converters.*

in regard to pressures exerted on them or by them. In a **hydrostatic system,** pressure is applied to liquid confined in the system. Pressure transmitted through the liquid serves as a means for applying force at some other point in the system. Hydrostatic principles are used in the control systems of automatic transmissions and transaxles. Most of this section will focus on the hydrostatic system and the individual components that go into a typical system. A simple hydrostatic system is shown in **Figure 6-2.**

Hydraulic Pressure

A common application of hydraulic principles involves the transmission of **hydraulic pressure** by applying force on an enclosed liquid. The force pressurizes the liquid so it can perform *work.* The pressure is transmitted undiminished (ideally) to every point in the fluid, throughout the system. The reaction is quite different compared to applying force on a solid object, wherein the force is only transmitted in the same direction as original force.

Hydraulic pressure is applied in many areas of daily life. The systems that deliver water to your home and oil to the critical areas of your engine use hydraulic pressure to perform their jobs. Other vehicle systems, such as the windshield washer and power steering systems, work because of hydraulic pressure.

Hydraulic pressure is also used to operate automatic transmissions and transaxles. The pressure produces a pushing force on some output device—a transmission *servo piston,* for example.

Pascal's Law

To understand hydraulic pressure, you must know that liquids, for all practical purposes, cannot be compressed. They are not like gases, such as air, which can be squeezed down to occupy a smaller space. (*Compressed air* is familiar to anyone who has filled a tire.) Since liquids, such as water or oil, cannot be compressed, any pressure placed

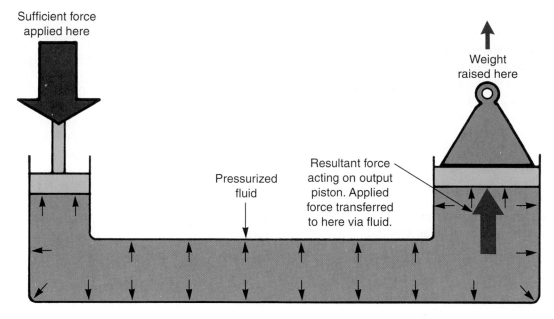

Figure 6-2. *A simple hydrostatic system. Force applied to piston on left creates pressure in the closed system. Pressure created in the system can cause the weight on the right to be raised. Modifications of this simple system are used in all automotive hydraulic systems.*

on them is immediately transferred to all parts of the liquid. As a result, liquids are very useful for transmitting motion.

In **Figure 6-3,** a cylinder contains a piston on each end. The space between the pistons is filled with fluid. Pushing on either piston will cause movement of the fluid, moving the other piston. The piston that starts the movement is called the ***apply piston.*** The other piston—the piston that is moved as a result of the apply piston—is called the ***output piston.*** This demonstrates that motion may be transmitted by a liquid.

The same principle can be applied to transmit motion from one cylinder to another. In **Figure 6-4,** two cylinders are connected by a hydraulic line. When the apply piston is moved, the liquid transfers the motion to the output piston. If the cylinders have the same diameter, the distances the pistons travel will be equal.

The underlying principle in the examples just discussed is covered by Pascal's law. ***Pascal's law*** basically states that a liquid that has been pressurized transmits the same amount of pressure to every part of its container. In **Figure 6-5,** the pressure is equal at all points and acts equally in all directions, on all areas. This principle was formulated by 17th century physicist *Blaise Pascal.*

Hydraulic Pressure Calculations

Hydraulic pressure is measured in pounds per square inch (psi) or pounds per square foot (psf) in the English, or customary system. In the metric system, it is measured in kilopascals (kPa). Mathematically, pressure is force per unit area. In other words:

$$P = \frac{F}{A}$$

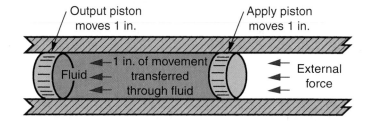

Figure 6-3. *Note how motion is transferred through a liquid. Since liquid is incompressible, an external force to the apply piston will cause a simultaneous movement of the output piston. The output piston will travel the same distance as the apply piston.*

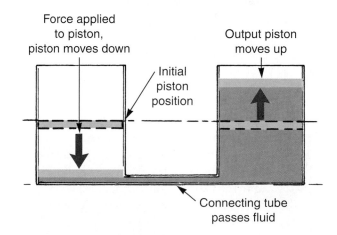

Figure 6-4. *Hydraulic pressure can be transmitted through hydraulic lines. Force applied to the piston on the left created pressure in the system, causing the output piston to move up.*

where, in customary units, *P* is hydraulic pressure in psi (or psf), *F* is force in pounds acting on a given surface, and *A* is the area of the surface in square inches (or square feet).

Pressure in a simple hydraulic system, then, can be easily determined if you know the value of the applied force and the area of the apply piston. Simply divide the force applied by the area of the apply piston.

In **Figure 6-6,** a 20-lb (89-N) force is applied to an apply piston with an area of 2 in² (12.9 cm²). Dividing the force by the apply piston area to get pressure gives:

$$P = \frac{F}{A} = \frac{20 \text{ lb}}{2 \text{ in}^2} = \frac{10 \text{ lb}}{\text{in}^2}$$

The hydraulic pressure in the system is equal to 10 psi (69 kPa).

To find the force exerted on or by a piston, the formula for pressure is rearranged as follows:

$$F = P \times A$$

Putting in values from **Figure 6-6,** we can easily determine the value of force at the output piston:

$$F = P \times A = \frac{10 \text{ lb}}{\text{in}^2} \times 20 \text{ in}^2 = 200 \text{ lb}$$

The force on the output piston is 200 lb (890 N).

Hydraulic Leverage

Compare input versus output forces in both **Figures 6-5** and **6-6.** You will note, in **Figure 6-6,** that a smaller force develops the same hydraulic pressure (10 psi, or 69 kPa) and this pressure applied to a larger output piston resulted in a greater output force. Note that in obtaining this advantage, output distance (and speed) is sacrificed. For example, in **Figure 6-6,** the smaller piston must move 10" (254 mm) to move the larger piston 1" (25.4 mm).

What you find by comparing these two examples illustrates how Pascal's law can be used to develop a hydraulic system that will increase force. This *hydraulic leverage* resembles the mechanical leverage gained from using a simple lever to lift a heavy object, **Figure 6-7.** However, where mechanical advantage is obtained by varying placement of a pivot point in a lever system, it is obtained by varying the size of the input and output pistons in a hydraulic system.

The hydraulic jack in **Figure 6-8** is a simple example of hydraulic leverage. Due to hydraulic leverage, a person can raise an automobile by hand.

Another application of hydraulic leverage that you may be familiar with is the vehicle brake system. A simple brake system is shown in **Figure 6-9.** Pressure is developed in the system when the driver steps on the brake pedal,

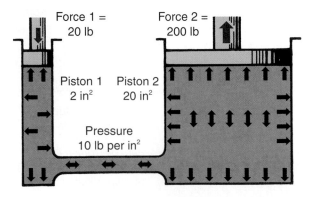

Figure 6-5. *An external force applied to a fluid in a closed system creates a pressure that is equal at every spot in the system. This principle is known as Pascal's law. Note that the pressure acts with equal force on equal areas.*

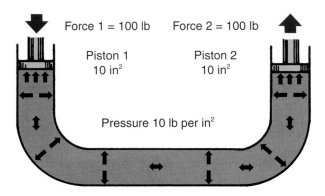

Figure 6-6. *Pascal's law, applied to systems with unequal pistons, is an easy way to multiply force. In this example, a 20-lb force is applied to the 2-in² piston on left, creating a 10-psi pressure in the system. This pressure acts on the 20-in² piston on right to produce an output force of 200 psi.*

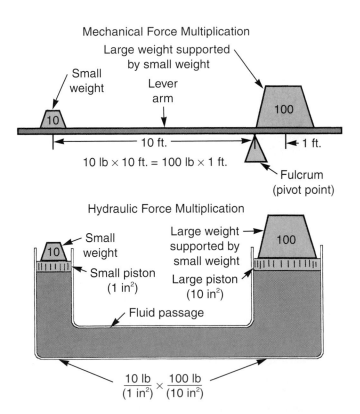

Figure 6-7. *The force multiplication in a hydraulic system can be compared to a lever. In the lever system, mechanical advantage is obtained by varying the placement of the pivot point. In a hydraulic system, it is obtained by varying the size of the input and output pistons. In both systems, the increase in force is accompanied by a loss of distance. For any given input, respective outputs travel a shorter distance.*

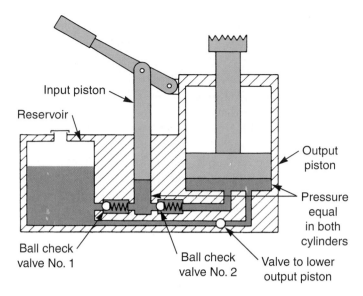

Figure 6-8. *A simplified drawing of a hydraulic jack. The piston operated by the jack handle is the input piston. Pressing down on the handle will raise the output piston. Mechanical advantage is obtained because the input piston is smaller than the output piston.*

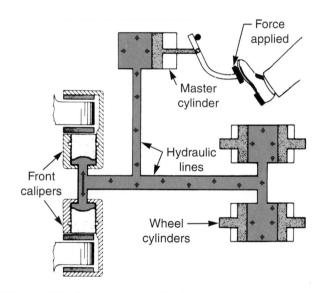

Figure 6-9. *A common application of hydraulic principles is found in the vehicle brake system. Pressure created in the master cylinder is transferred to the calipers and wheel cylinders to apply the brakes. The master cylinder piston is smaller than the brake pistons; therefore braking force is increased.*

which is connected to the master cylinder. The pressure is transferred from the master cylinder to the front calipers and rear wheel cylinders by hydraulic lines. Force is multiplied due to the difference in size between the master cylinder piston and the brake-actuating pistons.

Flow Rate

Fluid movement, or *flow,* is what produces component movement in a hydraulic system. *Hydraulic pumps* (discussed next) actually produce flow, *not* pressure.

Pressure develops as a result of the restriction to flow downstream. If there were no restrictions, then pressure at the pump discharge would be very low.

Volume flow rate is typically measured in *gallons per minute (gpm)* or *liters per minute (l/m).* Sometimes, it is measured in *cubic feet per minute (cfm).* Flow rate determines how far and how fast an output device, such as a servo piston, can move.

Increasing (or decreasing) volume flow rate impacts hydraulic pressure. The higher the flow rate is in a given hydraulic system, the greater the system pressure is, but also, the greater the pressure drop is through the system. (The increased pressure drop is due to increased frictional losses from increased velocity through the system.) In an automatic transmission, high pressures are needed, and the flow rate is large. A *high-capacity* (gpm) pump is needed to generate required system pressure and compensate for pressure loss.

Hydraulic Pumps

Hydraulic systems rely on **hydraulic pumps** to move fluid and pressurize the system. There are many different kinds of hydraulic pumps. They fall into one of two major classes: *positive displacement pumps* and *nonpositive displacement pumps.*

Nonpositive displacement pumps are used for lower-pressure, high-volume applications—primarily, for transporting fluids from one location to another. Pump output varies with pump speed and system internal resistance. An example of a hydrodynamic pump is the engine coolant pump.

Positive displacement pumps always put out the same volume of fluid during any rotation, no matter what the pump speed or internal resistance. Output from this type of pump is the same during every rotation. This type of pump can generate very high pressure. Therefore, it is very useful in fluid power systems. Positive displacement pumps will be the focus of the remainder of this section.

Three kinds of hydraulic pumps are commonly used in automatic transmissions and transaxles. They are called *gear, rotor,* and *sliding-vane pumps.* These pumps have four common features, which are:

❑ An *inlet port,* where liquid is drawn in from a non-pressurized *fluid reservoir.*
❑ An *outlet port,* where liquid is discharged into the pressurized system.
❑ Internal elements—gears, *rotors,* or *vanes,* to draw in and generate hydraulic pressure in the system. The area containing these parts is often called the *pumping chamber.*
❑ A power source to operate the pump.

In addition to these common features, these pumps all generate system pressure in similar ways. The internal elements rotate in the pumping chamber within the housing. On the inlet side of the chamber, the gear teeth, rotor lobes, or sliding vanes move apart as they rotate. On the outlet side, they move together. The side where the elements move apart is connected by an inlet passageway to the system oil reservoir. As the gears or vanes move apart, they create a vacuum. The vacuum draws in hydraulic fluid from the system reservoir. (Actually, fluid is *pushed* into the low-pressure region by atmospheric pressure.) The fluid is carried in spaces between the teeth, lobes, or vanes. As the gears or vanes come back together, the fluid is forced out of the chamber, through an outlet passageway, and into the hydraulic system. The flow encounters resistance in the hydraulic system, which causes the buildup of pressure in the system.

Gear Pumps

As mentioned, a **gear pump** is commonly used in automatic transmissions. The simplest form consists of two meshed gears inside a housing, **Figure 6-10.** The inlet and outlet ports are on opposite sides of the gear meshing point. The *driving gear* is normally driven by the engine, through the torque converter. Since the two gears are meshed, the driving gear causes the other gear to turn. As the gears move apart on the inlet side, they create a vacuum that draws in fluid. The fluid that enters is carried, between the teeth, around the pumping chamber. It is then discharged as the teeth move together on the outlet port side.

One version of the gear pump commonly used in automatic transmissions is the **internal gear pump,** or **crescent pump, Figure 6-11.** This compact design saves space and reduces system complexity. The internal gear pump consists of a spur gear that meshes with an internal gear inside the pump housing. The spur gear, which is driven by the engine through lugs on the rear of the torque converter, drives the internal gear. A crescent-shaped seal extends into the pumping chamber. It seals off the inlet and outlet sides. As the gears rotate, they move apart at the inlet,

creating suction that draws in fluid. The fluid is carried by the teeth to the outlet port and is discharged.

Rotor Pumps

A **rotor pump, Figure 6-12,** is used in some automatic transmissions. It consists of closely fitting *inner* and *outer rotors* and resembles the internal gear pump. The inner rotor is usually driven by the torque converter drive lugs. It causes the outer rotor to turn. As the rotors move apart, fluid is drawn in at the inlet port. The fluid is carried to the outlet port and is discharged when the rotors move together. A close fit between the lobes of the rotors in about the middle of the pumping chamber seals off the inlet and outlet sides.

Sliding Vane Pumps

A **sliding-vane pump,** or more simply, a **vane pump,** is used in many late-model automatic transmissions. It uses less engine power than the previously described pumps, improving gas mileage. **Figure 6-13** shows one type of vane pump called a *balanced vane pump.* It differs from other vane pumps in that it has two inlet and two outlet ports. This feature eliminates side loads due to pressure imbalance.

In general, the sliding-vane pump mainly consists of an inner *rotor body,* driven by engine power, and a series of *vanes.* The vanes slide in and out of slots in the rotor body. The pump also has an outer *rotor ring,* with an oval-shaped inner surface.

As the rotor turns, the vanes are thrown outward against the surface of the rotor ring. As the rotation continues, the

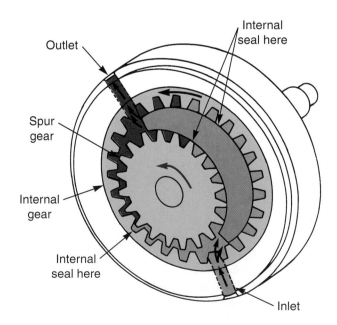

Figure 6-11. *The internal gear pump is used on many automatic transmissions. The gears move apart on the inlet side, drawing in fluid. The fluid is carried in spaces between the teeth. When the teeth come together on the outlet side, the fluid is discharged. The crescent section forms a seal between the inlet and outlet sides. (Deere & Co.)*

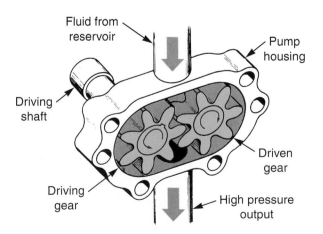

Figure 6-10. *The gear pump is a simple hydraulic pump. The gears move apart on the inlet side, creating a vacuum. The vacuum draws in fluid from the system reservoir. The fluid is carried around the pumping chamber between the gear teeth. When the gears come together on the outlet side, the fluid is discharged.*

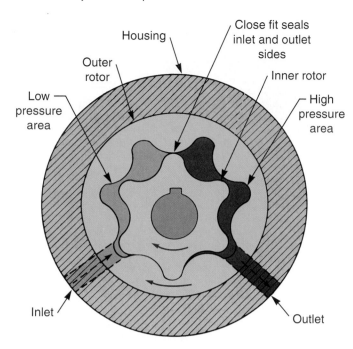

Figure 6-12. *The rotor pump is used in some automatic transmissions. The arrangement of the internal and external rotor resembles the crescent pump, only without a crescent seal. Instead, the close fit of the inner and outer rotors forms a seal at the center of the pumping chamber.*

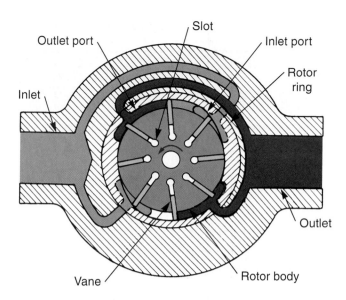

Figure 6-13. *The sliding-vane pump is often used in modern transmissions. Fluid is drawn in on the inlet side of the pump. The fluid is carried between the vanes and is discharged at the outlet side of the pump. The vane slots are pressurized to keep the vanes in close contact with the rotor ring. (Deere & Co.)*

vanes are forced back into the rotor body by the ring. As the vanes follow the contour, compartments are formed, which continually expand and contract in volume. During this process, fluid from the reservoir is drawn, by suction, through the inlet port into the expanding compartments. It is then carried around and discharged at the outlet port, as the trapped volume of fluid is squeezed out.

One unique feature of the vane pump is that once hydraulic pressure is developed, the pressurized fluid is fed to the vane slots, or the space behind the vanes. The pressurized fluid helps force the vanes against the housing wall and form a tight seal.

Hydraulic Valves

A pump that would generate enough pressure for a vehicle in idle would create too much pressure at high speeds. To control system pressure, *hydraulic valves* can be used. In addition, these valves can be used to determine where flow will be used in the rest of the hydraulic system. *Hydraulic valves* are devices used in a hydraulic circuit to control pressure and, also, direction and rate of fluid flow. In general, these devices help control the entire operation of a hydraulic system. Various types of valves can be used in an automatic transmission or transaxle.

Check Valves

A *check valve* permits flow in only one direction. The most basic of check valves is a steel ball, or *check ball,* held within the fluid passageway of a *valve body*

(discussed shortly). Flow in one direction through the system and passageway causes the check ball to unseat, permitting flow. Flow in the other direction pushes the ball against the valve seat, blocking flow. A check ball is shown in **Figure 6-14.** Placement of the check balls in a typical modern transmission is shown in **Figure 6-15.**

Check valves are considered a type of *directional control valve,* discussed shortly. Directional check valves are used to prevent or restrict fluid flow to parts of the hydraulic system. This is done to more precisely control the action of the hydraulic system.

Pressure Relief Valves

The *pressure relief valve,* **Figure 6-16,** is essentially a check valve that opens to exhaust excess pressure. It keeps pressure from exceeding some preset maximum value. Its purpose is to prevent excess pressure from damaging the hydraulic system. The pressure relief valve is designed to work as a *two-position device*—that is, to be either *open* or *closed.*

The internal valve element is held against its seat by spring pressure. When hydraulic pressure becomes greater than the given spring pressure, the element unseats, as shown by the ball in **Figure 6-16.** Fluid then passes through the valve body and flows back to the reservoir, reducing pressure. Pressure relief valves are primarily used to relieve excess system pressure when the *pressure regulator valve,* discussed in upcoming paragraphs, is overloaded or malfunctioning.

Control Valves

Hydraulic system **control valves** are used to regulate the operation of the system and of other hydraulic

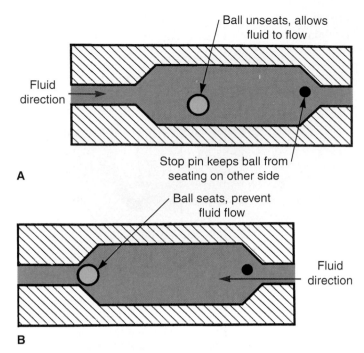

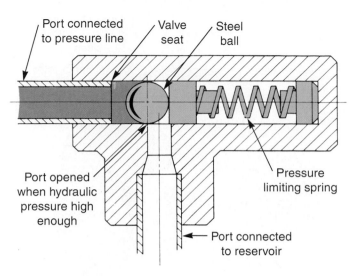

Figure 6-14. *The check ball allows fluid to flow in one direction only. A—When fluid is flowing in the direction shown, the ball unseats. This allows the fluid to flow through the valve passageway. B—If fluid attempts to flow in reverse, the ball seats and fluid flow is blocked. Directional check balls are widely used in automatic transmissions.*

Figure 6-16. *The ball in this pressure relief valve is held against the seat by a spring. When fluid pressure becomes greater than spring pressure, the ball is pushed back, allowing fluid to return to the reservoir. Fluid pressure is then reduced to less than spring pressure, and the ball reseats. Some springs are adjustable, making relief pressure adjustable.*

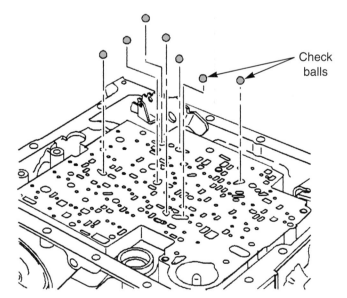

Figure 6-15. *Placement of check balls in a modern rear-wheel drive transmission. (General Motors)*

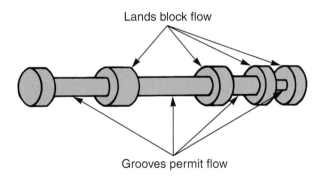

Figure 6-17. *Spool valves are so named because they resemble ordinary spools used to hold wire or thread. The valve's larger-diameter segments are machined to closely fit the bore of the valve body. These segments of the valve are called lands. The grooves between the lands allow fluid flow.*

components. They are used in automatic transmissions and transaxles. One type of control valve in particular is commonly called a **spool valve.** This type of valve is so named because of the spool-like flow-controlling element, hereinafter called the *valve.* See **Figure 6-17.**

Spool valves are installed in an automatic transmission **valve body,** which consists of an aluminum or iron

casting with internal passageways. Holes are drilled into the valve body to receive the spools. These holes, or **valve bores,** are a very close fit with the valves themselves.

The movement of the valves in the valve body directs pressurized fluid to other parts of the hydraulic system. The valve body is often referred to as the "brain" or "control center" of the hydraulic system. A valve body may contain only a single valve. Most automatic transmission valve bodies contain between 5 and 30 valves. **Figure 6-18** is a sectional view of a simple valve body containing a single valve. Note the relationship of the valves to the bores and passageways.

Pressure Regulator Valves

Some control valves are used to control system pressure; these are called **pressure regulator valves.** Pressure regulator valve operation is similar to that of a pressure

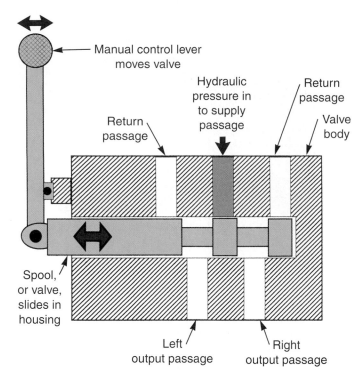

Figure 6-18. *Study sectional view of a simple valve body with one valve. Note the relationship of the valve and valve body passageways. Most transmission valve bodies contain many valves. Some valves are manually activated; others are automatic.*

relief valve. The difference is the pressure *regulator* tries to maintain a constant pressure, and it does this by *modulating,* or moving back and forth. The relief valve, on the other hand, works as a two-position device to keep system pressure from exceeding a specified maximum value.

In operation, the pressure regulator valve, which is a sliding piston, works against a spring. When pressure is low, the valve is positioned so that a port, which is connected to a line returning to the oil reservoir, is closed off. Hydraulic pressure above a certain value causes the valve to begin to move and uncover the port. Some of the fluid is exhausted back to the reservoir, reducing pressure. The valve piston modulates back and forth in the cylinder bore, bleeding off more or less oil in the system to maintain constant output pressure. The valve movement, then, allows a constant pressure to be maintained from a variable pressure source. The pressure setting of the pressure regulator valve is controlled by the valve spring, some of which are adjustable.

Throttle Valves

Whereas pressure regulator valves control pressure in response to internal pressure conditions, other valves control system pressure based on some external input. For example, **throttle valves** can modify system pressure according to engine load and vehicle speed. This information may be sent to the throttle valve mechanically, by a linkage from the carburetor or throttle body, or pneumatically, by engine manifold vacuum. This input causes modulation of the valve, which compensates for the varying conditions.

Directional Control Valves

Directional control valves are used to direct the path of flow in a hydraulic circuit. See **Figure 6-19.** These valves are used in automatic transmissions to control the application of different hydraulic output devices. They control transmission shifts, shift quality, and other transmission actions.

The valves themselves may be actuated either by external sources or internally. External sources include the shift lever moved by the driver, engine vacuum operating a vacuum modulator, and electric solenoids. Internal actuation is accomplished by different hydraulic pressures from within the hydraulic system acting on the valve.

Actuators

Hydraulic system output devices are driven by **actuators.** These may be pistons actuated by hydraulic pressure, **Figure 6-20,** or they may be electric solenoids (discussed later). Most hydraulic actuators are held in the *released,* or *normal* (unpressurized), position by one or more springs, called *release,* or *return, springs.* Hydraulic pressure overcomes spring pressure to move the piston. Hydraulic actuators used in automatic transmissions include servo pistons, used to apply friction *bands,* and clutch apply pistons, used to apply friction *clutches.* (These components are discussed in detail in later chapters.) Other hydraulic actuators, such as hydraulic motors, are seldom used on automatic transmissions and transaxles.

Pneumatics

Pneumatics is the study of the *mechanical properties* (reactions to applied forces) and physical properties of air and other gases. The study is concerned with gases both at rest and in motion. **Pneumatic systems** typically use air for a power source.

Since air is compressible, some power must be used to pressurize the air before a compressed-air system can do any work. With a hydraulic system, on the other hand, no power is wasted in pressurizing the system; any pressurizing force applied translates directly into useful work. Therefore, a pneumatic system cannot transfer power as efficiently as a hydraulic system. In addition, pressure and flow in pneumatic systems are affected by air temperature and humidity. These are all disadvantages of a pneumatic system.

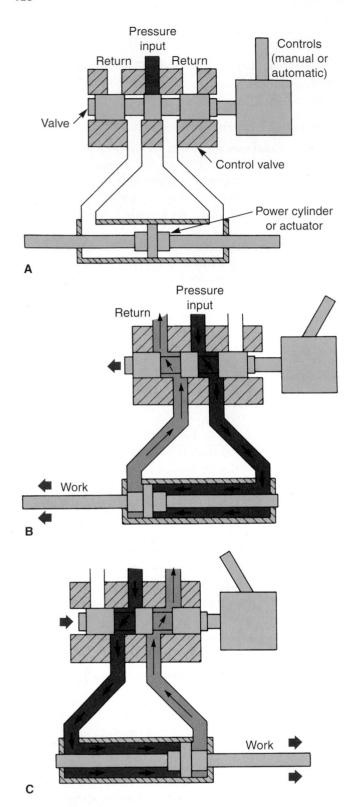

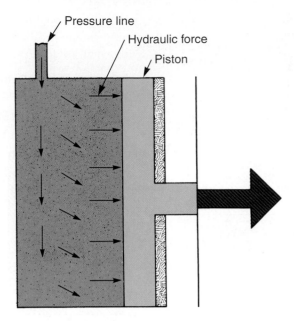

Figure 6-20. *Actuators change hydraulic pressure into motion and force. Two types of hydraulic actuators are used in automatic transmissions. The servo piston is used to apply transmission bands. The clutch apply piston is used to apply clutch packs. (DaimlerChrysler)*

Figure 6-19. *Actions of a directional control valve can be summarized by this series of illustrations. A—Valve is in position to prevent flow to the power cylinder. B—In this position, the valve has moved to allow fluid to flow to the right side of the power cylinder. The piston moves to the left. C—In this position, the valve has moved to allow fluid to flow to the left side of the power cylinder. The piston moves to the right.*

One advantage of a pneumatic system is that air, used to transmit power, is always available at no cost. Also, a pneumatic system is an *open system.* This means that after the air has been used, it is exhausted to the atmosphere. In this respect, it is unlike a hydraulic system, which is a *closed system,* requiring special reservoirs and no-leak designs. In a pneumatic system, air exhausting out of the system does not create a mess, as oil would in a hydraulic system. Further, slight leaks do not affect system performance. The pneumatic system does not need to be as tightly sealed, and components and piping can be lighter and less complex.

Although we seldom think of it, the surrounding air is pressurized by its own weight to about 14.7 psi (101 kPa) at sea level. This is called *atmospheric pressure.* The operating principle of any pneumatic system is the difference in pressure between air inside of the system and outside air. This difference in pressure, often called a *pressure differential,* is used to create the force that moves an actuator or spins an air motor.

Types of Pneumatic Systems

There are two types of pneumatic systems: *compressed air systems* and *vacuum systems.* If the pressure inside of the system is greater than the pressure of the outside air, the system is a *compressed-air system.* If the pressure inside of the system is a *vacuum*—that is, if it is less than that of the outside air, or atmospheric pressure—it is a *vacuum system.*

Compressed-Air Systems

A common example of a compressed-air system is the air compressor and piping found in most repair shops. The compressor is driven by an electric motor. It draws in and compresses air to a higher pressure than atmospheric pressure. This high-pressure air is then sent through piping to operate impact wrenches and hammers, clean and dry parts, and inflate tires.

Vacuum Systems

An example of a device operated by a vacuum system is the **vacuum modulator valve** found on some automatic transmissions. The modulator is a chamber, or container, with an internal *diaphragm*, **Figure 6-21.** The diaphragm is a flexible partition that divides the chamber into two regions, sealing one off from the other. One side of the diaphragm is at atmospheric pressure. The other side is connected to engine vacuum by tubing that leads to the intake manifold.

When the engine is running, downward-moving pistons on their intake strokes produce suction in the intake manifold. This is where the vacuum originates. The vacuum will vary according to the load placed on the engine by changes in throttle opening or weight changes to the vehicle.

Together, atmospheric pressure and engine vacuum act to force the diaphragm in the vacuum modulator toward the vacuum source. However, the modulator is *spring loaded,* and the force of the spring opposes the force caused by the pressure differential acting on the diaphragm. As engine vacuum goes up and down, the diaphragm, acting against the spring, moves back and forth, causing the transmission valve connected to it to move. This, in turn, causes throttle oil pressure to vary as

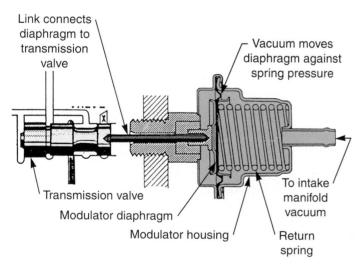

Figure 6-21. *The vacuum modulator used on some automatic transmissions is an example of a pneumatic device. Engine vacuum moves the modulator diaphragm against spring pressure. The modulator diaphragm is connected to a valve that modifies transmission fluid pressure. Changes in engine vacuum caused by load changes on the engine result in pressure changes in the transmission.*

changes in engine load occur, in order to match transmission shift points to engine loading.

Summary

Hydraulics is the science of how power is transmitted through liquids. Hydraulic systems are a simple way to transmit and multiply power. Hydraulics is composed of hydrodynamics—the study of moving liquids—and hydrostatics—the study of liquids under pressure.

Liquids, such as hydraulic fluids, cannot be compressed. This principle can be put to use so that hydraulic fluid will transmit pressure and motion.

Pascal's law states that pressure on a fluid in a closed system is equal at every place in the system. Pascal's law can be put to use to multiply force. If the area of the input piston of a hydraulic system is smaller than the area of the output piston, force will be increased at the output piston. This is the principle used in the hydraulic jack and on vehicle brake systems.

Automatic transmissions must develop high pressure and allow for high fluid flow rates. To do this, they require engine-driven pumps. The three major types of pumps are gear pumps, rotor pumps, and sliding-vane pumps. They all operate by creating a vacuum that draws in fluid and compressing the fluid so it can be discharged.

Valves are used to control the flow of fluid in the hydraulic system. Check valves are steel balls that control fluid flow. They are used with a spring to control system pressures or in a special chamber to ensure fluid flows in one direction only.

Spool valves resemble thread spools. They are installed in a valve body to control fluid pressure and flow. Spool valves are operated by external controls or by fluid pressures from other valves.

Actuators change fluid pressure into force. The force and movement they exert are the end result of hydraulic system operation. Examples of actuators are transmission servos and clutch apply pistons.

Pneumatics is the study of the physical properties of air and other gases. Pneumatic systems can be compressed air or vacuum systems. The automatic transmission vacuum modulator is an example of a pneumatic system.

Review Questions—Chapter 6

Please do not write in this text. Write your answers on a separate sheet of paper.

1. Explain the basic difference between hydrodynamics and hydrostatics.

2. _____ _____ states that liquid under pressure transmits the same amount of pressure to every part of its container.

3. Describe the basic system that produces hydraulic leverage.

4. List the four common features of hydraulic pumps and state their function.

5. Describe the operation of gear, rotor, and vane pumps.

6. _____ are used to help control the operation of a hydraulic system by controlling pressure and, also, direction and rate of fluid flow.

7. How does a pressure relief valve work?

8. _____ is the study of the physical and mechanical properties of air and other gases.

9. Explain why pneumatic systems are classified as open systems.

10. What part of the engine produces the vacuum used to operate vacuum devices?

ASE-Type Questions—Chapter 6

1. Technician A says that a torque converter is a hydrodynamic device. Technician B says that hydrostatic principles are used in the control systems of automatic transmissions. Who is right?
   (A) A only.
   (B) B only.
   (C) Both A and B.
   (D) Neither A nor B.

2. Each of the following is a type of hydraulic device or system except:
   (A) residential water supply.
   (B) power steering.
   (C) windshield washers.
   (D) vacuum modulators.

3. Pascal's law is given by which of these statements?
   (A) Pressurized liquid will burst from its container.
   (B) Pressurized liquid will distribute no further pressure.
   (C) Pressurized liquid transmits equal pressure to all points.
   (D) Pressurized liquid transmits unequal pressure to all points.

4. Hydraulic leverage is gained from varying the _____ of the input and output pistons in a hydraulic system.
   (A) length
   (B) weight
   (C) area
   (D) force

5. Technician A says hydraulic pumps produce hydraulic pressure, not hydraulic flow. Technician B says that flow produces component movement in a hydraulic system. Who is right?
   (A) A only.
   (B) B only.
   (C) Both A and B.
   (D) Neither A nor B.

6. The output of a _____ displacement pump decreases considerably as system pressure increases.
   (A) single
   (B) nonpositive
   (C) multiple
   (D) positive

7. Each of the following is commonly used in an automatic transmission except:
   (A) internal gear pumps
   (B) rotor pumps
   (C) hydrodynamic pumps
   (D) sliding-vane pumps

8. Technician A says hydraulic valves are used in a hydraulic circuit to control pressure. Technician B says hydraulic valves are used in hydraulic circuits to control the direction and rate of fluid flow. Who is right?
   (A) A only.
   (B) B only.
   (C) Both A and B.
   (D) Neither A nor B.

9. All of these involve the study of pneumatics except:
   (A) oil pressure.
   (B) gases at rest and in motion.
   (C) physical properties of gases.
   (D) mechanical properties of gases.

10. The two types of pneumatic systems are:
    (A) open and closed systems.
    (B) vacuum and compressed-air systems.
    (C) hydrostatic and hydrodynamic systems.
    (D) pressure regulation and pressure relief systems.

Chapter 7

Basic Electricity and Electronics

After studying this chapter, you will be able to:
- ❑ Explain the electron theory of electricity.
- ❑ Describe basic electrical circuits.
- ❑ Identify basic electrical measurements.
- ❑ Explain the purpose of vehicle wiring and connectors.
- ❑ Explain the purpose of common vehicle electrical devices.
- ❑ Explain the purpose of semiconductor devices.
- ❑ Describe the construction and operation of the automotive computer.
- ❑ Explain the operation of control loops.

Technical Terms

Conductors
Insulators
Circuit
Continuity
Current
Amperes
Ammeter
Voltage
Volts
Voltmeter
Resistance
Ohms
Ohmmeter
Ohm's law
Direct current
Alternating current

Series circuit
Parallel circuit
Series-parallel circuit
Short circuit
Open circuit
Magnetic
Electromagnet
Induction
Harnesses
Color coding
Ground
Wire gauge
Plug-in connector
Schematic
Circuit protection devices
Switch

Solenoid
Relay
Motors
Resistors
Capacitor
Semiconductors
Diodes
Transistor
Integrated circuit (IC)
Microprocessors
Control loop
Input sensors
Output devices
Central processing unit (CPU)
Memory
Self-diagnostic system

Introduction

Without a thorough understanding of basic electricity and electronics, the technician cannot diagnose problems on electronically controlled transmissions and transaxles. This chapter contains a brief overview of electricity, electronics, and computer operation. In addition, it explains the purpose of various electrical and electronic components. Emphasis is placed on the components used on electronically controlled automatic transmissions and transaxles.

Electrical Basics

The following section discusses fundamental electrical principles that govern the operation of all automotive electrical equipment. The information presented here applies to any modern vehicle and will help when troubleshooting electronic transmissions and transaxles.

Atoms and Electricity

Everything is made of **atoms.** Every atom has a center, or **nucleus,** that consists of **protons** and **neutrons.** The neutrons have no charge, and the protons have a positive charge, making the nucleus of the atom positively charged. Negatively charged **electrons** revolve around the nucleus. See **Figure 7-1.** In the ideal atom, the number of electrons is equal to the number of protons in the nucleus. In reality, the number of electrons varies, and there is some movement of electrons between atoms. Electricity is the movement of large numbers of electrons from atom to atom.

Conductors and Insulators

Some atoms easily give up or receive electrons. Elements made of these atoms are good electrical **conductors.** Examples are copper and aluminum. Materials made of atoms that resist giving up or accepting atoms are **insulators.** Glass and plastic are good insulators. Some materials can alternate between conducting and insulating. These materials will be discussed later in this chapter.

Electron Flow

As previously mentioned, electricity is the movement, or flow, of electrons. The path through which the electrons move is called a **circuit.** Electrons will not flow in a circuit unless two conditions are met:
- ❑ There are more electrons in one place in the circuit than in another. In a vehicle, the battery and alternator create this difference in the number of electrons.
- ❑ There is a connection between the two places. This connection is composed of the vehicle wiring and the electrical unit being operated. It must be made of materials that are good conductors.

A simple automotive electrical circuit is shown in **Figure 7-2.** In this circuit, electricity flows from the battery through the positive battery cable, fuse, brake light switch, brake lightbulbs, frame, and negative cable, and then back to the battery. This is called a **complete circuit,** since the electricity makes a loop through the battery, cables, fuse, switch, bulbs, frame, and back to the battery. The path for the electrons to return to the battery is as important as the path from the battery to the electrical device. When a circuit or component forms a complete path for electricity, the circuit is said to have **continuity.**

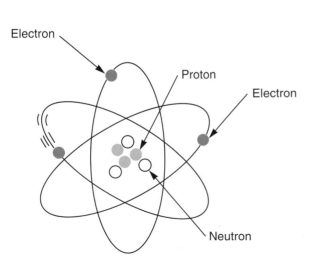

Figure 7-1. *In an atom, the electrons travel around the nucleus, which is made up of protons and neutrons.*

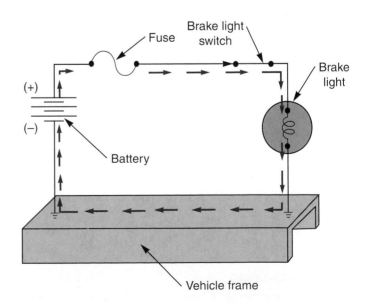

Figure 7-2. *A simple automotive electrical circuit.*

Electrical Measurements

The flow of electricity through a circuit depends on three electrical properties, all of which can be measured. These electrical properties, as well as the meters and techniques used to measure them, are explained below.

Current

Current is a measure of the number of electrons flowing past any point in a circuit. Current is measured in **amperes,** or **amps.** The higher the amps, the more electrons are moving in the circuit.

Current is measured with an **ammeter.** To measure current in a circuit, the meter must be connected as shown in **Figure 7-3.** When the meter is connected in this way, it is said to be in series with the circuit. This allows the current flowing through the circuit to travel through the meter.

Voltage

Voltage is the electrical pressure created by the difference in the number of electrons between two terminals. It provides the push that makes electrons flow and is measured in **volts.**

Voltage is measured with a **voltmeter.** The voltmeter is connected in parallel with the circuit. See **Figure 7-4.**

Resistance

Resistance is the opposition of the atoms in a conductor to the flow of electrons. All conductors, even copper and aluminum, have some resistance to giving up their electrons. Resistance is measured in **ohms.** An **ohmmeter** is used to measure resistance. Power to the circuit must be disconnected before hooking up the meter. The ohmmeter is then connected to each end of the component or wire that is being checked. See **Figure 7-5.**

Ohm's Law

Sometimes, none of the electrical values described in the previous section are known. However, if you know two of these values, you can calculate the third using **Ohm's law. Figure 7-6** is a graphic representation of Ohm's law, which is sometimes called the Ohm's law triangle. To use the Ohm's law triangle, cover the unknown property and perform mathematical calculations using the other two. Using Ohm's law requires no more than simple multiplication or division.

$$\text{amps} = \text{volts} / \text{ohms}$$
$$\text{ohms} = \text{volts} / \text{amps}$$
$$\text{volts} = \text{amps} \times \text{ohms}$$

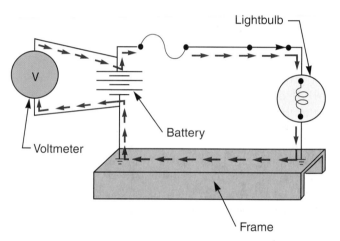

Figure 7-4. *When using a voltmeter, it must be connected in parallel with the circuit.*

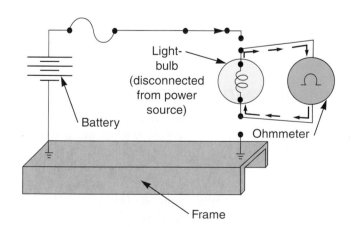

Figure 7-5. *To measure the resistance of a component, disconnect the power source before connecting the meter. If the ohmmeter is used in an energized circuit, the meter will be damaged. Connect the meter as shown.*

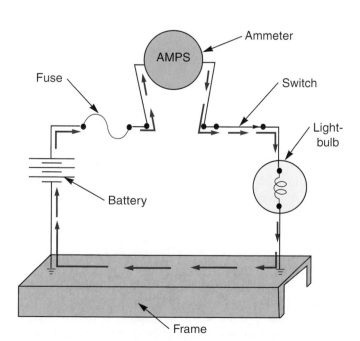

Figure 7-3. *An ammeter is used to measure current in a circuit. Note that an ammeter is always connected in series, as shown.*

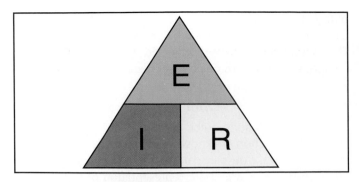

Figure 7-6. *A graphic representation of Ohm's law. To find an unknown value, cover the unknown value and perform the necessary calculations.*

When amperage is unknown, divide voltage by resistance. When resistance is unknown, divide voltage by current to obtain the unknown value. When voltage is unknown, multiply amperage and resistance to get voltage.

Direct and Alternating Current

The vehicle's battery and alternator always have positive and negative terminals. There is always a shortage of electrons at the positive terminal and an excess of electrons at the negative terminal. Current flows in only one direction, from negative to positive. This is called a **direct current,** or **dc,** system.

In the electrical systems used in homes, schools, and factories, the flow of electrons changes direction many times each second. These are **alternating current,** or **ac,** systems.

A vehicle's battery produces direct current. Therefore, most automotive electrical systems operate on direct current. However, a vehicle's alternator produces alternating current, which must be changed to direct current, or rectified, before being used by the vehicle's electrical system.

Alternating current is sometimes used as a signal device in a vehicle system. For instance, speed sensors produce an alternating current that is proportional to engine or vehicle speed.

Types of Electrical Circuits

There are three types of automotive circuits. Every wire in a car or truck is part of one of these types of circuits. The three types of circuits are series circuits, parallel circuits, and series-parallel circuits.

The circuit in **Figure 7-7** is a **series circuit**. This is the simplest type of automotive circuit. This particular series circuit consists of the battery, a switch, a lightbulb, and connecting wiring. Electrons flow through the wiring from the battery, through the switch and bulb, and back to the battery. The wire in the bulb, called a *filament*, is made of

a special type of resistance wire that glows as the electrons pass through it. The other wiring is made of metals that allow the electrons to pass with very little resistance. The same amount of current (number of electrons) flows through every part of the series circuit. Common examples of the series circuit are the various dashboard warning lights for oil pressure, coolant temperature, and charging system condition.

The circuit in **Figure 7-8** is a **parallel circuit.** In a parallel circuit, current flow is split so that each electrical component has its own current path. Different amounts of current will flow in different parts of the circuit, depending on the resistance of each part.

The **series-parallel circuit** has some components that are wired in series and others that are wired in parallel, as shown in **Figure 7-9.** All the current flows through some parts of the circuit, while the current path is split in other parts.

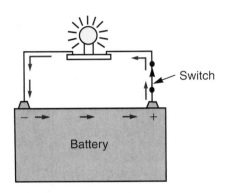

Figure 7-7. *A simple series electrical circuit.*

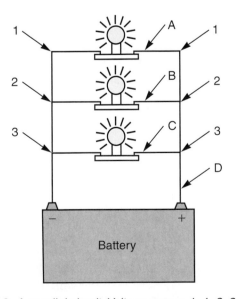

Figure 7-8. *A parallel circuit. Voltage across 1–1, 2–2, and 3–3 would be equal. Current flow in amperes at A, B, and C would depend on the resistance of the units. Current flow at D would equal the sum of current flowing through A, B, and C. Total resistance of A, B, and C would be less than any single one.*

Circuit Defects

There are two types of circuit defects: short circuits and open circuits. A **short circuit** occurs when a wire's insulation fails or is removed and the bare wire contacts the frame, body, or other grounded part of the vehicle, as in **Figure 7-10.** Not only is a needed part of the circuit bypassed, but the current flow through the rest of the circuit is too high. If a fuse or circuit breaker does not protect the circuit, excess current can damage wiring and components, or start a fire.

An **open circuit** is a circuit that is not complete. Current cannot flow in an open circuit, **Figure 7-11.** Common causes of open circuits are loose or corroded connections, disconnected wires, and defects in electrical components, such as switches, bulbs, and fuses. A related problem is a high-resistance electrical connection, which is usually caused by corrosion or overheating. The high resistance may cause the circuit to stop operating or, in a high amperage circuit, catch fire at the connection.

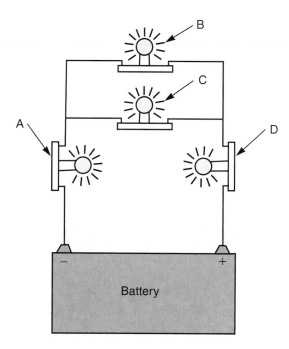

Figure 7-9. *A series-parallel electrical circuit. Lights A and D are in series. Lights B and C are in parallel.*

Electromagnetism

There is a well-defined relationship between electricity and magnetism. When a material is **magnetic,** the electrical charges of its electrons are aligned to create a force that extends outward from the material.

Magnetic fields have definite North and South poles. This property of magnetic fields is called *polarity.* Like poles repel each other and unlike poles attract. It is usually not necessary for the technician to determine the North and South poles when servicing a magnetic unit.

A magnetic material attracts iron and metals that contain iron. Some materials are naturally magnetic, while others can be magnetized by electricity. When current flows in a wire, the electrons start moving in the same direction. Electron movement creates a magnetic field around the wire as long as the current is flowing. When the wire is wound into a coil, the magnetic fields of each wire loop combine to create a very strong magnetic field. The combination of the coil winding and a metal core is called an **electromagnet, Figure 7-12.** The core helps to increase magnetic field strength and may be moveable. The magnetic field created may be used to move linkages or open electrical contacts. Electromagnets are the basic components in solenoids, relays, and starters.

When a wire moves through a magnetic field or a magnetic field moves through a wire, the electrons in the wire begin moving, creating current flow in the wire. This process is called **induction.** Current is produced in many vehicle speed sensors by induction. The sensor wires are stationary and a toothed wheel moves with the camshaft or crankshaft. The teeth on the wheel create a magnetic fluctuation, or movement, in a permanent magnet built into the sensor. This fluctuation induces ac current in the sensor wiring.

The relationship between electricity and magnetism is called **electromagnetism.** Electromagnetism is used to operate many electrical devices on the vehicle. These devices will be discussed later in this chapter.

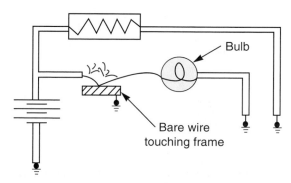

Figure 7-10. *A short circuit. Note that electrical current is flowing directly to ground (vehicle frame). (Chevrolet)*

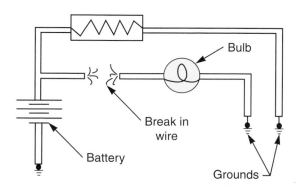

Figure 7-11. *An open circuit. This one is caused by a break in the wiring. (Chevrolet)*

Vehicle Wiring and Electrical Components

The modern automotive electrical system is a complex arrangement of wiring and electrical components. The electrical system must produce electricity and deliver it to the proper places; protect circuits from damage; reduce or increase voltage; change electricity into light, motion, or heat; and use electricity to control the movement of liquids and gases. The construction, operation, and use of the devices that accomplish these tasks are discussed in the following sections. The operation and function of electronic devices will also be discussed.

Wiring

Most automotive wiring is made of copper, aluminum, or copper-coated aluminum. Wires are plastic coated and are generally installed into wrapped assemblies called *harnesses,* **Figure 7-13.** However, harnesses used inside the transmission or transaxle are usually not wrapped. See **Figure 7-14.** For easier circuit tracing, the insulation of every vehicle wire is given a specific color. This is called *color coding.* Modern vehicles have many wires, and it is necessary to increase the number of color coding variations by adding a stripe or stripes of contrasting colors to the original insulator color. Service literature contains schematics that show all wires and their colors.

On most vehicles, there is no return wiring from various electrical units back to the battery. Instead, the vehicle's frame or body forms the return, or *ground,* as shown in **Figure 7-2.** On all modern vehicles, the negative

terminal is the ground terminal. The battery, charging system, and all other electrical devices have their negative terminal wired to a common negative ground connection, the vehicle's frame and body. On most vehicles, the negative battery cable is attached to the engine block, **Figure 7-15.** The body and chassis may be grounded directly to the battery through a smaller ground cable attached to the negative post, or grounded to the engine block by one or more ground straps, **Figure 7-16.**

Wire Size

All wires have some resistance and lose some electrical power as heat. It is important that the wire in a circuit be large enough to carry the rated amperage without overheating. At the same time, the wire should not be larger than necessary. *Wire gauge* is the rating system for wire diameter. The larger the gauge number, the smaller the wire. Wire gauge is measured in AWG (American Wire Gauge) sizes or metric sizes. Note that the gauge refers

Figure 7-13. *Instead of installing many single wires, manufacturers combine them into a wiring harness. The harness protects the individual wires and makes electrical connections easier.*

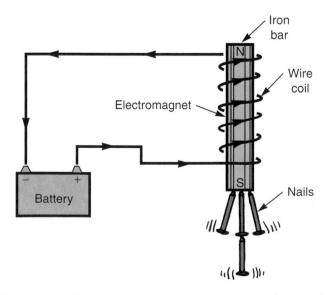

Figure 7-12. *An electromagnet. As current passes through the wire coil surrounding the iron bar, the bar becomes a magnet. The nails will adhere to the bar until current flow stops. The nails will then drop off the bar.*

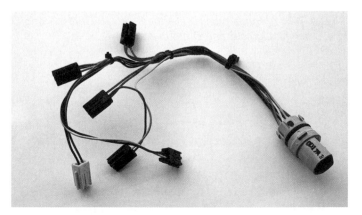

Figure 7-14. *The wiring harness shown here is used inside a common transaxle. Note that the wires are simply tied together instead of being wrapped with tape.*

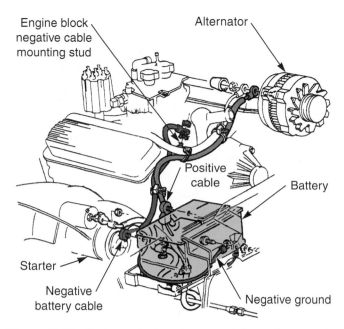

Figure 7-15. *Body and chassis ground from the negative battery terminal to the engine block. (General Motors)*

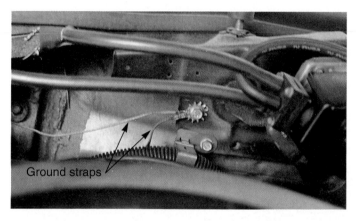

Figure 7-16. *Ground straps are commonly used to ensure a good electrical connection between the vehicle body or frame and the engine block.*

American Wire Gauge	Wire diameter in inches	Cross-sectional area in circular mils
0000	.4600	211600
000	.40964	167800
00	.3648	133100
0	.32486	105500
1	.2893	83690
2	.25763	66370
3	.22942	52640
4	.20431	41740
5	.18194	33102
6	.16202	26250
8	.12849	16510
10	.10189	10380
12	.080808	6530
14	.064084	4107
16	.05082	2583
18	.040303	1624
20	.031961	1022
22	.025347	642.4
24	.0201	404.0
26	.01594	254.1
28	.012641	159.8
30	.010025	100.5

A

Metric wire sizes (mm²)	AWG sizes American Wire Gauge
.22	24
.35	22
.5	20
.8	18
1.0	16
2.0	14
3.0	12
5.0	10
8.0	8
13.0	6
19.0	4
32.0	2

B

Figure 7-17. *A—American Wire Gauge chart showing some common sizes. B—American Wire Gauge sizes compared to metric wire sizing. (Belden, Chevrolet)*

only to the thickness of the conductor itself, not the thickness of the insulation. Typical wire gauges are shown in **Figure 7-17.** The smallest gauges (heaviest wires) are used to connect the battery to the vehicle's starter. To minimize resistance losses, wires carrying high amperages are designed to be as short as possible.

Plug-In Connectors

Most connectors used on modern vehicles are plug-in types. A *plug-in connector* has male and female ends that are plugged into each other. See **Figure 7-18.** Connectors having more than one wire are called multiple connectors. Many service manuals often refer to a specific connector by the number of wires it contains, such as a 12-wire connector or a 23-wire connector. Modern connectors are shaped so that they cannot be assembled incorrectly. Some have aligning lugs and slots on each side of the connector.

Figure 7-18. *Plug-in connectors can be found in almost every part of the vehicle.*

Modern plug-in connectors are often thoroughly sealed to keep out moisture and corrosion, **Figure 7-19.** Since many automotive electronic components operate on very low voltages, a small increase in resistance due to moisture and corrosion can affect circuit operation.

A few connectors may be screw terminals, or bolts, which pass through a terminal eye to form a connection, **Figure 7-20.** These connectors are generally used to connect a ground strap to the vehicle body or to connect circuits with very high current loads, such as those in the starting and charging systems.

Wiring Schematics

A *schematic,* or *wiring diagram,* is a drawing showing electrical devices and the wires connecting them. Schematics also show wire colors and terminal types. Use of the schematic allows the technician to trace defective components in the wiring system. Many vehicle manufacturers break down the overall vehicle wiring into separate circuit diagrams, as shown in **Figure 7-21.**

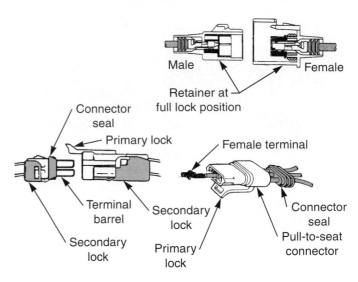

Figure 7-19. *Three types of plug-in connectors that are sealed to prevent moisture from entering and causing corrosion. (General Motors, Toyota).*

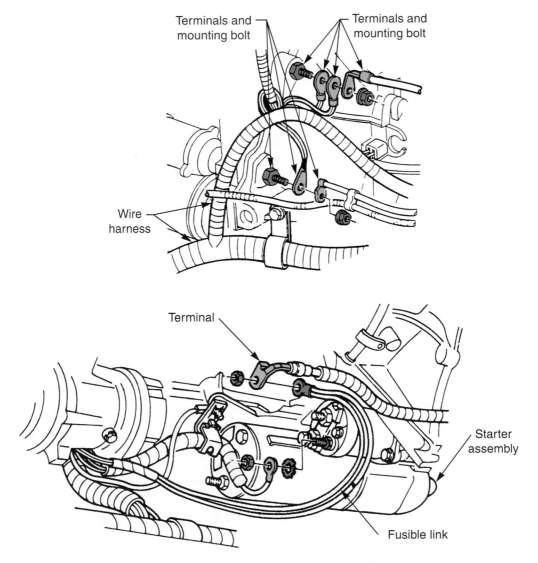

Figure 7-20. *Typical screw terminals. The bolts pass through the wire terminal eyes. (General Motors)*

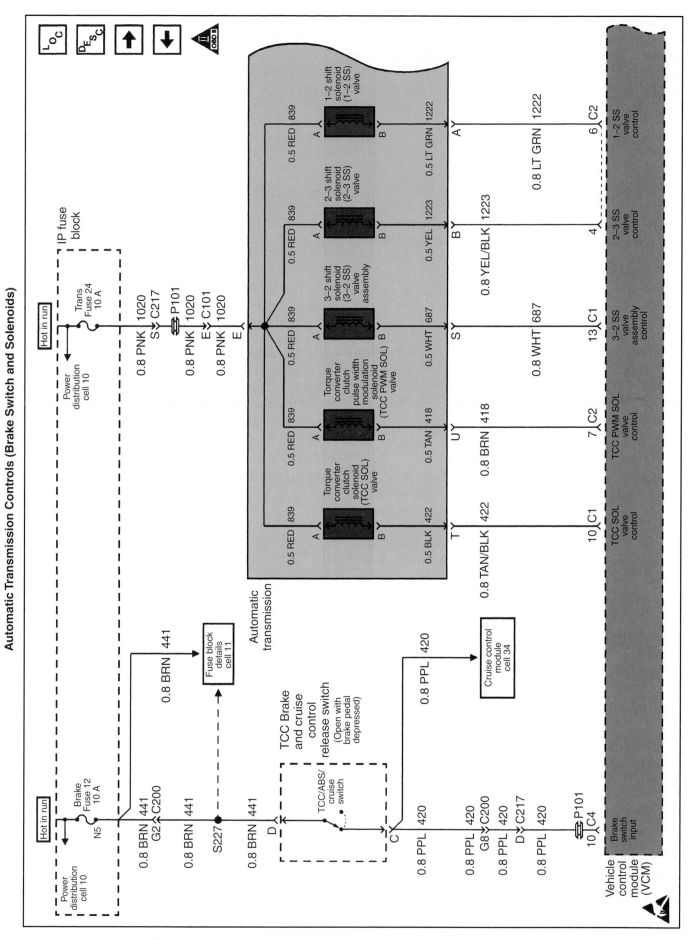

Figure 7-21. *A wiring diagram for an automatic transmission control circuit. (General Motors)*

Schematics use symbols to represent electrical devices. There is some variation in the use of these symbols. Some schematics have a combination of company-specific symbols and standardized symbols. **Figure 7-22** illustrates some symbols commonly used in automotive electrical diagrams.

Circuit Protection Devices

Short circuits or defective components can cause excessive current flow in a circuit. To protect circuits from damage due to excessive current, **circuit protection devices** are used. These devices include fuses, fusible links, and circuit breakers. All electrical circuits except the starter cables and the alternator output wire have some type of circuit protection device.

Fuses

A **fuse** is made of a soft metal that melts when excess current flows through it. The metal melts before the current can damage other components or circuit wiring. Most fuses are installed in a **fuse block** located under the dashboard or in the glove compartment. A melted, or blown, fuse must be replaced. See **Figure 7-23.**

Fusible Links

A **fusible link** is a length of wire made of soft metal. It operates in the same manner as a fuse, melting when excess current flows through it. Fusible links are usually installed in the wiring leading from the battery or starter solenoid to the main electrical circuits. See **Figure 7-24.** Both a fusible link ahead of the fuse box and a fuse in the fuse box may be used to protect a circuit.

Circuit Breakers

Circuit breakers consist of a contact point set attached to a bimetallic strip. The bimetallic strip will bend as it heats up. When it becomes hot enough, it bends enough to open the point set, breaking the circuit. When the strip cools off, it straightens out and allows the point set to close. The advantage of the circuit breaker is that it can reset itself. System circuit breakers are usually installed in the fuse panel, **Figure 7-25.** Individual circuit breakers may be installed in or near the device being protected.

Switches

To control the flow of electricity through a circuit, a **switch** is used. A switch is a device containing internal electrical contacts that are positioned by some external force. Closing the contacts allows current to flow through the switch. Opening the contacts causes current flow to stop. Some switches have simple on-off positions, while others, such as windshield wiper and blower motor switches, have several positions to place the circuit in various operating modes.

The vehicle's driver manually operates switches such as the headlight, ignition, windshield wiper, and air conditioner/heater switches. Other switches, such as the backup light switches, are operated as a result of other driver actions. Engine or transmission operating conditions activates some switches. Examples are oil pressure and coolant temperature switches.

Solenoids and Relays

Some electrical components are electromagnetic control devices. Electricity creates a magnetic field, which causes movement of a metal part. In **Figure 7-26,** a solenoid is being used to move a plunger that opens or closes an oil pressure passage. In a **solenoid,** the magnetic field performs a mechanical task, such as opening or closing a transmission valve. **Figure 7-27** shows a typical transmission shift solenoid.

In a **relay,** the magnetic field closes one or more sets of electrical contacts, allowing electrical flow in a circuit. **Figure 7-28** shows a typical relay used on a modern vehicle. This is useful when switching high-current devices, such as motors or resistance heaters, without using excessive lengths of heavy wire.

Motors

To turn electricity into rotation, **motors** are needed. Most motors consist of a central **armature** made of many loops of wire. **Field windings** surrounding the armature produce a magnetic field. By controlling the direction of current flowing through the armature windings, the armature can be made to turn inside the field windings. Current direction is usually controlled by the use of a commutator and brushes. Refer to **Figure 7-29.**

Electric motors are found throughout the vehicle. Typical motors include the starter motor; heater blower motor; windshield wiper motor; power window, seat, and antenna motors; etc.

Resistors

Resistors are placed in a circuit to reduce current flow. They are made of carbon or various metals that create extra resistance to the flow of electrons, reducing current flow. Resistors are commonly used to reduce motor speeds and protect other circuit components from excess current flow. Some resistors are **variable resistors.** The resistance of a variable resistor can be changed by mechanical linkage.

Capacitors

A **capacitor** can be used to damp out voltage fluctuations or control electronic frequencies. The capacitor

Legends of Symbols Used on Wiring Diagrams

Symbol	Description	Symbol	Description
+	Positive		Connector
−	Negative		Male connector
	Ground		Female connector
	Fuse		Denotes wire continues elsewhere
	Gang fuses with bus bar		Denotes wire goes to one of two circuits
	Circuit breaker		Splice
	Capacitor	J2 2	Splice identification
Ω	Ohms		Thermal element
	Resistor	TIMER	Timer
	Variable resistor		Multiple connector
	Series resistor		Optional Wiring with / Wiring without
	Coil		"Y" windings
	Step up coil	88:88	Digital readout
	Open contact		Single filament lamp
	Closed contact		Dual filament lamp
	Closed switch		L.E.D.–Light Emitting Diode
	Open switch		Thermistor
	Closed ganged switch		Gauge
	Open ganged switch		Sensor
	Two pole single throw switch		Fuel injector
	Pressure switch	#36	Denotes wire goes through bulkhead disconnect
	Solenoid switch	#19 STRG COLUMN	Denotes wire goes through steering column connector
	Mercury switch	INST PANEL #14	Denotes wire goes through instrument panel connector
	Diode or rectifier	ENG #7	Denotes wire goes through grommet to engine compartment
	Bi-directional zener diode		Denotes wire goes through grommet
	Motor		Heated grid elements
	Armature and brushes		

Figure 7-22. *Various electrical symbols on one vehicle manufacturer's wiring diagrams. Learning the various symbols will improve your ability to use and understand wiring diagrams. (DaimlerChrysler)*

Autofuse

Autofuse

Current Rating	Color
3	Violet
5	Tan
7.5	Brown
10	Red
15	Blue
20	Yellow
25	Natural
30	Green

Maxifuse

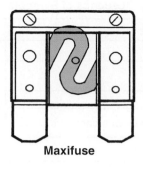

Maxifuse

Current Rating	Color
20	Yellow
30	Green
40	Amber
50	Red
60	Blue
70	Brown
80	Natural

Minifuse

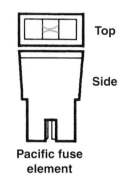

Minifuse

Current Rating	Color
5	Tan
7.5	Brown
10	Red
15	Blue
20	Yellow
25	White
30	Green

Pacific Fuse Element

Top

Side

Pacific fuse element

Current Rating	Color
30	Pink
40	Green
50	Red
60	Yellow

Figure 7-23. *An assortment of common fuses. Note their shapes and sizes. Each fuse type is color-coded according to its current rating. The autofuse is the most common type of fuse found in late-model vehicles. The minifuse is smaller than the autofuse, but it provides the same level of circuit protection. The maxifuse is sometimes used instead of a fusible link and is designed to protect cables that run between the battery and fuse block. It will protect a circuit from a direct short. The Pacific fuse element is also designed to be used in place of a fusible link. This type of fuse is easier to inspect and service than a standard fusible link.*

Fusible link

A

Broken circuit inside insulation

Cut wire here

Fusible link after short circuit

B

Figure 7-24. *A—A fusible link is used to power vehicle systems. B—Damaged fusible link. Never replace a fusible link, fuse, or circuit breaker using a piece of wire. A fire may result.*

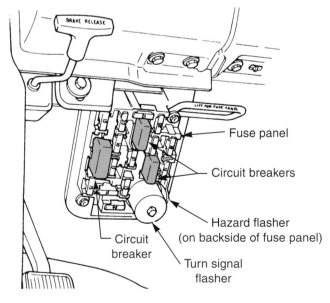

Fuse panel

Circuit breakers

Hazard flasher (on backside of fuse panel)

Circuit breaker

Turn signal flasher

Figure 7-25. *Circuit breakers are commonly located in the fuse box.*

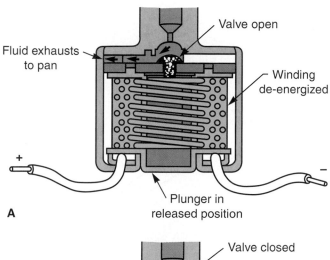

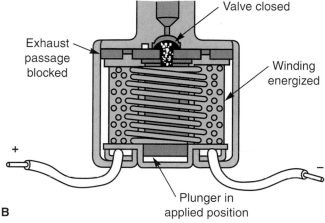

Figure 7-28. *An electrical relay used on a late-model vehicle. The relay controls the flow of current to an electrical device. Note that the electrical schematic is printed on the side of this relay.*

Figure 7-26. *Operation of a typical transmission/transaxle pressure control solenoid. A—The solenoid is de-energized. The return spring keeps the plunger and valve assembly in the retracted position. In the retracted position, the valve is unseated, and pressure entering from the center passage is exhausted to the oil pan. B—The solenoid is energized. The plunger and valve are pulled upward and the valve seals off the center pressure passage. Since pressure is not being exhausted, pressure in the center passage rises. This rise in pressure is used to move a valve elsewhere in the hydraulic system.*

serves as a trap for voltage surges (sometimes called spikes) before they can affect electrical or electronic circuits. Modern capacitors are very small and have the same capacity as older models. They are called **chip capacitors.**

Semiconductor Devices

All modern automotive computers and other electronic devices depend on the use of materials known as **semiconductors**. Semiconductors are made of silicon or germanium. Small amounts of impurities are added to these materials to cause them to act as either conductors or insulators, depending on how voltage is directed into them. Semiconductors change at the atomic level, depending on how their electrons are affected. Common semiconductor devices are diodes and transistors. These components are extremely small and can be combined to make compact computers.

Diodes

Diodes are semiconductor devices that act as one-way check valves for electricity: they allow current to flow in only one direction. The diode acts as a conductor or an insulator, depending on which way the current tries to flow through it. Diodes are used in alternators to rectify, or change, alternating current into the direct current needed to charge the battery.

Diodes are also used to prevent voltage surges, either by being installed inside the computer or in the wiring harness to some high-current devices. Diodes in these applications prevent reverse current surges when the device is turned on or off.

A special type of diode, called a *Zener diode,* will block current flow until a certain voltage is reached. When

Figure 7-27. *A solenoid used in a common electronically controlled automatic transmission. Note the coil winding that surrounds the plunger. This valve is used to open and close a pressure passage as necessary.*

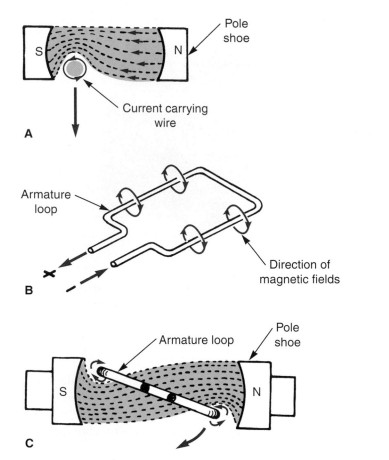

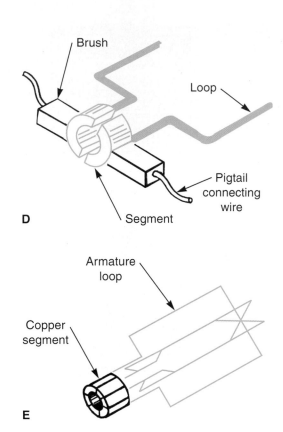

Figure 7-29. *An electric motor in various stages of construction. A—Current-carrying conductor in a magnetic field. The field around the wire (see small arrows) will oppose that of pole field. This will cause wire to be forced out in the direction of arrow A. B—Armature loop. Note direction of magnetic field on each side of loop. C—The armature loop will rotate. Each side of loop has a field moving in opposite directions. These fields oppose the pole shoe field causing the loop to rotate. D—Commutator segments. In an actual starter motor, many loops and segments are used. Each segment is insulated from the shaft and the other segments. E—Many loops are required. Additional loops will increase power or torque of the starter motor.*

this voltage is reached, the Zener diode will open and allow current to flow. Zener diodes are commonly used in voltage regulators and to limit voltage to sensitive electronic circuits.

Transistors

A **transistor** is a semiconductor device that can be used as a switch or an amplifier. Transistors can carry heavy current, but they are operated by very low currents. When a small amount of current is sent to one part of the transistor, the transistor becomes a conductor and much heavier current can flow through another part of the transistor. When this small amount of current is removed, the transistor becomes an insulator and the heavy current flow stops. Transistors are much more reliable than mechanical contacts or vacuum tubes. Therefore, they have largely replaced these older components. Transistors used in computers are usually too small to be seen. Large transistors designed to carry heavy current loads are called **power transistors,** or **drivers.**

Vehicle Computers

The modern automotive computer is a complex device that can operate many vehicle systems. It is made up of a collection of electronic parts and circuits. All modern cars and trucks have one or more computers. All vehicle computers have common operating features, which are discussed in the following sections.

Computer Circuits

Many diodes and transistors, as well as chip capacitors, resistors, and other parts are combined onto a large complex electronic circuit by etching the circuitry on small pieces of semiconductor material. A circuit made by this process is called an **integrated circuit (IC).** The IC also contains resistors, capacitors, and other electronic devices. Modern ICs, sometimes called **chips,** or **microprocessors,** can control many vehicle functions formerly controlled by mechanical devices. Microprocessors have made some systems, such as electronic transmission controls, not only possible but almost universally used.

Control Loops

The microprocessors that make up a modern computer can perform control operations based on their ability to process input messages and issue output commands. This ability allows them to operate as part of a *control loop,* **Figure 7-30.** A control loop can be thought of as a continuous circle of causes and effects used to operate part of the vehicle. When the control loop is operating, the *input sensors* furnish information to the computer, which makes output decisions and sends commands to the *output devices.* The operation of the output devices affects the operation of the vehicle system, causing changes in the readings furnished by the input sensors. There is a continuous loop of information, from the input sensors, to the computer, to the output devices, to the vehicle system, and back to the input sensors. Some computers operate many control loops at the same time.

Main Computer Sections

All automotive computers contain two main sections: the central processing unit and the memory. The construction and operation of these two sections is explained below.

Central Processing Unit

The *central processing unit (CPU)* is the section of the computer that receives the input sensor information, compares this information with the information stored in memory, performs calculations, and makes output decisions. The CPU may contain several microprocessors, as well as other electronic parts, arranged to perform needed calculations. The CPU also contains other electronic parts to change incoming signals to a form the CPU can understand, control high current that is too heavy for the microprocessor circuits, and protect CPU components from voltage and amperage overloads.

Memory

The types of *memory* commonly used in automotive computers are *read only memory (ROM), programmable read only memory (PROM),* and *random access memory (RAM).* ROM contains permanent operating instructions and specifications for the system. This information is installed in ROM at the time of manufacture. ROM is a permanent memory that will be retained if the battery is disconnected or the system fuse is removed. The information in PROM is similar to that in ROM. However, the information in PROM can be changed. On older vehicles, PROM is a removable chip, which can be replaced if necessary. See **Figure 7-31.** On newer vehicles, a variation of the PROM called EPROM (erasable programmable read only memory) or a flash EPROM is used. Some of the information in EPROM can be replaced by erasing and reprogramming.

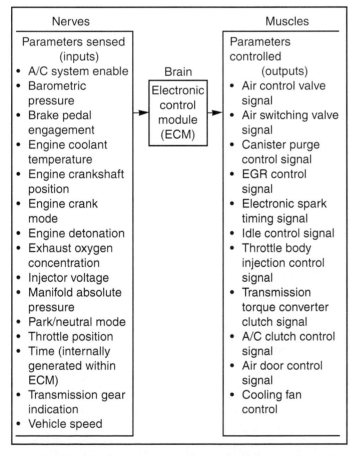

Nerves		Muscles
Parameters sensed (inputs)		Parameters controlled (outputs)
• A/C system enable	Brain	• Air control valve signal
• Barometric pressure	Electronic control module (ECM)	• Air switching valve signal
• Brake pedal engagement		• Canister purge control signal
• Engine coolant temperature		• EGR control signal
• Engine crankshaft position		• Electronic spark timing signal
• Engine crank mode		• Idle control signal
• Engine detonation		• Throttle body injection control signal
• Exhaust oxygen concentration		• Transmission torque converter clutch signal
• Injector voltage		
• Manifold absolute pressure		• A/C clutch control signal
• Park/neutral mode		• Air door control signal
• Throttle position		• Cooling fan control
• Time (internally generated within ECM)		
• Transmission gear indication		
• Vehicle speed		

Figure 7-30. *The computer operating input devices are the sensors (nerves). Information from the sensors goes to the ECM (electronic control module), or the brain. The brain (ECM) controls the output devices (muscles). (AC-Delco)*

Figure 7-31. *This replaceable PROM was used on some vehicles between 1987 and 1996. Indentations in the PROM body prevent reverse installation.*

Random access memory (RAM) is a temporary storage place for data from the input sensors. As information is received from the sensors, it is temporarily stored in RAM, overwriting any old information. The computer constantly receives new signals from the input sensors as the vehicle is driven. RAM is a temporary, or volatile, memory. If the battery cable or system fuse is removed, all the data in RAM will be erased.

Computer Diagnostics

In addition to operating the output devices, the computer's *self-diagnostic system* provides a *diagnostic output* to the technician. The diagnostic system consists of the computer itself, a dashboard-mounted warning light, and a diagnostic connector. When a defect occurs in the system monitored by the computer, the computer stores information about the defect in its memory. This information takes the form of a *trouble code.* A trouble code is a number or a series of letters and numbers that corresponds to a specific defect. A dashboard-mounted light is illuminated whenever the diagnostic system detects a problem. To retrieve the trouble codes from the computer's memory, the technician must generally use a *scan tool.* The scan tool is plugged into the diagnostic connector, or data link connector, located in the vehicle's engine or passenger compartment. On-board diagnostics and trouble code retrieval will be discussed in more detail in Chapter 14.

Summary

Every atom has a positively charged nucleus. Negatively charged electrons revolve around the nucleus. Electricity is the movement of electrons from atom to atom. Materials whose atoms easily give up or receive electrons are good conductors. Materials whose atoms resist giving up or accepting atoms are insulators.

The path through which the electrons move is called a circuit. Electrons will not flow in a circuit unless there are more electrons in one place than in another, and there is a path between the two places. The three basic electrical properties are amperage, voltage, and resistance. Unknown electrical properties can be calculated using Ohm's law.

The two types of current flow are alternating and direct. Automotive electrical systems are direct current systems. Alternating current is used in some sensors. The three types of automotive circuits are series, parallel, and series-parallel. Magnetism can be used to make electricity, and electricity can be used to create magnetism. This relationship between electricity and magnetism is called electromagnetism.

Automotive wiring can be copper, aluminum, or copper coated aluminum. This wiring is plastic coated, color coded, and installed into harnesses. Wire gauge is the rating system for wire diameter. Most electrical connectors are plug-in types. A connector with more than one wire is called a multiple connector. A schematic, or wiring diagram, is a drawing of electrical units and connecting wires. Schematics allow the technician to trace out defective components in the wiring system.

Circuit protection devices include fuses, fusible links, and circuit breakers. Switches control the flow of electricity through a circuit. Relays and solenoids are electromagnetic control devices. Relays control electrical flow, while solenoids cause physical movement. Motors turn electricity into rotation. Resistors are used to reduce current flow through circuits.

All modern automotive computers depend on the use of materials known as semiconductors. Semiconductors can be conductors or insulators, depending on how voltage is directed into them. Diodes and transistors are common semiconductors. Diodes allow current to flow in one direction only. Transistors are used as switches or amplifiers. Transistors carry large amounts of current, but they are operated by low currents. Transistors can perform switching operations with no moving parts. Transistors, diodes, capacitors, and resistors can be etched onto a small piece of semiconductor material to form an integrated circuit, or IC. ICs are also called chips or microprocessors.

Computers are arrangements of microprocessors and other electronic components. They form control loops, using the inputs and outputs of the system and internal controls. The computer processes inputs from sensors and issues commands to output devices. Computer internal controls are the central processing unit (CPU) and the memory circuits. The three types of memory are ROM, PROM, and RAM. In addition to controlling the operation of a system, computers can produce diagnostic codes to aid in troubleshooting problems.

Review Questions—Chapter 7

Please do not write in this text. Write your answers on a separate sheet of paper.

1. Every atom has a center that consists of _____ and _____, around which the _____ rotate.

2. If the atoms of a substance easily give up or receive electrons, the substance is a good _____.

3. Amperage is the number of _____ flowing past a point in a circuit.

4. The same amount of current flows through all parts of a(n) _____ circuit.

5. All the current flows through some parts of a(n) _____ circuit.

6. Current flow is _____ as it flows through a parallel circuit.

7. In a(n) _____ circuit, the current does not flow where it is intended.

8. In a(n) _____ circuit, there is no current flow.

9. Current can flow only one way through a _____.

10. Explain how a Zener diode works.

Match each of the following to the type of memory it describes.

11. Performs calculations. _____

12. Permanent memory storage section. _____

13. Makes output decisions. _____

14. Receives inputs from sensors. _____

15. Temporary memory storage section. _____

(A) RAM
(B) ROM
(C) CPU

ASE-Type Questions—Chapter 7

1. Technician A says that glass and plastic are good conductors. Technician B says that aluminum and copper resist giving up electrons. Who is right?
   (A) A only.
   (B) B only.
   (C) Both A and B.
   (D) Neither A nor B.

2. If there are a lot of electrons in one part of a complete circuit and very few electrons in another part, what will happen?
   (A) Atoms will flow through the circuit.
   (B) Protons will flow through the circuit.
   (C) Electrons will flow through the circuit.
   (D) Neutrons will flow through the circuit.

3. The volt is a measure of electrical _____.
   (A) heating
   (B) light
   (C) pressure
   (D) resistance

4. If amperage and resistance are known, the technician can use Ohm's law to find _____.
   (A) horsepower
   (B) heat
   (C) wattage
   (D) voltage

5. In a _____ circuit, all the current flows through every part of the circuit.
   (A) series
   (B) parallel
   (C) short
   (D) open

6. The measurement of wire diameter is called wire _____.
   (A) length
   (B) material
   (C) gauge
   (D) resistance

7. Each of the following is a circuit protection device *except:*
   (A) fusible links.
   (B) resistors.
   (C) circuit breakers.
   (D) fuses.

8. Solenoids use magnetism to turn electricity into _____.
   (A) sound
   (B) movement
   (C) heat
   (D) light

9. Technician A says that capacitors are used to reduce voltage spikes. Technician B says that capacitors are used to prevent damage to other electronic parts. Who is right?
   (A) A only.
   (B) B only.
   (C) Both A and B.
   (D) Neither A nor B.

10. Each of the following statements about semiconductors is true *except:*
    (A) semiconductors are made from aluminum and glass.
    (B) semiconductors can act as conductors.
    (C) semiconductors can act as insulators.
    (D) transistors and diodes are common semiconductor devices.

11. A power transistor is a large transistor designed to carry heavy _____.
    (A) voltage
    (B) heat
    (C) current
    (D) resistance

12. The modern IC or chip may contain all the following parts *except:*

    (A) transistors.

    (B) capacitors.

    (C) diodes.

    (D) solenoids.

13. Each of the following statements about control loops is true *except:*

    (A) a control loop is a complete circle of causes and effects.

    (B) control loops are used to operate vehicle systems.

    (C) during control loop operation, the input sensors make decisions and send commands to the output devices.

    (D) some computers can operate several control loops at the same time.

14. Which of the following computer components is a permanent, factory-installed memory section?

    (A) RAM.

    (B) ROM.

    (C) Controller.

    (D) CPU.

15. The computer self-diagnostic system consists of all the following parts *except:*

    (A) scan tool.

    (B) data link connector.

    (C) warning light.

    (D) computer.

Chapter 8

Automatic Transmission Mechanical Components

After studying this chapter, you will be able to:
- ❏ Identify the major mechanical parts of an automatic transmission.
- ❏ Discuss the purpose of each major mechanical part of an automatic transmission.
- ❏ Explain the operation of a torque converter.
- ❏ State various outcomes resulting from driving certain planetary gears and holding others.
- ❏ Diagram power flow through a typical automatic transmission.
- ❏ Summarize operations of planetary holding members.
- ❏ Identify and explain the purpose of stationary mechanical parts.

Technical Terms

Torque converter	Bounce-back effect	Multiple-disc clutch
Automatic transmission shafts	Torque multiplication	Clutch packs
Planetary gearsets	Overrunning clutch	Clutch drum
Planetary holding members	Coupling phase	Drive discs
Converter housing	Stall speed	Clutch plates
Automatic transmission case	Lockup torque converter	Clutch apply piston
Extension housing	Converter clutch torque converter	Clutch pressure plate
Oil pan	Input shaft	Clutch piston return springs
Fluid coupling	Automatic transmission output shaft	Clutch hub
Slippage	Stator shaft	Drive shell
Impeller	Planetary gearsets	Sun-gear shell
Turbine	Compound planetary gearset	Bands
Stator	Ravigneaux gear train	Servos
Converter cover	Simpson gear train	Engine braking
Converter pilot hub	Wilson gear train	Bell housing
Flywheel	Lepelletier gear train	Sump
Pump drive hub	Planetary holding members	

Introduction

In this chapter, you will study the mechanical components of rear-wheel drive automatic transmissions. These parts are used to transfer power from the engine to the drive shaft. Many mechanical components, such as shafts, gears, drums, and clutch plates, rotate. Other parts, such as the case, housings, and bands, are stationary and act as supports or housings for the rotating parts. The moving and stationary parts sometimes act together to stop some parts of the gear train while causing other parts to rotate, creating a particular gear ratio. Other times, the moving parts create the gear ratios, and the stationary parts serve as supports or are simply inactive.

Studying this chapter will give you the information that you need to understand how the hydraulic control system discussed in the next chapter interacts with the mechanical parts to produce vehicle movement. This, in turn, will lead to an understanding of the electronic controls covered in Chapter 10. Much of the information contained in this chapter also applies to the mechanical components used in automatic transaxles.

Construction Overview

Automatic transmissions are made of many separate parts and systems. **Figure 8-1** shows the relative position of the mechanical parts in the transmission. Also, refer to **Figure 8-2,** which is an exploded view of an automatic transmission. The major mechanical parts in an automatic transmission include:

❏ *Torque converter*—a hydrodynamic device that uses transmission fluid to transmit and multiply power. The torque converter is mounted directly behind the engine and is turned by the engine crankshaft.

❏ *Automatic transmission shafts*—transmit power from the torque converter to the driveline. Another transmission shaft provides a stationary support for the torque converter stator.

❏ *Planetary gearsets*—constant-mesh gears arranged in such a manner that they resemble the solar system. They create different forward gear ratios or provide reverse operation when certain members are held stationary and others are driven. Most transmissions use at least two planetary gearsets.

❏ *Planetary holding members*—allow planetary gearset operation to occur. The holding members function by either holding a planetary member stationary or applying drive power to it.

❏ *Converter housing*—a stationary enclosure that surrounds and protects the torque converter. It is located at the front of the transmission, directly behind the engine. Some converter housings cast as an integral part of the transmission case.

❏ *Automatic transmission case*—encloses the planetary gearsets and holding members, as well as the inner ends of the transmission input and output shafts. Passageways in the case allow hydraulic fluid to travel between hydraulic components. The case is bolted to the rear of the converter housing, or it may be an integral part of the converter housing casting.

❏ *Extension housing*—supports and encloses the tail end of the transmission output shaft. This housing bolts to the rear of the transmission case.

❏ *Oil pan*—installed on the bottom of the transmission case. It fits over the valve body and serves as a reservoir for transmission fluid.

Torque Converter

Before torque converters, a simpler device called a *fluid coupling* was used in automatic transmissions. A *fluid coupling* is a hydrodynamic device designed to

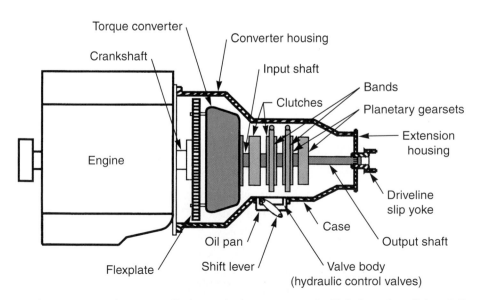

Figure 8-1. *This diagram shows most major automatic transmission components. Note how they fit in relation to each other.*

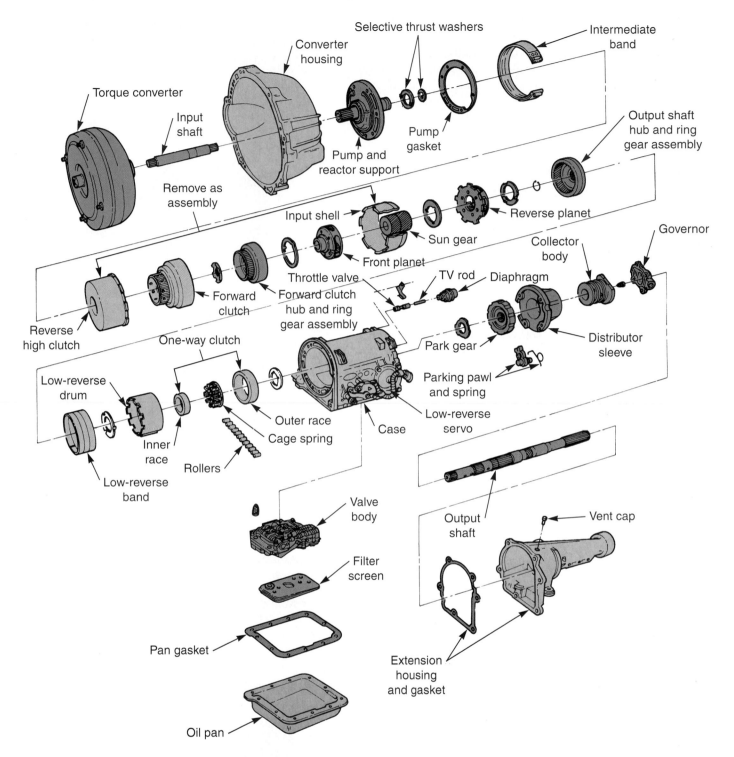

Figure 8-2. *Exploded view of a typical automatic transmission. (Ford)*

transmit power through a fluid. This device essentially consists of a *drive member* (impeller) and a *driven member* (turbine), **Figure 8-3.**

Fluid couplings have several shortcomings. For instance, the impeller and the turbine do not link together well at low speeds. The impeller must build up considerable speed before the turbine will turn. Therefore, the drive member rotates faster than the driven member and fails to transmit 100% of its power. This problem is known as ***slippage.*** Another shortcoming is a result of the *bounce-back effect* (described later in this section). This effect causes torque reduction through the coupling.

To reduce slippage and multiply torque, a simple fluid coupling is modified to create a torque converter, which is used in modern automatic transmissions. The major advantage of fluid couplings and torque converters is that they do not have to be disengaged when the vehicle is stopped and the engine is running.

Torque Converter Construction

A torque converter basically consists of three separate bladed elements located inside a fluid-filled housing. See **Figures 8-4** and **8-5.** The basic components of a torque converter include an impeller, a turbine, a stator, and a converter cover. These components, as well as associated parts, are discussed in the following section.

Impeller

The **impeller,** or pump, is composed of a set of curved vanes welded to the inside of a shell that forms the rear half of the converter. The impeller is attached to the converter cover. Since the converter cover is attached to the flywheel, the impeller turns whenever the engine turns. The impeller is the torque converter drive member.

Turbine

The **turbine** consists of a set of curved vanes welded to a *turbine shell.* The turbine is driven by the impeller through fluid contained within the converter. The turbine is the torque converter driven member. The central hub of the turbine contains internal splines that fit the external splines of the transmission input shaft, or turbine shaft. The turbine drives the turbine shaft.

Stator

The **stator,** or guide wheel, fits between the impeller and turbine and is smaller than either. It consists of a set of vanes that are curved in the opposite direction of the turbine vanes. The stator redirects fluid coming from the turbine to the impeller's direction of rotation, helping the torque converter multiply power. The stator is the component that makes a torque converter out of a simple fluid coupling.

The stator is mounted on an *overrunning clutch,* which allows the stator to move under some conditions and keeps it stationary under others. The overrunning clutch is splined to a hollow stationary shaft, called the stator shaft, which is connected to the oil pump housing.

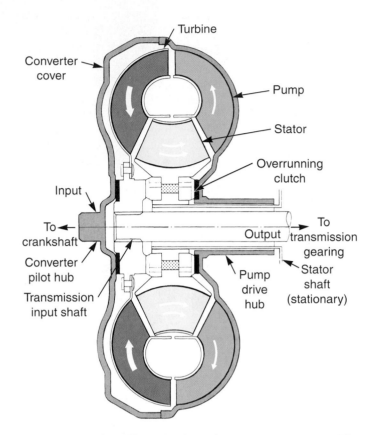

Figure 8-4. *Simplified drawing of a torque converter. The engine crankshaft is fastened (indirectly) to the converter cover and the impeller. When the crankshaft turns fast enough, fluid movement rotates the turbine element and the transmission input shaft. Note that the stator is mounted on a special type of clutch, which in turn, is mounted on a stationary shaft. (Sachs)*

Figure 8-3. *This cutaway shows a fluid coupling in action. Oil is thrown from the impeller into the turbine. Note that member positions are usually the opposite of what is shown here.*

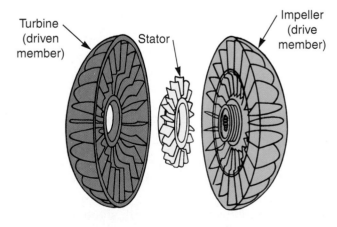

Figure 8-5. *Illustration shows the three main elements of a torque converter. The impeller is driven by the engine. The turbine, which drives the transmission input shaft, is driven by the impeller. The stator redirects fluid flow from the turbine to multiply torque. Note the curvature of the vanes is not shown. (Texaco)*

Torque Converters with Two Stators

A few torque converters have two stator assemblies, **Figure 8-6.** The use of two stators increases the ability of the converter to multiply torque while maintaining good fuel economy at higher vehicle speeds. The stator nearest the impeller has more pitch (greater angle of blade curvature) than the other, and therefore, it causes the fluid to strike the impeller blades with more force. When the vehicle begins moving, both stators are locked by the one-way clutch. Fluid strikes the impeller with great force, increasing the torque multiplication of the converter.

As vehicle speed increases, turbine speed begins to catch up to impeller speed. The stator nearest the impeller, which has more pitch and greater fluid flow, causes its one-way clutch to unlock sooner. The stator freewheels, and fluid is redirected by the second, less pitched, stator. This provides less torque multiplication than when both stators are locked, but still provides some torque multiplication. When the impeller and turbine speeds are almost the same, the second stator unlocks and the converter essentially becomes a fluid coupling.

If the vehicle is accelerated, the low-pitch stator will lock first and then the high-pitch stator will relock to provide additional torque multiplication.

Converter Cover

The **converter cover** fits over the turbine, attaches to the impeller, and indirectly attaches to the engine crankshaft. In some designs, a ring gear is welded to the perimeter of the cover. The ring gear engages with the starter motor during cranking.

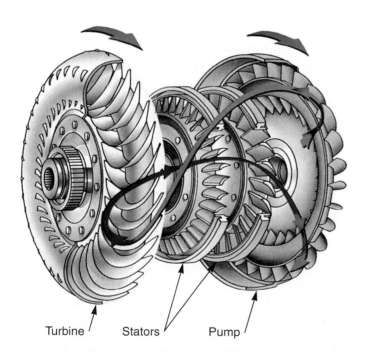

Figure 8-6. *Using two stators increases the ability of the converter to multiply torque while maintaining good fuel economy at higher vehicle speeds. (General Motors)*

Turbine   Stators   Pump

A cylindrical metal extension called the **converter pilot hub** is located at the center of the converter cover. The pilot hub fits into a mating hole in the engine crankshaft flange. The close fit of the pilot hub and crankshaft bore maintains the alignment of the converter and crankshaft centerlines. The converter cover may also have balance weights. Some covers have oil drain plugs for removal of fluid during maintenance.

Flywheel

The converter cover has tapped holes or threaded studs for connection to a flywheel. The flywheel is attached to the crankshaft at the rear of the engine. The **flywheel** is a lightweight disc with a ring gear that is used to engage the starter motor. If the ring gear is installed on the converter cover, the flywheel will not have a ring gear and is often called a flexplate. The flywheel is attached to the engine crankshaft with bolts. The flywheel and converter turn with the engine whenever it is running.

Although the true definition of a flywheel is a heavy disc that smoothes engine crankshaft rotation on manual transmission vehicles, industry practice is to call all discs containing the ring gear flywheels. Vehicles with automatic transmissions do not require heavy flywheels. Instead, the torque converter acts as a flywheel since it is heavy and contains several quarts of fluid, each quart weighing about 1.5 pounds (0.675 kg). The automatic transmission flywheel is a lightweight disc that does nothing to smooth crankshaft rotation.

Pump Drive Hub

Modern torque converters are sealed units that are welded together after all internal elements have been installed. However, the torque converter has a hollow rear shaft, called the **pump drive hub,** that fits over the stator and turbine shafts and into the oil pump at the front of the transmission case. Oil is pumped into the converter through one of the internal shafts, and it exits between the stator shaft and the inside of the hub. Drive lugs on the rear of the hub turn the transmission oil pump. An *oil pump bushing* within the pump housing supports the pump drive hub.

Thrust Washers

The torque converter contains several thrust washers, which are generally located between all parts that can move in relation to each other. Thrust washers are usually placed between the cover and turbine, the turbine and stator, and the stator and impeller.

Torque Converter Operation

Torque converter operation depends on the movement of transmission fluid. When the engine is running, the converter cover is turning. This causes the impeller to turn, as it is solidly connected to the cover. Since the torque converter is completely filled with transmission fluid, the turning impeller moves the fluid that is caught between the vanes. This fluid rotates at the same speed and in the same relative direction as the impeller. Engine power is transferred from the impeller to the rotating fluid.

As the impeller rotates, the fluid is thrown outward and, following the curvature of the impeller shell, is redirected toward the turbine. The fluid leaves the impeller, striking the turbine. The fluid presses forward on the vanes of the turbine, as shown by the small arrows in **Figure 8-7.** As a result of this force, the turbine begins to rotate. The moving fluid transmits motion to the turbine, and engine power is transferred from the fluid to the turbine.

At *idle,* with the vehicle stopped and the transmission in gear, the engine continues to run. If the vehicle had a manual transmission, the clutch would have to be disengaged or the transmission would have to be placed in Neutral; otherwise, the engine would stall.

With an automatic transmission, the impeller is turned by the engine, causing the fluid to rotate and strike the turbine blades. At low engine speeds, the force of the fluid hitting the turbine is relatively small, and the driver can overcome it by keeping the brakes applied. The turbine does not move.

Note that the energy of the fluid hitting the stationary turbine causes friction instead of motion. This friction changes the energy in the fluid into heat. The heat generated can become intense. For this reason, vehicles with automatic transmissions should not be idled in gear for long periods. Idling in gear for more than 10 minutes is considered excessive, especially if the air conditioner is being used.

During *acceleration,* the force of the fluid leaving the impeller turns the turbine, which turns the automatic

transmission input shaft. Turning the input shaft, and the resultant operation of the other transmission parts, causes the vehicle to move. However, there is some slippage between the impeller and the turbine.

At *cruising speeds* (maintained speed above 35 mph), the impeller and turbine spin at almost the same speed. There is very little slippage, and *torque multiplication* is minimal.

Stator Action

As a result of turbine vane curvature, fluid coming from the impeller will reverse direction before it returns to the impeller. This reversal is greatest at conditions of high torque multiplication, and it decreases as torque multiplication declines.

When the impeller is moving much faster than the turbine, the fluid still has considerable forward motion as it bounces back to the impeller. Without a stator, return flow would oppose impeller rotation and would hit the forward (oncoming) faces of the impeller vanes with a good deal of force. This action, seen in fluid couplings and known as the **bounce-back effect,** reduces the effectiveness of the impeller.

To counteract this problem, the stator reverses the direction of fluid coming off the turbine. Fluid leaving the turbine must pass through the stator before reaching the impeller. The stator blades are curved so that they turn the fluid around. Fluid leaving the stator strikes the impeller in the direction of impeller rotation. Instead of losing engine power in working to reverse the direction of the fluid coming from the turbine, which is what happens without the stator, the impeller is now assisted in rotating. The stator not only eliminates torque loss due to the bounce-back effect, but it also improves efficiency by providing torque multiplication.

Torque Multiplication

Just as a small gear can multiply torque by driving a large gear, a torque converter can also multiply torque. Unlike the gearset, however, the torque converter operates through a *continuous range* of "gear" ratios, and torque multiplication varies accordingly.

Torque multiplication refers to the ability of a component, such as a torque converter or gearset, to increase output torque above input torque. Specifically, for a torque converter, it refers to the amount of torque at the transmission input shaft above that which the engine puts out.

Torque multiplication basically occurs as a result of the redirection of fluid by the stator. The stator causes the fluid to re-enter the impeller in the direction of rotation, helping the impeller to turn. The fluid passes through the impeller, and when it re-enters the turbine, it imparts additional force (torque) to the turbine.

The amount of torque multiplication changes with the difference in rpm between the impeller and turbine—the greater the difference, the greater the multiplication. It follows that the greatest torque multiplication for any given

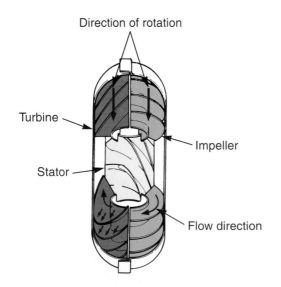

Figure 8-7. *Cutaway of torque converter. Heavy arrows show direction of rotation of impeller and turbine. The medium arrows show how the fluid circulates between the impeller and turbine. The smaller arrows depict the driving force with which the fluid strikes the turbine vanes. Note the curvature of impeller, turbine, and stator vanes. (Subaru)*

engine speed occurs at initial acceleration. The point of maximum slippage occurs when the turbine shaft is stopped, or *stalled,* and the difference in speed between the two elements is the greatest. Maximum *converter ratios* can be as high as 2.5:1 on some converters; however, most ratios are about 2:1.

Torque multiplication and output (turbine shaft) rpm have an inverse relationship; that is, for a constant engine speed, torque multiplication drops off as turbine speed increases. Thus, as the speed of the turbine is brought up to the speed of the impeller, torque multiplication is continuously reduced, and as the vehicle reaches cruising speed, the converter ratio drops to about 1:1. This means that as the impeller and turbine approach the same speed, torque multiplication drops off to almost zero.

Overrunning Clutch Operation

The stator is installed on the stator shaft over an **overrunning clutch,** or a **one-way clutch.** When the vehicle is accelerating, fluid deflects off the turbine blades and reverses direction. Fluid is deflected into the stator, striking the front faces of the vanes. The direction of applied force acts to lock the one-way clutch and, therefore, the stator. When held stationary, the stator can return fluid to the impeller in a *helping* direction, and it can multiply torque.

At cruising speeds, the turbine and impeller begin turning at almost the same speed. During this condition, the fluid enters the turbine at the same speed the turbine is moving, and there is no fluid deflection. The turbine is no longer reversing the fluid that passes through it. Therefore, when turbine speed approaches about 90% of impeller speed, which is called the **coupling phase,** the stator is not needed. In fact, it begins to restrict the flow of fluid from the turbine back to the impeller.

When the coupling phase is reached, the flow leaving the turbine strikes the back of the stator blades. As a result, the direction of applied force causes the overrunning clutch and, therefore, the stator to unlock and freewheel. The stator can then rotate with the impeller and turbine, and it does not restrict the movement of the fluid through the impeller and turbine. The action of the one-way clutch is shown in **Figure 8-8.**

Converter Stall Speed

Torque converter **stall speed** may be the most misunderstood concept of automatic transmission theory. Stall speed is the maximum engine speed that can be obtained when the converter turbine is stationary. To hold the turbine stationary, the vehicle is placed in gear and the vehicle's brakes are applied. Then the engine throttle is opened until the maximum engine speed is obtained. At this point, the engine is said to *stall,* or be unable to increase its speed no matter how large the throttle opening.

Converter stall should not be confused with the engine stalling, which occurs when the engine stops running.

At stall speed, internal converter fluid flow and slippage is so great that it creates massive internal friction between the moving fluid and the internal parts. This fluid friction causes the converter to hydraulically lock up, preventing any increase in engine speed. At stall speed, torque multiplication is at its maximum. Three internal converter factors and three external factors affect stall speed. The internal factors are turbine and stator blade angles, internal converter clearances, and converter size. The external factors include engine power, vehicle weight, and transmission design. Of these factors, turbine and stator blade angles are the most important. However, the other factors also play a part. These factors are discussed in the following sections.

Turbine and Stator Blade Angles

Stall speed varies with the shape of the turbine and stator blades. Remember from earlier sections that the turbine and stator blades redirect the fluid behind the turbine blades to help the engine turn the impeller. The amount of help the turbine and stator blades give the engine determines the speed at which internal fluid friction finally overcomes the engine's ability to turn the impeller. Stall speed is the result of an internal converter conflict between friction and torque multiplication.

The turbine and stator blades of a low stall converter have a relatively low, or shallow, angle (sometimes called pitch). Therefore, they return the fluid to the impeller with less assisting force than blades on a conventional converter. At a relatively low speed, the engine will run out of power to turn the impeller against fluid friction. In a high-stall converter, the turbine and stator blades have a higher, or more acute, angle. They deliver the fluid to the impeller with more force, giving the impeller extra help in turning. With more help from the returning fluid, the engine can turn the impeller at a higher speed before it stalls. The higher the blade angle, the more torque the converter can deliver over a greater range of engine speeds.

Internal Converter Clearances

Another internal converter factor that affects stall speed is the clearance (space) between the impeller, turbine, and stator. This clearance allows some of the fluid to bypass these parts instead of circulating through them. Large clearances mean that fluid friction will be less and stall speed will be higher. The action of the converter, however, will be less efficient because not all the fluid is being used to multiply power. Converters with large internal clearances are often used in extreme competition situations, such as drag racing, because bypassing some of the fluid increases stall speed and allows the engine speed to climb to near its maximum torque output. Large clearances, however, are not useful in normal driving. Older converters had more clearance than modern converters.

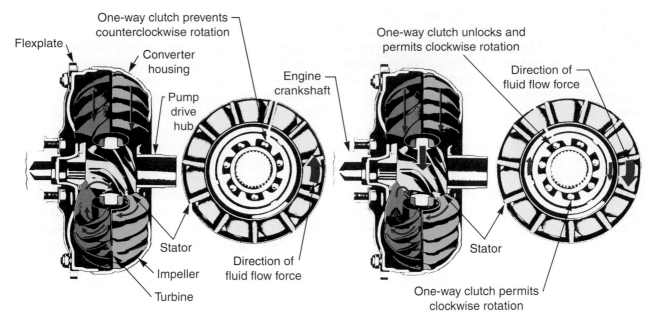

A—Converter in torque multiplication phase   **B—Converter in fluid coupling phase**

Figure 8-8. *Study the action of an overrunning clutch mounted between the stator and stator shaft. A—In the torque multiplication phase, fluid is deflected off the turbine vanes onto the front side of the stator vanes. This action locks the one-way clutch and holds the stator stationary. B—In the coupling phase, deflection (or lack thereof) of fluid is such that the fluid strikes the back side of the stator blades. This causes the stator to spin forward, preventing it from interfering with the fluid flow. (Ford)*

Larger clearances allowed the converter to absorb some engine vibrations. Large clearance converters were also easier and cheaper to manufacture. Modern converters have extremely tight clearances for maximum efficiency in both performance and economy.

Converter Size

The size of the torque converter also affects stall speed. A relatively small converter has less fluid to create friction, without greatly affecting torque multiplication. To compensate for their size, small converters usually have greater blade angles than large converters. A smaller converter also has less mass for the engine to turn. A disadvantage of small high-stall converters is that they get extremely hot, and additional cooling must be provided, usually in the form of a large add-on cooler.

Engine Power

An often-overlooked external factor that affects stall speed is the amount of power (in the form of torque) the engine produces. The same converter will have different stall speeds when attached to engines of different horsepower. This is because a powerful engine can overcome much more internal fluid friction than a lower power engine. Regardless of the angle of the blades, converter stall speed is determined by engine torque.

Vehicle Weight

Another external factor that affects stall speed is the weight of the vehicle. An engine and transmission combination in a large vehicle requires a higher stall speed than the same combination in a smaller vehicle. The higher stall speed forces the engine to climb higher in its torque range to more efficiently move the vehicle.

Transmission Design

The design of the transmission also affects converter stall speed. The number of transmission gears and their ratios affect how torque is multiplied after leaving the converter. As a rule, the more gears a transmission has, the lower the stall speed should be. Rear axle ratio also affects stall speed, but it can usually be disregarded. An exception to this occurs when an axle with a much lower ratio is installed for racing or towing.

Why Is Stall Speed Important?

While stall tests are never performed as part of everyday driving, stall speed is important. Stall speed is a way to measure the angles of the turbine and stator blades, and determine how they relate to the available engine power. It is, therefore, a way to measure how the converter will process the power delivered to it by the engine. A converter with an excessively low stall speed for the engine and vehicle will cause slow acceleration and a lack of power.

A converter with an excessively high stall speed will cause lower fuel economy. A high stall speed converter may also cause a loose, slipping feeling at cruising speeds. For these reasons, the converter must be carefully matched to the engine and vehicle.

Whenever a technician is considering changing to a converter of a different stall speed, he or she should consider how the change would affect overall vehicle operation. Engine power, vehicle weight, and the type of transmission should be carefully considered. A slightly higher stall speed (300–400 rpm) often increases performance without affecting drivability and fuel economy. For greater changes in stall speed, the technician should contact the converter manufacturer for more information.

Lockup Torque Converter

A *lockup torque converter,* or *converter clutch torque converter,* has an internal friction-clutch mechanism. This mechanism mechanically locks the turbine to the impeller and converter cover when the transmission is in high gear. It is designed to eliminate the 10% slippage that takes place between the impeller and turbine during the coupling phase of operation. When the turbine is locked to the impeller, the transmission input shaft will travel at the same speed as the engine crankshaft and efficiency will be 100%.

The lockup torque converter offers several advantages over a conventional torque converter. Fuel economy is improved because engine power is not wasted. In addition, engine wear is reduced because the engine can now be operated at lower rpms to attain the same vehicle speed. Another advantage is that converter's operational heat is reduced.

Modern lockup torque converters are equipped with an internal clutch, which locks the input shaft to the converter cover. **Figure 8-9** shows a cutaway of a torque

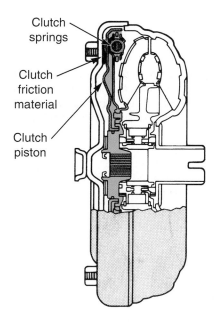

Figure 8-9. *Lockup torque converters use an internal clutch to make a direct connection between the engine crankshaft and transmission input shaft. The clutch is splined to the input shaft. When applied by hydraulic pressure, the clutch makes frictional contact with the converter cover. (DaimlerChrysler)*

Clutch springs

Clutch friction material

Clutch piston

converter with a lockup, or converter, clutch. Typically, the clutch assembly consists of a hydraulic clutch piston (or pressure plate), torsion springs, and a clutch friction material. The piston is splined to the input shaft. The friction material is attached to the converter cover. The torsion springs dampen out engine firing impulses and absorb shock loads that occur during lockup. **Figure 8-10** shows an exploded view of a typical hydraulic-type lockup torque converter.

A few transmissions use a viscous clutch. The viscous clutch consists of two sets of blades that are placed close together but do not touch. One set of blades is attached to the lockup clutch and the other is connected to the transmission input shaft. The space between the blades is filled with a very thick, or viscous, liquid. Power from the clutch must pass through one set of blades, through the viscous fluid, and onto the other set of blades. The fluid is thick enough to transfer power but thin enough to absorb shocks from clutch engagement and changes in load.

Operation of a typical hydraulic converter clutch is controlled through a *lockup module* and a *switch valve.* Both are housed in the transmission valve body. These mechanisms control the flow of hydraulic fluid to and from the clutch piston.

In the lower gears, the lockup torque converter operates in the same manner as a conventional torque converter, **Figure 8-11A.** When the transmission is shifted into high gear, or direct drive, transmission fluid is typically channeled through the input shaft into the space between the turbine and the clutch piston. Fluid pressure causes the piston to be pushed into the friction material on the converter cover, **Figure 8-11B.** This action locks the turbine to the impeller. When the transmission shifts out of high gear, fluid pressure and, therefore, the clutch piston are released. The converter returns to conventional operation.

> **Note:** There is another type of lockup torque converter that works by centrifugal force. In this type, shoes made of a friction material are thrown outward against the converter cover, forming a direct mechanical connection between the impeller and the turbine.

Automatic Transmission Shafts

The two main shafts in the automatic transmission are the *input shaft* and the *output shaft.* These shafts transfer power from the torque converter to the driveline. They are straight shafts made of hardened steel.

The *stator shaft* is a third type of shaft found in the automatic transmission. It is a hollow, stationary shaft that extends from the transmission oil pump. It supports the stator overrunning clutch. Refer to **Figure 8-12** as you read the following sections.

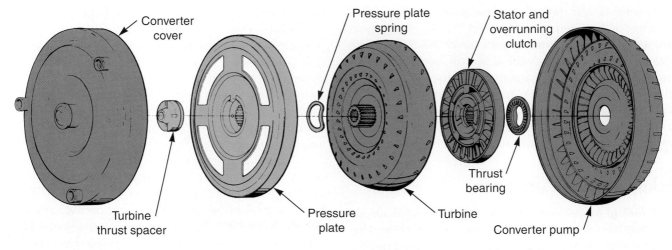

Figure 8-10. *This exploded view of a converter clutch torque converter shows the relative positions of the pressure plate and associated spacers and washers. (General Motors)*

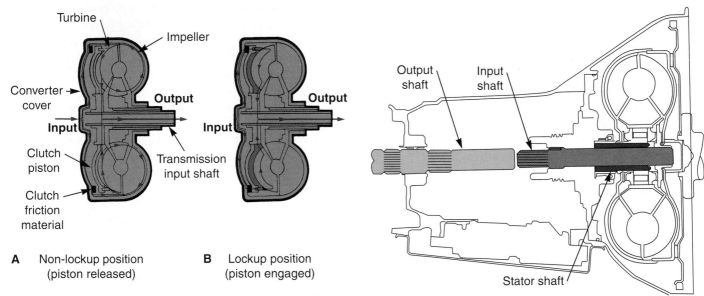

A Non-lockup position
(piston released)

B Lockup position
(piston engaged)

Figure 8-11. *Typical lockup torque converter operation. A—In lower gears, no fluid pressure acts on the clutch piston. The torque converter operates like a conventional unit: the impeller drives the turbine. B—In high gear, fluid is transferred into the piston chamber. The clutch piston is forced into the clutch friction material on the converter cover. The turbine is mechanically locked to the cover and impeller. The crankshaft drives the transmission input shaft directly, without slippage. Note that the arrows depict power flow. (DaimlerChrysler)*

Figure 8-12. *The main transmission shafts are illustrated here. The input shaft turns whenever the engine is running. The output shaft turns whenever the vehicle is moving. The stator shaft is stationary at all times. The input and output shafts turn on bushings.*

clutch apply pistons. The shaft rides on bushings, which are lubricated by transmission fluid.

Input Shaft

The front end of the automatic transmission *input shaft,* or *turbine shaft,* has external splines that fit into mating internal splines on the turbine of the torque converter. Power is transferred from the turbine via the input shaft directly (or indirectly) into the front planetary gearset of the transmission. Hydraulic fluid from the valve body to the transmission clutch apply pistons often travels through passageways in the input shaft. The input shaft may contain rings that seal in the fluid traveling to the

Output Shaft

The tail end of the *automatic transmission output shaft* has external splines that fit into mating splines on the driveline slip yoke. The front end of the output shaft is splined to one or more planetary gearsets. The output shaft centerline aligns with the input shaft centerline. The front end of the output shaft almost touches the input shaft.

Most output shafts contain the speedometer drive gear. Some speedometer drive gears are machined into the shaft, while others are clipped or pressed onto the shaft. In addition, the *governor,* which will be discussed later in this chapter, may be installed on the output shaft. However, in some designs, the shaft has a gear, similar to the speedometer drive gear, for driving the governor. Other units have a toothed wheel for use with a speed sensor. The output shaft may also have internal passageways and seal rings for hydraulic purposes.

Stator Shaft

The **stator shaft** is solidly mounted on the front of the transmission oil pump. It provides a stationary support for the torque converter stator and overrunning clutch. The overrunning clutch must be solidly supported on a stationary part to work properly. The stator shaft has external splines that fit into mating internal splines on the overrunning clutch, preventing rotation of the inner race of the clutch. The stator shaft is hollow, and the transmission input shaft passes through it. Bushings in the stator shaft support the input shaft. The stator shaft is sometimes called the reaction shaft.

Planetary Gearsets

The **planetary gearsets** used on modern automatic transmissions consist of a sun gear, several planet pinions, a planet-pinion carrier, and a ring gear. These constant-mesh gears are so named because their design somewhat resembles the solar system. Just as planets, such as Earth and Mars, revolve around the Sun, the planet gears revolve around the sun gear. Planetary gearsets were discussed in Chapter 4. A simple planetary gearset is shown in **Figure 8-13.** An exploded view of common planetary gears is shown in **Figure 8-14.**

Planetary Gearset Power Flow

Power is transferred through planetary gearsets by holding one part of the gearset stationary and driving another. Power exits the third part of the gearset. For example, holding the sun gear and driving the planet carrier causes the ring gear to rotate.

With regard to the direction of rotation, it is helpful to remember two points. The first point is that two gears with external teeth in mesh will rotate in opposite directions. The second point is that two gears in mesh, one with internal teeth and one with external teeth, will rotate in the same direction. Understanding these two points will help you analyze planetary gearset operation.

Different combinations of gears that are driven and gears that are held stationary produce different gear ratios or actions. Thus, by holding or driving the gears of a planetary gearset, it is possible to obtain:
- ❏ Gear reduction.
- ❏ Overdrive.
- ❏ Direct drive.
- ❏ Reverse gear.
- ❏ Park or Neutral.

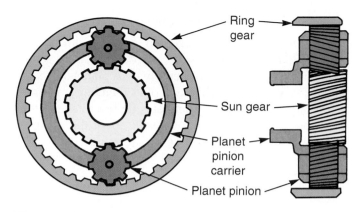

Figure 8-13. *A simple planetary gearset consists of a central sun gear surrounded and meshed with planet gears in a planet carrier. A ring gear surrounds and meshes with the planet gears. Different gear ratios can be established, in design, by varying the number of teeth on meshing gears, just as with conventional gearsets. Note that planetary gearsets are compact and tough, and since the gears are constantly engaged, gear clash and broken teeth are not a problem. (Ford)*

Figure 8-14. *A typical planetary gearset. Note the ring gear, planet gears, and sun gear. Almost all automatic transmissions have these parts.*

Gear Reduction

Automatic transmissions can achieve gear reduction through planetary gearsets. Gear reduction provides increased torque and reduced output speed. One method of obtaining a gear reduction is to turn the sun gear while holding the ring gear stationary. Conversely, another method is to hold the sun gear stationary while driving the ring gear. In both methods, the planet carrier is the output member. The first method provides the maximum reduction (torque multiplication) that can be achieved in one planetary gearset. The second method produces minimum gear reduction. See **Figure 8-15.**

Overdrive

Automatic transmissions can also achieve overdrive through planetary gearsets. Overdrive provides increased output speed while reducing torque. When the sun or ring gear is held stationary and the carrier is driven, the remaining member (output) will be driven at a faster speed than the planet carrier. This will produce overdrive in a planetary gearset, with maximum speed increase occurring as a result of holding the ring gear stationary. See **Figure 8-16.**

Direct Drive

Automatic transmissions can achieve direct drive through planetary gearsets. In direct drive, the gearset serves as a solid unit to transfer power. A planetary gearset will act as a solid unit when two of its members are turned at the same speed. The planet gears do not rotate on their axes; they can neither idle nor walk about the sun gear. The entire unit is locked together, forming one rotating unit. See **Figure 8-17.**

Reverse Gear

A planetary gearset is capable of providing output rotation that is opposite of input rotation. This action causes the vehicle to move in the reverse direction. The carrier is held stationary. The input shaft drives either the sun gear or the ring gear. A driven sun gear yields torque multiplication at ring gear output. A driven ring gear yields torque reduction at sun gear output. The planet gears act as idler gears. They reverse the direction of rotation between the sun gear and the ring gear. See **Figure 8-18.**

Park or Neutral

A planetary gearset can be allowed to freewheel, stopping power flow. When all the planetary members are free, the gearset cannot transfer power. This freewheeling condition occurs when an automatic transmission is placed in Park or Neutral.

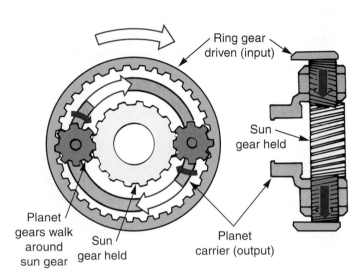

Figure 8-15. *Gear reduction is produced with the ring gear as the input and the sun gear held stationary. Planet gears walk around the sun gear, taking the planet carrier (output) with them. The planet carrier rotates in same direction as the ring gear but at a slower speed. Although this arrangement produces good torque, maximum torque is produced with the sun gear as input and the ring gear held stationary. (Ford)*

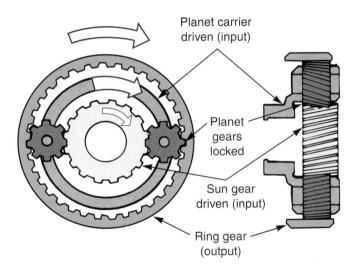

Figure 8-17. *In direct drive, driving any two of the gearset elements at the same speed will lock the entire gearset together, making it turn as a unit. (Ford)*

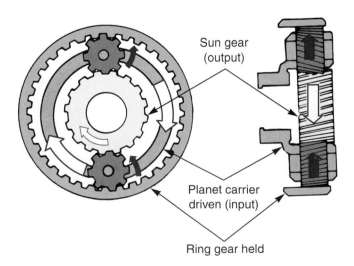

Figure 8-16. *Overdrive is produced with the planet carrier as input, the ring gear held stationary, and the sun gear as output. As planet gears walk around the ring gear, they drive the sun gear. The sun gear rotates in the same direction as the planet carrier but at a faster speed. (Ford)*

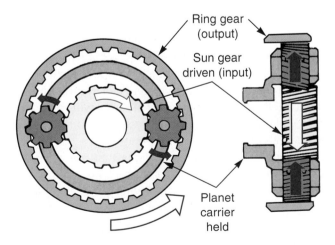

Figure 8-18. *Reverse is produced here by holding the planet carrier and driving the sun gear. The ring gear, as output, turns in the reverse direction of input. (Ford)*

Compound Planetary Gearsets

A modern three- or four-speed automatic transmission uses two or more planetary gearsets that are linked together to make up what is called a **compound planetary gearset.** This arrangement can provide more forward gear ratios than the basic planetary gearset described previously. Common variations of compound planetary gearsets are the *Ravigneaux gear train* and the *Simpson gear train.*

Ravigneaux Gear Train

The **Ravigneaux gear train** was one of the first compound planetary gear designs, used on torque converter automatics in the late 1940s. It was used on single speed (nonshifting) and two-speed automatics until the 1960s. Ravigneaux gear trains were used on some three-speed automatic transmissions until the 1990s. One transmission obtained overdrive directly from a Ravigneaux gearset by using a second input shaft. Ravigneaux gear trains are currently used in combination with other gears in many late model four-, five-, and six-speed overdrive transmissions.

There are several versions of the Ravigneaux gear train, but they all have the same basic design. All versions have two sun gears, one ring gear, and one planet carrier. The sun gears are installed together but can turn independently. The sun gears are usually referred to as the front and rear, or the forward and reverse, sun gears. The planet carrier contains two sets of planet gears. These are usually called *short planet gears* and *long planet gears,* although other names may be used. The short planet gears are meshed with the rear sun gear and the long planet gears. The long planet gears are in mesh with the front sun gear, the short planet gears, and the ring gear. The ring gear is attached to the output shaft. The Ravigneaux gear train shown in **Figure 8-19** is commonly used and is an example of the general design of all Ravigneaux gear trains. The major components of this gear train are identified in **Figure 8-19A.** Refer to **Figures 8-19B** through **8-19E** as you read the following paragraphs.

Drive Range, First Gear

In first gear, **Figure 8-19B,** the rear sun gear is driven and the planet carrier is held stationary. This causes the rear sun gear to turn the short planet gears. The short planet gears turn the long planet gears, which turn the ring gear and output shaft. Note that the power is reversed by the short planet gears and re-reversed by the long planet gears, turning the ring gear in the same direction as the rear sun gear, but at greatly reduced speed. The front sun gear is not held and is turned freely by the long planet gears.

Drive Range, Second Gear

In second gear, the front sun gear is held stationary while the rear sun gear continues to be driven. See **Figure 8-19C.** Since the long planet gears are still driven by the rear sun gear through the short planet gears, the long planet gears walk around the front sun gear, carrying the planet carrier with them. The long planet gears also carry the ring gear with them. The ring gear is moved in the same direction as the rear sun gear but at a reduced speed. Note that both sun gears and all the planet gears are involved in the creation of second gear.

Drive Range, Third Gear

In third gear, the front and rear sun gears are locked together and driven as a unit. Since the planet gears are connected between the two sun gears, they are also locked and cannot rotate. The entire gear assembly, including the ring gear and output shaft, is turned as a unit. This provides a direct (1:1) gear ratio, **Figure 8-19D.**

Reverse

In reverse, the front sun gear is driven and the planet carrier is held stationary.

The front sun gear drives the long planet gears, which turn the ring gear and output shaft. The long planet gears act as reverse idler gears, causing the ring gear to turn in the opposite direction of the front sun gear, **Figure 8-19E.** The rear sun gear and the short planet gears rotate freely with no effect.

Simpson Gear Train

Another compound planetary gearset is the **Simpson gear train, Figure 8-20.** The Simpson gear train consists of two separate ring gears, two separate planet carrier sets, and one common sun gear. The sun gear meshes with both sets of planet gears. The front planet carrier and rear ring gear are splined to the output shaft.

In the gearset shown in **Figure 8-20,** note the components for handling thrust loads. Since the sun and planet gears are helical, their operation places thrust loads on the moving parts of the transmission. To prevent wear of the moving parts, thrust washers (and plates) are used. The washers absorb some of the thrust loads and keep moving parts from touching. The average transmission has 6–10 thrust washers at various locations. Note the location of the thrust washers in the transmission shown in **Figure 8-21.**

To track power flow through a Simpson gear train, refer to **Figure 8-22.** It shows how power moves from input shaft to output shaft in a typical three-speed transmission. **Figure 8-22A** identifies major components. Study each illustration carefully.

Drive Range, First Gear

Power flow through the Simpson gear train in first gear with the selector lever in *Drive* is shown in **Figure 8-22B.** Power flows through the front ring gear, turning the gear at engine speed and direction of rotation. This movement drives the front planet gears, which are in mesh with the front ring gear, turning them in the same direction as the ring gear. The movement of the planet gears drives the planet carrier. The carrier drives the output shaft, as it is

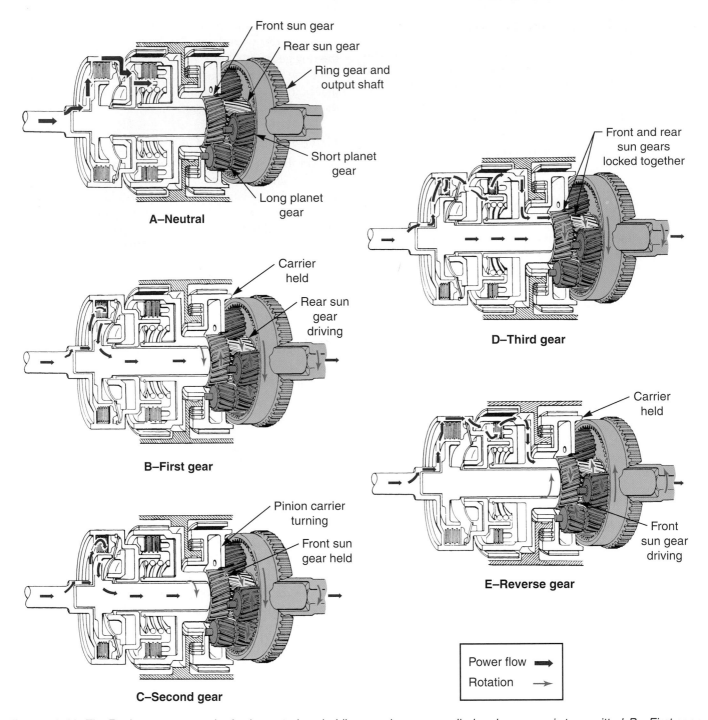

Figures 8-19. *The Ravigneaux gear train. A—In neutral, no holding members are applied and no power is transmitted. B—First gear. C—Second gear. Note that the rear band has been released, while the front band is applied. D—Third gear. Both bands are off and both clutches are applied. E—Reverse. Note that the long planet gears are being used as reverse idler gears.*

splined to the shaft. The planet gears also cause the sun gear to rotate. Note that there is gear reduction through the front gearset.

The sun gear rotates opposite engine rotation, causing the rear planet gears to rotate in the same direction as engine rotation. The rear planet carrier is held stationary. The planet gears cause the rear ring gear to rotate in the same direction as engine rotation. Torque is then trans-ferred through the rear ring gear, which is splined to the output shaft. The arrangement produces an additional gear reduction through the rear gearset, and the total gear reduction through both gearsets is compounded. Since both gearsets are used, a maximum speed reduction and torque multiplication is achieved. Gear ratio in this particular design is 2.45:1, meaning the input shaft must turn 2.45 times to turn the output shaft once.

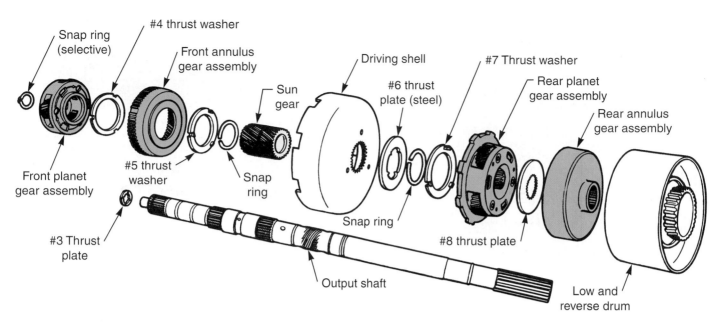

Figure 8-20. *The Simpson gear train, shown here in an exploded view, is a type of compound planetary gearset used on modern transmissions. This design consists of a single long sun gear, two sets of planet gears and carriers, and two ring gears. Note the components for handling thrust loads. (DaimlerChrysler)*

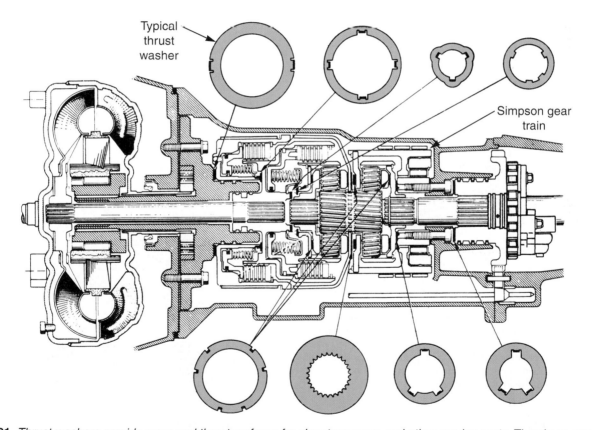

Figure 8-21. *Thrust washers provide wear and thrust surfaces for planetary gears and other moving parts. They keep moving parts from rubbing on each other or on the transmission case. Note the Simpson gear train in this transmission. (Ford)*

Drive Range, Second Gear

Figure 8-22C shows power flow through the Simpson gear train in second gear with the selector lever in *Drive*. Power enters the front ring gear, which drives the front planet gears and planet carrier. The sun gear is held stationary. The front planet carrier is splined to the output shaft and drives the shaft. Notice that power does not go through the rear gearset. This gives a reduction gear ratio that is not as great as in first gear (1.45:1) but it still multiplies power.

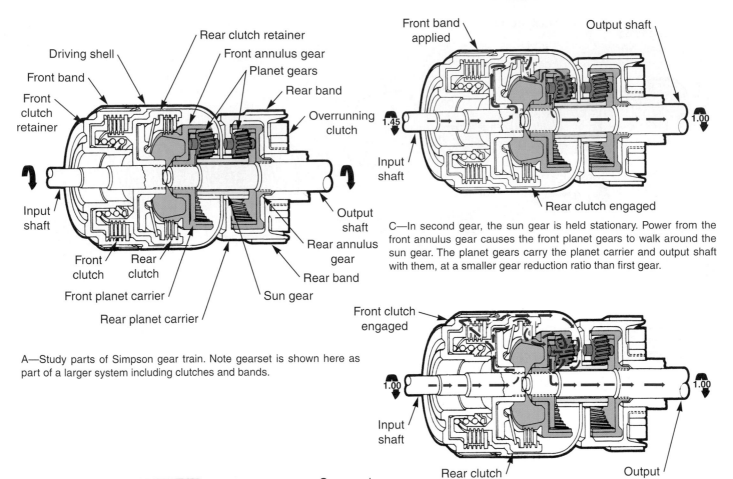

A—Study parts of Simpson gear train. Note gearset is shown here as part of a larger system including clutches and bands.

C—In second gear, the sun gear is held stationary. Power from the front annulus gear causes the front planet gears to walk around the sun gear. The planet gears carry the planet carrier and output shaft with them, at a smaller gear reduction ratio than first gear.

D—In third gear, the sun gear and ring gear are both driven at the same speed, and the gears rotate as a unit. Power is transmitted through the planet gears, to the planet carrier, to the output shaft, splined to the carrier. The rear gearset rotates as a unit with the output shaft, but does not transmit torque.

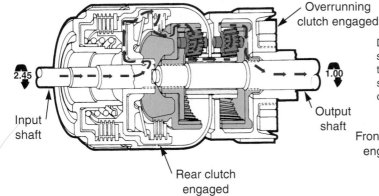

B—In first gear, power flows through front ring gear into planet gears. Planet gears reduce input speed and cause sun gear to rotate opposite of engine direction. Sun gear drives rear planet gears, which further reduce speed. Planet gear rotation is same direction as engine rotation. From planet gears, power enters rear annulus gear, which is splined to output shaft. The double reduction provides a low gear.

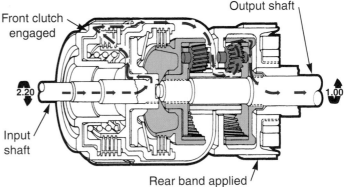

E—In reverse gear, the sun gear is driven by the engine, and the rear planet carrier is held stationary. The rear planet gears are driven by the sun gear and act as reverse idler gears. Power exits the rear annulus gear and output shaft in opposite direction of engine rotation. Front gearset is in idling state and does not transmit power.

Figure 8-22. *Power flow through a typical automatic transmission. (DaimlerChrysler)*

Drive Range, Direct Drive

In third gear, **Figure 8-22D,** power enters the front ring gear and sun gear. Both gears are driven at the same speed, locking the planet gears in place. The gearset rotates as a unit. The front planet gears transmit torque to the output shaft through the front carrier, which is splined to the shaft.

The rear ring gear is splined to the output shaft. It rotates at the same speed as the sun gear. The rear planet gears are locked to the ring gear and sun gear. The rear gearset rotates as a unit with the output shaft but does not transmit torque. Gear ratio through the compound gearset is 1:1.

Reverse Gear

Power flow in reverse gear is shown in **Figure 8-22E.** Power enters the sun gear, turning it in direction of engine rotation. The rear carrier is held stationary, and the rear planet gears act as reverse idler gears. Torque is transmitted directly to the rear ring gear, which is splined to the output shaft. Rotation of the ring gear and output shaft is opposite engine rotation. The front planetary is in an idling state and does not transmit power.

Four-Speed Planetary Gear Operation

The last section explained how a Simpson gear train is used to obtain three forward speeds and reverse. This design was widely used on three-speed automatic transmissions and is still used to some extent. However, three-speed automatic transmissions have been replaced by four-speed transmissions that have a fourth gear overdrive. Some four-speed transmissions were made by adding an extra set of planetary gears to an existing three-speed transmission design. **Figure 8-23** shows a common three-speed automatic transmission that has been converted into a four-speed transmission by adding an extra set of planetary gears to the output shaft area at the rear of the transmission.

In other designs, the overdrive planetary gears are placed at the front of the transmission, directly behind the pump. A typical example of this type of four-speed transmission is shown in **Figure 8-24.**

Most four-speed automatic transmissions are completely new designs that use a compound gearset. The following section illustrates how a late-model four-speed transmission obtains four forward gears and reverse using a compound gearset.

Study **Figure 8-25** as you read the following sections. Note that this gearset resembles the one shown in **Figure 8-22.** Although slightly different in design, it contains the same major parts: two separate sun gears, two sets of planet gears, and two ring gears. The front planet gearset is called the input gearset. The rear gearset is the reaction gearset. The input ring gear and reaction planet carrier are connected to each other. The input planet carrier and reaction ring gear are also connected.

Drive Range, First Gear

In **Figure 8-25A,** the gearset is in first. The input sun gear is driving the input carrier planet gears. The front ring gear is splined to the reaction carrier assembly, which is

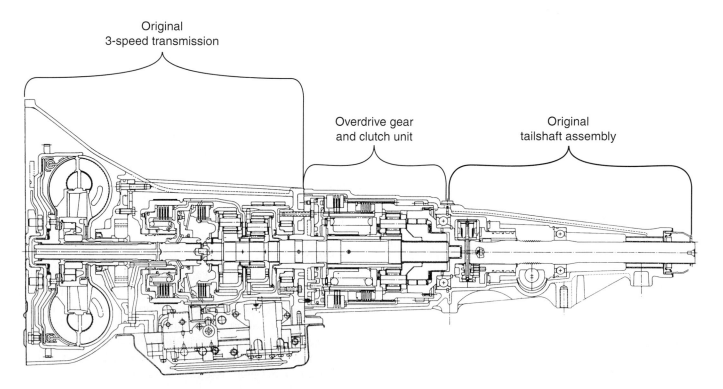

Figure 8-23. *This three-speed transmission was updated to a four-speed transmission by installing a fourth gear, or overdrive, unit at the rear of the original transmission. (DaimlerChrysler)*

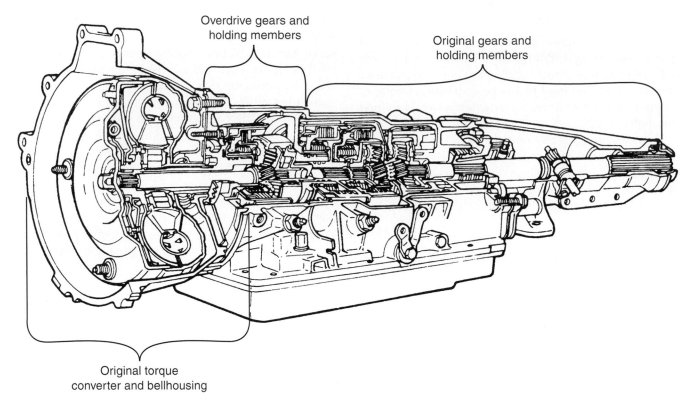

Figure 8-24. *The four-speed transmission in this illustration was created by adding an overdrive unit to the original three-speed transmission. Later versions of this transmission have five speeds. (Ford)*

held stationary. The combination of sun gear input and stationary ring gear causes the input planet carrier to rotate at a reduction of about 3:1. Since the output shaft is splined to the input planet carrier, the output shaft also turns at 3:1. Note that the movement of the rear ring gear turns the rear planet gears, driving the reaction sun gear and the sun gear shell. At this time, however, the reaction sun gear is not connected to any other part of the transmission and has no effect on first gear operation.

Drive Range, Second Gear

The gearset is in second gear in **Figure 8-25B.** The input sun gear continues to drive the input carrier. However, the reaction sun gear is now being held stationary and the reaction carrier assembly can now turn. In second, the power flows through the sun gear to the input carrier, the reaction ring gear, the reaction planet pinions, the input ring gear, the input planet pinions, the input carrier, and the output shaft. The input sun gear drives the input carrier assembly, which is splined to the reaction ring gear. The reaction ring gear drives the reaction carrier, which is splined to the input ring gear. The input ring gear drives the planet pinions and the input carrier. The combination of ratios produced by both planetary gearsets is about 1.6:1. Since the output shaft is splined to the input carrier, it also rotates at a 1.6:1 ratio.

Drive Range, Third Gear

In third gear, **Figure 8-25C,** the input sun gear and the input ring gear are both turning at engine speed, forcing the input carrier to turn with them. This produces direct drive, or a 1:1 ratio. Note that the rear planet gears are inactive in direct drive. The reaction sun gear, the planet carrier, and the ring gear are not held or driven.

Drive Range, Fourth Gear

In fourth gear, **Figure 8-25D,** the reaction sun gear is held stationary. The reaction carrier assembly is driving the reaction ring gear at a reduced ratio. The reaction ring gear is splined to the output shaft, and both are turning at a ratio of about 0.7:1. In fourth gear, the front planet gear assembly is not being used. The input sun gear, planet carrier, and ring gear are turning but have no effect on operation.

Reverse

To obtain reverse gear, **Figure 8-25E,** the reaction carrier is held stationary. The reaction sun gear drives the reaction planet gears. The reaction planet gears reverse the sun gear rotation before driving the reaction ring gear. This causes the reaction ring gear and output shaft to turn in the reverse direction at a ratio of about 2.3:1. In reverse, the front planet gear assembly is not used.

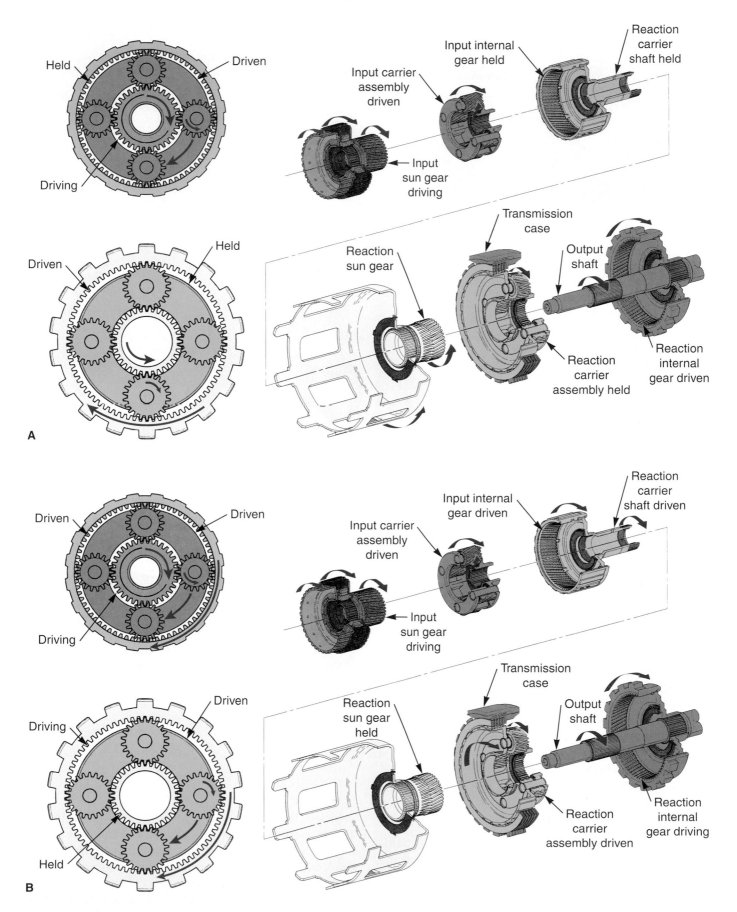

Figure 8-25. *A—A four-speed planetary gearset in first gear. Can you determine what kind of planetary gear train is being used? B—Second gear.*

(Continued)

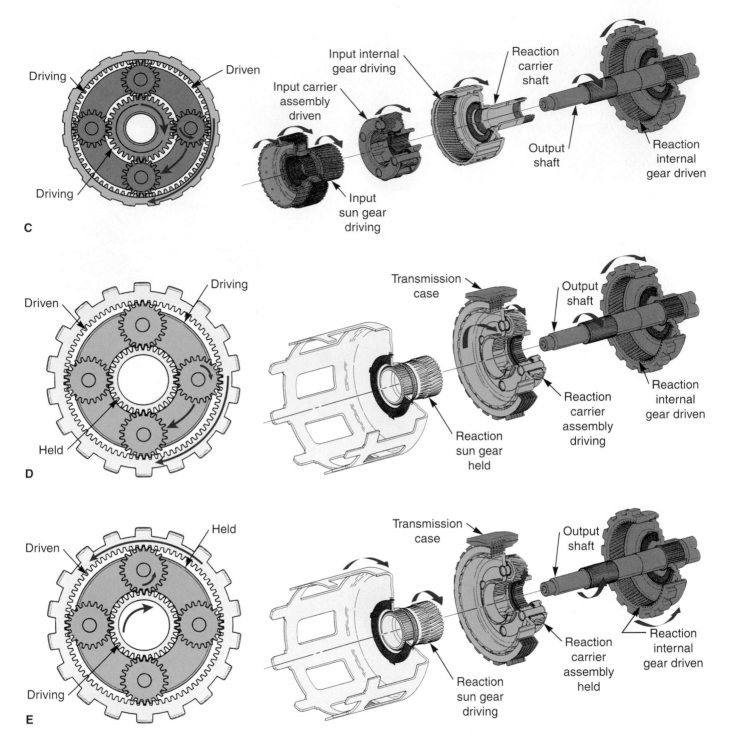

Figure 8-25. *(Continued) C—Third gear, or direct drive. Can you tell from the illustration why the planetary gear is in direct drive? D—Fourth gear, or overdrive. E—Reverse. (General Motors)*

Recent Gear Train Designs

Within the last few years, two new types of gear trains have been used. They are the *Wilson gear train* and the *Lepelletier gear train*.

Wilson Gear Train

Wilson gear trains are used on five-speed, rear-wheel drive transmissions installed in some late-model European vehicles. The Wilson gear train is a combination of three simple gearsets that are connected and located at the rear of the transmission case. Each of the three gearsets has a sun gear, a planet gear, and a ring (annular) gear. The center and rear sun gears are connected to each other, as are the front ring gear, the rear ring gear, and center planet carrier. The front planet carrier and the center ring gear are connected as well. These connections, plus three driving clutch packs, three holding (braking) clutch packs, and one overrunning (one-way) clutch, provide five forward gears and reverse. Gears one through three are reduction gears. Fourth gear is direct drive (1:1), and fifth gear is overdrive. **Figure 8-26** is a cross section of a Wilson gear train.

Lepelletier Gear Train

The *Lepelletier gear train* creates six forward speeds with less complexity and approximately half the parts of typical four- and five-speed automatic transmissions. The basic design of the Lepelletier gear train can be modified for various drive train configurations, including transverse-engine front-wheel drive, longitudinal-engine front-wheel drive, and rear-wheel drive.

The Lepelletier gear train in **Figure 8-27** consists of one double gearset located at the rear of the transmission case and one gearset located at the front of the case. The front gearset is a simple planetary gear with a sun gear, planet gear, and ring gear. The sun gear is connected to the transmission case through an overrunning (one-way) clutch. The front gearset forms two possible paths for engine power into the rear, double gearset.

 Note: Some variations of the Lepelletier gear train use two front gearsets, one for each power path.

The rear gearset assembly consists of two simple planetary gearsets, with an extra sun gear attached to the front planetary sun gear. This gearset is similar to Simpson and Ravigneaux gear trains. The main difference is in how the double gearset is attached to the front gearset. Interconnecting the planetary gears allows the double gearset to work with the front planetary gearset to obtain six forward speeds and reverse. Since the front gearset always produces a speed change, there is no direct (1:1) gear ratio. Gears one through four are reduction gears. Gears five through six are overdrives. The fourth and fifth gears are designed to be close to a 1:1 ratio.

Six holding members control all transmission operations. These holding members consist of three driving clutch packs, two holding (or braking) clutch packs, and an overrunning clutch. All the holding members except the overrunning clutch are located between the front and rear planetary gear assemblies. The overrunning clutch is located at the front sun gear. The driving clutch packs transfer engine power, and the holding clutch packs lock certain parts of the rear planetary gearset to the case.

Applying the drive and brake clutches operates the rear planetary gearset and also determines which of the front planetary gear power paths will be used to deliver power to the rear gearset. Since the overrunning clutch controls the front gearset, only two holding members are needed to obtain each gear.

Planetary Holding Members

Planetary gearsets are controlled by planetary holding members. As previously discussed, *planetary holding members* are the transmission components that apply force to hold or drive other parts. They make the gearsets transfer power. Typical holding members are *multiple-disc clutches, bands and servos,* and *overrunning clutches.* **Figure 8-28** shows a simple arrangement of clutches and bands.

Multiple-Disc Clutches

A *multiple-disc clutch* is used to hold gearset members in place and to lock, or couple, two elements together for power transfer. The coupled elements may be the input shaft and a planetary member or two planetary members.

Figure 8-26. *Although not as popular as the Ravigneaux and Simpson gear trains, the Wilson gear train has been used in various imported transmissions. The Wilson gear train in this illustration is used in a popular transmission installed on various European cars. (ZF)*

Figure 8-27. *The Lepelletier gear train is a new design that provides six forward speeds with less complexity than four- and five-speed designs. (ZF)*

Multiple-disc clutches are often called *clutch packs.* A typical automatic transmission has two or three clutch packs.

The clutch apply piston puts pressure on alternating sets of friction *drive discs* and driven *clutch plates* held within the *clutch drum.* This action locks the clutch elements together and, thereby, locks transmission elements attached to the clutch. Depending on the circumstance, power will be transferred through the clutch, or the clutch will hold a planetary member stationary. A detailed description follows.

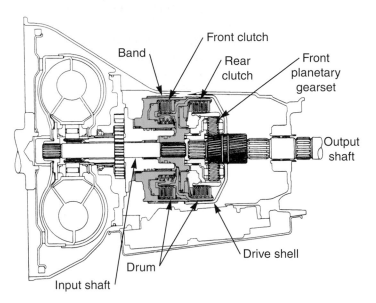

Figure 8-28. *This shows the positions of multiple-disc clutches and band in a common automatic transmission. This arrangement is only one of several used in automatic transmissions. Note the position of the input shaft, the front clutch, the rear clutch, the front planetary gearset, and the drive shell. The rear gearset is not shown. (Ford)*

Multiple-Disc Clutch Construction

The multiple-disc clutch is made up of several parts. Major components include the clutch drum, drive discs, and clutch plates. Also included are the clutch apply piston, clutch pressure plate, and clutch return spring(s). In addition, the multiple-disc clutch has an associated hub that may or may not be part of the assembly itself. There may also be a drive shell associated with the clutch pack. Although designs vary, the parts are similar on all makes of transmissions. **Figure 8-29** is an exploded view of a multiple-disc clutch assembly.

Clutch Drum

The *clutch drum* is the housing for all the other multiple-disc clutch components. The outer surface of the clutch drum may be a holding surface for a band (discussed shortly). The inner surface contains large splines, or channels, that mate with teeth on the clutch plates or drive discs.

The clutch drum is also called a clutch cylinder or a clutch retainer. A modern clutch drum is shown in **Figure 8-30.** This drum is made of stamped sheet metal. Compare it with the cast iron drum in **Figure 8-29.**

Drive Discs

The *drive discs,* or *friction discs,* are metal plates covered with a friction lining. The friction lining is made of a combination of paper and various heat-resistant materials, such as Kevlar and graphite, bound into a mixture with various resins and other adhesives.

The lining is similar to the friction material used on a manual clutch, but it is designed to operate when immersed in transmission fluid. The drive discs have internal teeth, or

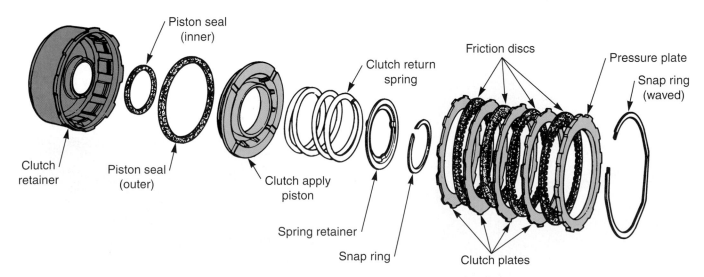

Figure 8-29. *This front clutch is typical of a multiple-disc clutch assembly. Hydraulic pressure moves the piston to apply the clutch. The large return spring holds the clutch apply piston against the back of the clutch drum (retainer) when pressure is released. Some clutches use many small coil springs to hold the apply piston. The clutch plates have external teeth to engage the drum. The drive discs have internal teeth to engage a splined hub. The hub for this clutch is shown in following illustration. (DaimlerChrysler)*

tangs, that engage splines on the clutch hub, as shown in **Figure 8-31.** Drive discs are sometimes called composition plates. Some discs have external splines. These splines engage clutch splines inside the clutch drum.

Clutch Plates

The driven **clutch plates,** or **steel plates,** look like the drive discs, but they are not covered with a friction lining. Also, they have external teeth rather than internal teeth. These teeth lock into channels on the inside of the clutch drum. The drive discs and clutch plates form a sandwich of alternating layers. Their purpose is to lock together the drum and hub, as well as the parts connected to each (input shaft or planetary gearset members), when the clutch is engaged. A steel plate is shown in **Figure 8-32.**

Clutch Apply Piston

The **clutch apply piston** is an aluminum or steel disc that fits into a piston bore in the lower portion of the clutch drum. The piston is moved by hydraulic pressure. Movement of the apply piston clamps the drive discs and clutch plates together. Outer and inner piston seals fit around the respective diameters of the piston to prevent fluid loss and resulting pressure loss when the clutch is applied. **Figure 8-33** shows a typical clutch apply piston.

Seals can be either square-cut types or lip seals, such as those shown in **Figure 8-34.** On many modern transmissions, the seals are bonded to the piston. A bonded seal is either glued to the piston with oil- and temperature-resistant glues, or formed on the piston by heat that melts the seal material onto the piston metal. **Figure 8-35** compares bonded and nonbonded piston seals.

Many clutch apply pistons contain a check ball. The check ball has two purposes: to allow air trapped behind the clutch piston to escape when the clutch is applied and to bleed off extra fluid when the clutch is not applied. If the extra fluid was not allowed to exit, centrifugal force could

Figure 8-32. *Clutch steel plates usually have external teeth. These teeth engage the grooves or channels on a drum.*

Figure 8-30. *The clutch drum shown here is typical of drums used in late-model transmissions and transaxles. It is made of heavy sheet steel, which is stamped into the proper shape. Drums like this one may contain two or more clutch pistons.*

Figure 8-31. *Clutch friction plates usually have internal teeth. The teeth engage the grooves or channels on a hub. The friction material is bonded to a flat steel plate.*

Figure 8-33. *All clutch apply pistons resemble the one shown here. Note the inner and outer grooves, which hold the seals in position. The small air bleed hole near the edge of the apply piston contains the check ball.*

cause it to slightly apply the clutch pack while other clutches were in operation. This drag could damage the clutches. Clutch piston check ball operation is shown in **Figure 8-36.** When the clutch is applied, pressure seats the check ball, preventing fluid loss. When the clutch is released, centrifugal force moves the check ball to one side, allowing fluid to exit instead of moving the clutch apply piston. Since air is much thinner than transmission fluid, any trapped air can exit through the check ball seat, even if the ball is seated.

Clutch Pressure Plate

The **clutch pressure plate** serves as a stop for the set of drive discs and clutch plates when the piston is applied. The pressure plate is held to the drum or case by a large snap ring. The pressure plate is always installed at the end of the clutch pack opposite the apply piston.

Clutch Piston Return Springs

The **clutch piston return springs**, or release springs, push the apply piston away from the drive discs and clutch plates after hydraulic pressure is released from the piston. Return springs may be one of three general designs, as shown in **Figure 8-37.** The multiple spring assembly on the left is a set of several small springs held in a cage. This design is common on late-model units. Some older transmissions used spring assemblies that were open on one

Figure 8–34. Clutch apply piston seals used on some late-model vehicles are bonded to the piston.

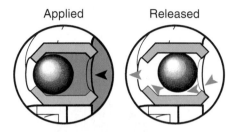

Figure 8-36. The clutch apply piston check ball is shown in the applied and released positions. In the applied position, pressure seats the check ball to prevent fluid loss. In the released position, centrifugal force moves the check ball to one side, allowing fluid to escape. (General Motors)

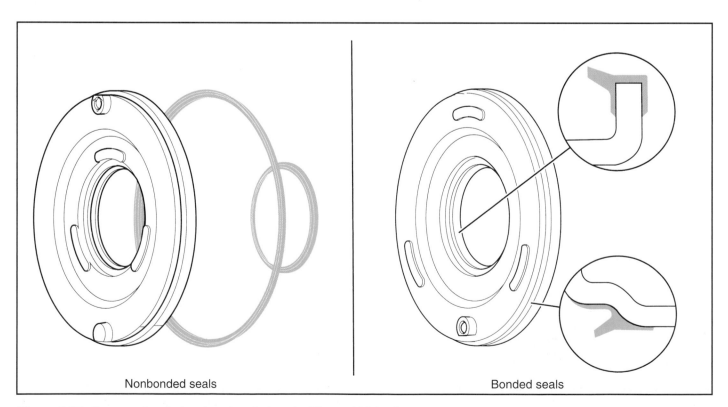

Nonbonded seals   Bonded seals

Figure 8-35. Compare bonded and nonbonded seals. (General Motors)

end. Other transmissions used separate springs that were not attached to each other in an assembly.

The middle return spring is a single large spring. The spring on the right is a flexible dished plate that deforms when the piston moves upward. When the piston pressure is released, the spring returns to its original shape. As a general rule, multiple springs are used on shifting clutches (those that apply to change gears), while single springs and dished plates are used on nonshifting clutches, such as those that are applied to place the transmission or transaxle in drive or reverse.

All piston return springs are held in place by a flat retainer. The retainer is usually secured by a snap ring that connects it to the drum or housing. Some return springs used on clutch pistons installed in the case are held in place by a bolted on retainer. The springs are still under some tension in the released position, and the retainers must be carefully removed.

Clutch Hub

The **clutch hub** may be a part within the clutch pack, or it may be part of another component, such as the planetary ring gear. The clutch hub fits into the inside diameter of the set of drive discs and clutch plates. External splines on the hub engage the internal teeth of the drive discs. The clutch hub may also be splined to the transmission input shaft or to a part of a gearset.

Some clutch packs form a connection between rotating transmission parts and the transmission case. The purpose of these clutch packs is to lock the rotating parts to the case. These clutch packs serve the same purpose as bands, but provide smoother application. **Figure 8-38** shows a typical clutch pack installed in the case.

Drive Shell

A **drive shell,** or **input shell,** is commonly used to transfer power to the planetary sun gear. It is a bell-shaped part made of sheet metal, **Figure 8-39.** Tangs on the shell fit into notches on the drum of the front clutch, or direct clutch. The shell surrounds the neighboring rear clutch (some manufacturers call this the forward clutch) and the front planetary gearset. The shell is splined to the sun gear; and the clutch drum, shell, and sun gear turn together.

In a typical design, the shell drives the sun gear when the front clutch is engaged. When the front clutch is not engaged, the shell and front clutch drum are turned by the sun gear, but they do not transfer power. Also, a band may be tightened around the drum to hold the sun gear stationary. The drive shell is sometimes called the **sun-gear shell.**

Clutch Operation

During clutch engagement, pressurized fluid enters the area behind the clutch apply piston. The fluid forces the piston into the clutch plates and drive discs. The plates and discs lock together, forming a unit. The hub drives the discs, the discs drive the plates, and the plates drive the drum; the clutch rotates as a unit to transfer power through the assembly. A simplified illustration is shown in **Figure 8-40.**

When the hydraulic fluid is exhausted from behind the piston, the return spring pushes the piston away from the clutch plates and drive discs. The plates and discs are no longer locked together, and power is no longer transferred

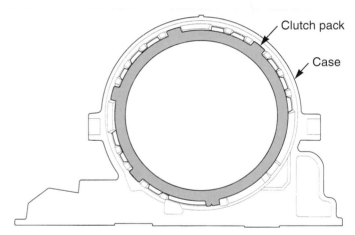

Figure 8-38. *Some steel clutch plates are splined to the transmission case. This type of clutch pack holds a part of the planetary gear to the case, causing it to be stationary. (General Motors)*

Figure 8-37. *The three main types of clutch piston return springs are shown here. The multiple spring type at left is usually used on clutches that apply to change gears. The large spring and dished plate springs are used on clutch packs that are placed in operation only one time during a drive cycle.*

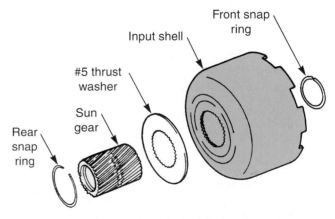

Figure 8-39. *The drive shell, or input shell, is used to transfer power from the front clutch drum to the sun gear. It is used on all transmissions using the Simpson gear train. (Ford)*

through the clutch. The drive discs and clutch plates can turn independently. The hub drives the discs, but no power is transferred to the plates and drum. See **Figure 8-41.**

Bands and Servos

Automatic transmission **bands** are flexible metal friction devices designed to hold planetary gearset members stationary. The bands are wrapped around clutch drums and can be tightened to stop drum rotation. Sometimes bands are wrapped directly around a planetary gearset member. For example, the ring gear sometimes has a smooth outer surface for accepting a band. Modern transmissions have one or two bands. (Some modern transaxles have as many as three.) Bands are applied and released by hydraulic devices called **servos,** which essentially consist of a piston and cylinder.

Note that clutches can be attached to the case to hold a planetary gear stationary, replacing a band used for the same purpose. Such a clutch does not connect two rotating members, but one rotating member and one stationary member. The clutch plates are splined to the case, and when engaged, the clutch prevents any movement of the unit to which it is connected.

Band and Servo Construction

A band is made of flexible-steel strap with a friction lining on its inner surface. The friction material is a combination of paper and heat-resistant materials, such as Kevlar or graphite. The material used is able to operate in transmission fluid and withstand high temperatures.

Bands on modern transmissions attach to the band servo on one end and to the transmission case on the other. The ends of the band are usually called *band ears.* **Figure 8-42** shows a *kickdown band* and linkage. **Figure 8-43** shows a *double-wrap band* and linkage.

The *servo apply piston* is a metal plunger that operates in a cylinder. The cylinder is machined in the transmission case, or it is a separate assembly mounted inside the oil pan. Rubber or Teflon™ seals on the outside diameter of the piston prevent fluid leakage and resulting pressure loss. The apply side of the piston usually contains a large return spring.

The servo *apply piston pin* actuates the *band lever* attached to the band ear. The servo piston and spring assembly are held in the cylinder by a large snap ring or a bolted cover. Sectional and exploded views of a servo are shown in **Figure 8-44.**

The band ear across from the servo is attached to a band *anchor pin,* which holds the ear stationary. The anchor pin may contain a *band adjustment screw.* This screw provides a means of adjusting the *band-to-drum clearance* by moving the band closer to the drum as the friction material wears. See **Figure 8-45.**

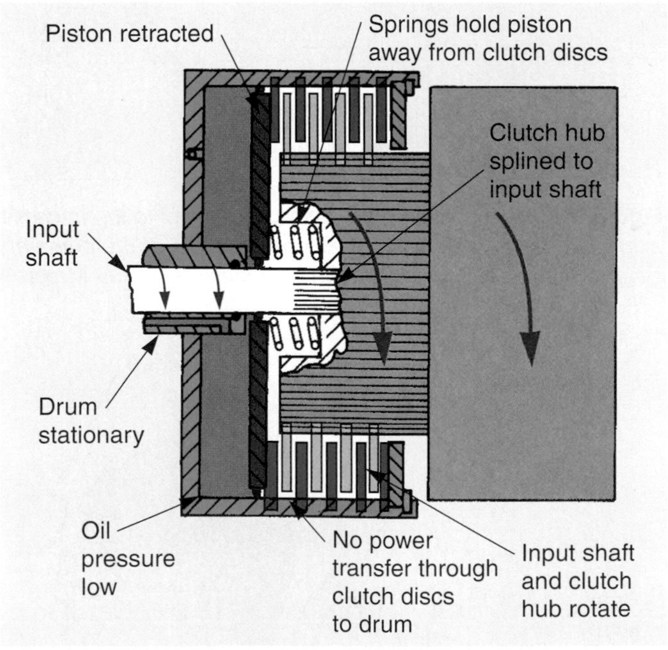

Figure 8-40. *Study a multiple-disc clutch being applied. Pressurized fluid enters the space behind the apply piston, pushing it into the clutch plates and drive discs. The input shaft drives the hub. When the plates and discs are jammed together, power flows from the center hub to the drum.*

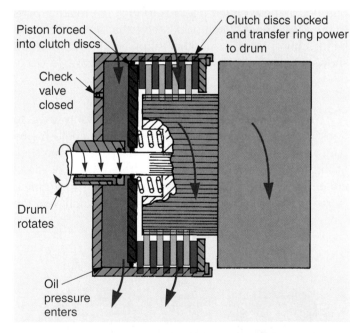

Figure 8-41. *In this illustration, the clutch pack is released. When hydraulic fluid is exhausted, the return spring pushes the piston away from the clutch plates and drive discs. Since the plates and discs are no longer engaged together, power does not flow to the drum. Note that the check ball is open, allowing fluid to escape for quick release of pressure. This is to prevent the plates and discs from dragging when the clutch is released.*

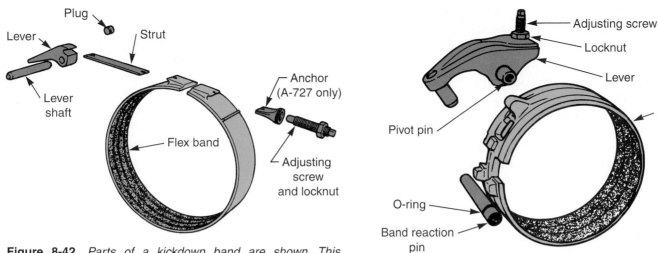

Figure 8-42. *Parts of a kickdown band are shown. This particular band is positioned around the front clutch. (DaimlerChrysler)*

Figure 8-43. *Parts of a double-wrap band are shown. This type of band gives more uniform clamping action and has greater holding power than a conventional band. It is used only where necessary. (DaimlerChrysler)*

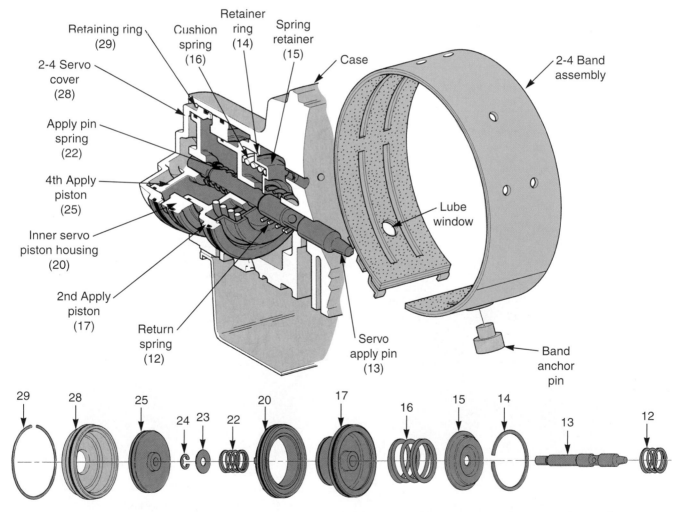

Figure 8-44. *Typical band servo shown in both sectional and exploded views. This servo contains extra pistons that act as an accumulator to cushion application of the band. (General Motors)*

Band and Servo Operation

Transmission fluid is sent into the apply side of the servo cylinder. Hydraulic pressure moves the apply piston toward the band, compressing the return spring and pushing on the apply piston pin and band ear. Since the other side of the band is anchored to the case, the band tightens around the drum (or planetary member). The friction material locks the band to the drum and stops the drum from turning. Stopping the drum keeps one of the planetary components from revolving. When the oil pressure to the servo is relieved, the return spring pushes the

apply piston and piston pin away from the band. The band then releases the drum and/or planetary gearset member. Some servos have an *accumulator piston,* which is applied before the servo apply piston. The accumulator piston provides an initial application that is fast and soft. It is followed immediately by the application of the larger apply piston, which provides greater pressure for firm engagement. In this arrangement, the apply piston works against a spring, which serves to delay its application.

Overrunning Clutches

Overrunning clutches are mechanical holding members. They can be used to hold a planetary gearset member to improve shift quality and timing. These devices can also be used with torque converter stators to improve efficiency, as discussed earlier in this chapter.

The overrunning clutch prevents backward rotation of certain planetary gearset members during shifting. The action of a typical overrunning clutch is illustrated in **Figure 8-46.** The outer race is usually held to the transmission case, and the inner race is connected to a planetary member. When turned in one direction, the clutch rotates freely, allowing the attached member to rotate freely, too. When turned in the other direction, the clutch locks up and acts as a holding member.

There are two main types of overrunning clutches: the roller clutch, shown in **Figure 8-46,** and the sprag clutch. A typical *roller clutch* consists of inner and outer races, rollers, and springs. The clutch performs its holding function when the rollers become jammed between the inner

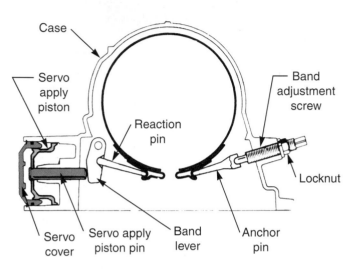

Figure 8-45. *This servo is a simple design with one piston. The band, linkage, and adjustment screw are also shown. (Ford)*

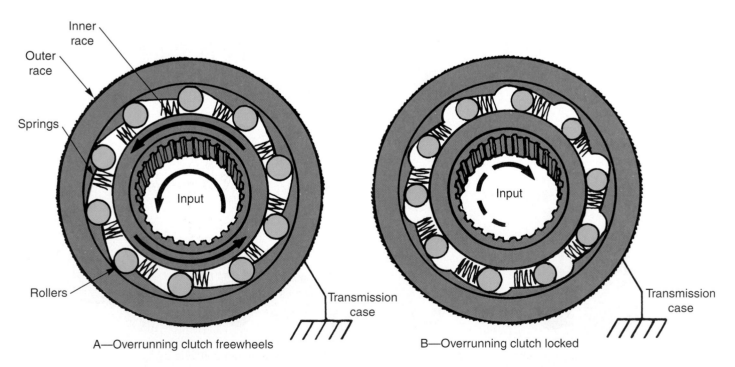

A—Overrunning clutch freewheels

B—Overrunning clutch locked

Figure 8-46. *The overrunning, or one-way, clutch shown here locks or unlocks according to the direction it is turned. A—When the planetary gearset member (input) tries to turn counterclockwise, the inner race and attached planetary member turn freely. B—When the planetary member tries to turn clockwise, rollers become jammed between the inner and outer races. The inner race and attached planetary member are prevented from turning, as the outer race is attached to the transmission case.*

and outer races. The *sprag clutch* uses small metal pieces known as *sprags* to perform the holding action instead of rollers. See **Figure 8-47.**

A less common type of overrunning clutch is the *mechanical diode*. Mechanical diodes use small struts in place of rollers or sprags. The struts are small, flat pieces of metal. The struts engage notches in the races to prevent rotation when power is applied in one direction. When power is applied in the other direction, the struts lie against the inner hub and do not interfere with rotation. This design is stronger than roller- or sprag-type clutches, but has not been widely adopted.

Typically, the one-way clutch is locked only when the transmission is in drive low gear and manual low gear. Note, however, that the clutch can *overrun* (freewheel) in *drive* low, permitting one planetary member to turn faster than another. This happens when the vehicle goes faster than the engine, as when coasting downhill. In contrast, selecting *manual* low will prevent this overrunning by hydraulically applying an extra holding member. As a result, the engine can be used to slow down the vehicle when traveling downhill.

In other words, manual low (and manual second) provides engine braking. ***Engine braking*** means using the engine (via engine resistance) to slow down the vehicle, rather than using the brakes. Engine braking does not happen on a coastdown in drive low because the one-way clutch freewheels; the car wheels are, in effect, disconnected from the engine.

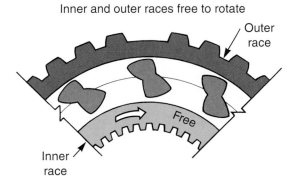

Inner and outer races free to rotate

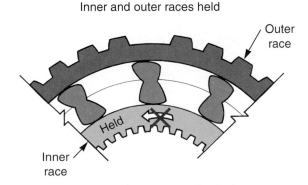

Inner and outer races held

Figure 8-47. *A sprag clutch uses small metal pieces called sprags, which lock the inner and outer races in one direction only. The inner and outer races are circular. (General Motors)*

Converter Housing

The converter housing, **Figure 8-48,** is a stationary component that covers and protects the transmission torque converter. It is often referred to as the **bell housing** because of its bell-like shape. The housing is bolted to the rear of the engine. Many converter housings are combined with the transmission case into a one-piece, or integral, casting. Others are separate and bolt to the front of the transmission case.

Older converter housings were made of cast iron, but most modern housings are cast from aluminum to reduce weight. Most of these have cast ribs at the top for added strength. The front of the housing is machined flat for alignment with the mounting surface on the engine. Holes for the *converter housing attaching bolts* (converter housing-to-engine bolts) and for aligning dowels are often drilled into the front of the housing. Bolts pass through the housing to thread into the rear of the engine block.

Aligning dowels are usually pressed into the rear of the engine block. They fit into the matching holes in the converter housing. The dowels make transmission installation easier and ensure that the centerlines of the transmission input shaft and the engine crankshaft align. The rear of the converter housing is accurately machined and drilled to mate with the front of the transmission case (for nonintegral castings).

Converter housings may be either fully or partially enclosed. The fully enclosed housing completely surrounds the torque converter, **Figure 8-49.** It is usually bolted to the engine through a steel *stiffener plate*. The stiffener plate adds strength to the engine-transmission mounting. The starter motor is often attached to the

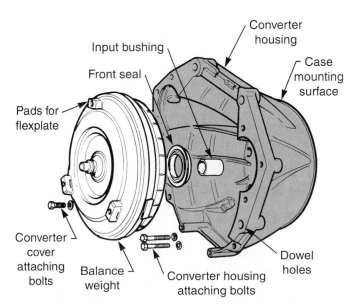

Figure 8-48. *The converter housing encloses the torque converter and mounts on the back of the engine. This converter housing is separate from the transmission case. (Fiat)*

converter housing through the stiffener plate. **Figure 8-50** shows a typical stiffener plate.

An *inspection cover* is usually installed at the bottom half of the converter housing, directly behind the engine oil pan. The inspection cover can be removed to reach the *converter cover attaching bolts* (converter cover-to-flexplate bolts). When a stiffener plate is used, the inspection cover is usually smaller and is attached directly to the stiffener plate with machine screws.

The partially enclosed housing, or *half housing,* is open at the bottom. See **Figure 8-51.** The bottom of the housing is covered by a *dust cover,* which is made of sheet metal. The dust cover protects the converter from road debris and water. It is removed to reach the converter cover attaching bolts. When a half housing is used, the starter motor is usually mounted on the engine block. The starter drive pinion reaches the flexplate ring gear through a hole in the dust cover. For added strength, this kind of converter housing is usually cast as an integral part of the transmission case. On some vehicles, a *spacer plate* is installed between the engine and converter housing. The spacer plate is designed to maintain the correct distance between the torque converter and the transmission oil pump. On these vehicles, if the spacer plate were not used, the moving parts of the converter and pump would bind, causing severe damage to the transmission.

Automatic Transmission Case

The transmission case is the central part of the transmission. Modern transmission cases are made of aluminum. Many older transmissions had cast iron cases.

The front, rear, and bottom of the transmission case are machined flat and drilled to accept the converter housing, extension housing, and oil pan. The inside of the case is machined to provide a mounting for other transmission parts, such as band anchors, servo pistons, accumulators,

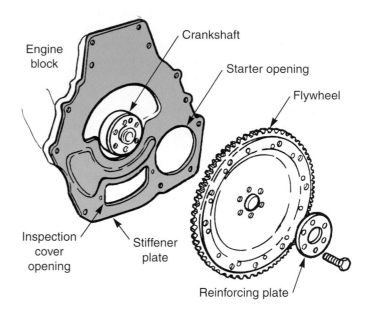

Figure 8-50. *When a fully enclosed converter housing is used, a steel stiffener plate is often installed between the engine block and the housing. The stiffener plate will have openings for the starter and inspection cover. Not all enclosed converter housings have a stiffener plate. Some designs have only a bottom-mounted cover plate to keep road debris from striking the flywheel and converter.*

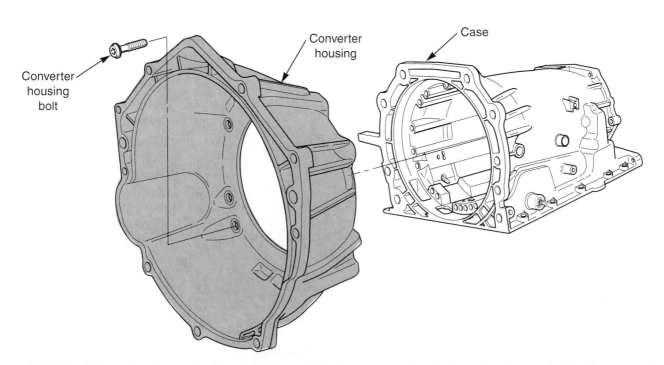

Figure 8-49. *The fully enclosed converter housing surrounds the torque converter. Most enclosed converter housings are bolted to the transmission case, as shown here. Some enclosed housings an integral part of the case. (General Motors)*

clutch plates, center supports, and the oil pump. The bottom of the case is also machined to hold the valve body and, sometimes, servos or accumulators. The rear of the case usually has a pressed-in bearing to support the output shaft. Many transmission accessories, such as shift levers, electrical connectors, the filler tube, and pressure-relief vents, are installed in openings drilled and threaded into the case.

1. Converter assembly complete
2. Retainer and ball assembly, 3rd accumulator
3. Plug, cup (lube hole) (8.05 dia.)
4. Bushing, stator shaft
7. Pump assembly, complete
9. Washer, thrust (housing to pump)
10. Seal, ring (pump to case)
11. Gasket, pump cover to case
12. Case and bushing assembly
13. Pipe, vent
14. Connector, inverted flare (brass)
15. Ring, servo cover retaining
16. Cover, intermediate servo
17. Seal, O-ring (intermediate servo cover)
18. Ring, oil seal (inner)
19. Piston, intermediate servo (outer)
20. Ring, oil seal (outer)

21. Piston, intermediate servo (inner)
22. Ring, oil seal piston (inner)
23. Spring, intermediate servo cushion
24. Ring, snap (apply pin/retainer)
25. Retainer, servo spring
26. Spring, intermediate servo (inner)
27. Pin, intermediate band apply (selective)
28. Ring, oil seal (intermediate band apply pin)
29. Plug, hex head 1/8″ pipe (2)
31. Nameplate
32. Bushing, case rear
33. Seal assembly, rear oil
35. Plug, cup (4th accumulator or servo) (2)
36. Plug, cup (orifice)
37. Ring, oil seal
38. Washer, thrust (governor driven gear/case)
39. Connector, electrical
40. Seal, O-ring (electrical connector)

41. Governor assembly, complete
42. Gasket, governor cover
43. Cover assembly, governor
46. Seal, reverse oil (case to housing)
47. Pin, band anchor
49. Piston, 3-4 accumulator
50. Ring, oil seal (3-4 accumulator piston)
51. Spring, 3-4 accumulator
52. Seal, O-ring (solenoid)
53. Solenoid assembly
55. Ball, .25 diameter
56. Plate, valve body spacer
57. Plate, accumulator
58. Gasket, accumulator hsg. to accum. plate
59. Spring, 1-2 accumulator piston
60. Ring, oil seal (1-2 accumulator piston)

61. Piston, 1-2 accumulator
62. Housing and pin assembly, 1-2 accumulator
64. Screw, conical washer assembly
65. Pan, transmission oil
66. Gasket, pan
67. Seal, O-ring (filter)
68. Filter assembly, transmission oil
69. Lever and bracket assembly, throttle
70. Spring, lifter (T.V. exhaust valve)
71. Link, throttle lever to cable
72. Lifter, T.V. exhaust
73. Switch, pressure 4-3
74. Valve assembly, control
75. Switch assembly, pressure (3rd or 4th)
76. Pin, accumulator piston
82. Clip, filter retainer
83. Pipe, signal oil
84. Retainer, signal oil pipe
85. Magnet, chip collector
86. Gasket, spacer plate to case
87. Gasket, valve body to spacer plate

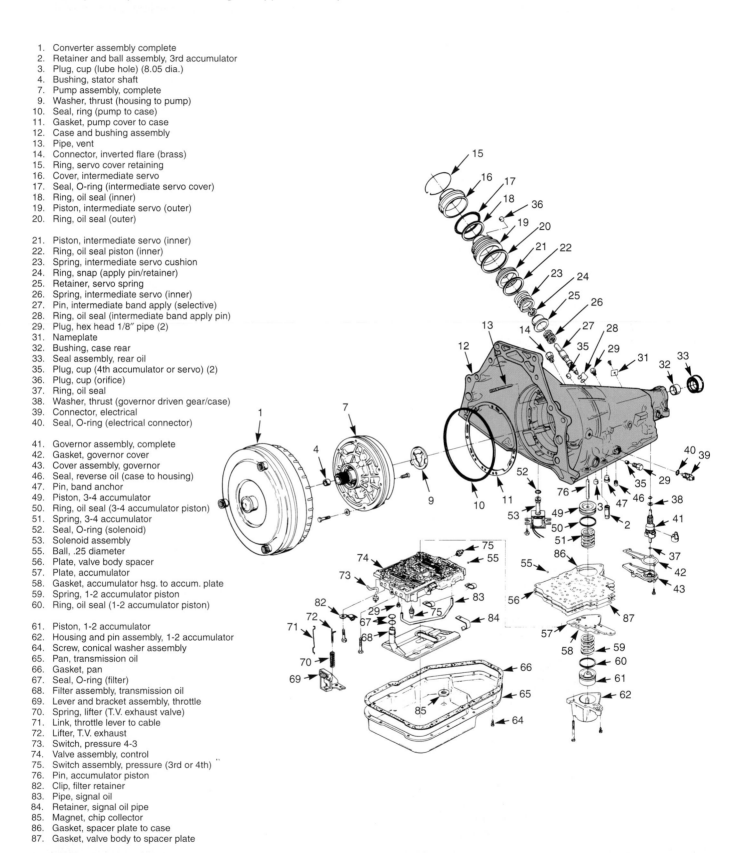

Figure 8-51. *This transmission design features a partially enclosed converter housing. (General Motors)*

Oil passageways are cast or drilled in most automatic transmission cases, **Figure 8-52.** These passageways direct the flow of transmission fluid between other parts of the hydraulic system, such as the oil pump and valve body or the valve body and transmission servo pistons. One passageway directs hot transmission fluid from the torque converter to the transmission oil cooler. Cooled fluid is directed through another passageway to components requiring continuous lubrication and cooling, such as the rear transmission bushings. Some passageways also lead out from the case to allow for testing of the automatic transmission. These passageways are sealed with pipe plugs.

Extension Housing

The extension housing is used to support and protect the tail end of the output shaft. It also contains other internal transmission parts, such as the governor, parking gear and linkage, and speedometer gears. Extension housings are usually made of aluminum. However, cast iron is sometimes used on heavy-duty applications. **Figure 8-53** shows an extension housing.

The rear of the extension housing contains an extension bushing. The bushing is pressed into the housing bore. The bushing supports the driveline slip yoke, which in turn, supports the transmission output shaft. The output shaft turns the slip yoke, which rides on the extension bushing. Some extension housings also use an antifriction bearing to support the output shaft. The extension housing is sealed at the tail end by the transmission rear seal.

Transmission Oil Pan

The oil pan is the transmission fluid reservoir. It is from here that fluid is drawn, to be distributed throughout the transmission. It is also where it ultimately returns. The oil pan is installed at the bottom of the transmission and encloses the valve body, filter, and other internal parts. The pan holds enough extra fluid to allow for changes in fluid level caused by temperature and road-angle variations, as well as leaks. The oil pan is sometimes called the *sump.*

Transmission oil pans are wide and relatively shallow. Their shape allows them to transfer heat from the fluid to air passing under the vehicle. A few pans have cast aluminum fins to aid cooling even more. **Figure 8-54** shows a typical oil pan.

Since it is one of the lowest components on the vehicle, the oil pan is subject to damage from gravel and other road debris. As a result, oil pans are often made from

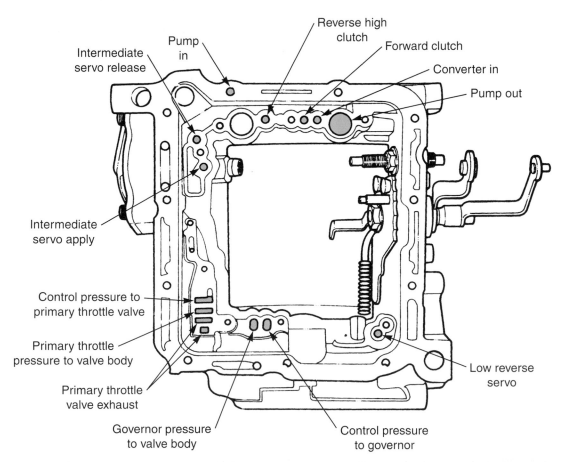

Figure 8-52. *Bottom view of a transmission case (pan and valve body removed). Surfaces are machined flat and the fluid passageways are drilled. (Ford)*

heavy sheet metal, stamped into the proper shape. However, they are sometimes made of cast aluminum.

The oil pan is attached to the transmission case by bolts. A gasket is always used between the oil pan and the case. Some oil pans contain a drain plug. A few pans are protected by a stone guard, **Figure 8-55.** The stone guard also helps dissipate heat.

The transmission oil filter is installed in the oil pan. Other transmission components are often installed on or in the oil pan, as well. The filler tube is sometimes welded or bolted to the side of the pan. A few used on late-model vehicles have an attached electrical connector that connects a sensor to the computer system. Some pans contain a small magnet to pick up stray metal shavings. This keeps the shavings from entering the hydraulic system.

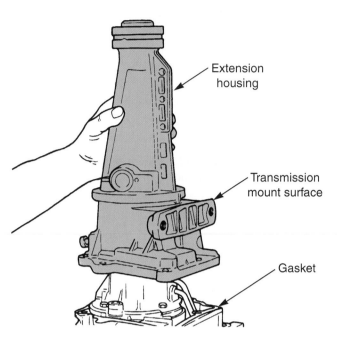

Figure 8-53. *The extension housing supports and protects the output shaft. It is bolted to the rear of the transmission housing. (Ford)*

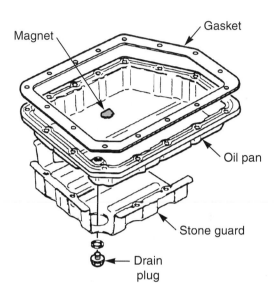

Figure 8-55. *Many oil pans contain a small magnet to pick up metal particles before they can enter the transmission's hydraulic system. The stone guard under the pan helps prevent damage and remove heat. (General Motors)*

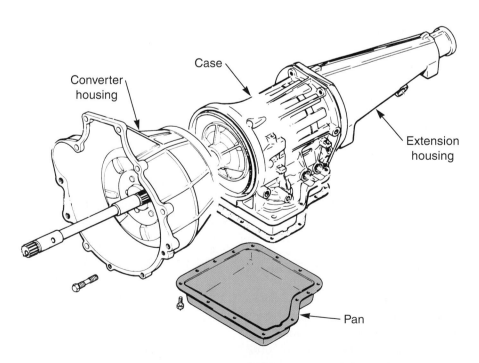

Figure 8-54. *This exploded view shows the relationship of the oil pan to the transmission case. The oil pan seals and protects the bottom of the transmission and serves as a fluid reservoir. (General Motors)*

Summary

The mechanical parts of a rear-wheel drive automatic transmission transfer power from the engine to the drive shaft. The torque converter transfers power and multiplies torque through a fluid coupling. It is installed at the front of the transmission. The rotating impeller moves the fluid outward and in the same direction as engine rotation. The turbine receives the fluid and reverses its direction by action of its curved blades. As a result, power is transferred from the impeller to the turbine through the fluid. The stator reverses the fluid direction so that the fluid assists the impeller in turning.

On many modern transmissions, a torque converter lockup clutch bypasses the fluid coupling to save energy. The converter clutch is splined to the transmission input shaft. The clutch contacts the inside of the converter cover through a friction surface. When the clutch is applied by hydraulic pressure, it locks the cover and input shaft together.

The input and output shafts transfer power into and out of the transmission. They connect to the planetary gearsets. The stator shaft is solidly mounted on the oil pump. It provides a stationary support for the torque converter stator and the overrunning clutch.

Planetary gears are constant-mesh gears arranged in such a way that they resemble the solar system. The gearset consists of a sun gear, planet-pinion gears and carrier, and a ring gear. Different gear ratios are obtained by holding one of the gears stationary and driving another. Power exits through the third member.

Simple planetary gears are combined into compound planetary gearsets for use in automatic transmissions. The Ravigneaux gear train has two sun gears, two sets of planet gears, one planet carrier, and one ring gear. The Simpson gear train has one sun gear, two sets of planet gears, two planet carriers, and two ring gears.

Planetary gearsets are controlled by holding members called multiple-disc clutches, bands, and overrunning clutches. Clutches and bands are operated by hydraulic pressure. Clutch operating pistons are called apply pistons. Bands are operated by pistons called servos. When engaged, multiple-disc clutches, splined to certain planetary gears, hold parts of the gearset stationary or transfer power through parts of the drive train. Bands wrap around drums, holding them stationary when applied. One-way clutches are mechanical devices that either hold or freewheel, depending on what direction power is driven through them.

The converter housing, transmission case, extension housing, and oil pan are stationary parts of the transmission. They provide mounting surfaces for other transmission components. They also seal in transmission lubricant and protect the internal components from dirt. The converter housing and case are often cast into one unit. The case usually contains drilled or cast oil passageways.

Review Questions—Chapter 8

Please do not write in this text. Place your answers on a separate sheet of paper.

1. The three-bladed elements of the torque converter are the _____, the _____, and the _____.

2. Briefly summarize the operation of a torque converter.

3. What does a lockup torque converter do and how does it work?

4. List the five actions that can be accomplished by holding or driving the members of a planetary gearset.

5. A _____ gear train may combine two planetary gearsets with two separate sun gears or a single common sun gear.

6. Components used to hold gearset members in place or to couple two elements together for power transfer are referred to as _____.
   (A) torque converters
   (B) bands
   (C) servos
   (D) multiple-disc clutches

7. Which of the following components is used to operate a band?
   (A) oil pump
   (B) clutch disc
   (C) servo
   (D) clutch apply piston

8. A(n) _____ _____ is a mechanical holding member that can be used to hold a planetary gearset member when rotated in one direction.

9. The converter housing is bolted to the rear of the _____ and may be bolted to the _____.

10. The bottom of the transmission case is machined to accommodate the _____ _____.

ASE-Type Questions—Chapter 8

1. The torque converter is mounted:
   (A) within the engine.
   (B) directly behind the engine.
   (C) directly in front of the engine.
   (D) on the pump drive hub.

2. Planetary gearsets do all the following *except:*
   (A) provide reverse operation.
   (B) resemble the solar system.
   (C) provide different forward gear ratios.
   (D) drive the torque converter.

3. Each of the following is part of a torque converter *except:*
   (A) a one-way clutch.
   (B) a clutch disc.
   (C) a lockup clutch.
   (D) a turbine.

4. Thrust washers are usually placed between all the following torque converter parts *except:*
   (A) converter cover and turbine.
   (B) turbine and stator.
   (C) stator and impeller.
   (D) stator and overrunning clutch.

5. Technician A says the impeller turns whenever the engine is running. Technician B says the turbine turns whenever the engine is running. Who is right?
   (A) A only.
   (B) B only.
   (C) Both A and B.
   (D) Neither A nor B.

6. As a result of turbine vane curvature, fluid coming from the impeller will do which of these before it returns to the impeller?
   (A) Overheat.
   (B) Increase flow.
   (C) Decrease flow.
   (D) Reverse direction.

7. In a torque converter, fluid leaving the turbine must pass through the _____ before reaching the impeller.
   (A) stator
   (B) flexplate
   (C) drive shell
   (D) accumulator

8. Torque multiplication refers to a component's ability to:
   (A) decrease output torque below its input torque.
   (B) increase output torque above its input torque.
   (C) increase input torque above its output torque.
   (D) decrease input torque below its output torque.

9. Which of the following is used to mechanically lock the turbine to the impeller and converter cover when the transmission is in high gear?
   (A) Multiple-disc clutch.
   (B) Lockup clutch.
   (C) Overrunning clutch.
   (D) Apply springs.

10. An automatic transmission can achieve overdrive through:
    (A) a clutch pack.
    (B) the converter cover.
    (C) the coupling phase.
    (D) a planetary gearset.

11. Two or more planetary gearsets can be linked together to form a:
    (A) multiple planetary gearset.
    (B) compound planetary gearset.
    (C) combination planetary gearset.
    (D) main planetary gearset.

12. Technician A says that a Ravigneaux gear train is common in transmissions using overdrive. Technician B says that a Ravigneaux gear train contains two sun gears, two planet carriers, and one set of planet gears. Who is right?
    (A) A only.
    (B) B only.
    (C) Both A and B.
    (D) Neither A nor B.

13. Each of the following is a planetary holding member *except:*
    (A) pawl clutch.
    (B) band.
    (C) overrunning clutch.
    (D) multiple-disc clutch.

14. Multiple-disc clutches are also referred to as:
    (A) combination units.
    (B) disc packs.
    (C) clutch packs.
    (D) overrunning clutches.

15. Technician A says bands are applied and released by servos. Technician B says servos are metal plungers that operate in cylinders. Who is right?
    (A) A only.
    (B) B only.
    (C) Both A and B.
    (D) Neither A nor B.

Hydraulic valves help direct fluid to various parts of the transmission or transaxle. The valves shown above fit into the valve body and are secured by a retaining clip.

Chapter 9

Automatic Transmission Control Components

After studying this chapter, you will be able to:
- ❏ Explain how transmission oil pumps develop hydraulic pressure.
- ❏ Describe the purpose of the transmission oil filter.
- ❏ State the two locations of transmission oil coolers.
- ❏ Explain the purpose and design of valve bodies and control valves.
- ❏ Describe the operation of the manual valve and related linkage.
- ❏ Explain the two ways of operating throttle valves.
- ❏ Describe the types of transmission governor valves.
- ❏ Describe the operation of accumulators.
- ❏ Explain the operation of the park lock mechanism.
- ❏ Describe the purpose and operation of electrical switches.
- ❏ Describe the purpose and operation of electrical solenoids.

Technical Terms

Transmission oil pump
Front pump
Oil filter
Oil strainer
Main oil cooler
Auxiliary oil coolers
Valve body
Auxiliary valve bodies
Oil transfer tubes
Hydraulic valves
Pressure regulator valve
Line pressure

Manual valve
Shift valves
Throttle valve
Throttle pressure
Throttle valve (TV) linkage
Vacuum modulator
Governor valve
Governor pressure
Detent valve
Passing gear
Check ball
Accumulator

Shift linkage
Selector lever
Neutral start switch
Parking gear
Parking pawl
Shift-brake interlock
Switches
Solenoids
Detent solenoid
Lockup solenoid

Introduction

In the last chapter, you studied the mechanical parts of the automatic transmission. These parts are used to transfer power from the engine to the drive shaft in the most efficient manner. In this chapter, you will study the hydraulic parts that make the mechanical parts operate. The hydraulic parts control the mechanical parts to get maximum power transfer through the transmission with the least use of fuel. If the hydraulic parts do not operate properly, the maximum benefit of the mechanical parts is lost. Knowledge of hydraulic control systems is vital to transmission diagnosis and service.

This chapter describes the transmission devices that create, direct, and control hydraulic pressure. Since all hydraulically controlled transmissions use the same basic devices, studying this chapter will give you a fundamental understanding of how all automatic transmissions operate. This chapter covers driver-controlled valves and related linkage. It also briefly describes the electrical components used by all new and many older transmissions. Some sections review information that was covered in earlier chapters.

Studying this chapter will give you the grounding needed to understand the transaxle controls covered in the next chapter, as well as electronic controls systems detailed in Chapter 11. Studying this chapter will also enable you to move on to the diagnosis and service chapters later in this book.

Construction Overview

Automatic transmission control systems are made of many separate components. The major automatic transmission control parts include:

❑ Transmission oil pump—produces the fluid pressure that operates other hydraulic components and lubricates the moving parts of the transmission. It is driven by lugs on the torque converter.

❑ Transmission oil filter—removes particles from the automatic transmission fluid before the fluid is circulated by the oil pump.

❑ Transmission oil cooler—needed to dissipate heat from the transmission fluid, which is generated by the transmission and picked up from the engine. The transmission oil cooler is often mounted in the engine cooling-system radiator. Separate external oil coolers, or auxiliary oil coolers, may also be installed on the vehicle.

❑ Valve body—a complex casting that serves as the control center for the transmission's hydraulic system. It contains many internal passageways and components, including hydraulic valves.

❑ Hydraulic valves—used in the valve body to control transmission fluid pressure, as well as the direction and rate of fluid flow. Specifically, they control the application of clutches and bands. In general,

hydraulic valves help control the entire operation of the automatic transmission.

❑ Shift linkage—operates the manual valve, which is one of the hydraulic valves. This forms a direct mechanical connection between the driver and the transmission. Shift selector levers can be mounted on the floor or on the steering column. The shift linkage also operates the parking gear mechanism.

❑ Electrical components—used in the transmission to send signals to other vehicle components or to partially control the transmission's hydraulic system.

Transmission Oil Pump

The *transmission oil pump,* sometimes called the *front pump,* has several basic functions. It generates *system pressure* for the entire hydraulic control system and keeps the moving parts of the transmission lubricated. It also keeps the torque converter filled with transmission fluid for proper operation. (Most of the fluid in an automatic transmission is inside the converter.) Finally, it circulates transmission fluid to and from the oil cooler for heat transfer.

The transmission oil pump is a positive displacement pump. It is driven by the engine, through lugs on the back of the torque converter. The relative position of the oil pump in the transmission is shown in **Figure 9-1.**

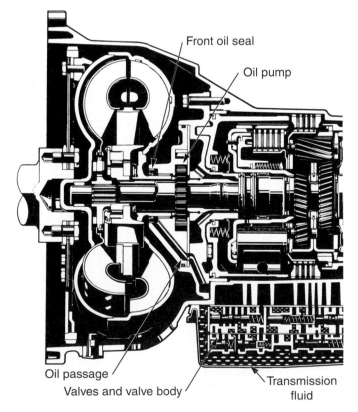

Figure 9-1. *The oil pump is often located at the front of the transmission case. It is driven by lugs on the rear of the torque converter. Bolts hold the pump to the case. Some older transmissions also have a rear oil pump, which is driven by the output shaft. (Mercedes-Benz)*

There are several types of transmission oil pumps, all of which consist of elements—gears, rotors, or vanes—that rotate inside a housing, **Figure 9-2.** (The types of hydraulic pumps used in automatic transmissions were discussed in Chapter 6.) As the elements rotate, they come together on one side of the pump housing and move apart on the other.

As they move apart, a low-pressure region is created on the inlet side of the pump. As a result, transmission fluid is drawn from the *sump,* or oil pan, through the transmission filter and pump inlet, and into the pumping chamber. The fluid is then carried around the pumping chamber to the pump outlet. The fluid is discharged into the hydraulic

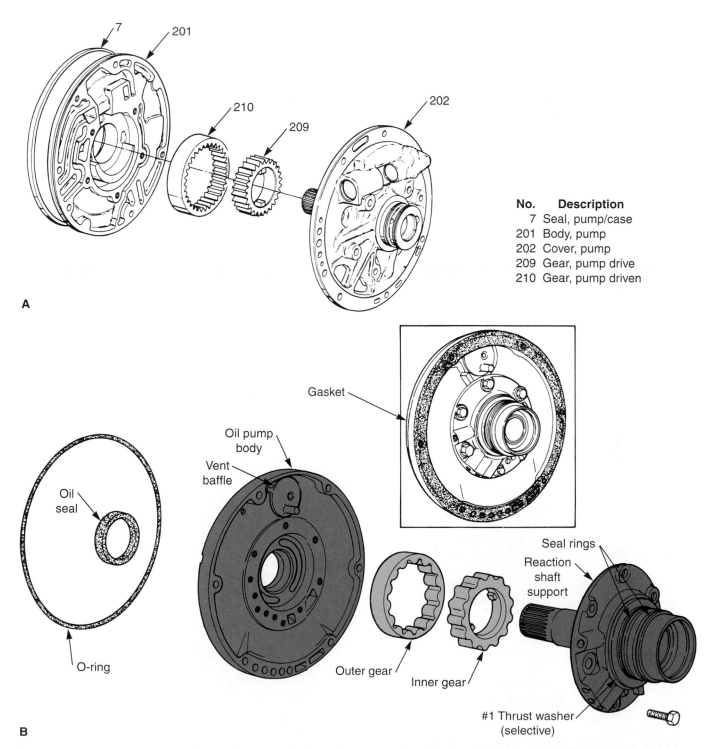

No.	Description
7	Seal, pump/case
201	Body, pump
202	Cover, pump
209	Gear, pump drive
210	Gear, pump driven

Figure 9-2. *Three types of transmission oil pumps are shown here. Note the stator shaft, which extends forward from oil pump. It supports the torque converter stator. The shaft extending back from oil pump is the support and fluid feed for some transmission clutches. A—Gear-type oil pump. The pump housing also contains built-in converter clutch and pressure regulator valves. The pressure regulator valve lowers pump pressure by decreasing pump capacity. B—Rotor-type oil pump. The pump cover contains the stator and clutch support shafts. (continued)*

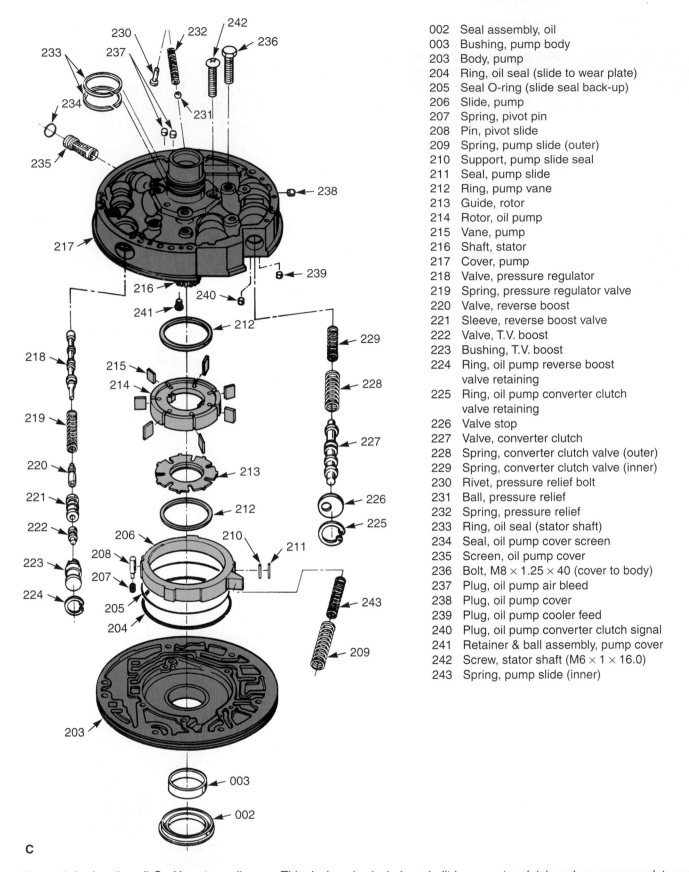

002	Seal assembly, oil
003	Bushing, pump body
203	Body, pump
204	Ring, oil seal (slide to wear plate)
205	Seal O-ring (slide seal back-up)
206	Slide, pump
207	Spring, pivot pin
208	Pin, pivot slide
209	Spring, pump slide (outer)
210	Support, pump slide seal
211	Seal, pump slide
212	Ring, pump vane
213	Guide, rotor
214	Rotor, oil pump
215	Vane, pump
216	Shaft, stator
217	Cover, pump
218	Valve, pressure regulator
219	Spring, pressure regulator valve
220	Valve, reverse boost
221	Sleeve, reverse boost valve
222	Valve, T.V. boost
223	Bushing, T.V. boost
224	Ring, oil pump reverse boost valve retaining
225	Ring, oil pump converter clutch valve retaining
226	Valve stop
227	Valve, converter clutch
228	Spring, converter clutch valve (outer)
229	Spring, converter clutch valve (inner)
230	Rivet, pressure relief bolt
231	Ball, pressure relief
232	Spring, pressure relief
233	Ring, oil seal (stator shaft)
234	Seal, oil pump cover screen
235	Screen, oil pump cover
236	Bolt, M8 × 1.25 × 40 (cover to body)
237	Plug, oil pump air bleed
238	Plug, oil pump cover
239	Plug, oil pump cooler feed
240	Plug, oil pump converter clutch signal
241	Retainer & ball assembly, pump cover
242	Screw, stator shaft (M6 × 1 × 16.0)
243	Spring, pump slide (inner)

C

Figure 9-2. *(continued) C—Vane-type oil pump. This design also includes a built-in converter clutch and pressure regulator valves. (General Motors, DaimlerChrysler)*

system through the outlet as the internal elements come back together.

Note that modern transmissions use only one pump, which is driven from the engine. However, many older transmissions used two oil pumps. The front pump was driven by the engine through the converter. The *rear pump* was driven by the transmission output shaft. The rear pump turned whenever the vehicle was moving. It developed pressure for push starting and for providing lubrication to the transmission if the vehicle was being towed.

Pushing a vehicle with a rear pump caused the pump to develop enough pressure to apply the holding members and turn the engine in the same manner as pushing a manual shift vehicle.

Push starting became less of a factor in transmission design when alternators were introduced in the early 1960s. Alternators require some current input to begin charging. Therefore, alternator-equipped vehicles cannot be push started when the battery is completely dead. Safety and vehicle damage factors were also considered, and after 1964, new transmissions were designed without rear pumps. In the late 1960s, older transmissions were redesigned to eliminate the rear pump.

Transmission Oil Filter

The transmission *oil filter,* or *oil strainer,* is used to remove particles from the automatic transmission fluid. Fluid being drawn into the oil pump must first pass through the filter. This keeps fluid contaminants from entering the pump or the other hydraulic components. Most filters are made of a disposable filter paper or felt. Some filters are fine mesh screens, which can be cleaned and reused.

The filter is located inside the oil pan. It must be positioned where it will always be submerged in the automatic transmission fluid. If the filter were not submerged with fluid, air would be drawn into the hydraulic system, affecting operation. Usually, the filter is attached to the bottom of the valve body. It may be held in place by screws, clips, or bolts. Some filters have a built-in pickup tube that goes directly up to the oil pump. Filters used with deep transmission pans may have an extended pickup tube.

Transmission Oil Cooler

The action of the transmission—the torque converter, in particular—heats the transmission fluid. In addition, the transmission's close coupling with the engine serves to add heat to the fluid. This heat must be removed so the fluid and the transmission do not become overheated. Fluid subjected to overheating will quickly lose its lubricating ability. Further, it will break down into sludge and varnish, which will plug passageways and destroy the transmission.

In addition, excessive heat destroys the nonmetallic materials used in the clutch and band friction linings and in the seals. As a result, transmission fluid that is

overheated can damage these materials or fail to cool the friction materials properly. Then, not only are the materials destroyed, but the transmission fluid ends up with particulates from the damaged materials. These particulates can further plug transmission passageways and destroy the transmission.

In general, fluid temperature should be kept below 275°F (135°C). Running at elevated temperatures greatly reduces the life of the transmission fluid. If the transmission is being worked hard, such as in trailer towing, temperatures could exceed 300°F (150°C). Under such conditions, the fluid and filter *must* be changed frequently.

Main Oil Cooler

Fluid in older transmissions was sometimes air-cooled. The converter housing was open, so air could pass through and absorb heat directly from fins on the torque converter. This design has been abandoned.

Fluid in modern automatic transmissions is cooled in the oil cooler. The **main oil cooler** consists of a heat exchanger built into a side or bottom tank in the engine cooling-system radiator. Cooler lines connect the transmission to the oil cooler, as shown in **Figure 9-3.** Some vehicles have a direct air cooler installed in front of the air conditioner condenser and do not use a radiator cooler.

The oil cooler is immersed in cooled engine coolant. In operation, hot transmission fluid is pumped from the transmission to the oil cooler, usually directly from the torque converter. As the fluid passes through the cooler, its heat is transferred to the engine coolant. The cooled transmission fluid then returns to the transmission. The cooled fluid may be routed through the transmission lubrication system before returning to the oil pan.

Auxiliary Oil Cooler

When a vehicle is used to tow a trailer or subjected to other extreme operating conditions, **auxiliary oil coolers,** or external oil coolers, may be installed. Auxiliary coolers consist of direct, air-cooled heat exchangers that connect to the existing fluid cooling system. They may be installed to completely replace the main oil cooler in the radiator, or they may be helper units that further cool fluid that has passed through the radiator oil cooler. Auxiliary coolers are usually mounted ahead of the radiator.

Valve Body

The **valve body** is a complex casting made of cast iron or aluminum. It contains many internal passageways and components, including hydraulic valves. The valve body serves as the control center for the transmission's hydraulic system. It is precisely machined, and valve body components are manufactured to exact tolerances. Some tolerances are as small as 1 ten-thousandth of an inch (0.0001"), or 25 ten-thousandths of a millimeter (0.0025 mm).

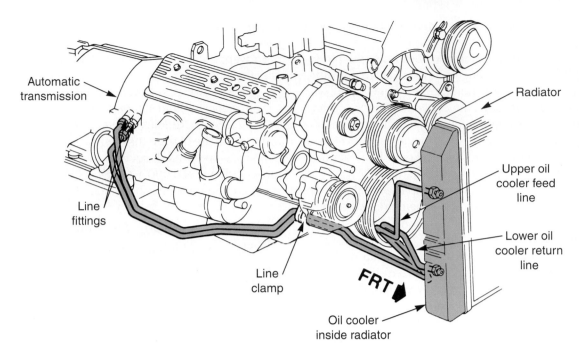

Figure 9-3. *Steel tubing connects the oil cooler in the radiator to the transmission hydraulic system. Oil is pumped from the torque converter outlet to the cooler. Cooled oil is directed through the transmission to the components requiring continuous lubrication and cooling, such as the rear transmission bushings. Excess runs to oil pan, where it is drawn in by the oil pump and distributed elsewhere through the system. (General Motors)*

Oil passageways, which are sometimes called worm tracks, are cast into the valve body during the casting process. After casting, holes are drilled for valves, and one side of the valve body is machined flat. This flat area is intended to closely fit a similarly machined area on the transmission case and, in some cases, other transmission parts. The close fit prevents oil leaks that could lead to faulty transmission operation.

A *one-piece valve body* is shown in **Figure 9-4.** This component would bolt directly to the bottom of the transmission case. Some valve bodies are made in two sections. Such *split valve bodies* must be assembled before being installed in the transmission. **Auxiliary valve bodies** are often bolted to the main valve body or to the transmission case. Auxiliary valve bodies are separated from the main valve body because of design considerations, such as clearance restrictions.

Valve bodies are used with *spacer plates,* which are sometimes referred to as *separator plates* or *transfer plates.* Refer to **Figure 9-5.** These steel plates seal off passageways in the valve body and in the transmission case. Spacer plates are also used between the halves of split valve bodies. Holes are drilled in the spacer plate to connect certain passageways of mating components.

The spacer plate holes are sized to give a desired flow rate. In other words, the size of the hole in the spacer plate dictates how quickly oil can flow through a passageway. This is to obtain a desired type of shift. Large holes give quick, hard shifts. Small holes restrict oil flow to lengthen shift time and cushion shifts. Check balls are often used to block spacer plate holes, allowing oil to flow in only one direction. Spacer plates are usually installed with gaskets on both sides.

Oil Transfer Tubes

Many valve bodies are equipped with *oil transfer tubes.* Oil transfer tubes are used to connect valve body pressure chambers when it would be impractical to cast the needed passages in the valve body or the case. Smaller oil transfer tubes are usually made of steel tubing, while larger transfer tubes are made of aluminum. **Figure 9-6** shows how oil tubes are used to connect parts of the valve body of a typical transmission.

Hydraulic Valves

Hydraulic valves, which were discussed in Chapter 6, are moving parts of the automatic transmission that act as pressure and flow controls, directing fluid to various parts of the transmission. Some hydraulic valves are actuated manually; others are actuated automatically. For instance, some hydraulic valves are operated by pressure inputs from other hydraulic valves, which themselves are operated by outside factors, such as vehicle speed. Some valves are located in the valve body, **Figure 9-7,** while others are positioned elsewhere in the transmission.

Valves located in the valve body are installed in drilled holes called *valve body bores.* The clearance between the valves and bores is very small, usually less than 2 thousandths of an inch (0.002"), or 50 thousandths of a millimeter (0.050 mm). Valves are held in the valve body bores by snap rings, retaining pins, or retainer plates. Retainer plates are attached to the end of the valve body bores by machine screws or small bolts. In addition,

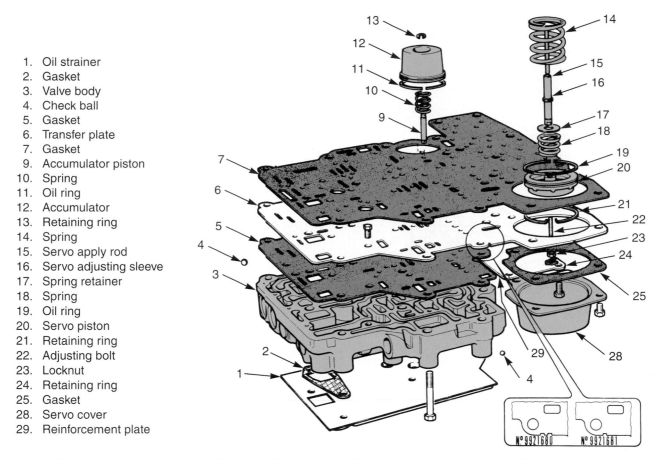

1. Oil strainer
2. Gasket
3. Valve body
4. Check ball
5. Gasket
6. Transfer plate
7. Gasket
9. Accumulator piston
10. Spring
11. Oil ring
12. Accumulator
13. Retaining ring
14. Spring
15. Servo apply rod
16. Servo adjusting sleeve
17. Spring retainer
18. Spring
19. Oil ring
20. Servo piston
21. Retaining ring
22. Adjusting bolt
23. Locknut
24. Retaining ring
25. Gasket
28. Servo cover
29. Reinforcement plate

Figure 9-4. *Valve bodies contain many of the hydraulic valves found in an automatic transmission. The valve body shown here is a one-piece design. Note the position of the spacer plate and gaskets in the assembly. Also note the accumulator and band servo incorporated into this valve body and the position of oil filter (strainer). (Fiat)*

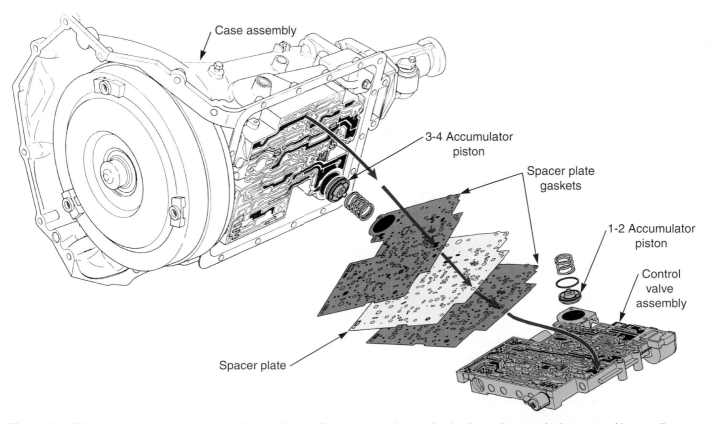

Figure 9-5. *This illustration shows the relative positions of the spacer plate, valve body, and transmission case. Almost all spacer plates have a gasket on each side. Many transmissions have two-piece valve bodies, with a spacer plate between the two halves. (General Motors)*

springs made of heat-resistant steel are used in the valve body to hold valves in position when they are not in operation.

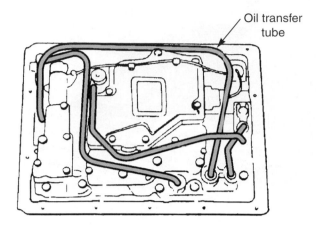

Figure 9-6. *The tubes in this illustration connect pressure passages in the valve body. Using tubes is an alternative to casting complex extra passages in the valve body or case. (General Motors)*

Springs have other functions as related to hydraulic valves. In some cases, the springs are used to provide constant resistance that hydraulic pressures work against, such as in a pressure regulator valve. In other cases, the springs are used for an initial loading of a valve, to which a varying hydraulic *control pressure* is added. Such valves produce a varying output pressure. In either case, valve springs are carefully sized to achieve a specified pressure. Specific springs are used with each valve, and they cannot be interchanged. For this reason, care should be taken so springs are not mixed up when working on the valve body.

Types of Valves

There are many types of hydraulic valves used in the transmission. These include *pressure regulator, manual, shift, throttle, governor, detent,* and *check valves.* A working knowledge of these valves will help you understand the internal operation of different transmissions. A brief discussion on each type is presented in the following sections. (The different valves are explained here on an individual basis. The operation of these valves as integrated into the complete hydraulic system is presented in Chapter 12.)

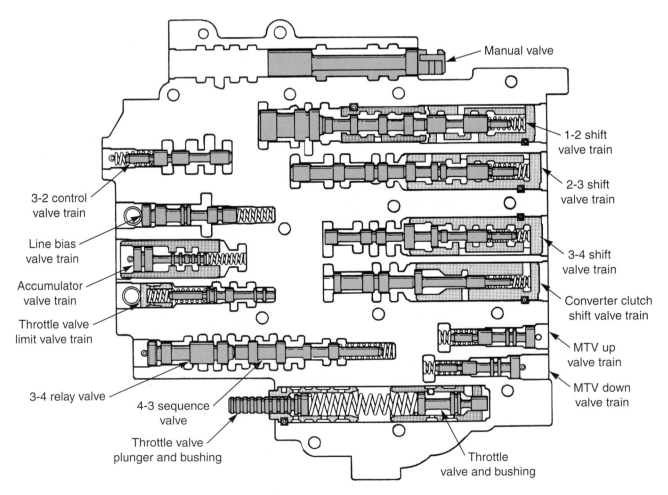

Figure 9-7. *Note the hydraulic valves in this sectional view of a typical valve body. Notice the difference in valve size and configuration and in the placement of the springs. (General Motors)*

Pressure Regulator Valve

The **pressure regulator valve** controls the pressure of the overall transmission hydraulic system, or the system's **line pressure.** This valve is sometimes located on the pump or in an auxiliary valve body, but it is most often found in the main valve body. A few transmissions have a separate regulator valve to control pressure in the torque converter. A schematic of a pressure regulator valve is shown in **Figure 9-8.** The pressure regulator is often installed on the pump body, **Figure 9-9.**

In operation, pressurized fluid at the oil pump discharge is directed to the pressure regulator valve before going to the rest of the transmission hydraulic system. A heavy spring holds the valve closed against oil pressure from the pump. At idle, the regulator valve is usually closed.

When the engine begins running faster, it turns the oil pump at a faster rate. Fluid pressure rises. When hydraulic pressure exceeds some predetermined value, it forces the valve to move against spring pressure. This opens a passageway back to the oil pan or pump intake. Hydraulic fluid is diverted back through this passageway, and pressure falls back within the limit. This action keeps line pressure constant.

Note that if the pressure were allowed to exceed the set maximum, it would cause hydraulic system malfunctions and damage to parts. Also note that the pressure limit is established by the spring; thus, the stiffer the spring, the higher the line pressure.

On many vehicles with slipper-type pumps, the pressure regulator controls line pressure by increasing or decreasing pump output. Refer to **Figures 9-10** and **9-11.** Note the pressure decrease line from the pressure regulator to the left side of the pump. Also note that the pump assembly can pivot on a pin located at the center top of the pump assembly. A spring called the priming spring pushes the pump to the left. This is the full output position.

If high pressures are needed, or if the pump is turning slowly, the pressure regulator remains closed and no oil enters the pressure decrease line. See **Figure 9-11.** The priming spring holds the pump in the full output position. In the full output position, the pump intake side chambers are much larger than the output side chambers and the pump creates full pressure.

If pump pressure becomes excessive, the pressure moves the regulator valve against spring pressure, allowing line pressure to flow into the pressure decrease line. This pressure pushes the pump against priming spring pressure, moving the pump assembly to the right. In this position, the pump intake side chambers are only slightly larger than the output side chambers, and the pump output is decreased.

Manual Valve

A **manual valve,** located in the valve body, is operated by the driver through the shift linkage. This valve allows the driver to select *Park, Neutral, Reverse,* or different *Drive* ranges. When the shift selector lever is moved, the shift linkage moves the manual valve. As a result, the valve routes hydraulic fluid to the correct components in the transmission, including other hydraulic valves, clutches, and band servos.

Note that when *Overdrive, Drive,* or *Second* is selected, the transmission takes over, shifting automatically to meet driving conditions. In *Low* and *Reverse,* however, the transmission is locked into the selected gear.

Shift Valves

Shift valves, which are located in the valve body, control transmission upshifts and downshifts. Hydraulic

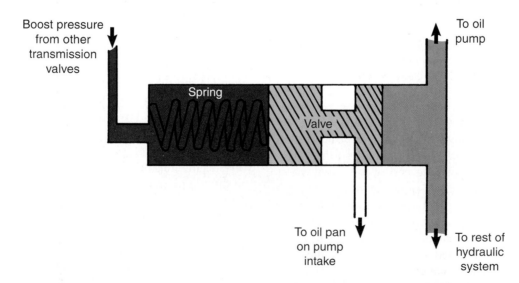

Figure 9-8. *The main pressure regulator valve controls hydraulic system pressure by exhausting fluid to the pan or pump intake. Basic pressure settings are determined by a spring that holds the regulator valve closed until a certain pressure is reached. The valve setting is modified by boost pressure from the manual valve and throttle valve.*

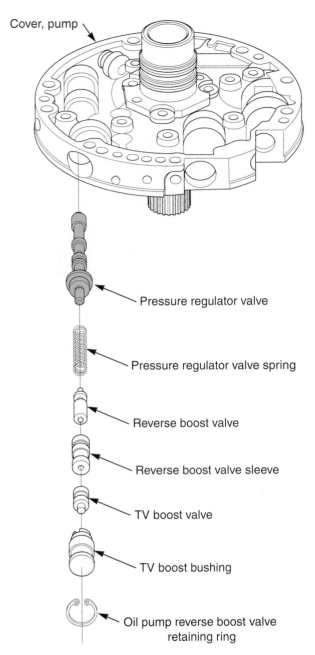

Cover, pump

Pressure regulator valve

Pressure regulator valve spring

Reverse boost valve

Reverse boost valve sleeve

TV boost valve

TV boost bushing

Oil pump reverse boost valve retaining ring

Figure 9-9. *The main pressure regulator valve is often installed in the pump housing. Some regulators are installed in the valve body or a separate auxiliary body. (General Motors)*

pressures established by the other transmission valves act on the ends of each shift valve. In addition, a spring located at one end of the shift valve puts a preload on the valve. Hydraulic pressure acting on one end of the valve works against hydraulic pressure *plus* spring pressure acting on the other end. The valve moves back and forth in the bore as pressures on each end change. When pressures on each end are equal, the valve is balanced and does not move.

Line pressure is directed to the center of the shift valve. It is used to apply a planetary holding member. As the passageway leading to the holding member is uncovered by movement of the shift valve, line pressure will pass through the valve's annular groove and be routed to the holding member.

Every transmission will have one fewer shift valve than it has forward gears. For example, a 3-speed transmission will have two shift valves, a 4-speed transmission will have three, and so on

The sequence of shift valve operation must be carefully controlled so the transmission upshifts through all available gears. This is done during design by proper sizing of the springs and valve lands. By designing lands of different sizes, or different **effective face areas,** hydraulic forces on valves acted on by the same pressure will differ.

Throttle Valve

The **throttle valve's** role, in effect, is to sense how hard the engine is working and to delay upshifting, as necessary. The throttle valve may be located in the valve body, or it may be installed in a separate bore in the transmission case.

Movement of the throttle valve modifies line pressure. The resulting pressure, or **throttle pressure**, is then transmitted to the shift valve (and a few other valves) at some pressure less than line pressure. Throttle pressure modulates, or changes, with engine load. As load increases, throttle pressure in the line to the shift valve increases, providing greater opposition to the counter (governor) pressure and delaying the upshift. The throttle valve may be operated by the *throttle valve linkage* or by the *vacuum modulator*.

Throttle Valve Linkage

The **throttle valve (TV) linkage** is a mechanical connection between the throttle valve in the transmission and the *throttle lever* on the vehicle's throttle body or carburetor, **Figure 9-12.** The throttle lever, in turn, is connected to the *throttle plate(s),* which are also located in the throttle body or carburetor. Engine power increases with the opening of the throttle plate and decreases with its closing.

The TV linkage consists of several levers connected by a series of rods or a cable. Movement of the vehicle's accelerator causes the throttle lever (and throttle plate) to move via the *throttle linkage,* which connects the throttle lever to the accelerator. The movement is transferred from the throttle lever to the throttle valve.

The action of the linkage on the throttle valve causes TV output pressure to change. Throttle pressure, then, varies as the throttle plate opening is varied, increasing as the throttle plate is opened and decreasing as the throttle plate is closed.

TV linkage always has some provision for adjustment. The rod-type linkage is adjusted at either the throttle lever or where the throttle rod enters the transmission case. The cable-type linkage is usually adjusted at the throttle lever.

Vacuum Modulator

The **vacuum modulator,** discussed in Chapter 6, moves the throttle valve with changes in engine load,

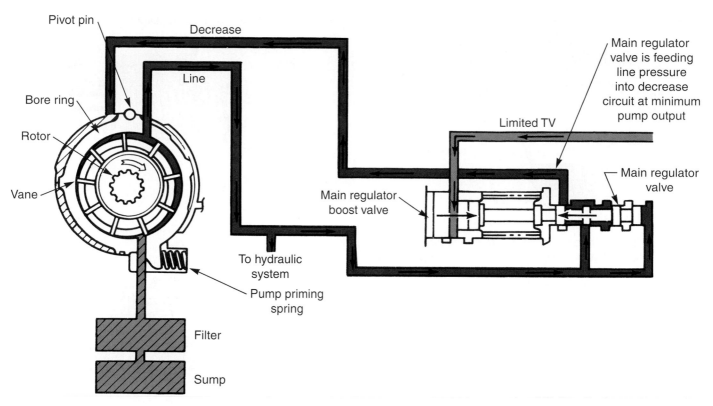

Note: Bore ring moved to left and pump priming spring is compressed.

Figure 9-10. *At minimum pump output (low engine speeds), the pressure regulator valve keeps pressure from moving the pump. Spring pressure keeps the pump in position to produce maximum output.*

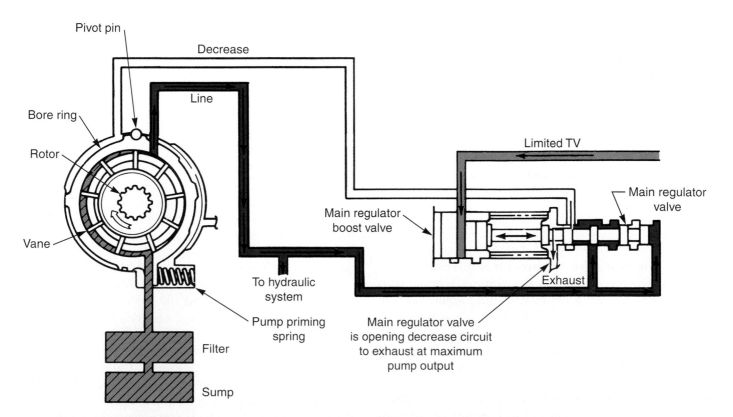

Figure 9-11. *At maximum pump output (high engine speeds), the pressure regulator valve allows pressure to work against spring pressure to move the pump to its lowest output position.*

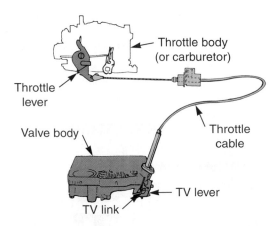

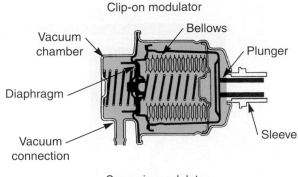

Figure 9-12. *The throttle valve on many automatic transmissions is operated by a cable from the engine throttle plate. Changes in throttle plate position change the position of the throttle valve and, therefore, the throttle pressure. Throttle pressure is sent to the shift valves, pressure regulator valve, and sometimes, other valves. (General Motors)*

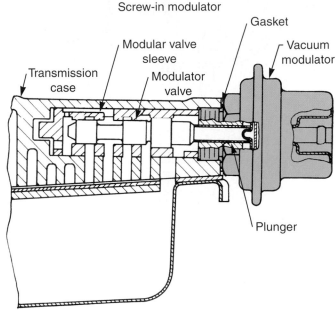

Figure 9-13. *Two different kinds of vacuum modulators are shown here. Both use a vacuum-operated diaphragm and spring to affect valve position. Note that the internal bellows, which is evacuated, and the smaller spring in the one design are used to compensate for power loss that occurs at reduced atmospheric pressures, such as at higher elevations. (General Motors, Fiat)*

causing the throttle pressure to vary. The basic modulator is a sealed container with an internal, vacuum-operated diaphragm. The modulator is threaded into the transmission case or is held to the case with a bolt or clip, **Figure 9-13.**

The diaphragm divides the container into two regions, sealing one off from the other. One side of the modulator is connected to engine vacuum by tubing connected to the intake manifold; the other side is at atmospheric pressure.

In operation, the vacuum side of the modulator contains a spring that pushes on the diaphragm, opposing manifold vacuum. Changes in manifold vacuum cause the diaphragm, working against spring pressure, to move back and forth. The other side of the diaphragm is connected to the throttle valve, either directly or through a plunger. The throttle valve is moved by the diaphragm as engine vacuum changes.

Engine vacuum will vary with changes in throttle opening and engine load. If the load is high, the manifold vacuum will be low and throttle pressure will be high. If engine load is low, vacuum will be high and throttle pressure will be low.

Governor Valve

The **governor valve,** or **governor** as it is often called, senses output shaft speed (and, therefore, vehicle speed) to help control shifting. It works with the throttle valve to determine *shift points,* or the vehicle speeds at which the shifts will occur. A typical assembly consists of a driven gear, weights, springs, a hydraulic valve, and a shaft.

The governor, which is driven by the transmission output shaft, takes line pressure at the valve and modifies it according to vehicle speed. Greater vehicle speed increases the governor valve output pressure, or **governor pressure.** Lower vehicle speed decreases governor pressure. Governor pressure opposes throttle pressure acting

on the shift valves. The higher of the two pressures—governor or throttle—overrides and controls the shift.

As stated, the governor is driven by the transmission output shaft. The assembly may be mounted on the output shaft, or it may be mounted in the transmission case and driven through gears similar to speedometer drive and driven gears. Either type takes line pressure from the valve body and modifies it according to vehicle speed.

Governors vary in design. Aside from where they are mounted and how they are driven, they vary in other aspects of operation. The most common type of governor contains a valve that is moved by centrifugal force. The governor contains at least one valve that is held in the low-speed position by a spring or springs. The governor is also equipped with weights that can overcome spring tension. As the governor rotates, the weights are thrown outward by centrifugal force. This action overcomes spring tension and causes the valve to move. Movement of the valve partially opens the governor port to line pressure. Governor pressure builds up rapidly with increasing vehicle speed until about 20 mph (32 km/h). At higher speeds, pressure increases more gradually to prevent it from becoming too high in the high vehicle-speed ranges. Shaft-mounted and gear-driven versions of this type of governor are shown in **Figure 9-14.**

Note that a variation of this valve does not have *separate* weights; the weight is an integral part of the valve itself. The motion of the governor, aided by the weight, causes the valve to move outward and open.

Another kind of governor has two openings, or ports, that are served by check balls. Oil pressure pushes the balls away from the openings, and oil flows out. Weights are thrown outward as the governor rotates. This action moves the check balls to *close* the openings, causing the pressure in the circuit to rise. See **Figure 9-15.** Line pressure is fed to this governor through an orifice, which limits the amount of fluid that can flow in the governor circuit. At low speeds, the balls are positioned so transmission fluid leaks out through the openings. The governor pressure is zero. As the vehicle speeds up, the weights begin to move the check balls inward, allowing governor pressure to build. At maximum governor pressure, the ports are completely closed off.

Detent Valve

The **detent valve,** also called the **kickdown valve,** is used on some transmissions to aid downshifting. Downshifting is often needed when a vehicle that is already in a higher gear attempts to climb a hill or accelerate to pass. When the accelerator is pushed all the way to the floor, the valve provides a *forced downshift,* increasing torque to the drive wheels. The vehicle, in this circumstance, is said to be in **passing gear.**

The detent valve may be operated through a *kickdown linkage* connected to the engine throttle or by an electric solenoid. In operation, the valve increases pressure on the throttle side of the shift valves. The extra pressure

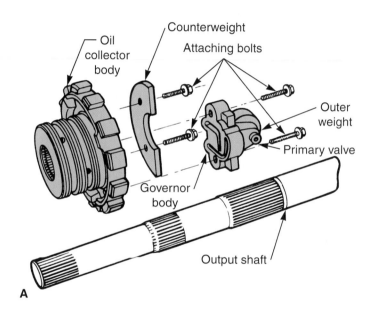

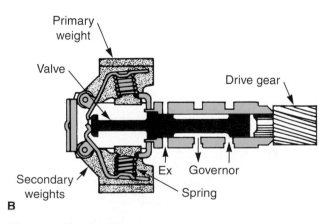

Figure 9-14. *This illustration shows two kinds of governors that control governor pressure by permitting restricted flow through the governor valve. A—Output shaft-mounted governor. B—Case-mounted, gear-driven governor. (Ford, General Motors)*

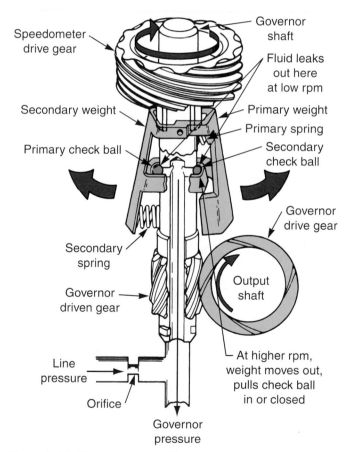

Figure 9-15. *The governor drive gear on the output shaft drives the governor. Transmission fluid leaks out past the check balls at low rpm. At higher rpm, the weights are thrown outward, against spring tension. The check balls are pulled inward, blocking the passageway and causing governor pressure to increase. When pressure is great enough to overcome the throttle pressure acting on the other end of the shift valve, upshift occurs. (General Motors)*

forces the shift valves to the downshifted position, putting the transmission in a lower gear.

Detent valves are necessary since changes in throttle pressure are not great enough to cause a downshift at highway speeds. The valve hastens the downshift, placing the vehicle in a lower gear before vehicle speed is slow enough to cause sufficient drop in governor pressure.

Note that some hydraulic circuit designs do not have a separate detent valve. In these designs, kickdown is achieved through the throttle valve. If the throttle valve is moved far enough, a *detent oil* passageway is uncovered and this oil is then routed to the appropriate shift valves.

Check Valves

A common type of check valve in automatic transmissions is the **check ball.** Check balls are one-way valves; they allow fluid to flow in one direction only. These steel balls are installed in chambers next to holes in the valve body spacer plate.

In operation, fluid flow in one direction will push the check ball into the spacer plate hole, sealing it. Fluid flow in the opposite direction will push the ball out of the hole, allowing fluid to flow.

Note that some transmissions use a *spring-loaded* check ball. These may function to control torque converter pressure, or they may serve in overpressure situations as pressure relief valves.

Accumulators

An **accumulator** is a hydraulic device used in the apply circuit of a band or clutch to cushion initial application. The result is a smoother shift. The accumulator essentially consists of a spring-loaded piston in a cylinder. The piston is pushed to one end of the cylinder by the spring. A hydraulic passageway connects the accumulator to the apply piston of a multiple-disc clutch or band servo.

As the clutch or servo is being applied, fluid in the passageway is received by the accumulator. The accumulator piston begins to move against spring pressure. The expanding cylinder chamber diverts some of the hydraulic fluid from the clutch or servo. As a result, pressure to the clutch or servo does not build up to *line pressure* immediately, and the initial application of the holding members will be soft. After the accumulator spring is fully compressed, full pressure goes to the apply piston and the clutch or band will be held tightly. **Figure 9-16** shows a typical accumulator installed in an auxiliary valve body.

Automatic Transmission Shift Linkage

The **shift linkage** is the only direct connection between the driver and the automatic transmission. It makes a mechanical connection between the driver and the transmission hydraulic control system. The shift linkage consists of a series of levers and linkage rods or cables. These transfer motion to the manual valve in the transmission valve body.

Selector Lever

The shift **selector lever** in the vehicle's passenger compartment can be mounted on the floor or on the steering column. With either arrangement, movement of the selector lever causes the manual valve inside the transmission to move. This action selects the driving range.

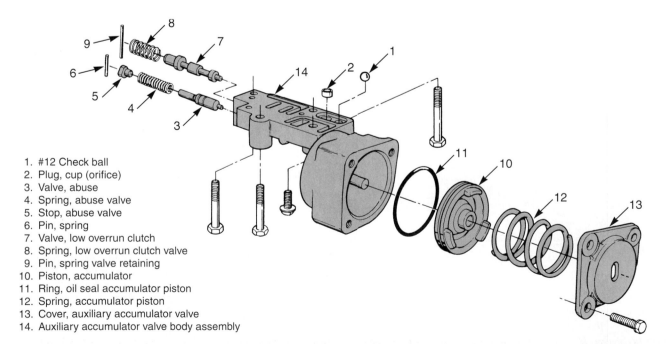

1. #12 Check ball
2. Plug, cup (orifice)
3. Valve, abuse
4. Spring, abuse valve
5. Stop, abuse valve
6. Pin, spring
7. Valve, low overrun clutch
8. Spring, low overrun clutch valve
9. Pin, spring valve retaining
10. Piston, accumulator
11. Ring, oil seal accumulator piston
12. Spring, accumulator piston
13. Cover, auxiliary accumulator valve
14. Auxiliary accumulator valve body assembly

Figure 9-16. *Accumulators are used to cushion the application of multiple-disc clutches and bands. The accumulator shown here is contained in a separate, or auxiliary, valve body. (General Motors)*

Floor-Shift Selector Lever

An assembly incorporating a typical floor-shift selector lever, **Figure 9-17,** is mounted directly on the floor or in a console between the front seats. Levers and rods (or cables) are used to transfer motion from the floor-shift selector lever to the transmission. The selector indicator, or *quadrant,* is usually built into the assembly; however, some are installed in the dash and connected to the transmission by a linkage arrangement.

To move the selector lever between certain positions, a lever release button must be depressed. Pressing the release button unlocks the selector lever latching mechanism. The latching mechanism typically consists of a latching spring pin (or lever) and a detent plate. The pin fits through a slotted hole at the bottom of the selector lever and into the detents in the plate, which is located on the mounting base of the shift mechanism. Pressing the release button lowers the pin so that it is no longer in the detent. The selector lever can then be moved. When the button is

released, the pin moves into the detent for the new selector position. The latching mechanism is usually designed to allow the selector lever to be moved from a lower gear to a higher gear and from drive to neutral, without pressing the release button. For other sequences, the button must be depressed.

Column-Shift Selector Lever

A shift linkage incorporating a column-shift selector lever uses a series of rods and levers to transfer movement from the steering column to the transmission. See **Figure 9-18.** The column-shift lever does not have a release button. Pulling the selector lever toward the driver disengages a latching pin from its slot, allowing the lever to be moved to the proper position.

Shift Linkage

Figure 9-19 shows typical shift linkage at a transmission case. Note the outer manual lever and shaft assembly.

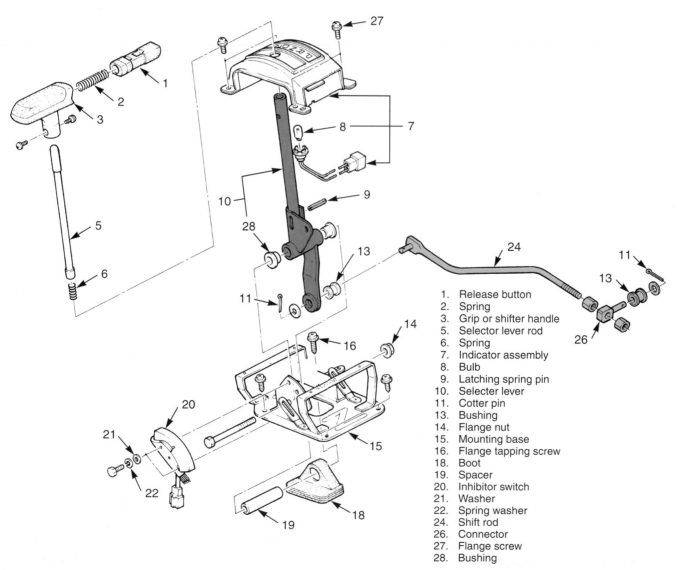

1. Release button
2. Spring
3. Grip or shifter handle
5. Selector lever rod
6. Spring
7. Indicator assembly
8. Bulb
9. Latching spring pin
10. Selecter lever
11. Cotter pin
13. Bushing
14. Flange nut
15. Mounting base
16. Flange tapping screw
18. Boot
19. Spacer
20. Inhibitor switch
21. Washer
22. Spring washer
24. Shift rod
26. Connector
27. Flange screw
28. Bushing

Figure 9-17. *A typical floor-shift selector lever is mounted directly on the floor or in a console between the front seats. The shift rod extends to the transmission. (Subaru)*

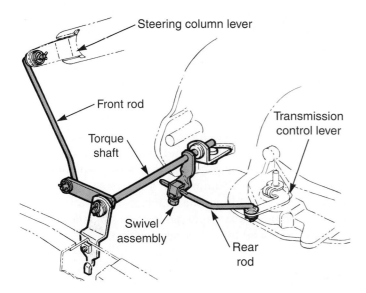

Figure 9-18. *Linkage transfers movement of the steering column-mounted gear selector lever to the manual valve in the transmission. Some vehicles use a cable instead of the rods and levers shown here. (DaimlerChrysler)*

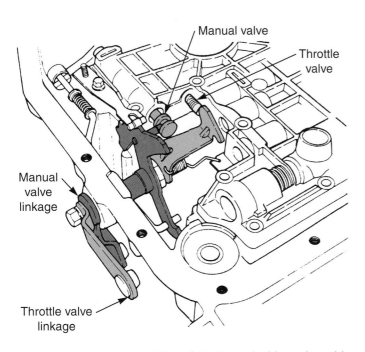

Figure 9-19. *Note the position of the levers inside and outside the case. The manual valve lever has a projection that fits between two lands of the manual valve. The throttle lever pushes on the spring-loaded throttle valve. (Ford)*

The outer manual lever is attached to the shift rod or cable from the shift selector lever. Movement of the outer manual lever is transferred through the shaft passing through the case to move the inner manual lever. This lever connects to the manual valve in the valve body. The manual valve opens and closes passageways that direct fluid to the different circuits in the transmission, as needed for the selected operating range.

Note that a hole runs through the shaft of the outer manual lever and shaft assembly shown in **Figure 9-19.** Through this hole passes another shaft having levers at each end. The outer throttle valve lever is connected, by linkage, to the engine throttle, and the inner throttle valve lever is connected to the throttle valve. The inner lever usually contacts the throttle valve directly. Seals at the case and throttle shaft prevent transmission fluid leaks.

The inner manual lever holds the manual valve in position. If the manual valve were not held in place, hydraulic pressure and vibration could cause the valve to move out of position. A series of detents, sometimes called a *cockscomb,* is part of the inner manual lever. This can be seen in **Figure 9-20.** A spring is used to hold a roller or ball against the detents. Spring pressure holds the lever in position; however, the spring is not strong enough to prevent movement of the lever when the driver moves the selector lever.

Neutral Start Switch

A ***neutral start switch*** is shown in **Figure 9-20.** Sometimes called a ***neutral safety switch,*** this switch prevents starting the engine when the selector lever is in any position but *Park* or *Neutral.* The switch may be transmission mounted or console mounted. (The column-shift safety mechanism is usually in the ignition switch. The ignition key will not turn to start the vehicle unless the selector lever is positioned in *Park* or *Neutral.*)

If the engine were to start when the selector lever was in any position but *Park* or *Neutral,* the vehicle would lurch forward, possibly causing an accident or injury. The switch is activated by movement of the shift linkage, and current is allowed only when the selector lever is in *Neutral* or *Park.* Otherwise, the switch contacts are open, and circuit current cannot flow. The vehicle's ignition switch is wired to the starter solenoid through the neutral start switch. If the ignition switch is operated when the selector lever is in any *Drive* gear or *Reverse,* current will not flow to the starter motor and the motor will not turn.

Note that in some cases, the neutral start switch has an integral backup-light switch. The shift linkage makes contact with a plunger within the switch when the selector lever is in *Reverse.* Moving the plunger causes a second set of contacts within the switch to close, completing the backup-light circuit and causing the lights to go on.

Parking Gear

Figure 9-20 also shows part of the linkage that locks up the ***parking gear,*** which is a toothed wheel splined to the output shaft. The linkage extends to the rear of the transmission and locks the output shaft to the case. A lever with a ***parking pawl, Figure 9-21,*** is attached to the transmission extension housing. The pawl moves in and out of engagement with the parking gear. When the driver shifts to the *Park* position, the shift linkage pushes on the parking

1. Screw—switch attaching
2. Lever—outer downshift
3. Nut—lever attaching
4. Washer—lever attaching
5. Switch—neutral start
6. Seal—downshift lever shaft
7. Lever—outer manual and shaft assembly
8. Seal—Manual lever shaft
9. Lever—inner manual
10. Nut—lever attaching
11. Lever—inner downshift
12. Rod—park pawl
13. Roller—lever actuating
14. Link—toggle operating
15. Clip—link retaining
16. Clip—link retaining
17. Clip—link retaining
18. Washer—link retaining

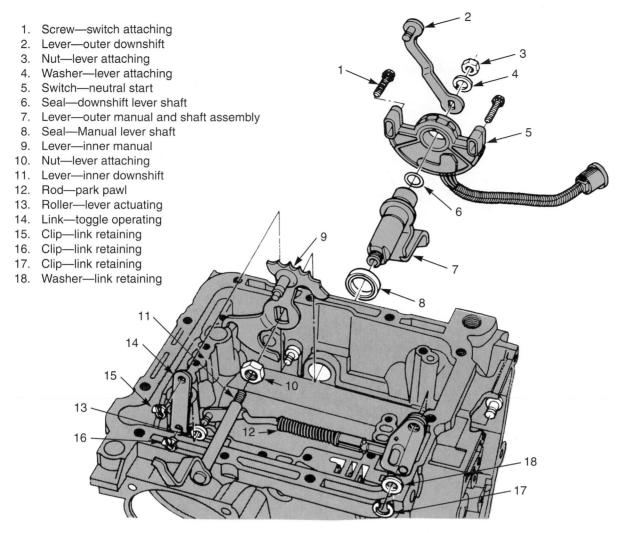

Figure 9-20. *This illustration shows the outer and inner, case-mounted shift linkage. Also shown are the parking pawl, throttle valve linkage, and neutral start switch. (Ford)*

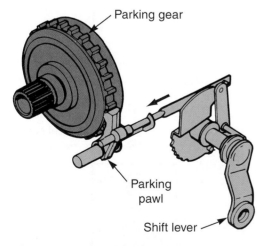

Figure 9-21. *The parking pawl is simply a latch that locks into large teeth on the parking gear. Since the pawl is mounted on the case, the output shaft is locked to the case when engaged and the vehicle will not roll. (Subaru)*

pawl. The pawl then engages the parking gear, locking the output shaft in place. The linkage also moves the manual valve to neutral.

Shift-Brake Interlocks

All modern vehicles have shift-brake interlocks. The purpose of the **shift-brake interlock** is to keep the driver from moving the gear selector out of *Park* without pressing on the brake pedal. This device keeps the vehicle from being placed in gear accidentally or at high engine speeds. Some older vehicles use a set of levers and rods connected between the brake pedal and shift lever. Newer vehicles have a solenoid on the shift linkage that applies a pin on the shift linkage to keep the linkage from being moved. **Figure 9-22** shows a shift-brake interlock solenoid attached to floor shift linkage. **Figure 9-23** illustrates a shift-brake interlock solenoid attached to column shift linkage. When the solenoid is energized, it withdraws the pin and the shift lever can be moved. A switch on the brake pedal linkage energizes the solenoid. Sometimes the brake light switch is used to provide an electrical signal to the body computer, which energizes the solenoid.

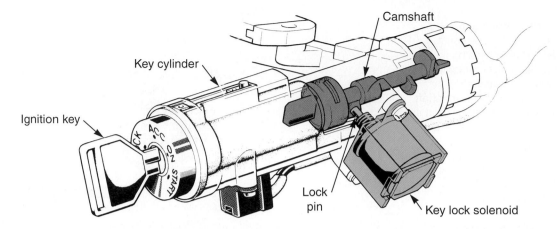

Figure 9-22. *This solenoid is installed on the ignition key assembly and is operated by a small computer. It prevents accidental starts in gear and reduces the chance of vehicle theft. (Lexus)*

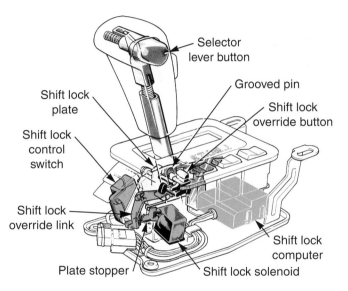

Figure 9-23. *This shift-brake interlock uses a small computer and a solenoid to prevent damage to the drive train and engine. The shift selector lever is connected to a small computer. The computer triggers the solenoid, which will not allow the lever to be moved out of gear unless the brake is applied. (Lexus)*

Electrical Components

Modern transmission hydraulic systems have electrical and electronic components. On newer vehicles, electrical and electrical components control all transmission functions, including gear changes. Some of the more common components are discussed below. These components and the devices that control them will be discussed in more detail in Chapter 12.

Switches

Switches use electrical contacts to energize or de-energize a solenoid or motor, or to send an electrical signal to another electrical device. Switches generally have two

contacts. When the contacts are touching, the switch is in the *on* position and current can flow through the switch. When the contacts are separated, the switch is in the *off* position and current cannot flow. Some pressure switches have one contact, and current flow is grounded through the switch body in the *on* position.

The driver activates some switches through shift linkage connected to the gear selector lever. Another type of driver-operated switch is installed on the accelerator linkage to operate the detent (passing gear) valve system. These types of switches are discussed in more detail later in this chapter.

Another commonly used switch is the pressure switch. Modern transmissions usually contain several pressure switches. Pressure switches are opened and closed by transmission hydraulic pressures, **Figure 9-24. Figure 9-25** illustrates a typical pressure switch and its general location on the valve body.

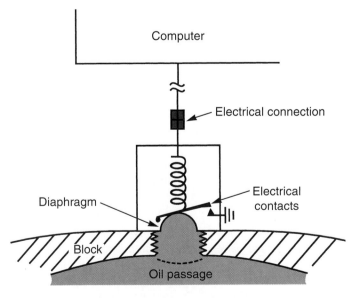

Figure 9-24. *The typical hydraulic pressure sensor is a switch that tells the computer that a hydraulic circuit is pressurized. Oil pressure acts on the switch diaphragm to open or close the electrical contacts.*

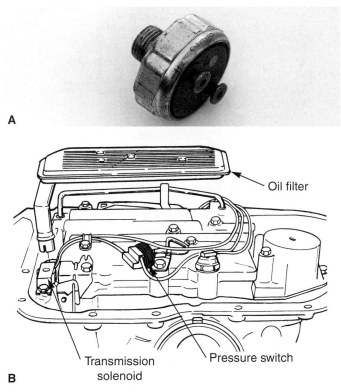

A

Oil filter

Transmission
solenoid

Pressure switch

B

Figure 9-25. *A—A typical pressure switch used on late-model transmissions. Note the two connections for current to flow in and out of the switch. Some switches have only one connection. B—A pressure switch installed on the transmission valve body. (General Motors)*

Solenoids

Electric **solenoids** control the operation of some transmission hydraulic components. Solenoids on modern transmissions are installed on the valve body. These are connected to the vehicle through a wiring harness having an electrical connector that is mounted on the case. See **Figure 9-26.** On a few older transmissions, the solenoids are threaded directly into the transmission case as shown in **Figure 9-27.**

A sectional view of a typical solenoid is shown in **Figure 9-28.** When energized, the solenoid valve functions by opening a small, self-contained valve or ball. This action causes transmission fluid in the connecting passageway to be exhausted. Loss of pressure in the passageway causes another valve to move, and action from this point is completely hydraulic.

An example of a transmission solenoid is the **detent solenoid** used on some automatic transmissions. This solenoid is also called the **kickdown solenoid** or the **downshift solenoid.** The detent solenoid is energized by closing the kickdown switch, which is located near the accelerator. **Figure 9-29** is a simplified diagram of a detent solenoid circuit.

Another common solenoid is the **lockup solenoid,** which is used to engage the torque converter lockup clutch, **Figure 9-30.** When the solenoid is energized, its

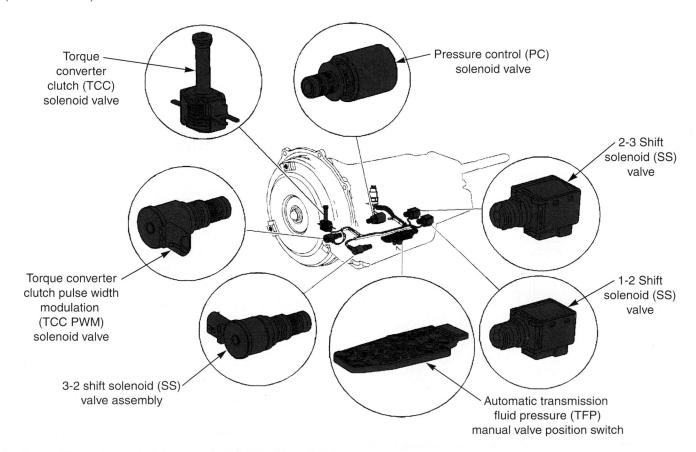

Torque
converter
clutch (TCC)
solenoid valve

Pressure control (PC)
solenoid valve

2-3 Shift
solenoid (SS)
valve

Torque converter
clutch pulse width
modulation
(TCC PWM)
solenoid valve

1-2 Shift
solenoid (SS)
valve

3-2 shift solenoid (SS)
valve assembly

Automatic transmission
fluid pressure (TFP)
manual valve position switch

Figure 9-26. *Most of the control solenoids in a modern transmission are installed on the valve body, inside the oil pan. This transmission has six solenoids and a pressure and temperature switch assembly. (General Motors)*

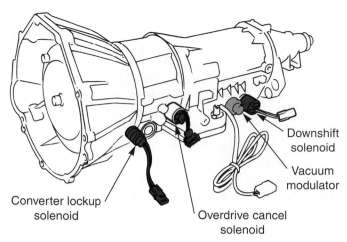

Figure 9-27. *Some older transmissions have solenoids installed on the outside of the case. (Nissan)*

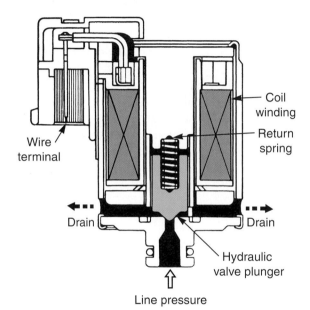

Figure 9-28. *A typical solenoid consists of a coil winding, a plunger, and a return spring. The winding, when energized, pulls the plunger against return spring pressure. This causes transmission fluid to be exhausted and activates the particular control circuit. (DaimlerChrysler)*

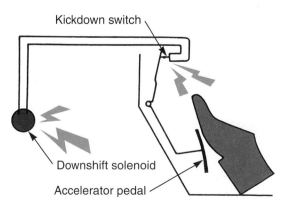

Figure 9-29. *Simplified diagram of a typical detent solenoid circuit is shown here. Pressing the accelerator to the floor closes contacts on the kickdown switch and energizes the solenoid. The valve opens and fluid is exhausted, reducing hydraulic circuit pressure. (Nissan)*

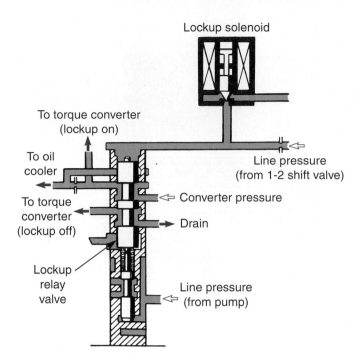

Figure 9-30. *The lockup solenoid operates the hydraulic portion of the torque converter lockup relay valve on many modern transmissions. (DaimlerChrysler)*

integral valve is unseated, causing transmission fluid in the connecting line to be exhausted. Loss of this fluid reduces circuit pressure. This causes the *lockup relay valve* to move in such a way as to apply the clutch. The first lockup solenoids were often operated by vacuum and pressure switches. Today, most lockup solenoids are operated by the on-board computer.

Summary

Transmission oil pumps use gears, rotors, or vanes that cause the hydraulic system to become pressurized. The oil pump is usually driven by lugs on the rear of the torque converter. Some older transmissions have a second oil pump that is driven by the transmission output shaft.

Transmission oil filters are used to remove particles from the automatic transmission fluid. Fluid being drawn into the oil pump must first pass through the filter. This keeps fluid contaminants from entering the pump or the other hydraulic components.

The transmission oil cooler removes heat from transmission fluid. Fluid that has been overheated does not lubricate properly. Further, it breaks down into sludge and varnish, which plug passageways. The oil cooler is located in the engine cooling-system radiator.

The valve body contains the manual valve, shift valves, and other hydraulic system valves. Valve bodies usually have several check balls, which are located next to the spacer plate and gaskets.

The pressure regulator valve may be in the valve body, in the back of the oil pump, or in an auxiliary valve body. It regulates line pressure according to spring pressure and inputs from other valves.

The manual valve is operated by the driver through the shift linkage. It allows the driver to select *Park, Neutral, Reverse,* or different *Drive* ranges. When the shift selector lever is moved, the shift linkage moves the manual valve. As a result, the valve routes hydraulic fluid to the correct components in the transmission.

Shift valves control transmission upshifts and down-shifts. Hydraulic pressures from the other transmission valves act on each end of each shift valve.

Throttle valves control the point at which shift valves move to the upshifted position. They can be operated by the TV linkage or by the vacuum modulator. Throttle pressure on the shift valves tries to keep them in the down-shifted position.

Governor valves produce pressure that varies with vehicle speed. As speed increases, governor pressure increases. Governors can be mounted on the output shaft or on the transmission case. Case-mounted governors are driven by a gear on the output shaft. Governor pressure opposes throttle pressure to upshift the transmission.

Accumulators cushion shifts by absorbing some of the fluid being sent to the holding members. The resulting pressure in the accumulator compresses a piston, expanding the chamber, causing a gradual pressure buildup in the circuit, and cushioning the shift.

The detent valve forces the transmission into a lower gear. This is often needed when a vehicle attempts to climb a hill or accelerate to pass. Detent valves are operated by linkage or by an electric solenoid.

Check balls are one-way valves. They allow fluid to flow in one direction only. The steel balls are installed in chambers next to holes in the valve body spacer plate.

Shift linkage connects the driver to the transmission through levers and rods or cables. The shift linkage also operates the neutral start switch, backup lights, and parking gear mechanism.

Review Questions—Chapter 9

Please do not write in this text. Place your answers on a separate sheet of paper.

1. Cite four functions of an automatic transmission oil pump.

2. Most of the fluid in an automatic transmission is inside the _____ _____.

3. All transmission oil pumps are _____ displacement pumps.

4. Rear pumps have not been used since the _____.
   (A) 1950s
   (B) 1960s
   (C) 1970s
   (D) They are still being used.

5. Why must the oil filter be located at the bottom of the oil pan?

6. Most transmission main oil coolers are installed in the vehicle _____.

7. Cooled fluid may be routed through the transmission _____system before returning to the oil pan.

8. Manufacturing tolerances for the _____ _____ must be exact.

9. Why must the holes in the spacer plate be carefully sized?

10. Oil transfer tubes connect _____ chambers in the valve body or case.

11. The _____ controls overall system pressure by exhausting extra fluid to the oil pan or pump intake.

12. The throttle valve changes position with changes in engine manifold vacuum. This means that a _____ is used to operate the throttle valve.

13. The following valve is used to modify pressure according to vehicle speed _____.
    (A) manual valve
    (B) throttle valve
    (C) governor valve
    (D) detent valve

14. Does governor pressure increase or decrease with road speed? _____ What effect does engine load have on governor pressure? _____

15. Some automatic transmissions use a _____ valve to aid downshifting.

16. The hydraulic device used in the apply circuit of a band or clutch to cushion initial application is the _____.
    (A) synchronizer
    (B) accumulator
    (C) limiter
    (D) regulator

17. The shift linkage transfers motion from the driver to the _____ valve in the valve body.

18. The _____ _____ switch prevents the engine from starting when the selector lever is in any position but *Park* or *Neutral.*

19. Pressure switches are opened and closed by transmission _____.

20. Solenoids are installed in the valve body or sometimes threaded into the transmission _____.

ASE-Type Questions—Chapter 9

1. Technician A says that the transmission oil pump is used to keep the torque converter filled with transmission fluid. Technician B says that the transmission oil pump is used to circulate transmission fluid through the oil cooler. Who is right?
   (A) A only.
   (B) B only.
   (C) Both A and B.
   (D) Neither A nor B.

2. Technician A says that excessive transmission fluid temperatures will reduce the life of the transmission fluid. Technician B says that temperature will have no effect on the fluid's lubricating ability. Who is right?
   (A) A only.
   (B) B only.
   (C) Both A and B.
   (D) Neither A nor B.

3. Technician A says that check balls are often used to block holes in the spacer plate. Technician B says that spacer plates may be installed between the valve body and transmission case. Who is right?
   (A) A only.
   (B) B only.
   (C) Both A and B.
   (D) Neither A nor B.

4. The basic control device for many transmission valves is:
   (A) the pressure developed by the pump.
   (B) centrifugal force.
   (C) oil viscosity.
   (D) strength of the valve spring.

5. All of the following are hydraulic valves used in automatic transmissions *except:*
   (A) pressure regulator valves.
   (B) manual valves.
   (C) shift valves.
   (D) recovery valves.

6. Every transmission will have one less shift valve than it has:
   (A) forward gears.
   (B) planetary gear sets.
   (C) servos.
   (D) clutch packs.

7. Technician A says that the throttle valve may be located in the valve body. Technician B says that the throttle valve may be located in a bore in the transmission case. Who is right?
   (A) A only.
   (B) B only.
   (C) Both A and B.
   (D) Neither A nor B.

8. Technician A says that throttle (TV) linkage is always adjustable. Technician B says that vacuum modulators are always adjustable. Who is right?
   (A) A only.
   (B) B only.
   (C) Both A and B.
   (D) Neither A nor B.

9. If the valve body does not contain a separate detent valve, what valve is used to obtain detent downshifts?
   (A) Main pressure regulator valve
   (B) Throttle valve
   (C) Governor valve
   (D) Manual valve

10. The switch that permits starting only when the selector lever is in the *Park* or *Neutral* position is called the:
    (A) restrictor switch.
    (B) neutral start switch.
    (C) ignition limiting switch.
    (D) transmission cutoff switch.

Chapter 10

Automatic Transaxle Construction and Operation

After studying this chapter, you will be able to:
- ❑ Identify and explain the purpose of the major parts of an automatic transaxle.
- ❑ Identify similar transaxle and rear-wheel drive transmission parts.
- ❑ Identify transaxle-specific parts.
- ❑ Identify the differences between transaxle and rear-wheel drive transmission parts.
- ❑ Explain the overall operation of an automatic transaxle.
- ❑ Trace power flow through an automatic transaxle.
- ❑ Compare automatic transaxle design variations.
- ❑ Describe the operation of a continuously variable transmission.

Technical Terms

Automatic transaxle case	Output shafts	Hypoid gear oil
Extension housings	Transfer shaft	Secondary pump
Oil pans	Oil pump shaft	Scavenger pump
Thermostatic element	Stator shaft	Oil scoops
Torque converter	Chain drive	Oil diverters
Transaxle shafts	Transfer gears	Oil cooler
Input shaft	Parallel shaft transaxles	Channel plate
Turbine shaft	Final drive unit	Continuously variable transmission (CVT)
Hub and shaft assembly	Wheel hop	

Introduction

Automatic transaxles are commonly used in domestic and imported automobiles, as well as a few light trucks. Automatic transaxle operating principles are similar to those for an automatic transmission and differential found in a rear-wheel drive vehicle. In a transaxle, however, the transmission and differential are usually housed in a single case. Design and layout of parts, while similar to those of a rear-wheel drive transmission, may vary in shape and position. Some parts are specific to transaxles and are not used on rear-wheel drive transmissions.

This chapter will describe the parts and operating principles of automatic transaxles. Where necessary, the differences between the design and placement of transaxle and rear-wheel drive transmission components will be called out. Parts that are specific to transaxles will be identified and explained. This chapter will also identify the major types of transaxles and explain how power flows through each type.

Before beginning this chapter, make sure you thoroughly understand the information in previous chapters. Understanding this information will make learning the material in this chapter much easier.

Transaxle Construction and Basic Operation

Automatic transaxles are made of many separate parts and systems. **Figures 10-1** and **10-2** show the relative positions of the major automatic transaxle parts. These parts include:

❑ Case.
❑ Oil pans.
❑ Torque converter.
❑ Transaxle shafts.
❑ Chain drive (some transaxles).
❑ Gear drive (some transaxles).
❑ Planetary gearsets (some transaxles).
❑ Parallel shaft gearsets (some transaxles).
❑ Holding members.
❑ Differential assembly.
❑ Oil pump.
❑ Oil filter.
❑ Valve body.
❑ Shift linkage.
❑ Electronic components.

Many of these parts are identical in construction and operation to their corresponding rear-wheel drive transmission parts and will not be presented in detail in this chapter. Other parts, such as transaxle shafts, drive chains, drive gears, and differential assemblies, are only used on transaxles and will be explained in detail. Still other parts,

such as valve bodies, oil pans, and the transaxle case, have the same operating characteristics as the corresponding rear-wheel drive components but are different in design and placement. These transaxle parts will also be covered in detail.

Transaxle Designs

Before studying individual transaxle parts, look at the transaxle designs shown in **Figure 10-3**. Most transaxles will use one of two designs. General Motors and Ford transaxles generally use a chain drive as shown in **Figure 10-3A**. The gear drive is used on almost all other domestic and imported transaxles. See **Figure 10-3B**. A variation of the gear-drive transaxle is shown in **Figure 10-3C**. This transaxle uses a chain in place of the gears but is laid out like a gear-drive transaxle. This design is used with longitudinal (front-facing) engine and transaxle layouts.

Automatic Transaxle Case

The *automatic transaxle case* is an aluminum casting that is machined to serve as a mounting and aligning surface for the moving parts of the transaxle. Most other transaxle parts are housed in the case.

Although most automatic transaxle cases are one-piece castings, the transaxle case can be divided into three main areas: the converter housing, the transmission case, and the final drive housing. See **Figure 10-4**. The converter housing covers and protects the torque converter. The transmission case contains the planetary gears, holding members, and other transmission parts. The final drive housing contains the final reduction gears, differential assembly, and the inboard ends of the CV axles.

The case is carefully machined to provide an accurate mounting to the engine and to provide a leakproof seal with the valve body and other hydraulic components. Cylinders are machined in the case to accept servo and accumulator pistons. Oil passageways are cast or drilled in the case to connect the hydraulic components of the transaxle. See **Figure 10-5**.

Extension housings or separate sections called channel plates, which contain some of the hydraulic passageways, are attached to some transaxle cases. Sheet metal covers and extension housings are installed at the end or on the sides of the case, as shown in **Figure 10-6**. The covers allow easier removal and installation of internal transaxle components. The covers enclose certain transaxle components, such as the valve body, and many serve as oil pans. A few transaxles have a cover mounted at the top of the case.

Many external parts, such as shift linkage, throttle linkage, vacuum modulators, and oil cooler lines, are installed on or enter through the transaxle case. Electrical wires pass through the case to operate solenoids or to

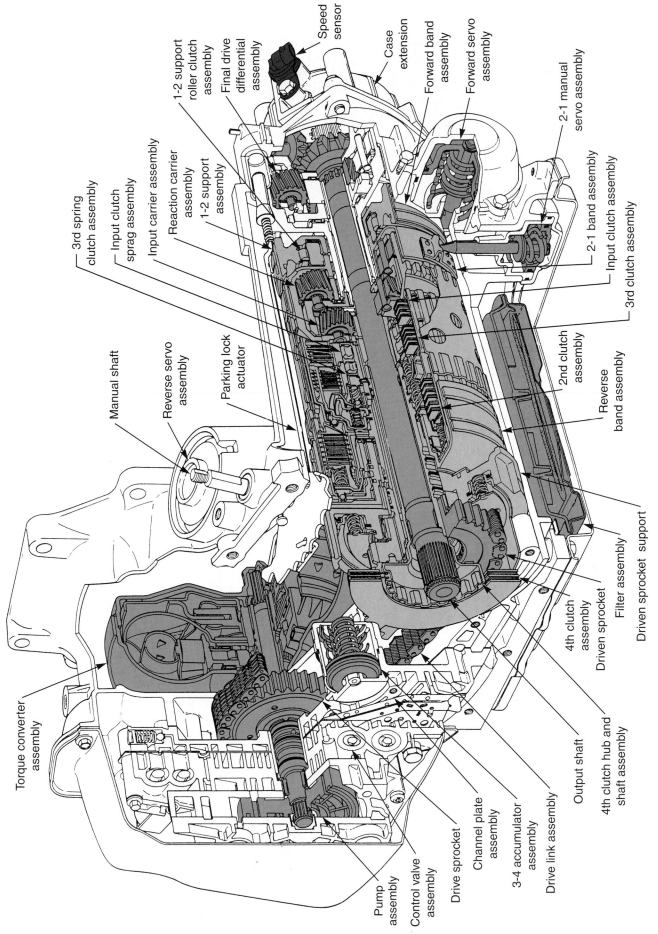

Figure 10-1. *This view shows the major components of one common type of automatic transaxle. Many of the components are identical in operation to those used in automatic transmissions on rear-wheel drive vehicles. Others are different and will be detailed in this chapter. (General Motors)*

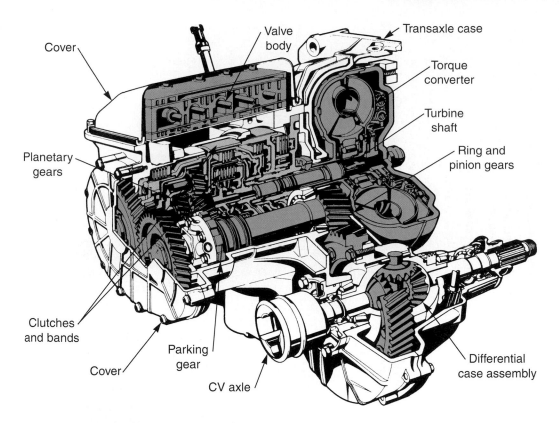

Figure 10-2. *This transaxle is obviously different from the one shown in Figure 10-1. However, many of the parts are similar. Various differences in design and parts placement will be discussed in this chapter. (Saab)*

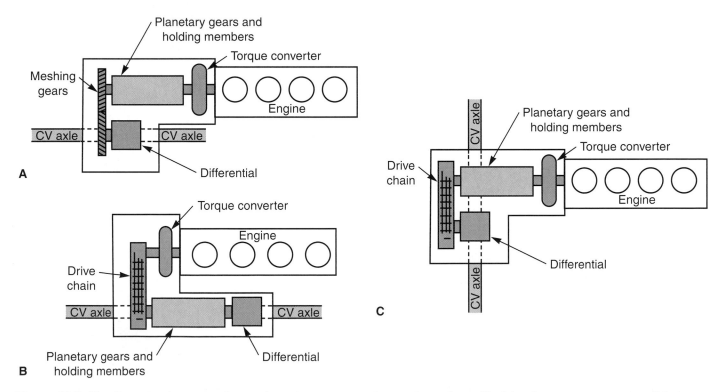

Figure 10-3. *The three most common transaxle parts arrangements are shown here. Studying these arrangements will help you understand the parts as they are discussed in this chapter. A—Chain drive transaxle. B—Gear-drive transaxle. C—Chain-drive transaxle used with a longitudinal engine.*

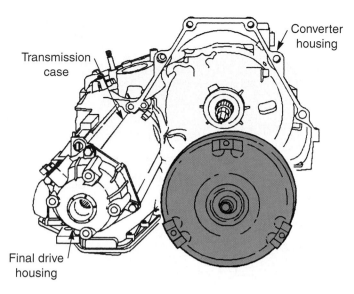

Figure 10-4. *Transaxles can be divided into three main areas: the converter housing, the transmission case, and the final drive. (General Motors)*

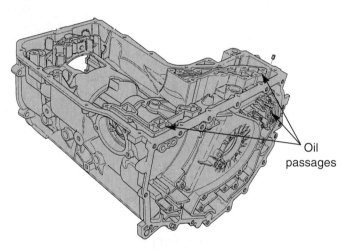

Figure 10-5. *The transaxle case is machined and drilled to create oil passages. (General Motors)*

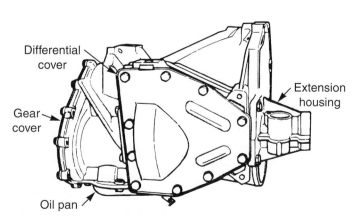

Figure 10-6. *Sheet metal covers and an extension housing are installed on the case. (DaimlerChrysler)*

connect to pressure sensors on the valve body. These wires are enclosed in special case connectors equipped with O-rings to prevent leaks. Speedometer assemblies are also installed on the case, as shown in **Figure 10-7.** A gear on the transaxle output shaft, the differential assembly, or one of the CV axles may drive the speedometer.

Transaxle Oil Pans

Most transaxles have two *oil pans.* One pan is installed on the bottom of the case, and the other is installed on one side of the case. **Figure 10-8** illustrates both pans. The bottom transaxle oil pan is the fluid reservoir. The bottom pan holds fluid that drains out of the operating parts of the transaxle. This oil pan is wide and relatively shallow. Some oil cooling occurs as air passes across this pan. The oil pan is attached to the transaxle case with bolts, and a gasket is always used between the pan and case. The transaxle oil filter is installed in this oil pan.

The side oil pan covers the chain or gears, and may cover the valve body. The side pan is also attached to the transaxle case using bolts and a gasket. Many side pans have an opening to permit the CV axles to exit, **Figure 10-9.**

A few transaxles have an oil pan at the top of the case. The top oil pan is usually a cover for a top-mounted valve body. A top-mounted oil pan is shown in **Figure 10-2.**

Thermostatic Elements

Transaxles with two pans use a *thermostatic element.* The thermostatic element allows the side pan to act as a transmission fluid reservoir when the transaxle is warmed up. This is necessary because the lower oil pan cannot hold enough fluid for cold operation without being over-filled when hot. Too high a level in the lower pan when the transaxle is hot would cause oil foaming if the rotating

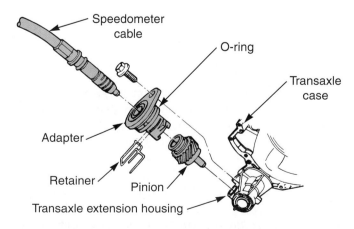

Figure 10-7. *This speedometer assembly mounts in the transaxle case. The pinion gear is typically driven by the output shaft. The gear drives the speedometer cable. (DaimlerChrysler)*

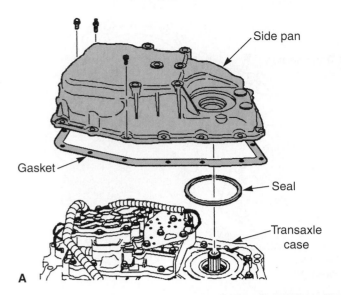

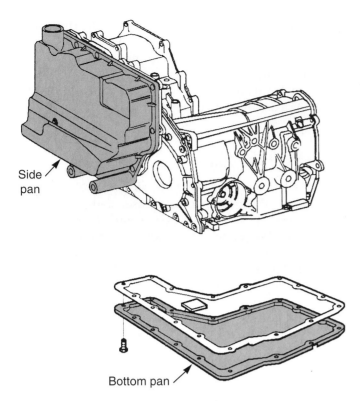

Figure 10-8. *Most transaxles have two oil pans. One is side mounted, and the other is installed on the bottom of the case. (General Motors)*

Figure 10-9. *A—This side oil pan has an opening for the CV axle to exit. B—The transaxle shown here is designed to allow the CV axle to exit directly from the case. (General Motors)*

planetary gears contact the oil. To prevent this, the thermostatic element keeps hot oil that drains from the side pan components in the side pan.

The thermostatic element consists of a bimetallic (two metals bonded together) strip and a flat metal valve. When the oil heats up, the bimetallic strip flexes, closing the valve and trapping oil in the side pan. When the transaxle cools down, the strip returns to its original shape. This causes the valve to open and allows fluid to return to the bottom pan. **Figure 10-10** illustrates the action of the thermostatic element. A unique feature of a transaxle with a thermostatic element is that, since the oil level is measured from the bottom pan, the level will be higher when the engine is cold than when it is hot.

Torque Converter

The *torque converter* used in an automatic transaxle is identical in general construction and operation to that used in a rear-wheel drive transmission. The torque converter is mounted directly behind the engine and is turned by the engine crankshaft. As in an automatic transmission, the engine flywheel is located between the engine and the converter. **Figure 10-11** shows the location of the torque converter in a transaxle. Some converters have a pump drive hub, with lugs to drive the transaxle oil pump. Others use a separate shaft to drive the oil pump. The torque converter drives the turbine shaft, which is called the transaxle input shaft in some designs. Most torque converters used in modern transaxles are lockup torque converters.

Magnetic Powder Clutch

A few transaxles have a *magnetic powder clutch* in place of the torque converter. The magnetic powder clutch has outer hub, an inner hub, and metallic powder (usually iron or steel particles) between the hubs, **Figure 10-12**. The outer hub forms the clutch housing and is driven by the engine. The inner hub is attached to the transaxle input shaft. An electromagnet is installed in the clutch and is energized through a set of brushes similar to those in starters and alternators. Some electromagnets are part of the inner hub as shown in **Figure 10-12**, while others are attached to the housing. The hubs, powder, and electromagnet are sealed in the housing. This housing is often finned for heat removal.

When the electromagnet is de-energized, the metallic powder particles rotate inside of the housing and have no effect. When the electromagnet is energized, the metallic powder particles are locked together by the magnetic field, forming a solid mass. When the particles lock, the hubs are also locked together. Power is transmitted from the outer hub to the inner hub.

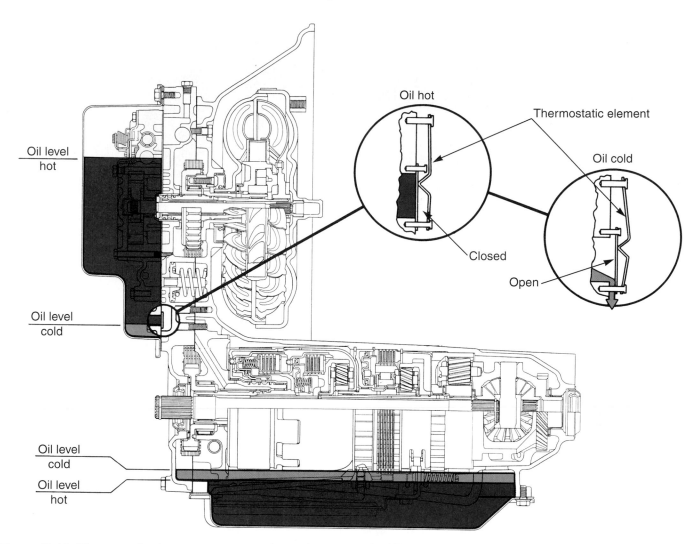

Figure 10-10. *Thermostatic elements are commonly used on transaxles. The element allows use of a shallow oil pan. When oil is cold, the valve is open and oil can flow into the bottom pan to ensure that the oil filter does not pick up any air. When the oil is hot, the valve closes to prevent a high oil level in the lower pan. (General Motors)*

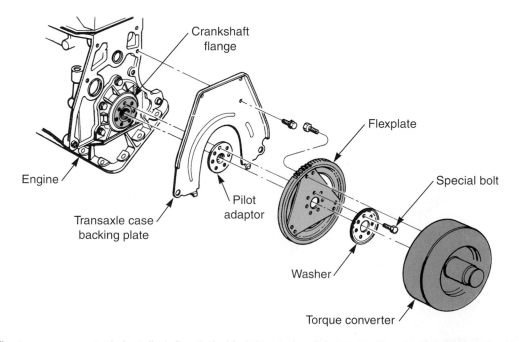

Figure 10-11. *The torque converter is installed directly behind the engine. It is attached to the flywheel or flexplate, which in turn, is attached to the engine crankshaft. The torque converter turns whenever the engine turns. (DaimlerChrysler)*

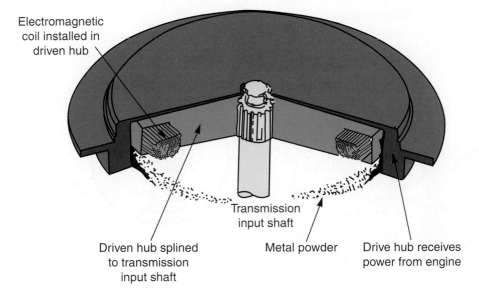

Figure 10-12. *The magnetic powder clutch is used on some transaxles, and may become more common in the future.*

Varying the amount of voltage to the electromagnet varies the amount of power transmitted by the powder clutch. Smooth engagement of the clutch can be obtained by sending a relatively low voltage to the electromagnet on first application. This reduces the locking force on the powder, allowing some slippage. More voltage can be applied quickly for slip-free power transfer.

Transaxle Shafts

The automatic *transaxle shafts* are made of steel that is hardened and tempered to withstand high loads and extreme temperatures. Transaxles may have several different shafts, including turbine shafts, input shafts, output shafts, and oil pump shafts. The type of shafts used varies with the type of transaxle. Refer to **Figure 10-13** as you read the following sections.

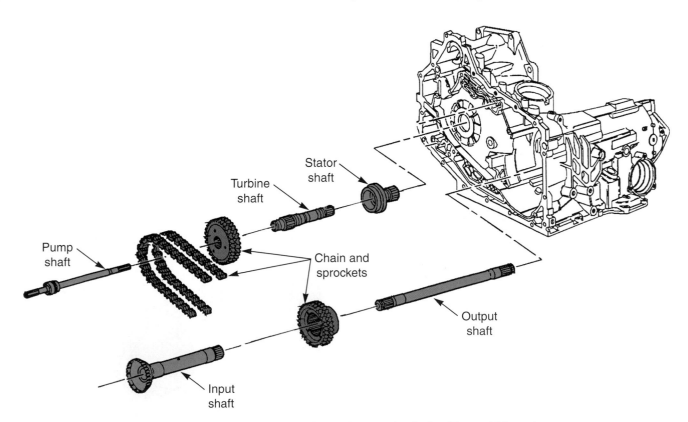

Figure 10-13. *This figure shows some of the most commonly used transaxle shafts. (General Motors)*

Input Shafts

The automatic transaxle **input shaft** has the same purpose as the input shaft on a rear-wheel drive transmission. Input shaft designs vary depending on whether the transaxle uses a chain or gear drive. These variations are discussed in the following sections.

Chain-Drive Input Shafts

If the transaxle uses a drive chain to transfer power between the converter and the other transmission parts, there will be two input shafts, one at each drive sprocket. The input shaft between the converter turbine and the drive sprocket is usually called the **turbine shaft.** The turbine shaft may be hollow to allow the pump drive shaft to pass through it. One end of the turbine shaft is splined to the converter turbine. The other end attaches to the drive sprocket or may be part of the drive sprocket assembly.

The input shaft between the driven sprocket and the other transmission parts may be called an input shaft or a **hub and shaft assembly.** This input shaft may be part of the driven sprocket and may not be called out as a separate part. On most chain-drive units, this input shaft is hollow and one of the output shafts passes through it.

Gear-Drive Input Shafts

Input shafts used on gear-drive transaxles closely resemble those used on rear-wheel drive transmissions. One end of the input shaft is splined to and driven by the torque converter turbine. The other end is splined to the input planetary gears or to an input drum.

Output Shafts

The exact placement and connection of automatic transaxle **output shafts** depends on the transaxle design. Most transversely mounted transaxles have output shafts that extend from the differential assembly to the CV axles. When the transaxle and engine are mounted longitudinally, a single output shaft transfers power from the planetary gearsets to the differential assembly. On other transaxles, the output shaft drives the gears or chain to power a transfer shaft, which will be discussed later in this chapter.

Oil Pump Shaft

An **oil pump shaft** drives the oil pump on most chain-drive transaxles. This shaft is usually splined or keyed to the converter cover and turns whenever the engine is running. Splines on the other end of the shaft are attached to the converter hub. The torque converter drives the oil pump via the oil pump shaft. See **Figure 10-14.** On some transaxles, the oil pump shaft assembly contains oil passageways to deliver oil to and from the torque converter.

 Note: Most gear-drive transaxles drive the pump with gears or lugs on the rear of the converter, in the same manner as a rear-wheel drive transmission.

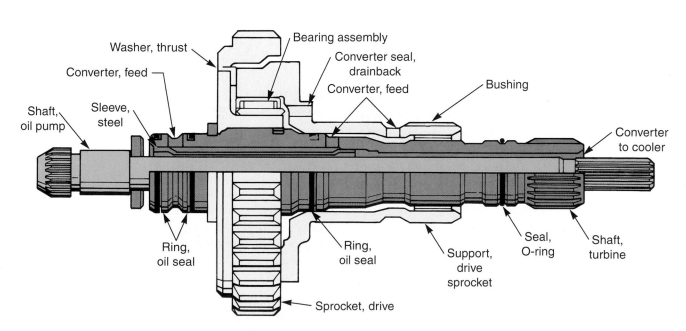

Figure 10-14. *An oil pump shaft passes through the turbine shaft, as shown here. The splines on the left side of the shaft drive the oil pump. The oil passages in this shaft assembly are used to fill and control the torque converter. Note that this assembly is used with a chain-drive automatic transaxle. (General Motors)*

Stator Shaft

The **stator shaft**, sometimes called the reaction shaft, provides a stationary support for the torque converter stator and the overrunning clutch. The stator shaft has external splines that mate with internal splines on the overrunning clutch. These prevent rotation of the inner race of the clutch. The stator shaft is hollow, and the transaxle input shaft passes through it. When used, the torque converter pump drive hub fits over the stator and input shafts. The stator shaft may be mounted on the front of the transaxle oil pumps, if the pump is located at the converter. When the pump is not close to the converter, such as on most chain drive transaxles, the stator shaft is installed on the case. A typical case-mounted stator support is shown in **Figure 10-13.** Case-mounted stator shafts may be called turbine shaft supports or input shaft supports.

Transfer Shaft

As mentioned, some transaxles have a **transfer shaft.** The transfer shaft transmits power from the transaxle output shaft to the differential assembly. The transfer shaft is positioned below the output shaft. A gear on the end of the output shaft drives a gear on the end of the transfer shaft. A helical pinion gear on the other end of the transfer shaft drives the differential ring gear. **Figure 10-15** shows a typical transfer shaft.

Chain Drives

A **chain drive** is a power-transfer system using a chain and two sprockets. It is used in some automatic transaxles to transmit motion in a small space, **Figure 10-16.** When a chain drive is used, the input and output sprockets turn in the same direction. The sprockets are mounted in bearing or bushing assemblies that provide support and alignment.

The drive chain is a link, or silent-chain, design. This type of chain meshes with its sprockets with a minimum of noise and friction. As shown in **Figure 10-17,** most chain drives are installed between the converter and the planetary gears and holding members. A few chain drives are placed at the rear of the transmission section and deliver power between the transmission output shaft and the final drive, **Figure 10-15.**

Gear Drives

Some transaxles use gear drives, which are sometimes called **transfer gears.** The gear drive consists of two large helical gears that are in constant mesh, **Figure 10-18.** Gear and chain drives serve the same purpose; they transfer power from the upper section of the transaxle to the lower section.

On a gear drive, the gears turn in opposite directions. A few manufacturers use an idler gear, **Figure 10-19,** but most design the final drive unit to re-reverse the direction of rotation. Most gear drives are installed after the planetary gears and holding members, but ahead of the final drive unit.

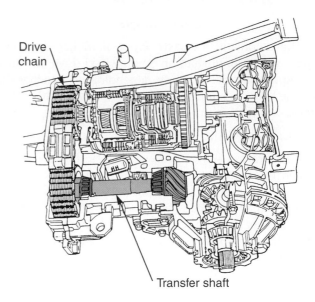

Figure 10-15. *The transfer shaft is used on some vehicles to transfer power from the helical gears to the differential. It is used on only a few transaxles. (DaimlerChrysler)*

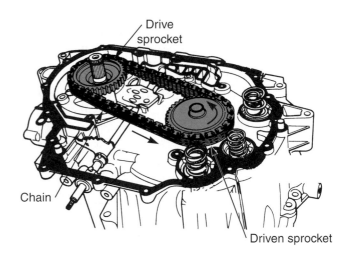

Figure 10-16. *Chain drive assembled view. The chain drive enables power transfer in a small space. (Ford)*

Planetary Gearsets

Planetary gearsets transmit power through the transaxle and change gear ratios. The planetary gearsets used in automatic transaxles are the same type found in rear-wheel drive automatic transmissions. Transaxles use Simpson, Ravigneaux, or Wilson planetary gears, and combinations of each. The operation of planetary gearsets was covered in Chapter 8.

Parallel Shaft Gearsets

The automatic transaxles used by two manufacturers have sliding gears that move in and out of mesh. They are sometimes called **parallel shaft transaxles.** The general layout of these transaxles resembles that of a manual transmission. Parallel shaft transaxles are controlled by hydraulically operated clutch packs. See **Figure 10-20.**

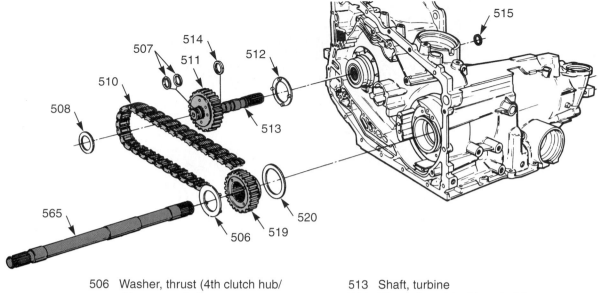

506 Washer, thrust (4th clutch hub/ driven sprocket)	513 Shaft, turbine
507 Ring, oil seal (shaft/sleeve) (2)	514 Ring, oil seal (turbine shaft/support
508 Washer, thrust (drive sprocket/ channel plate)	515 Seal, O-ring (turbine shaft/ turbine hub)
510 Link assembly, drive	519 Sprocket, driven
511 Sprocket, drive	520 Washer, thrust (driven sprocket/ sprocket support)
512 Washer, thrust (drive sprocket/ sprocket support)	565 Shaft, output
	566 Bearing, input sun gear/output shaft

Figure 10-17. *This exploded view of a chain drive assembly shows the relationship of the chain and sprockets to the automatic transaxle case. Sprocket size varies to match engine and vehicle needs. (General Motors)*

Figure 10-18. *Gears used on a common transaxle. Note the meshing of the gears. The drive and driven gears are usually the same diameter, with the same number of teeth.*

Holding Members

Holding members control transaxle planetary gearsets. These are the components of the transaxle that apply force to hold or drive other parts, causing the gearsets to transfer power. Holding members include multiple-disc clutches, bands and servos, and overrunning clutches. All these components were discussed in Chapter 8.

Final Drive Units

The major difference between transmissions and transaxles is that in a transaxle, the case contains the *final*

drive unit. The final drive unit corresponds to the rear axle gears and differential assembly of a rear-wheel drive vehicle. The housing enclosing the final unit is usually cast as an integral part of the transmission section of the case. There are usually one or more sheet metal access covers, which are removed for service. Some differential housings are separate assemblies. Many of these are made of cast iron for added strength. CV axle openings on the housing have seals to retain transmission fluid.

There are three major kinds of final drive units, as shown in **Figure 10-21.**

In **Figure 10-22,** the final gear ratio is obtained by using a planetary gear unit. Power from the transmission section flows into the sun gear. Since the sun gear is turning and the ring gear is splined to the case, the planet gears are forced to turn with the sun gear and carry the planet carrier with them. The ratios between the sun and planet gears cause the carrier to rotate in reduction. Power from the planet carrier then goes to the differential unit. This drive unit is found on many transaxles used with transverse (sideways-mounted) engines.

In **Figure 10-23,** gear reduction is through a set of helical gears. The gear ratio is determined by the number of teeth on the input, or pinion, gear and the number of teeth on the output, or ring, gear. The differential unit is installed on the ring gear. This drive unit is used on transaxles installed on transverse engines.

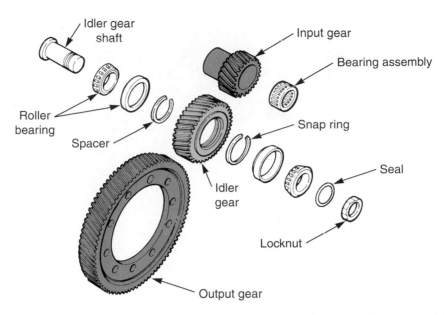

Figure 10-19. *On a gear drive, the gears generally turn in opposite directions. In a few cases, however, an idler gear is used between the input and output gears, so they turn in the same direction. (Ford)*

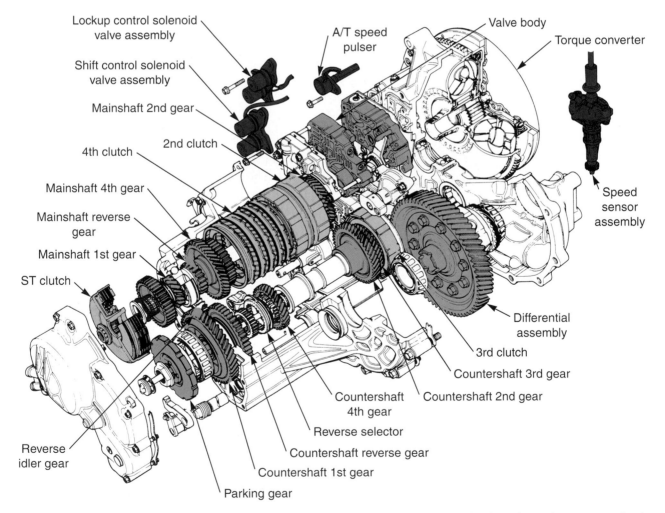

Figure 10-20. *This transaxle is unusual since it does not use planetary gearsets. Instead, clutch packs and a one-way clutch apply and release meshing helical gears. (Honda)*

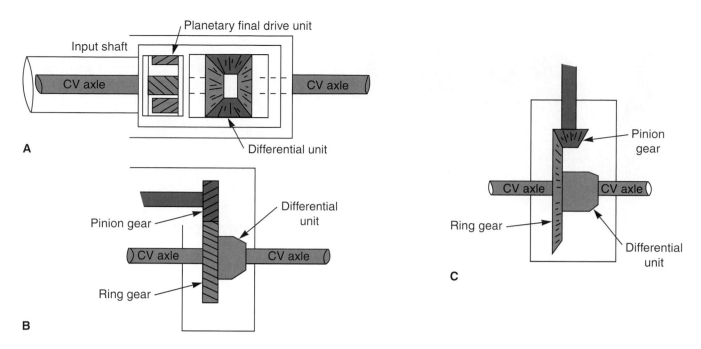

Figure 10-21. *The three kinds of final drive units are shown here. A—This final drive design, which is commonly used with chain-drive units uses a planetary gear to obtain reduction. B—The final drive in this illustration uses helical gears of different sizes to obtain reduction. It is most often used with transverse-mounted transaxles having transfer gears instead of a drive chain. C—The final drive unit in this illustration uses gears of different sizes to obtain reduction. The gears are a type of specially cut gears known as hypoid gears. This type of final drive is similar to the drives used in the rear axles found on rear-wheel drive vehicles.*

670 Ring, snap	679 Gear, governor drive
671 Spacer, final drive internal gear	680 Washer, selective thrust
673 Gear, final drive internal	681 Bearing, differential carrier/case thrust
674 Bearing, thrust sun gear/internal gear	682 Shaft, differential pinion
675 Gear, final drive sun	683 Pin, differential pinion shaft retaining
676 Bearing, thrust sun gear/carrier	684 Washer, pinion thrust
677 Ring, spiral retaining	685 Pinion, differential
678 Differential, carrier	686 Washer, differential side gear thrust
	687 Gear, differential side

Figure 10-22. *This exploded view shows planetary gears and differential gears installed into one compound carrier. The planetary gear is a final drive reduction gear. Combining parts reduces weight and complexity. (General Motors)*

The final drive unit in **Figure 10-24** is used on transaxles installed on longitudinal (front facing) engines. The ring and pinion provide a gear ratio in the same manner as those in transaxles used on transverse engines. However, these are hypoid gears, similar to those used on drive axles of rear-wheel drive vehicles. This type of final drive is usually lubricated by hypoid gear oil and installed in a separate housing. The housing may be part of the transaxle case, or it may be installed on the transmission section of the case.

Differential Assembly

All final drives contain a differential assembly, which allows a vehicle's driving wheels to rotate at different speeds when the vehicle is turning a corner. During a turn, the outer wheel has a longer distance to travel than the inner wheel. If a differential were not used, one wheel would break loose from the pavement every time the vehicle made a turn. This is called *wheel hop.* The differential assembly is installed in the final drive unit to prevent wheel hop.

Figure 10-23. *This photograph shows ring and pinion helical gears used on one popular transaxle.*

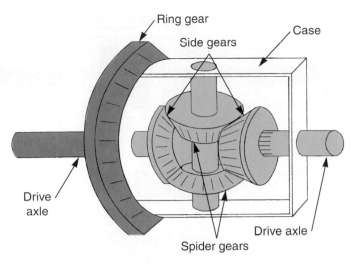

Figure 10-25. *Parts of a differential. The ring is not used on all final drive differentials.*

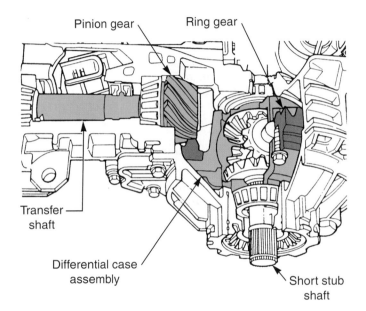

Figure 10-24. *The hypoid ring and pinion shown here are used on longitudinally mounted transaxles. This design is less common than other designs. (DaimlerChrysler)*

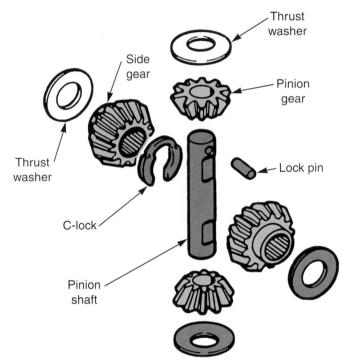

Figure 10-26. *Exploded view of the differential assembly. Notice the relationship of parts. (General Motors)*

Figure 10-25 illustrates the parts of a differential. These are the spider gears, the spider gear shaft, the side gears, and the case. The spider and side gears are bevel gears that are always in mesh. The spider gears are held in place by a steel shaft. The shaft is held to the case. The spider gears can turn on the shaft. Spider gears are sometimes called pinion gears, and shaft is called a pinion shaft. The side gears are held in the case but can turn in relation to the case. They are splined to the output shafts. **Figure 10-26** is an exploded view of the spider and side gears and the shaft.

Differential operation is shown in **Figure 10-27.** When the vehicle is being driven straight ahead, the spider and side gears rotate with the differential case but do not move in relation to it. Speed is transferred equally to both wheels.

When the vehicle makes a turn, the output shafts and side gears begin turning at different speeds. The difference in side gear speeds causes the spider gears to begin to rotate. Movement of the faster side gear causes the spider gears to walk around (rotate in relation to) the slower side gear. This allows the faster moving side gear to turn in relation to the slower moving side gear. Since the side gears are splined to the axle shafts and, therefore, to the related wheel and tire, the drive wheels can turn without wheel hop.

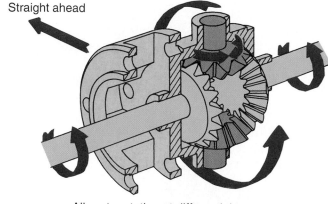

All parts rotating at differential
carrier speed

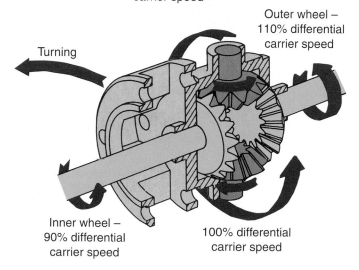

Figure 10-27. *Differential operation is shown in this illustration. When driving straight ahead, the spider and side gears turn as a unit. When the vehicle makes a turn, the spider gears walk around the slower moving side gear to divide speed according to individual wheel speed. (General Motors)*

Final Drive Lubrication

Many final drives are lubricated by gear oil. This is the same type of oil used in the rear axles of rear-wheel drive vehicles. The gear oil is sometimes called *hypoid gear oil.* Final drives using gear oil are housed in a sealed compartment, or sump. Seals and gaskets are used to keep the gear oil and transmission fluid separate. Some gear oil compartments have a separate dipstick to measure oil level. Oil level on other compartments can be checked by removing the fill plug.

Transaxle Oil Pump

Transaxles use the same kinds of oil pumps as rear-wheel drive transmissions. The transaxle pump generates hydraulic pressure to fill the torque converter, operate holding members, and lubricate moving parts. Transaxle oil pumps are turned by the converter whenever the

engine is running. The major difference between transaxle and transmission oil pumps is the location of the pump on certain transaxle designs and whether or not it is driven by a pump shaft.

Some oil pumps are driven directly by lugs on the pump drive hub of the torque converter. On other transaxles, the oil pump is mounted in a location that is remote from the torque converter, usually in the valve body. This arrangement is usually used on transaxles that use a chain drive to transfer power between the converter and transmission. Remote pumps are driven by the converter through the oil pump shaft, **Figure 10-28.** When the pump is installed in this manner, special oil passages must be drilled in the case or the shafts to allow the pump to send oil to other places in the hydraulic system.

To transfer oil from the bottom pan to the side pan, where it can be picked up by the primary oil pump, one automatic transaxle has a main pump, a *secondary pump,* and a *scavenger pump.* **Figure 10-29** is an exploded view of a valve body containing the secondary and scavenger pumps. The secondary and scavenger pumps are driven by the same shaft that drives the primary pump.

Transaxle Oil Filter

Transaxle oil filters are used to remove particles from the automatic transmission fluid. Fluid being drawn into the oil pump must first pass through the filter. This keeps fluid contaminants from entering the pump or the other hydraulic components. Most filters are made of a filter paper or felt and resemble rear-wheel drive transmission filters. The filter is located inside the bottom oil pan. One transaxle has the filter located in the lower part of the side pan. This transaxle also has two smaller debris screens in the bottom pan. Another transaxle has a spin-on oil filter that resembles an engine oil filter.

Oil Scoops and Diverters

Due to the layout of the transaxle, additional devices must sometimes be used to help lubricate the moving parts. Oil scoops and oil diverters are sometimes installed in the transaxle. *Oil scoops* are used to pick up oil being thrown from moving parts and deliver it to other portions of the transaxle, which would otherwise be starved for lubrication. *Oil diverters,* sometimes called weirs, keep excess oil away from parts that could be over lubricated. Diverters are also used to keep excess oil from sealing areas or vents that could leak lubricant. Scoops and diverters are usually made of plastic and fit into recesses in the transaxle case.

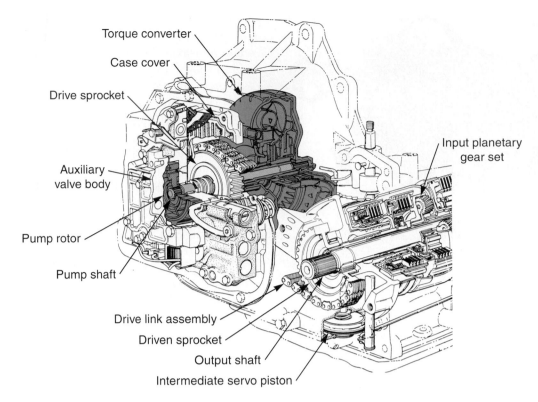

Figure 10-28. *Instead of being driven directly by the torque converter, this gear pump is driven by a shaft. Transaxles can use either method, depending on their design. (Ford)*

Transaxle Oil Cooler

Fluid in an automatic transaxle is cooled in the *oil cooler.* The cooler consists of a heat exchanger that is built into a side tank of the engine cooling-system radiator. Cooler lines connect the transaxle to the oil cooler. Transaxle oil coolers operate in the same manner as coolers in rear-wheel drive transmissions. Vehicles used for towing trailers or operated under other extreme conditions may benefit from direct, air-cooled external coolers.

Valve Body

The valve body is a machined casting that contains hydraulic valves and passageways. In conjunction with the hydraulic valves, it redirects and modifies oil flow to the transaxle holding members, converter, and cooling and lubrication systems. The valve body may also contain accumulators, check balls, and spacer plates. The driver operates the manual valve through linkage. The major difference between transaxle and transmission valve bodies is the placement and number of valve bodies in the transaxle. To make maximum use of limited space, some transaxles use more than one valve body.

It is common for the main valve body to be mounted on the side of the transaxle case, under the side oil pan. This valve body may contain the oil pump and is usually installed sideways in relation to the valve bodies used on rear-wheel drive transmissions. Sideways installation does not affect valve body operation.

Smaller auxiliary valve bodies are located in the lower part of the case, inside the lower oil pan, **Figure 10-30.** Each valve body operates a portion of the hydraulic control system. Oil transfer tubes are commonly used to connect various parts of the hydraulic system.

Some transaxles have an extra cast housing called a **channel plate.** The channel plate contains oil passages to direct the flow of oil pressure between components. A channel plate is shown in **Figure 10-31.**

Automatic Transaxle Shift Linkage

The shift linkage connects the shift selector lever, which may be mounted on the floor or on the steering column, to the automatic transaxle. It consists of a series of levers and linkage rods or cables. These transfer motion to the manual valve of the transaxle valve body to select the desired operating range. The design of the shift linkage is similar to that on a rear-wheel drive automatic transmission, including manual and throttle valve levers, parking pawl linkage, and a cockscomb on the inner manual lever.

Electronic Components

Most modern vehicles are controlled, at least in part, by an on-board computer. In addition to controlling the

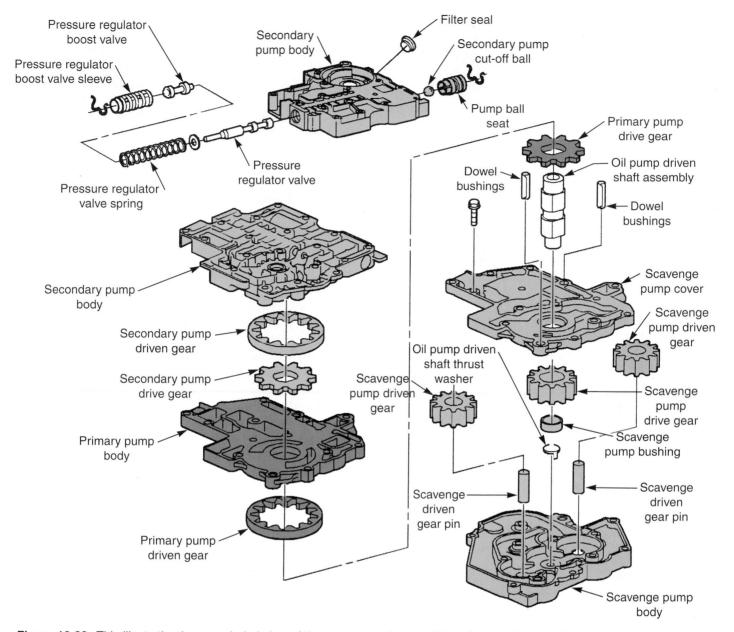

Figure 10-29. *This illustration is an exploded view of the pumps used on one kind of transaxle. Note that the valve body contains primary and secondary pumps, which draw fluid from the main filter, and a three-gear scavenger pump, which draws fluid though a filter screen in the lower oil pan. (General Motors)*

engine and emission systems, the on-board computer receives inputs from and sends commands to the automatic transaxle. Automotive computer systems include input and output devices, such as sensors and solenoids.

Through electronic hardware, the computer typically monitors engine speed, engine load, and throttle position, among other variables. It can then provide control of transaxle shift points, torque converter lockup, and other functions. This keeps the transaxle functioning at maximum efficiency. Chapter 12 contains more information on electronic control systems.

Automatic Transaxle Power Flow

Power flow through transaxles is similar for all transaxle designs. The main differences depend on whether the vehicle's engine is mounted longitudinally or transversely.

Figure 10-32 shows one make of transaxle used with a transverse engine. Power flows from the engine, through the torque converter and turbine (input) shaft, and on to the transaxle multiple-disc clutches. The planetary holding members match the proper gear ratio to conditions. Power flows from the clutches to the planetary gears. The planetary gears drive the output shaft. A gear on the output shaft drives a gear on the transfer shaft. The transfer shaft pinion gear, on the other end of the transfer shaft, drives the

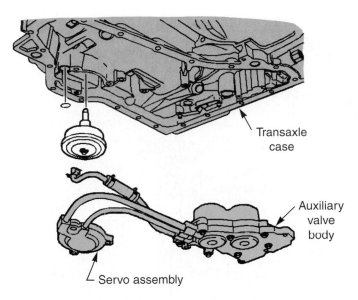

Figure 10-30. *This auxiliary valve body contains the accumulators. Other auxiliary valve bodies contain valves, thermostatic elements, and pressure sensors. (General Motors)*

differential ring gear and case assembly. Power leaves the differential assembly and drives the CV axles and front wheels.

Figure 10-33 shows a transaxle used with a longitudinal engine. Engine power enters the torque converter and flows through the turbine shaft to the chain drive. From the driven sprocket, power flows through the input shaft to the planetary gears. The hydraulic system and holding members select the proper gear ratio for driving conditions. Power from the output shaft turns the pinion gear. The ring and pinion turn the power flow 90°. (They are not hypoid gears, however, because their centerlines are not offset.) Power flows through the differential assembly and drives the CV axles and front wheels.

Power Flow Diagrams

The following figures show the power flow through a typical four-speed automatic transaxle. This unit uses a compound gearset consisting of two simple gearsets connected together. The front ring gear and rear planet carrier are connected to each other. The front planet carrier and rear ring gear are also connected. There are separate front and rear sun gears.

This transaxle uses a chain drive and sprockets to connect the torque converter and input shaft. Chains are used on most transaxles. Refer to **Figures 10-34** through **10-38** as you read the following paragraphs.

First Gear

In first gear, **Figure 10-34,** power flows through the engine crankshaft to the torque converter. From the torque

converter, power flows through the turbine shaft to the drive sprocket. The drive sprocket turns the driven sprocket through the drive chain. The front sun gear receives power from the driven sprocket by way of the forward clutch and low one-way clutch assembly. The front sun gear turns the front planet gears. The turning planet gears force the front planet pinions to walk around the front ring gear. Since the front pinion carrier and rear ring gear are a single piece, the rear ring gear drives the rear planet carrier. The rear ring gear causes the rear planet gears to walk around the rear sun gear. The rear sun gear is held stationary by the low/intermediate band. The rear planet carrier transmits power to the final drive unit. The action of the planetary gears results in a gear ratio of 2.77:1.

Second Gear

In second gear, **Figure 10-35,** the power flows through the torque converter and drive sprockets in the same manner as in first gear. However, in second, the intermediate clutch is applied. The intermediate clutch causes engine power to directly drive the front planet carrier and rear ring gear assembly. The rear ring gear drives the rear carrier and front ring gear around the stationary rear sun gear at a reduction of 1.5:1. The forward clutch remains applied, but it is ineffective because the low one-way clutch is overrunning.

Third Gear

In third gear, the power flows through the torque converter and sprockets to the front sun gear and the front carrier. All the components of both planetary gearsets turn as a single unit, **Figure 10-36.** This is accomplished by applying the front clutch and releasing the intermediate band. Third gear is a 1:1 ratio.

Fourth Gear

In fourth gear, the forward clutch is released and the overdrive band is applied. This causes the front sun gear to be held stationary. The intermediate clutch drives the front carrier and rear ring gear assembly directly from the driven sprocket. The front planet pinions rotate around the stationary sun gear and carry the planet carrier with them at a ratio of about 0.7:1. Since the front carrier and rear ring gear assembly is attached to the final drive gear, this is the fourth gear ratio. See **Figure 10-37.**

Reverse

In reverse, **Figure 10-38,** the forward and reverse clutches are applied. The reverse clutch holds the front carrier and rear ring gear assembly stationary. Power enters

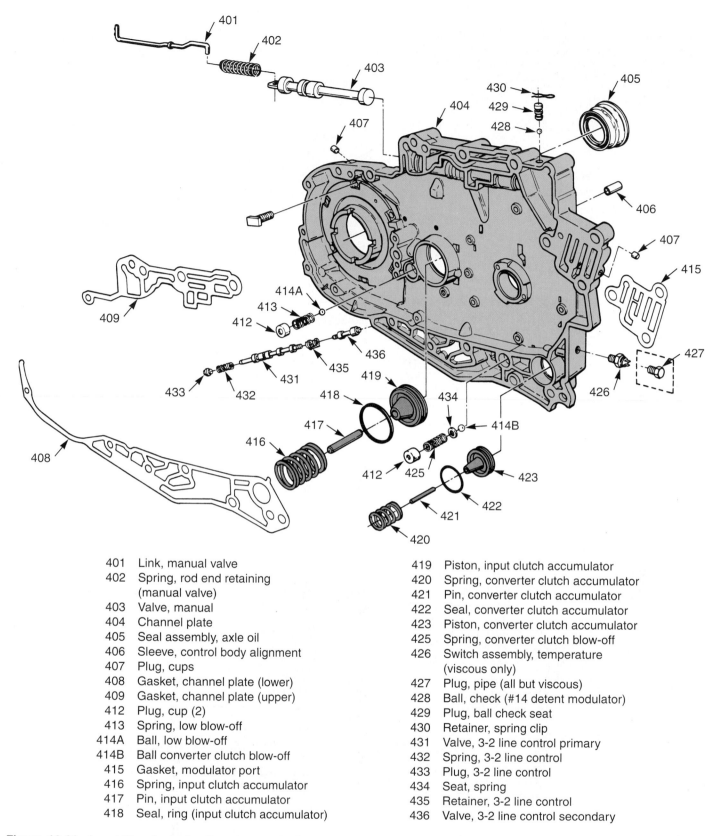

401	Link, manual valve	419	Piston, input clutch accumulator
402	Spring, rod end retaining (manual valve)	420	Spring, converter clutch accumulator
403	Valve, manual	421	Pin, converter clutch accumulator
404	Channel plate	422	Seal, converter clutch accumulator
405	Seal assembly, axle oil	423	Piston, converter clutch accumulator
406	Sleeve, control body alignment	425	Spring, converter clutch blow-off
407	Plug, cups	426	Switch assembly, temperature (viscous only)
408	Gasket, channel plate (lower)	427	Plug, pipe (all but viscous)
409	Gasket, channel plate (upper)	428	Ball, check (#14 detent modulator)
412	Plug, cup (2)	429	Plug, ball check seat
413	Spring, low blow-off	430	Retainer, spring clip
414A	Ball, low blow-off	431	Valve, 3-2 line control primary
414B	Ball converter clutch blow-off	432	Spring, 3-2 line control
415	Gasket, modulator port	433	Plug, 3-2 line control
416	Spring, input clutch accumulator	434	Seat, spring
417	Pin, input clutch accumulator	435	Retainer, 3-2 line control
418	Seal, ring (input clutch accumulator)	436	Valve, 3-2 line control secondary

Figure 10-31. *In addition to hydraulic valves and other components, this channel plate, which fits up against the automatic transaxle case, contains two servos. The servo springs are calibrated for each servo assembly and should not be interchanged. (General Motors)*

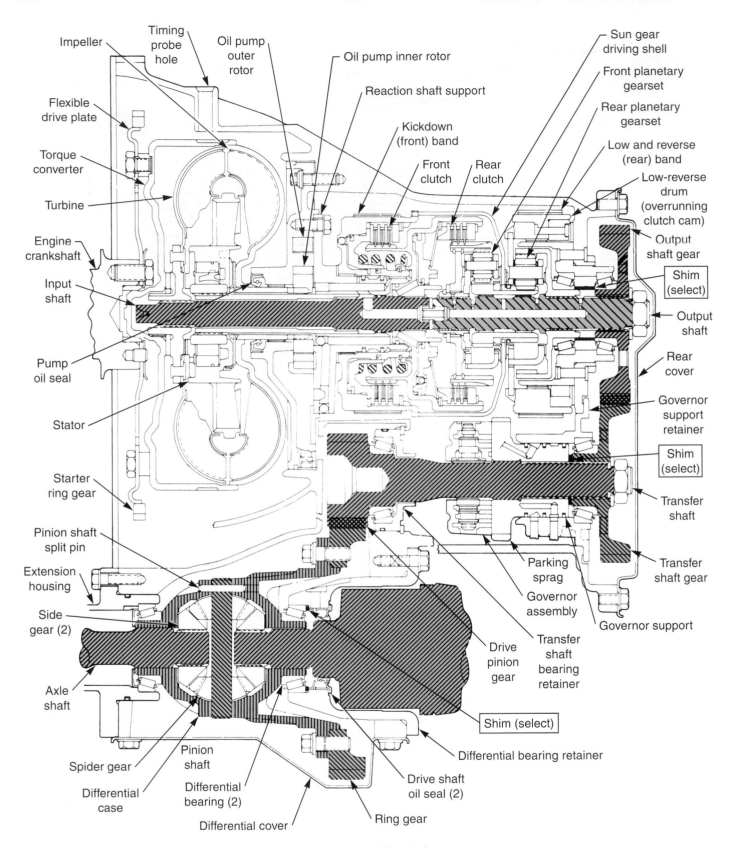

Figure 10-32. *Power flow through a transverse transaxle. (DaimlerChrysler)*

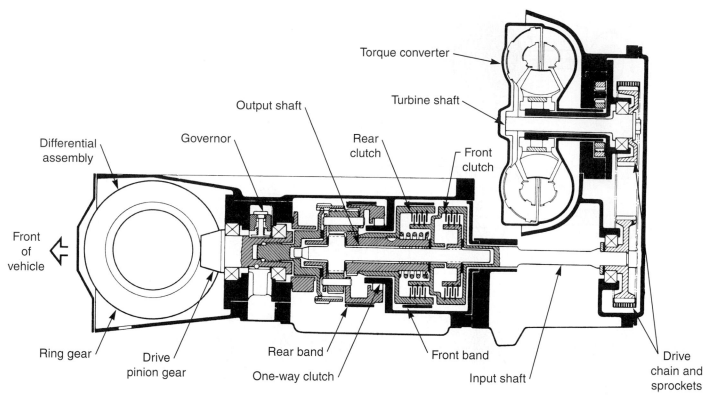

Figure 10-33. *This transaxle is used with a longitudinal engine. Power takes a 90° turn through the ring and pinion. (Saab)*

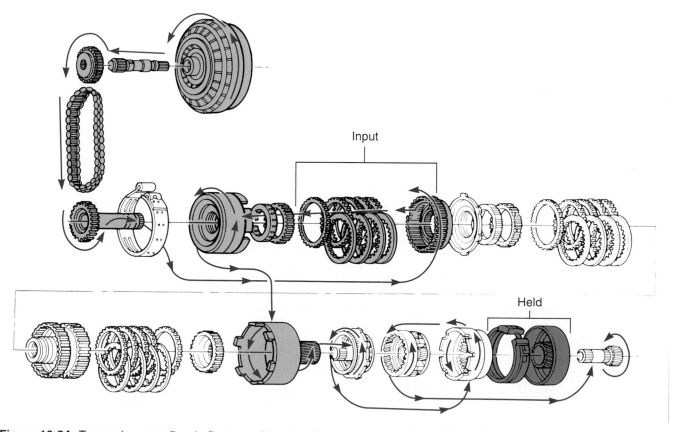

Figure 10-34. *Transaxle power flow in first gear. Note how the components operate to obtain reduction. (Ford)*

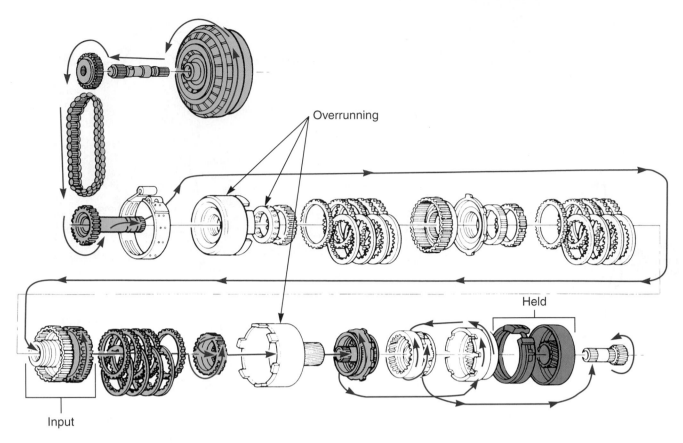

Figure 10-35. *Transaxle power flow in second gear. (Ford)*

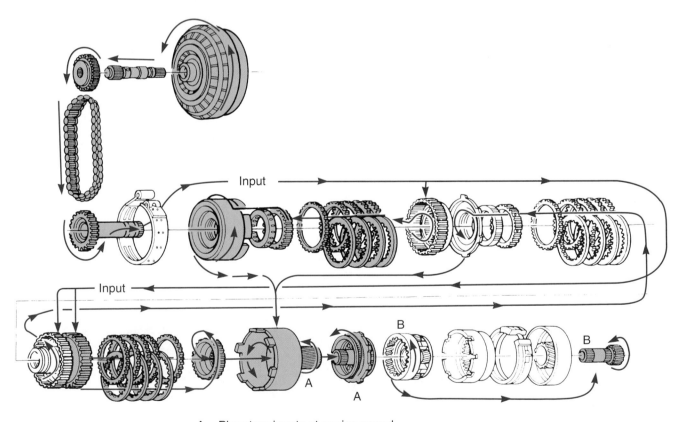

A = Planetary inputs at engine speed
B = Planetary output at engine speed

Figure 10-36. *Transaxle power flow in third gear. This arrangement of holding members and planetary gears is used to obtain direct drive. (Ford)*

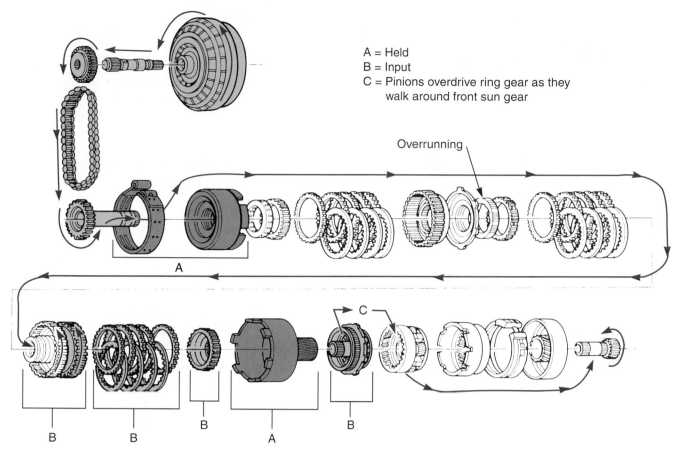

A = Held
B = Input
C = Pinions overdrive ring gear as they walk around front sun gear

Figure 10-37. *Transaxle power flow in overdrive gear. (Ford)*

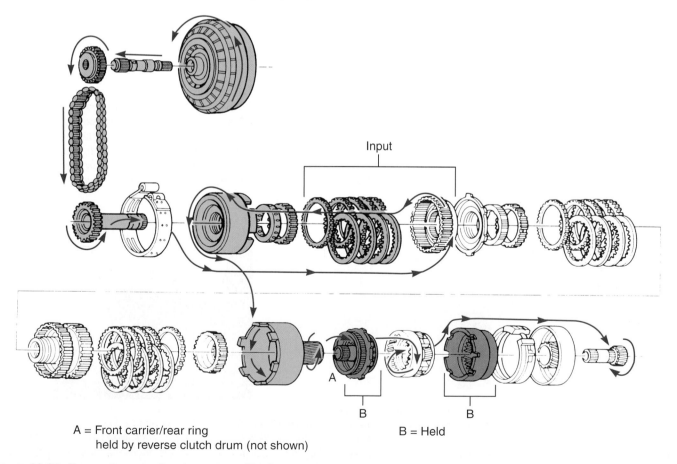

A = Front carrier/rear ring
held by reverse clutch drum (not shown)

B = Held

Figure 10-38. *Transaxle power flow in reverse. (Ford)*

the front sun gear by way of the forward clutch. With the front carrier stationary, the front planet pinions reverse the sun gear movement, causing the front ring gear to turn in the opposite direction of sun gear rotation. The front ring gear and rear carrier assembly carries the output shaft with them at a reverse gear ratio of about 2.3:1.

All Clutch Transaxle

An unusual type of transaxle is shown in **Figure 10-39.** This transaxle uses no bands or one-way clutches to hold the planetary gears. All gears are provided by applying and releasing clutch packs. There are two versions of this transaxle. The original design is used on vehicles with transverse engines. A later version of this transaxle is used on vehicles with front-facing, or longitudinal, engines.

The clutch-only application system allows this transaxle to use a simplified valve body with only five valves. Clutch packs are applied and released by solenoid-operated check balls. A body-mounted computer controls the operation of a solenoid pack, which in turn controls the hydraulic system. Computer operation of this transaxle is covered in more detail in Chapter 12.

Continuously Variable Transaxle

The *continuously variable transmission (CVT)* has been used for many years in lawn equipment, snowmobiles, and other small engine–driven equipment. Improvements in design and materials have made the CVT

practical for automotive use, particularly on front-wheel drive vehicles with engines having relatively low torque output.

The continuously variable transmission is an automatic transaxle using a belt and two variable-diameter pulleys to transfer engine power to the drive wheels. See **Figure 10-40.** The variable-diameter pulleys are made in two halves. One-half of the pulley is fixed; the other half is adjustable, **Figure 10-41.** The adjustable pulley half is operated by a hydraulic piston. The drive belt consists of several hundred steel segments that are held together by

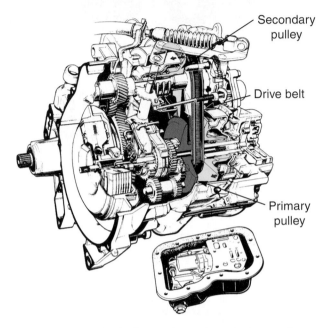

Figure 10-40. *Basic design and part layout of a continuously variable transmission (CVT). The most obvious difference between a CVT and a conventional transaxle is the belt and pulley system. Gears in the CVT assist in driving the belt in the forward and reverse directions. An electromagnetic clutch engages and disengages the engine and transaxle. The CVT also has a differential assembly, as in a conventional transaxle. (Subaru)*

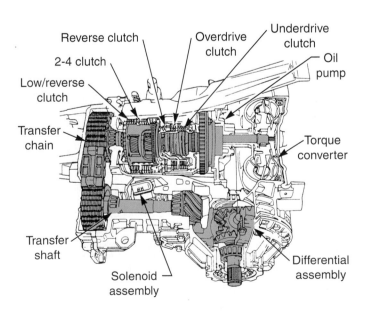

Figure 10-39. *The transaxle shown here uses clutch packs to obtain all gears. No bands or overrunning clutches are used. This transaxle relies on a simple valve body and a complex series of computer controls and sensors to operate. (DaimlerChrysler)*

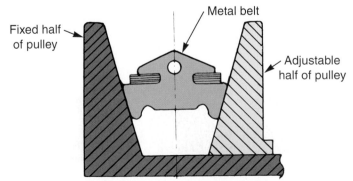

Figure 10-41. *A split pulley is used on a CVT. One-half of the pulley is stationary. The other half slides on the pulley shaft. Moving the pulley halves apart decreases the pulley's effective diameter. Moving the pulley halves closer together increases the pulley's effective diameter.*

overlapping steel bands. See **Figure 10-42.** The design of the belt allows it to be pushed by the drive pulley.

Varying the effective diameter of the pulleys by moving the adjustable pulley halves in and out causes the belt to ride higher or lower in the pulley grooves. **Figure 10-43** shows how changes in pulley diameter cause the belt to change position in the pulley groove. When the pulley halves are far apart, the belt rides very close to the center of the pulley (small effective diameter). When the pulley halves are close together, the belt rides farther from the center of the pulley (large effective diameter).

When the pulley diameter is changed, the ratio between the drive pulley and the driven pulley is also changed. This ratio change matches vehicle speed and engine power output. The design of the belt and the pulleys gives the CVT an unlimited number of drive ratios. Therefore, the transmission allows the engine to operate in its most efficient range at all times. This increases performance and reduces fuel consumption. Because the operation of the CVT is extremely smooth, the drive ratio changes are often referred to as stepless shifting.

Figure 10-44 shows a simple belt and pulley system. Engine power enters the primary pulley, passes through the belt, and exits the secondary pulley.

When a vehicle is accelerating, the primary pulley width is increased and the secondary pulley width is reduced. This causes the belt to ride close to the center of the primary pulley and near the outer part of the secondary pulley. See **Figure 10-45A.** Several revolutions of the small primary pulley are required to produce one revolution of

the secondary pulley. This results in a reduction ratio and provides maximum power multiplication.

As the vehicle approaches cruising speed, maximum power multiplication is no longer needed. The width of the primary pulley is decreased, and the width of the secondary pulley is increased. This causes the belt to move out on the primary pulley and closer to the center on the secondary pulley. Since the belt is approximately the same distance from the center of both pulleys, **Figure 10-45B,** the result is the same as driving a pulley with another pulley of the same size. A 1:1 ratio is achieved; the CVT is in direct drive.

When the vehicle is at cruising speed, the primary pulley width is at its smallest setting, causing the belt to

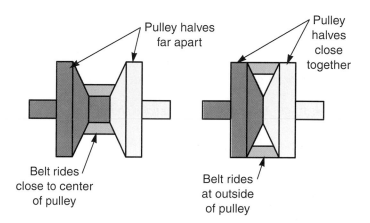

Figure 10-43. *Changing the effective pulley diameter can change the position of the belt in the pulley. A—The two sides of the pulley are far apart, and the belt moves close to the center of the pulley. B—The two sides of the pulley are close together, and the belt is pushed to the outside of the pulley.*

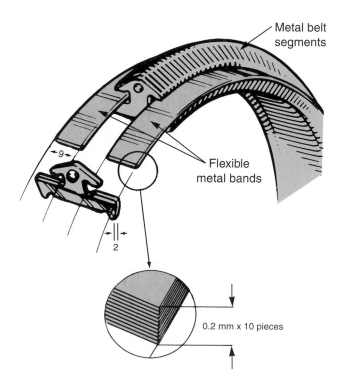

Figure 10-42. *The CVT drive belt is made up of many small metal segments, which are held together by a series of flexible steel bands. The belt is designed to be pushed, rather than pulled, by the drive pulley. (Subaru)*

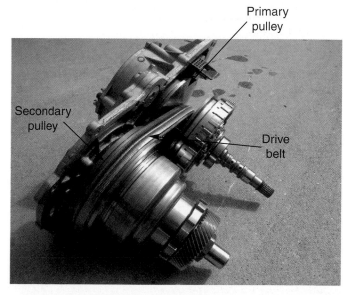

Figure 10-44. *The drive mechanism of a CVT is shown here. The effective diameters of the primary and secondary pulleys are changed to achieve different drive ratios.*

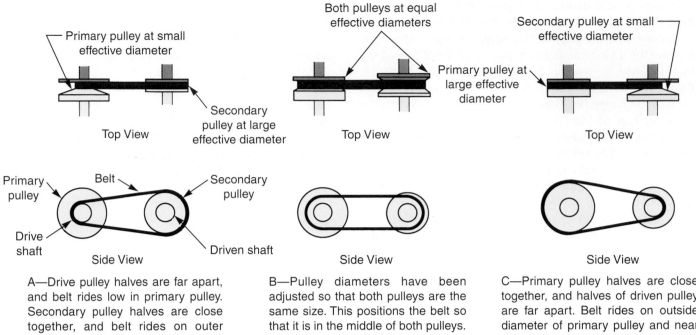

Figure 10-45. *This series of illustrations shows how changing the position of the belt on the pulleys can change the pulley ratio.*

ride on the outer portion of the pulley. The width of the secondary pulley is increased, causing the belt to ride near the center of the pulley. Since the belt is near the outside of the primary pulley and near the center of the secondary pulley, a small pulley is being driven by a large pulley. See **Figure 10-45C.** When the primary pulley turns through one revolution, it turns the driven secondary through more than one revolution. This is an overdrive ratio.

Rear-Wheel Drive Transaxles

Some older vehicles, as well as a few newer models, have rear-wheel drive automatic transaxles. These designs are generally used on rear-engine vehicles. The basic concept of the rear-wheel drive transaxle is the same as with front-drive models. The transmission and final drive are contained in a single unit. Within that basic design, however, rear-wheel drive transaxles vary. Some older transaxles have as many as three hollow shafts to transmit power and turn the oil pump. Some imports use a parallel shaft design similar to the ones discussed earlier in this chapter, while others resemble modern front-drive transaxles.

Summary

The automatic transaxle contains many systems and parts that are similar to those used in the rear-wheel drive automatic transmission and differential assembly. In a transaxle, however, both the transmission and the differential assembly are contained in one housing. The main difference between many transaxle and transmission parts is their location.

The transaxle case is made of aluminum, with passages for the flow of transmission fluid between other components. Sheet metal covers are used to enclose various parts of the transaxle case. The covers make transaxle disassembly easier. An oil pan is located at the bottom of the transaxle.

The torque converter is a fluid transfer device that can transmit and multiply engine power. It operates according to the same principles as the torque converters found on rear-wheel drive vehicles. Most transaxle converters use a lockup clutch in the torque converter. Some transaxles have a magnetic powder clutch instead of a torque converter.

Input and output shafts transfer power. The turbine shaft is a type of input shaft turned by the converter. It sends power to a drive chain or directly to the gear train. The output shaft is often the connection between the transmission and the differential. A transfer shaft is used to transfer power to the differential assembly in some transaxles. Some transaxles have an oil pump shaft to drive the oil pump.

Some transaxles use a chain drive to transfer power between the torque converter and the gear train. Transaxles use the same types of planetary gears as rear-wheel drive transmissions. The gearsets are controlled with clutch packs, bands, and one-way clutches.

The transaxle oil pump produces the fluid pressure that operates other hydraulic components and lubricates the moving parts of the transaxle. The transaxle oil filter removes particles from the automatic transmission fluid prior to circulation by the oil pump. Oil from the torque converter is pumped to an oil cooler at the front of the vehicle. The oil cooler is located in the radiator. Auxiliary coolers may be mounted ahead of the oil cooler.

Ring and pinion gears are sometimes used in automatic transaxles. Those used with transverse engines are straight-cut or helical gears. Those used with longitudinal engines must divert power flow by 90°, just as a hypoid ring and pinion on a rear-wheel drive vehicle.

Transaxle differentials operate in the same way as the differential in a rear-wheel drive rear axle. The interaction of the spider and side gears allows the vehicle to make turns. Some planetary and differential gears are combined into a single unit.

The continuously variable transmission, or CVT, is an automatic transaxle that uses a belt and two variable-diameter pulleys to transfer engine power to the drive wheels. Varying the diameter of the pulleys varies the gear ratio of the transaxle. This kind of transaxle has infinite gear ratios.

Rear-wheel drive transaxles are not commonly used. The basic design of the rear-wheel drive transaxle is the same as for front-drive units, but design details vary.

Review Questions—Chapter 10

Please do not write in this text. Place your answers on a separate sheet of paper.

1. The _____ serves as a mounting and aligning surface for the moving parts of a transaxle.

2. In most transaxles with a chain drive, the chain is located between the _____ and _____.
   (A) converter, final drive
   (B) converter, transmission assembly
   (C) transmission assembly, final drive
   (D) final drive, CV axles

3. In most transaxles with a gear drive, the gears are located between the _____ and _____.
   (A) converter, final drive
   (B) converter, transmission assembly
   (C) transmission assembly, final drive
   (D) final drive, CV axles

4. Do any transaxles with a chain drive also have a hypoid ring and pinion final drive?

5. The differential assembly allows the drive wheels to turn at different _____.

6. What two methods are used to drive the transaxle oil pump?

7. A side-mounted valve body does *not* contain the _____.
   (A) shift valves
   (B) oil pump
   (C) chain drive
   (D) accumulators

8. In a CVT, power is transmitted from the drive pulley to the driven pulley by a metal _____ made of many small _____.

9. How does a CVT provide an infinite number of ratios?

10. Rear-wheel drive transaxles were usually used with _____-mounted engines.

ASE-Type Questions—Chapter 10

1. All of the following statements about magnetic powder clutches are true *except*:
   (A) the clutch electromagnet is energized through brushes.
   (B) varying voltage to the electromagnet varies the amount of power transmitted.
   (C) the clutch housing may be finned for heat removal.
   (D) the magnetic powder is usually brass or bronze particles.

2. Each of the following shafts is used in automatic transaxles *except*:
   (A) countershafts.
   (B) input shafts.
   (C) turbine shafts.
   (D) oil pump shafts.

3. The turbine shaft is splined to the converter turbine and the:
   (A) oil pump.
   (B) drive sprocket.
   (C) driven sprocket.
   (D) transfer shaft.

4. Some automatic transaxles have a transfer shaft that transmits power from the transaxle output shaft to the:

   (A) input shaft.

   (B) engine crankshaft.

   (C) planetary gearsets.

   (D) differential assembly.

5. Each of the following is part of a planetary gear–type final drive assembly *except*:

   (A) side gears.

   (B) spider gears.

   (C) sun gear.

   (D) pinion gear.

6. Technician A says that automatic transaxles used with most longitudinal engines will have a drive chain. Technician B says that all automatic transaxles used with transverse engines will have a helical gear ring and pinion as part of the differential assembly. Who is right?

   (A) A only.

   (B) B only.

   (C) Both A and B.

   (D) Neither A nor B.

7. Technician A says that one function of the transaxle oil pump is to keep moving parts of the transaxle lubricated. Technician B says that keeping the torque converter filled with transmission fluid is one of the oil pump's functions. Who is right?

   (A) A only.

   (B) B only.

   (C) Both A and B.

   (D) Neither A nor B.

8. Technician A says that many transaxles have two oil pans. Technician B says that many transaxles have two valve bodies. Who is right?

   (A) A only.

   (B) B only.

   (C) Both A and B.

   (D) Neither A nor B.

9. The channel plate is installed between the valve body and the:

   (A) oil pump.

   (B) transaxle case.

   (C) final drive housing.

   (D) spacer plate.

10. The term *stepless shifting* refers to which of the following transaxles?

   (A) Transverse.

   (B) Longitudinal.

   (C) Continuously variable.

   (D) Linear.

Chapter 11

Transmission and Transaxle Circuits

After studying this chapter, you will be able to:
- ❏ Describe basic hydraulic circuit operation.
- ❏ Trace the path of oil flow through typical automatic transmission circuits.
- ❏ Read hydraulic circuit diagrams for automatic transmissions and transaxles.
- ❏ Explain what happens to cause an upshift or a downshift.

Technical Terms

Line pressure	Cut back valve
Throttle pressure	3-2 shift timing valve
Governor pressure	TV limit valve
Overrunning	Line bias valve
2-3 backout valve	Modulator TV valves
Throttle pressure booster	Overrun clutch

Introduction

This chapter details the operation of automatic transmission and transaxle hydraulic systems. It begins with a discussion of basic hydraulic circuit operation. Once this is understood, you will be prepared to follow the operation of a typical modern hydraulic control system.

Basic Hydraulic Circuit Operation

The components of a hydraulic circuit must work together to control the automatic transmission's mechanical system and, ultimately, produce movement and affect changes in gear ratios in response to engine and vehicle needs. This, then, is the primary function of the complete hydraulic circuit—to apply and release a band or to engage or disengage a multiple-disc clutch in order to accomplish these tasks. It is by this action that the hydraulic circuit regulates the shifting process. In addition to this primary function, the hydraulic circuit has secondary functions, such as routing fluid for cooling and lubrication purposes.

Simple Automatic Transmission

The hydraulic circuit of a simple automatic transmission is shown in **Figure 11-1.** The circuit consists of such items as the oil pump and pan, hydraulic valves, servos and clutches, and connecting passageways. The oil pump produces the system pressure. This pressure is regulated by the pressure regulator valve. From the pressure regulator valve, pressurized fluid, or *line pressure,* goes to the manual valve in the valve body. The driver controls the manual valve through the shift selector lever and shift linkage.

The transmission having the circuit shown in **Figure 11-1** has two planetary holding members that give it two speeds: low and high. The band is applied in low gear, and the clutch pack is used in high gear.

If the driver places the manual valve in drive and the vehicle is not moving, line pressure goes to the shift valve, as well as to the throttle and governor valves. The manual valve also directs fluid to the band servo apply side. Pressure to the apply side of the servo overcomes the pressure of the release spring and applies the band. This is shown in **Figure 11-2.**

The output pressure from the throttle valve, which is called *throttle pressure,* will vary, depending on engine load. The throttle valve in **Figure 11-2** is operated by a vacuum modulator. Some throttle valves are operated by a throttle valve linkage, or TV linkage. With either type, throttle pressure changes with engine load. Heavy engine load causes high throttle pressure; light engine load causes low throttle pressure.

Governor pressure is proportional to road speed. The governor is often driven by the transmission output shaft. It

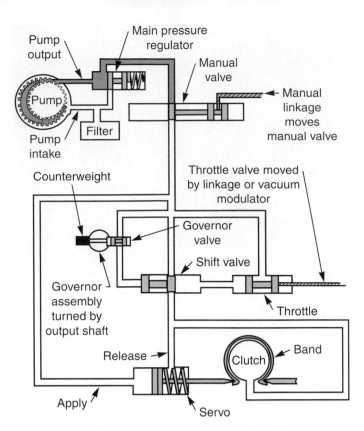

Figure 11-1. *Hydraulic circuit of a simple automatic transmission. All the parts shown here were discussed in earlier chapters.*

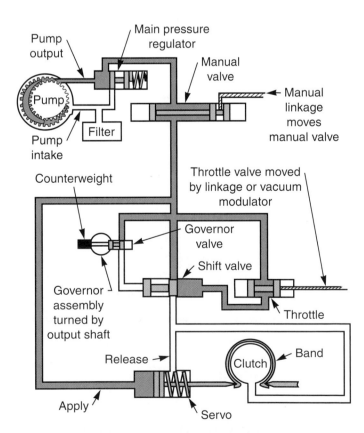

Figure 11-2. *The simple automatic transmission in the downshifted (low gear) position. The low band is applied. Note the valve positions and oil flows that create low gear.*

revolves faster as vehicle speed increases. As the governor speed increases, forces are exerted on the governor valve, causing it to move and increasing governor pressure. As a result, an increasing pressure signal is sent to the shift valve. When the vehicle slows down, forces exerted on the governor valve decrease, and it modulates accordingly. As a result, the governor valve puts out less pressure.

Governor pressure and throttle pressure, acting on opposite ends of the shift valve, oppose each other. At low speeds, governor pressure is low. Throttle pressure and spring pressure hold the shift valve in its normal position, to the left in **Figure 11-2.** With the shift valve in this position, line pressure from the manual valve is blocked from the clutch pack, and the clutch pack remains disengaged.

As the vehicle speeds up, governor pressure increases. At a certain vehicle speed, the governor pressure becomes greater than throttle pressure and spring pressure. This causes the shift valve to move to the right in **Figure 11-3.** The speed of the vehicle at which the upshift occurs is governed by the throttle pressure. Hard acceleration causes higher throttle pressure, and the governor has to spin faster to put out enough pressure to overcome throttle pressure. This means that the vehicle has to be moving at a higher speed before the upshift can occur. Light acceleration means lower throttle pressure, and a lower vehicle speed will cause enough governor pressure to move the shift valve.

With the shift valve pushed to the right by governor pressure, line pressure from the manual valve can flow through the shift valve to cause an upshift. Pressure goes to the clutch piston to apply the clutch and also to the release side of the band servo. Oil pressure equalizes on both sides of the servo piston. The release spring pushes the piston away from the band, releasing it. The vehicle is now in high gear.

If the vehicle slows down, governor pressure decreases. At some vehicle speed, depending on throttle position, governor pressure becomes less than the throttle and spring pressures. The combined throttle and spring pressures will then push the shift valve to the left, causing a downshift. An open throttle (low vacuum pressure) causes high throttle pressure, hastening the downshift. This action causes a forced downshift. A closed throttle (high vacuum pressure) causes low throttle pressure, delaying the downshift. A review of throttle valves, shift valves, and governor valves, presented in Chapter 9, may be helpful if you do not fully understand the operation just described.

3-Speed Automatic Transmission

The hydraulic system of a 3-speed automatic transmission is slightly different than that of the simple 2-speed unit just discussed. **Figure 11-4** shows a simple hydraulic control circuit in a 3-speed transmission. The most obvious

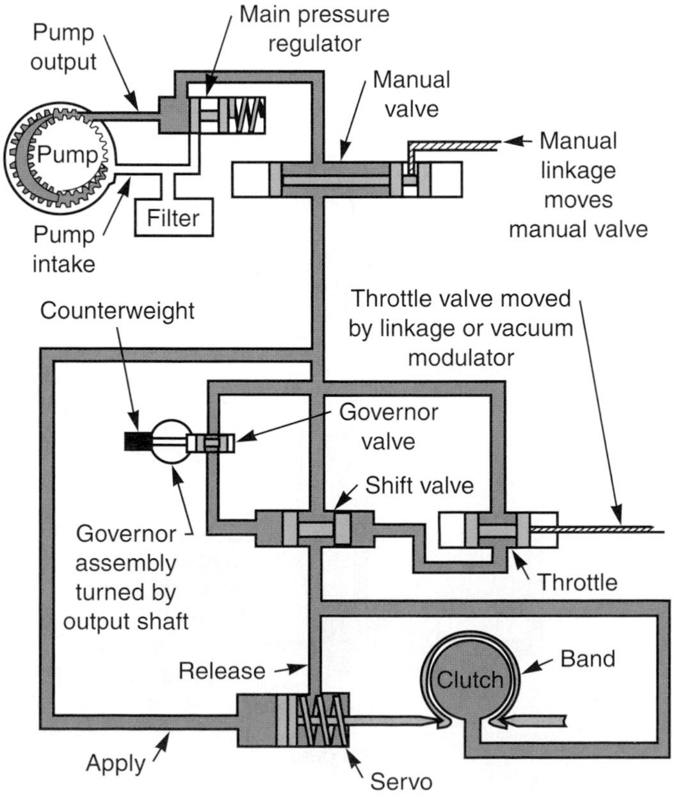

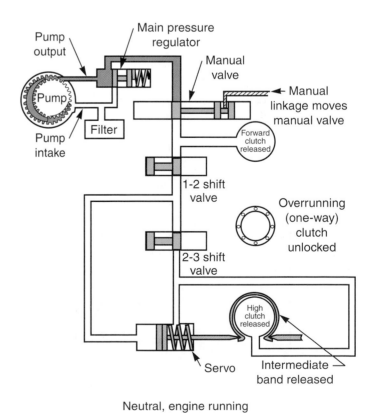

Figure 11-3. *The simple hydraulic circuit in the upshifted (high gear) position. The band has been released and the clutch has been applied. Compare the hydraulic circuit oil flows and valve positions with those in Figure 11-2.*

Figure 11-4. *This is a 3-speed version of the simple transmission that was shown in Figures 11-1 through 11-3. Compare the figures to identify the additional valves and holding members.*

feature of this transmission is that it has two shift valves. Note also that the transmission has a forward clutch and an overrunning (one-way) clutch. For clarity, the throttle and governor valves are not shown. In this figure, the transmission is in neutral with the engine running. None of the holding members is applied.

The transmission starts off from rest in drive, and the forward clutch is applied, **Figure 11-5.** The overrunning clutch locks, and the vehicle is in first gear. As vehicle speed increases, governor pressure overcomes throttle pressure to move the 1-2 shift valve to the upshifted position. See **Figure 11-6.** Moving the 1-2 shift valve to the upshifted position causes oil to flow to the apply side of the band servo, applying the band. Applying the band causes the overrunning clutch to release, placing the transmission in second gear. The forward clutch remains applied.

When the vehicle is moving fast enough, governor pressure overcomes throttle pressure and moves the 2-3 shift valve to the upshifted position, **Figure 11-7.** In the upshifted position, the 2-3 shift valve allows pressure to go to the high clutch and the release side of the band servo. Applying the high clutch and releasing the band puts the transmission in third gear. The forward clutch remains applied, and the overrunning clutch remains released.

When the vehicle slows down, governor pressure drops off and, at some point, the 2-3 valve moves to the downshifted position. In this position, pressure is blocked

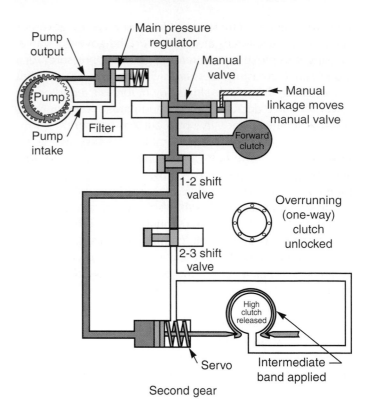

Figure 11-6. *In second gear, the 1-2 shift valve has moved to the upshifted position. This causes oil to flow to the apply side of the band servo, applying the band and causing the overrunning clutch to release.*

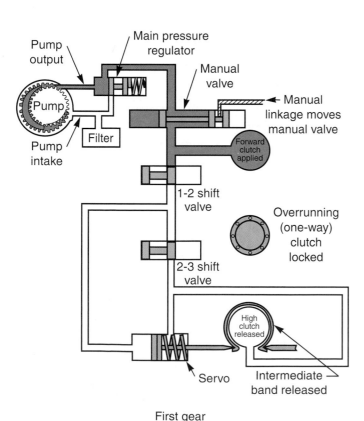

First gear

Figure 11-5. *In first gear, both shift valves are in the downshifted position. Notice that the forward clutch is applied with hydraulic pressure. The one-way clutch is locked mechanically by planetary gear rotation when the forward clutch is applied.*

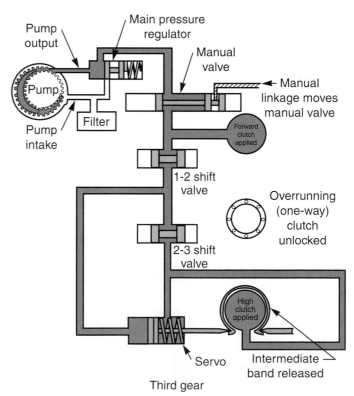

Third gear

Figure 11-7. *In third gear, both the 1-2 shift valve and the 2-3 shift valve are in the upshifted position. The 2-3 shift valve allows pressure to go to the high clutch and the release side of the band servo. Applying the high clutch and releasing the band puts the transmission in third gear.*

from the high clutch and the release side of the band servo. The high clutch releases, and oil on the apply side of the band servo is able to apply the band. The transmission returns to second gear.

As the vehicle slows down further, governor pressure drops even more. This causes the 1-2 shift valve to move to the downshifted position. Moving the 1-2 shift valve to the downshifted position blocks off oil pressure to the apply side of the band servo, and the transmission goes into first gear. If the vehicle is coasting to a stop, the overrunning clutch will be released. Should the driver accelerate in first, the overrunning clutch will apply.

Reverse is obtained when the manual valve is positioned to direct hydraulic pressure to the necessary holding members. Line pressure travels through the manual valve directly to the holding members. An accumulator may be used to prevent harsh engagement. For reverse, many transmissions apply the high clutch and a reverse band or clutch. The high clutch may be called the reverse-high clutch. The reverse band or clutch may also be used to prevent **overrunning** (release of the one-way clutch when the vehicle is coasting) in manual first gear and may be called the low-reverse band or clutch. Since reverse has only one gear ratio, the governor, throttle, and shift valves are not used. Many designs send reverse pressure to

the spring side of the pressure regulator valve. This extra pressure assists the pressure regulator spring and, therefore, raises line pressure in reverse.

A schematic of an actual 3-speed transmission is shown in **Figure 11-8.** The front and rear clutches correspond to the high and forward clutches previously discussed. The brake band is the same band explained in the last section. This band is used to obtain second gear. It is often called the second, or intermediate, band.

Notice that the 1-2 and 2-3 shift valves are constructed so that the throttle pressure side of the 1-2 shift valve is smaller than the throttle pressure side of the 2-3 shift valve. Therefore, less governor pressure is needed to overcome the throttle pressure on the 1-2 shift valve than on the 2-3 shift valve. This arrangement means that the 1-2 shift valve will always move to the upshifted position before the 2-3 shift valve.

Figure 11-9 is a more detailed diagram showing a different 3-speed automatic transmission. The hydraulic circuit pictured is obviously more complex than the one shown in **Figure 11-8.** Extra valves are added to more precisely control and cushion automatic and manual shifts.

The **2-3 backout valve** prevents a rough upshift if the driver lifts up on the accelerator just as the shift occurs.

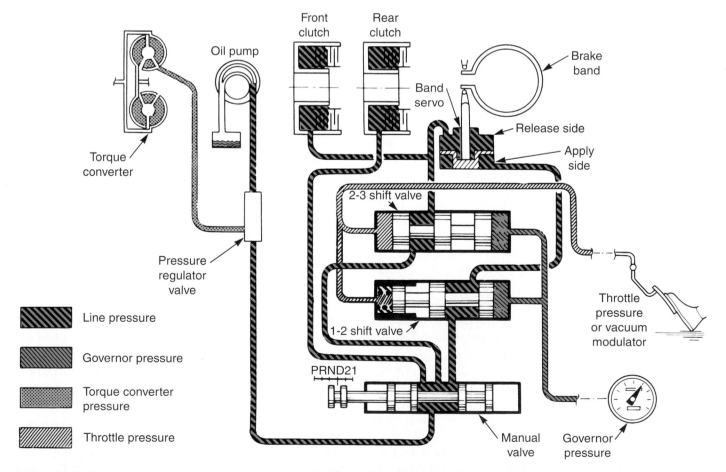

Figure 11-8. *Schematic of an actual 3-speed transmission. The shift valves are designed so the 1-2 shift valve will be moved before the 2-3 shift valve. (Nissan)*

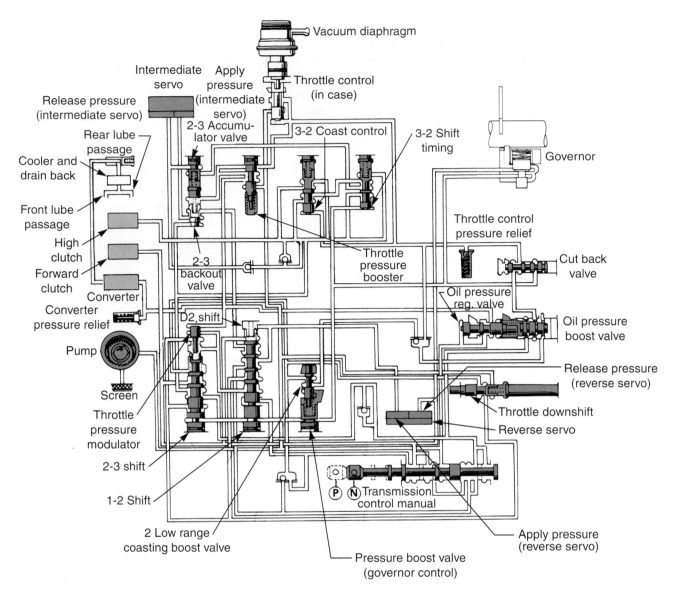

Figure 11-9. *This figure shows a somewhat more complex version of a three-speed automatic transmission. The components, however, are similar to those used in the transmissions discussed earlier. Most factory oil circuit diagrams resemble the one shown here.*

The ***throttle pressure booster*** works with the 2-3 accumulator valve to precisely time and cushion the shift during forced downshifts.

The ***cut back valve*** uses governor pressure to reduce throttle pressure. This allows the 2-3 shift valve to upshift when the engine speed becomes too high in passing gear. The ***3-2 shift timing valve*** causes all holding members to briefly disengage during detent (passing gear) downshifts. This allows the engine to speed up for more power in passing gear when the holding members reengage.

Other valves are used to control the torque converter lockup clutch, as shown in **Figure 11-10.** Oil pressure from the governor, throttle valve, and 2-3 shift valve ensure that the lockup clutch is engaged only at certain speeds and only when the transmission is in third gear. Other valves are used to regulate converter oil pressure, to boost pressure in reverse and in manual low or manual second gears,

and to ensure that the transmission bearings receive enough lubrication.

Transmission and Transaxle Oil Circuits

To this point, a simple transmission that identified the basic principles of hydraulic controls and simple transmissions used in the recent past have been briefly discussed. The following sections cover actual oil circuit diagrams used in modern transmissions and transaxles. You will note that some valves are used that have not been discussed before. Keep in mind that different transmission and transaxle manufacturers may use different valves and circuits to accomplish the same result. In many cases,

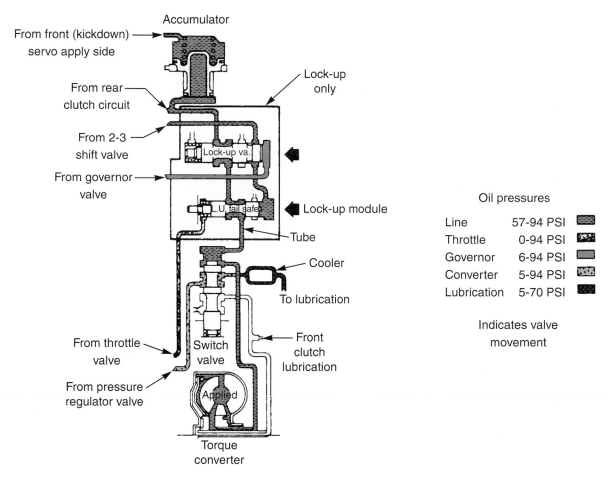

Figure 11-10. *The hydraulic circuit shown here is that of a lockup torque converter control system. (DaimlerChrysler)*

valves that have similar functions have different names in different transmissions.

The oil circuits presented here are for a common rear-wheel drive transmission and a transaxle used on a popular front-wheel drive car. These are probably the two most common hydraulically controlled units in current use. However, their oil circuit diagrams are not representative of those for other transmissions and transaxles, and they are to be used only to enhance your general understanding of how hydraulic controls work. Actual oil circuit diagrams will vary from one transmission or transaxle model to another, even on similar models. The diagrams used in this chapter should never be used in place of specific oil circuit diagrams for the unit you are servicing.

Transmission Hydraulic System Operation

The following section explains how the control valves, accumulators, oil passages, and holding members operate in a widely used rear-wheel drive automatic transmission with four forward gears. **Figures 11-11** through **11-22** show the hydraulic system operation in the Park, Neutral, Drive, Part Throttle Downshift, Detent Downshift, Manual Third, Manual Second, Manual First, and Reverse modes, as well as the application of the converter lockup clutch. This transmission was used on many rear-wheel drive cars and light trucks from 1982 until 1993. An electronic version, which will be discussed in Chapter 12, is widely used on modern vehicles.

Trace the oil circuits as you read the following paragraphs. Note how the shift valves are controlled by the action of the governor and throttle valves. In addition, notice how the other valves, as well as the accumulators, calibrated orifices, and check balls, operate to precisely time and cushion the shifts and provide maximum band and clutch life.

> **Note: Some check balls are used to bleed air from the hydraulic system. These are not described individually.**

Park, Engine Running

In Park with the engine running, **Figure 11-11,** the pump is turning and producing pressure. The main pressure regulator valve and spring are controlling the pump

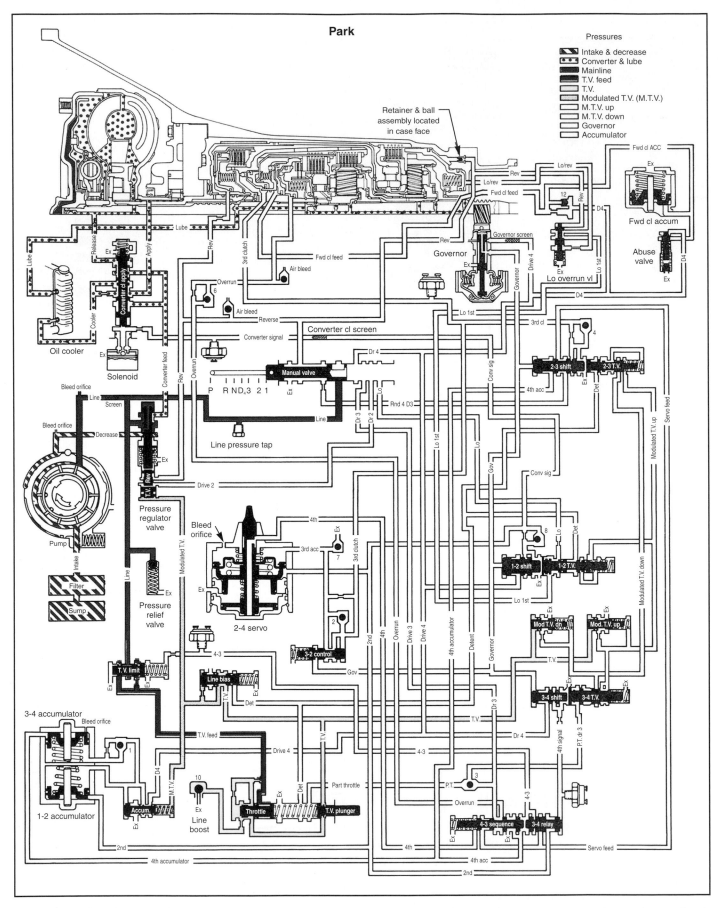

Figure 11-11. *This oil diagram is for a common 4-speed automatic transmission. The transmission is in park and the engine is running. Study the parts shown here, and notice how oil pressure is developed and controlled. (General Motors)*

output to produce regulated line pressure. Pressure is also sent to the pressure relief valve. If the pressure regulator sticks, the pressure relief valve will open to prevent damage to the other transmission components.

Some of the pump output is also diverted to the torque converter and transmission oil cooler. The converter clutch apply valve is positioned so that there is pressure on both sides of the converter lockup clutch. With pressure on both sides, the clutch is released.

Line pressure is directed to the throttle valve through the TV limit valve. The **TV limit valve** limits the amount of pressure that can reach the throttle valve. This prevents the development of excess throttle valve pressures, which could cause excessive line pressures.

From the TV limit valve, line pressure goes to the throttle valve and then to the TV plunger. The TV plunger is actually the main throttle valve, since it controls the delivery of oil to the shift valves. The throttle valve itself controls pressure to some extent, as it is acted on by the TV plunger through the spring.

Movement of the throttle valve plunger changes the amount of pressure sent to the TV plug (a small valve) on the underside of the main pressure regulator. Note that the TV plug is not connected to the TV rev. valve, although it presses the TV rev. valve to exert pressure on the main pressure regulator.

The TV plug pushes the rev. valve, which increases pressure regulator spring pressure. Increased throttle pressure provides more assist to the pressure regulator spring, raising line pressure. Less throttle pressure provides less assist to the spring, lowering line pressure. Increases in line pressure cause firmer holding member application, reducing slippage and increasing holding member durability. Throttle pressure also goes to the Modulator TV up and down valves, but has no effect on transmission operation at this time.

Throttle pressure is directed through the **line bias valve.** The line bias valve allows the throttle pressure to the main pressure regulator to increase quickly when the throttle valve is first moved, but slows the rate of increase after the first opening. This allows the line pressure increase to more closely match increases in engine torque.

Regulated line pressure is sent to the manual valve. In Park, the manual valve keeps oil from flowing to the other hydraulic system components and no holding members are applied.

Transmission Protection Device

Check ball #10 at the bottom of **Figure 11-11** is part of a fail-safe device that is designed to protect the transmission if the throttle valve linkage breaks or is adjusted too loosely. This setup is used only on linkage-controlled throttle valves. During normal operation, the throttle linkage holds the check ball open. Holding the check ball open exhausts oil pressure in the passages behind the throttle valve. Keeping pressures in these passages from

rising allows throttle valve pressures to be controlled normally by linkage movement. If the throttle linkage breaks or is adjusted too loosely, it can no longer hold the check ball open. Pressure will seat the check ball, causing a pressure rise in the passages behind the throttle valve. These pressures will move the throttle valve to its maximum output position. This results in high shifts and maximum line pressures. Check ball #10 ensures that the transmission will not be damaged by a broken cable and that the vehicle will be brought in for service as soon as possible.

Neutral, Engine Running

Neutral engine running, **Figure 11-12,** is similar to Park. In Neutral, the position of the manual valve causes pressure to go to the 2-3 shift valve, the 4-3 sequence valve, and the 3-4 relay valve. However, this is only because of the design of the manual valve and oil passages, and it has no effect on transmission operation. Other oil flows are the same as in the Park position, and no holding members are applied.

Drive, First Gear

When the manual valve is moved to the Drive position (sometimes called Overdrive), it allows oil to enter the forward clutch piston. This oil is called Drive 4, Dr 4, or D4 oil, depending on its location in the circuit. D4 oil causes the forward clutch to apply, exerting rotational force on the input and lo (low) overrunning clutches. This places the transmission in first gear. See **Figure 11-13.** The use of lo and input overrunning clutches allows the engine to freewheel when the vehicle is moving and the throttle is released. This increases fuel economy and reduces emissions and engine wear.

 Note: The input overrunning clutch is sometimes called the forward overrunning clutch.

When the forward clutch is first applied, some oil is diverted to the forward clutch accumulator. This oil is used to compress the spring in the accumulator. Compressing the spring reduces the oil pressure to the forward clutch, allowing it to apply softly. This prevents harsh drive engagement from Neutral or Park. When the accumulator spring is fully compressed, full oil pressure goes to the forward clutch. Notice that some D4 oil also goes directly to the forward clutch through a check ball and restriction. The check ball is seated, and only a small amount of oil flows to the clutch. The real purpose of the check ball and restriction is to allow quick release of the forward clutch when the vehicle is shifted into Park, Reverse, or Neutral.

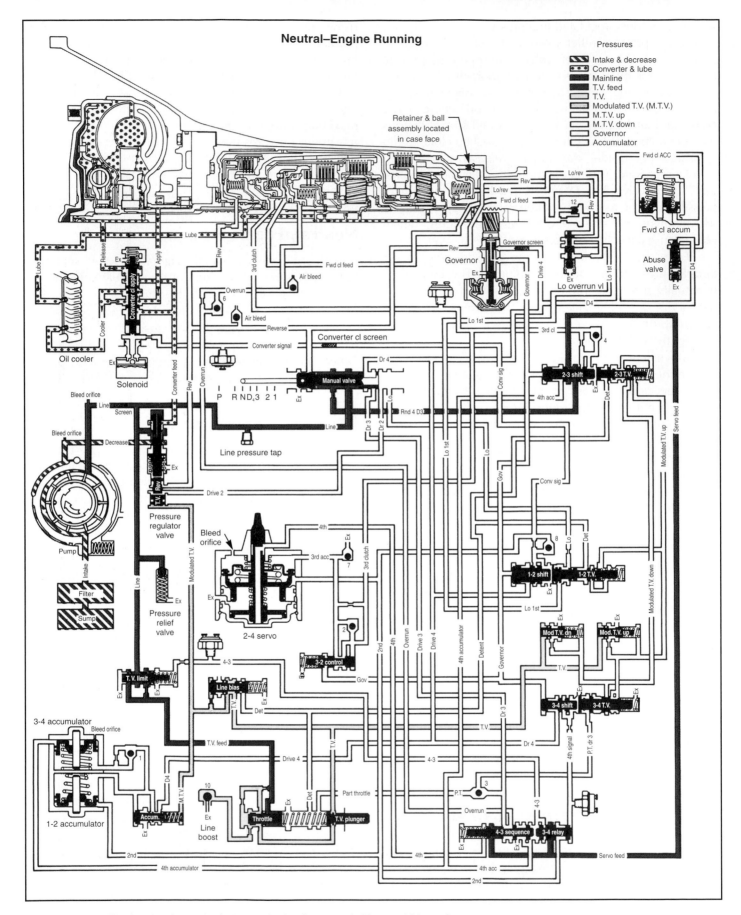

Figure 11-12. *The 4-speed automatic transmission in neutral. (General Motors)*

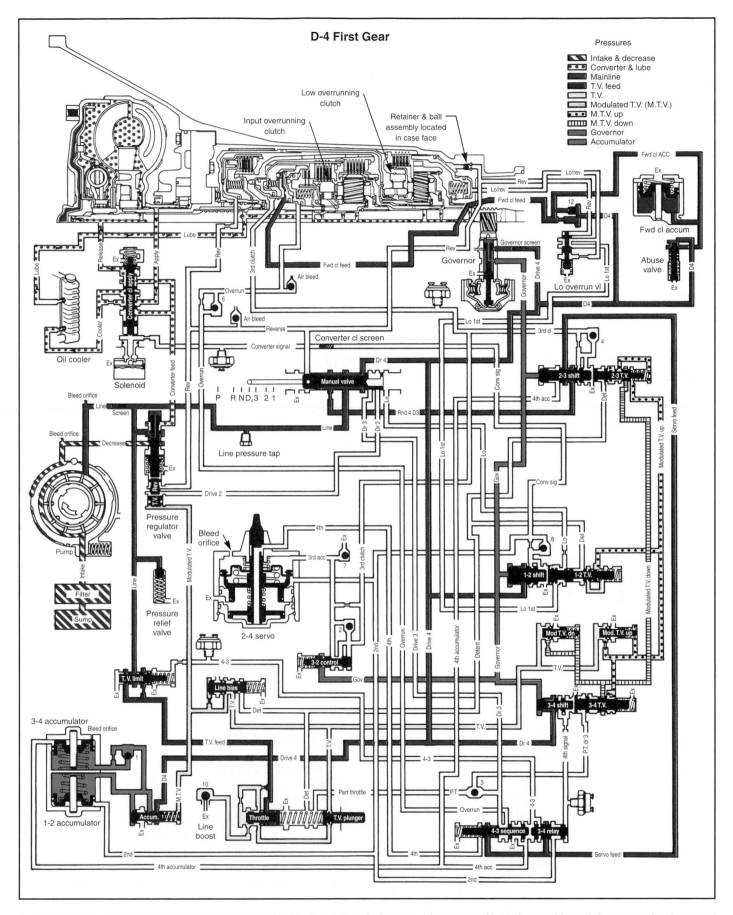

Figure 11-13. *The 4-speed automatic transmission in first gear of the overdrive range. Note the position of the manual valves and shift valves. Why is oil pressure directed to the accumulator in the upper-right corner? (General Motors)*

With pressure removed at the manual valve, pressure in the forward clutch unseats the ball and quickly exits the forward clutch passages.

Oil flows through the abuse valve before going to the forward clutch. Under normal operating conditions, the abuse valve spring holds the valve closed against D4 pressure. With the valve closed, oil flow into the accumulator and forward clutch is restricted, helping to provide a smooth forward clutch engagement. However, if the throttle position is above idle (such as when the vehicle is being rocked out of mud or snow), throttle pressure increases. This causes the line pressure and therefore the D4 pressure to increase. Increased D4 pressure overcomes the abuse valve spring and opens the abuse valve. With the abuse valve fully open, D4 oil can quickly apply the forward clutch, resulting in a firmer apply. This reduces drive and driven clutch slippage, reducing forward clutch wear when Drive is selected with the engine speed above idle.

The manual valve also sends oil to the 3-4 shift valve, but it is not affected at this time. Pressure is also directed to the accumulator valve, which opens to pressurize the two central chambers of the 3-4 and 1-2 accumulators. The line pressure directed to these valves and the forward clutch is called drive 4 oil.

In Drive position, the manual valve also directs oil to the governor valve. Line pressure enters the governor through a screen and orifice, and exits as governor pressure. Governor pressure is directed to the upshift side of the 1-2, 2-3, and 3-4 shift valves. Governor pressure also goes to the 3-2 control valve.

The manual valve sends oil to the 3-4 shift valve, but this valve is not affected at this time. The manual valve also directs oil to the 2-3 shift valve. This oil is called RND4-D3 oil since it is delivered to the 2-3 shift valve in every gear except manual first and second. The RND4-D3 oil passes through the 2-3 shift valve and goes to the 4-3 sequence and 3-4 relay valves as servo feed oil. This does not affect transmission operation in this gear.

When the vehicle is accelerated from stop, the throttle linkage moves the TV plunger and throttle valves. This sends enough pressure to move the modulator TV up and modulator TV down valves. Throttle pressure passes through the modulator TV up valve to the downshift side of the 1-2, 2-3, and 3-4 shift valves, and through the modulator TV down valve to the 2-3 and 3-4 shift valves.

The purpose of the **modulator TV valves** is to modify the pressure supplied by the throttle valve for more precise shifting. The modulator TV up valve allows the transmission to upshift at low speeds when throttle pressure is low, while holding the transmission in lower gears longer under heavy acceleration. The modulator TV down valve allows the transmission to downshift at higher speeds when necessary. The modulator TV up valve only affects a shift valve when it is in the downshifted position. The modulator TV down valve only affects a shift valve when the valve is in the upshifted position. To prevent engine overspeeding, modulator TV down oil is not sent to the 1-2 shift valve.

Note that the 1-2 shift valve is in the downshifted (first gear) position. Governor pressure is acting directly on the 1-2 shift valve. Throttle pressure through the modulator TV up valve acts on the shift valve through the 1-2 TV valves. Drive 4 (line pressure) oil from the manual valve is available at the 1-2 shift valve but is blocked at the valve.

Drive, Second Gear

As the vehicle is accelerated in the drive range, governor pressure increases with speed. When a certain speed is reached, the governor pressure at the 1-2 shift valve becomes greater than the throttle pressure (actually the Modulator TV up pressure). When this happens, the 1-2 shift valve is moved to the upshifted position. When the shift valve moves, drive 4 oil passes through the valve and applies the 2-4 band servo, **Figure 11-14.** This applies the band and stops the drum from turning. With the drum stationary, the lo overrunning clutch is unlocked and the transmission shifts into second gear. The input overrunning (forward) clutch is still holding and will allow the engine to freewheel if the throttle is released when the transmission is in second gear. The check ball directly above the shift valve (#8) seals the oil passage, forcing the oil to travel through the smaller orifice next to it. This restricts pressure to the servo and cushions the shift. On a downshift, the check ball unseats and oil can exhaust quickly.

To further cushion the shift, some servo oil pressure is diverted to the 1-2 accumulator. This pressure, called 2nd oil, moves the 1-2 accumulator piston against spring pressure and pressure that was delivered by the accumulator valve. As this happens, some of the pressure is diverted from the servo, causing the initial band apply to be smooth. After the servo piston is moved to the limit of its travel, no more oil can move into the accumulator and full pressure is applied to the band servo.

When the 1-2 shift valve moves, it also directs oil to the converter clutch apply valve. This oil, which is called converter signal oil, will be used in a later operation. Oil pressure is also sent to the 3-4 relay valve for use during a later shift.

Drive, Third Gear

As vehicle speed continues to increase, governor pressure continues to rise. At a certain speed, governor pressure becomes high enough to move the 2-3 shift valve to the upshifted position, **Figure 11-15.** Line pressure is available at the 2-3 shift valve directly from the manual valve. When the valve moves, this pressure, which is now called 3rd clutch oil, is directed to the 3-4 clutch and to the release side of the 2-4 band servo. The 3-4 clutch applies and the 2-4 band releases, placing the transmission in third gear. The combination of the forward clutch and the 3-4 clutch lock the planetaries into a single unit. All turning transmission parts are locked together, giving direct (1:1) drive.

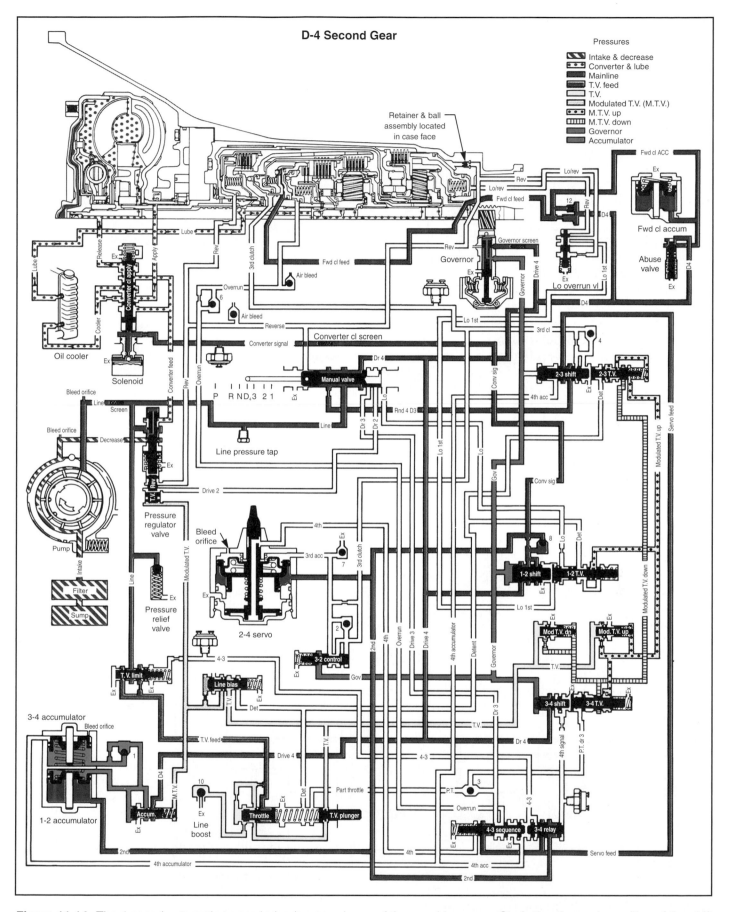

Figure 11-14. *The 4-speed automatic transmission in second gear of the overdrive range. Study the change in position of the shift valves, and note how oil is now directed to the 2-4 band, causing it to apply. Which holding member is released? (General Motors)*

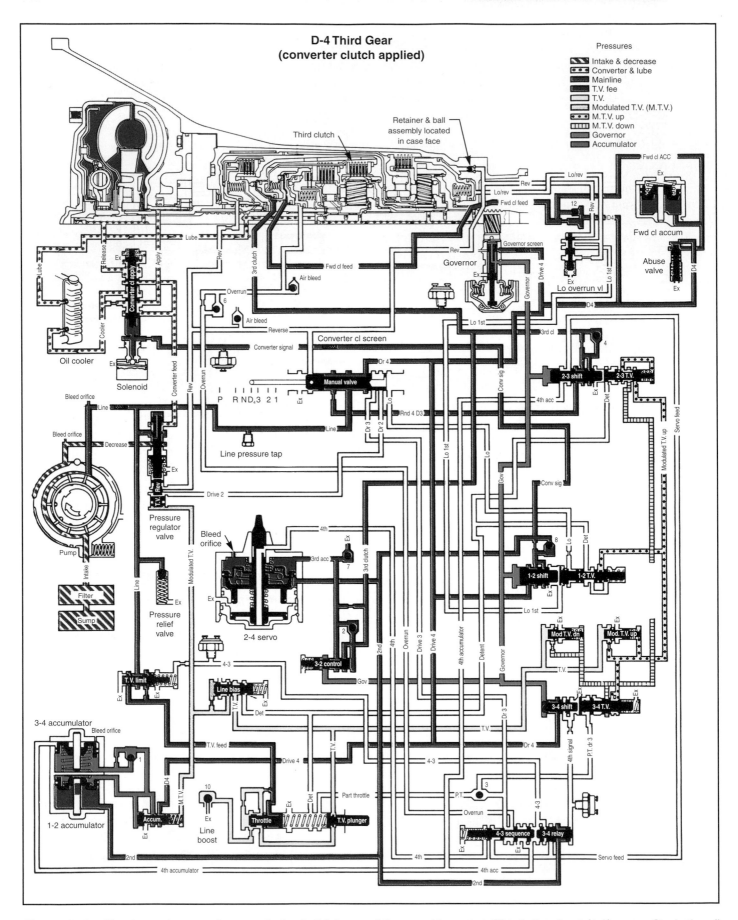

Figure 11-15. *The 4-speed automatic transmission in third gear of the overdrive range. This is the direct (1:1) range. Study the oil pressure flow to determine how the same oil that applies the direct clutch also releases the 2-4 band. This is similar to the operation of the hydraulic system in Figure 11-3. In this figure, the converter clutch has also been applied. (General Motors)*

The input (forward) overrunning clutch remains applied. The servo release side of the 2-4 servo acts as a 2-3 shift accumulator, diverting clutch apply pressure until the band is fully released. Pressure also goes to the 3-2 control valve for use when a 3-2 downshift is performed.

Drive, Third Gear – Converter Clutch Apply

For maximum fuel economy, the converter clutch is locked to eliminate converter slippage. **Figure 11-15** shows the clutch apply process in third gear. Converter clutch apply can also occur in second gear. Remember that when the 1-2 shift valve moved, it also directed oil to the converter clutch shift valve. This allows the converter clutch to apply.

As long as the converter clutch solenoid is de-energized, the oil will exhaust through the solenoid orifice and will not move the valve. Refer back to Figure 11-14 for the exhaust action.

When the solenoid is energized, an internal check ball seats and the orifice is closed. There is now enough pressure to move the converter clutch apply valve. When the valve moves, it exhausts the oil pressure on the lockup clutch release side. Pressure on the clutch apply side will then push the clutch plate into contact with the converter cover, locking the converter. Movement of the valve also forces oil to enter the oil cooler circuit through a small orifice. Oil flow through the oil cooler is reduced since there is no converter slippage, and therefore, less heat is produced.

Older converter clutch solenoids were operated by a combination of vacuum and speed switches. On most vehicles made since the 1980s, the on-board computer operates the solenoid. The converter clutch can be unlocked by one of two methods. If the solenoid is de-energized, the clutch will release. If the vehicle shifts back to first gear, there will be no oil available to the converter clutch apply valve and there will be no pressure to operate the converter clutch. This is done as a fail-safe measure, since a locked converter would cause the engine to stall when the vehicle is stopped.

Note: The converter clutch can be applied in any forward gear other than first. The solenoid is energized depending on speed and vacuum inputs to the computer. The clutch apply valve is moved depending on governor and throttle inputs. Therefore, the converter may or may not be applied in a particular gear.

Drive, Fourth Gear (Overdrive)

Increased vehicle speed causes governor pressure to increase. When governor pressure becomes high enough,

it moves the 3-4 shift valve against throttle pressure. See **Figure 11-16.** This causes the drive 4 oil available at the valve to be directed to the 3-4 relay valve. The 3-4 relay valve moves and pushes the 4-3 sequence valve. The 4-3 sequence valve directs servo feed oil through the 2-3 shift valve to a second servo apply area in the 2-4 band servo. The combination of the second and fourth apply sides overcomes the single release side and reapplies the band. The forward and 3-4 clutches remain applied. The input overrunning (forward) clutch overruns, placing the transmission in fourth gear.

The 3-4 relay valve also directs servo apply oil through the 2-3 shift valve to the 3-4 accumulator. This accumulator is opposite the 1-2 accumulator. Oil from the servo apply, now called 4th accumulator oil, compresses the 3-4 accumulator spring, diverting oil pressure until the spring is fully compressed.

Part Throttle Downshift to Third

For increased performance, the transmission will downshift from fourth to third when the throttle is partially depressed. In **Figure 11-17**, the TV plunger has been moved enough to open the part throttle passage. Oil pressure from the throttle valve then enters the 3-4 TV, located in the 3-4 shift valve bore. The combination of throttle pressure and 3-4 TV spring pressure will force the 3-4 shift valve into the downshifted position. Without oil pressure from the 3-4 shift valve, the 3-4 relay valve will return to the third gear position and oil will exhaust from the servo apply area and the 3-4 accumulator piston.

Detent Downshift (Passing Gear)

To obtain detent downshifts, **Figure 11-18,** the throttle valve plunger must be pushed far enough to open the detent passage. With the detent passage open, TV oil pressure is directed to the 1-2 and 2-3 throttle valves. Part throttle oil pressure will move the 3-4 shift valves to the downshifted position, as explained in the previous section.

Governor pressure dictates which shift valves will be moved by detent oil pressure. If vehicle speed is low enough, it is possible for the detent oil to move both the 1-2 and 2-3 shift valves to the downshifted position and place the transmission in first gear. At higher speeds, the governor pressure may be high enough to keep the 1-2 and possibly the 2-3 shift valves in the upshifted position. Therefore, even when the detent system is in operation, governor pressure will keep the transmission from downshifting into an excessively low gear.

If the downshift is down to first gear, movement of the 1-2 shift valve will cut off pressure to the converter clutch apply valve, releasing the converter lockup clutch. In most cases, the clutch will already be deactivated by the computer, based on the throttle position sensor signal.

Detent oil is also delivered to the line bias valve. This pressure assists TV pressure and causes the line bias valve

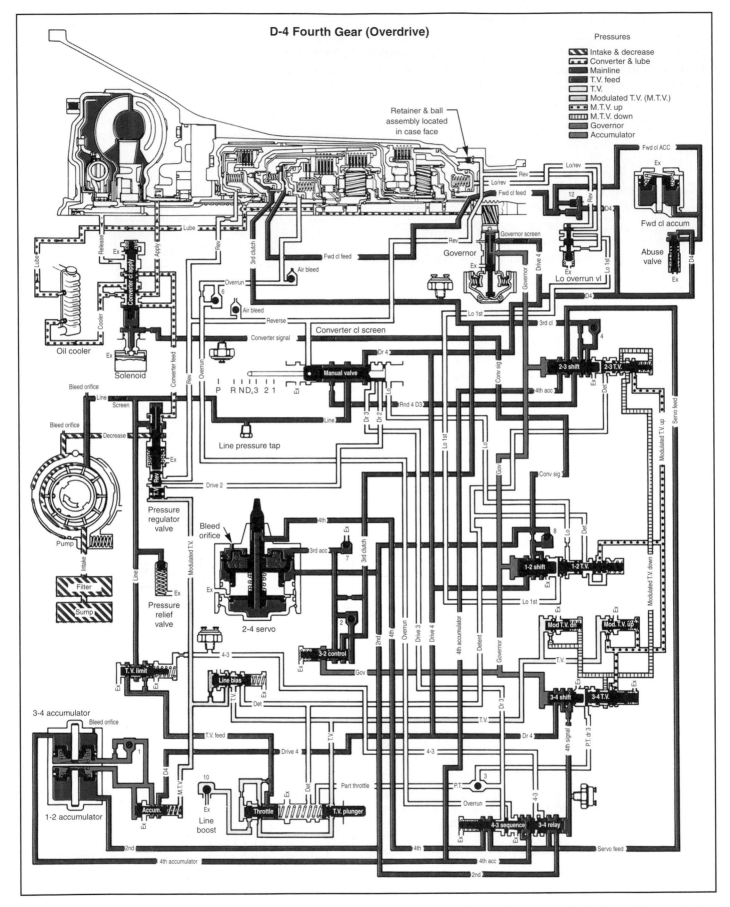

Figure 11-16. *The 4-speed automatic transmission in fourth (overdrive) gear of the overdrive range. Study the oil flow to the 2-4 band and note the way in which the band has been reapplied. (General Motors)*

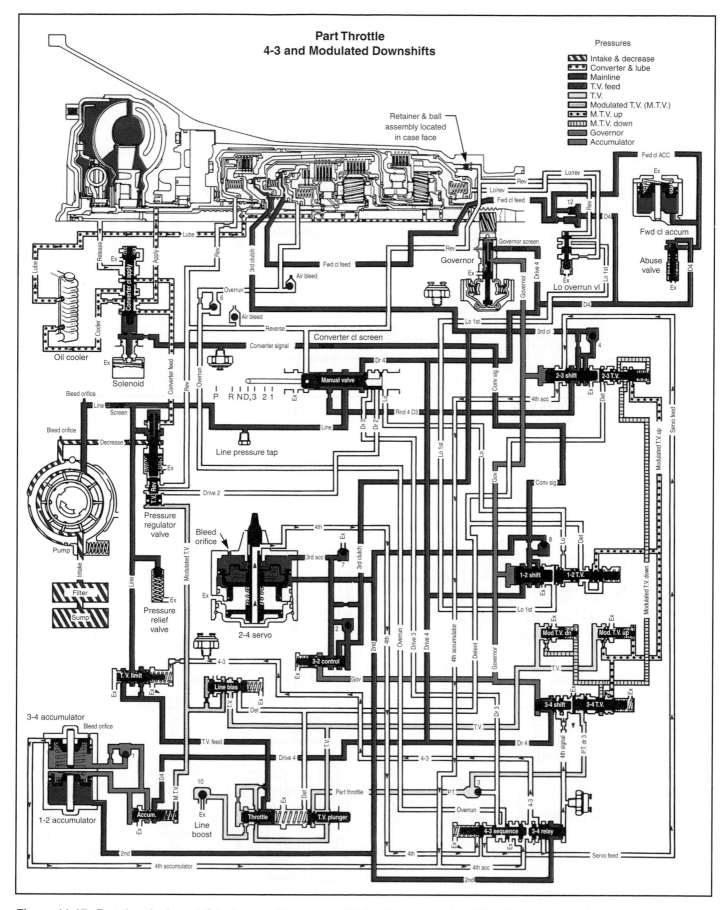

Figure 11-17. *Part throttle downshift in the overdrive range. Which valves cause the shift valves to move? (General Motors)*

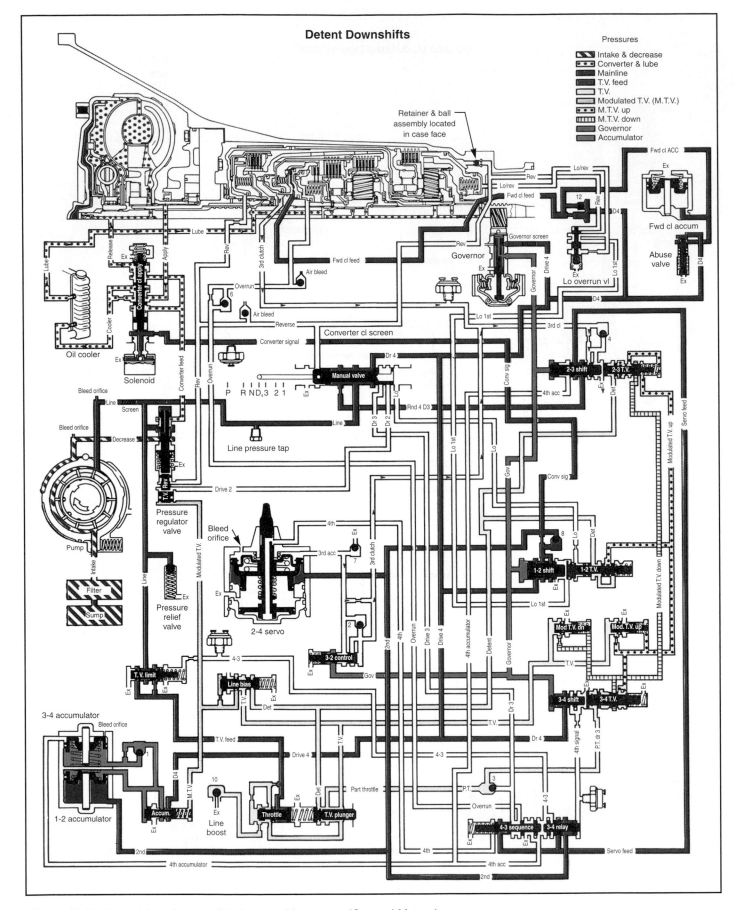

Figure 11-18. *Detent (passing gear) in the overdrive range. (General Motors)*

to increase pressure sent to the TV plug at the underside of the main pressure regulator. Line pressure is therefore increased to more firmly apply the holding members.

To prevent a rough downshift from third to second, the 3-2 control valve and its associated check balls (#2 and #7) control the rate of servo release. Operation of the 3-2 control valve is based on governor pressure to vary the release according to vehicle speed.

Manual Third

When the driver moves the shift lever to the manual third position, **Figure 11-19,** the manual valve directs oil pressure (called DR3 oil) to the 3-4 TV valve and the 4-3 sequence valve. DR3 oil at the 3-4 TV valve increases pressure on the throttle side of the 3-4 shift valve, and the 3-4 shift valve moves to the downshifted position. Sending oil to the 4-3 sequence valve causes it to close. This cuts off oil pressure to the second apply servo area and allows the oil to exhaust through the 4-3 sequence valve. The 2-4 band releases, placing the transmission in third gear. If the transmission was in fourth gear, the 3-4 accumulator oil is also exhausted. To keep throttle pressure from being affected by DR3 oil, a check ball (#3) seats to seal the part throttle passage to the throttle valve.

Oil pressure is also sent through the 4-3 sequence valve to the overrun clutch. The **overrun clutch** is a hydraulically applied clutch, which is separate from the mechanical overrunning clutches. The overrun clutch is applied, which keeps the input overrunning clutch from releasing. This provides engine braking. Oil passes through an orifice and check ball assembly (#6) to ensure a smooth 3rd gear overrunning clutch application. The check ball unseats to quickly exhaust oil pressure when the shift lever is returned to Drive.

 Note: Oil pressure is still available to the converter clutch apply valve, and the converter clutch may apply in manual third.

Manual Second

When the manual valve is moved to the manual second position, **Figure 11-20,** valve movement opens the RND4-D3 oil passage to exhaust. RND4-D3 oil pressure drops to zero, and oil exhausts from the third clutch and the servo release area. The clutch releases, and the band tightens around the drum to shift the transmission into second gear. The #2, #4, and #7 checkballs seat to cushion the shift. The 2-3 shift valve is held in the upshifted position by governor pressure, but it is ineffective since apply pressure has been exhausted. The overrun clutch remains applied through the 4-3 sequence valve to keep the overrunning clutch from releasing. This provides engine braking.

 Note: Oil pressure is still available to the converter clutch shift valve, and the converter clutch may apply in manual second.

Manual First

When the manual valve is shifted to manual first, **Figure 11-21,** it sends line pressure to the 1-2 TV valve. This pressure, called lo pressure, causes the 1-2 TV valve to push the 1-2 shift valve against governor pressure. The 1-2 shift valve moves to the downshifted position. Pressure in the servo apply area is exhausted through the 1-2 shift valve. With pressure gone, the servo release spring pushes the servo to its resting position, releasing the band. The transmission shifts into first gear. To keep the lo overrun clutch from releasing, lo pressure applies the lo and reverse clutch through the 1-2 TV valve. This oil, now called lo1st oil, passes through the lo overrun valve. The lo overrun valve consists of a spring that works against lo 1st oil pressure to control shift feel. More oil pressure causes more oil to flow through the lo overrun valve, causing a quicker, firmer apply. Lower oil pressure causes less oil to flow through the lo overrun valve. This results in a gentler clutch application.

To keep the input (forward) overrunning roller clutch from releasing, the overrun clutch remains applied through the 4-3 sequence valve. Since neither overrunning clutch can release, engine braking is provided in manual first.

With the 1-2 shift valve closed, no oil pressure can reach the converter shift valve and the converter lockup clutch cannot apply. If the driver tries to shift into manual first at high speeds, the governor pressure may be high enough to prevent a downshift to low, even with the extra pressure on the downshift side of the 1-2 shift valve. However, the transmission will downshift into manual second.

Reverse

When the driver shifts into reverse, the manual valve directs pressure to the lo and reverse clutch and the reverse input clutch. Reverse oil pressure also flows to the lo overrun valve, moving it against spring pressure.

Moving the lo overrun valve opens an oil passage to a second area behind the lo and reverse clutch piston, increasing the amount of pressure holding the lo and reverse clutches. The second area was used earlier to prevent lo overrunning clutch release in manual first. The transmission shifts into reverse. The reverse circuits are shown in **Figure 11-22.**

Since the reversal of directions requires that the clutches be held tightly, reverse pressure travels to the reverse valve, or rev. valve, in the main pressure regulator. The rev. valve assists spring pressure pushing on the main pressure regulator, causing line pressure to rise. Line pressure is also directed to the 2-3 shift and through it to the

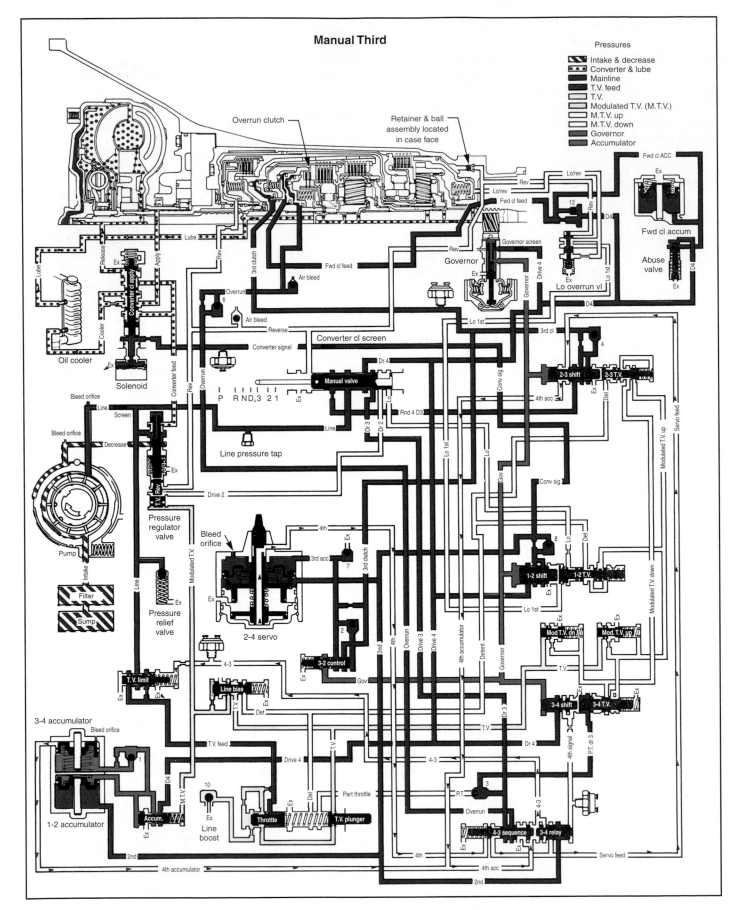

Figure 11-19. *The 4-speed automatic transmission in the manual third range. Note how the manual valve directs oil flow to prevent an upshift to fourth. (General Motors)*

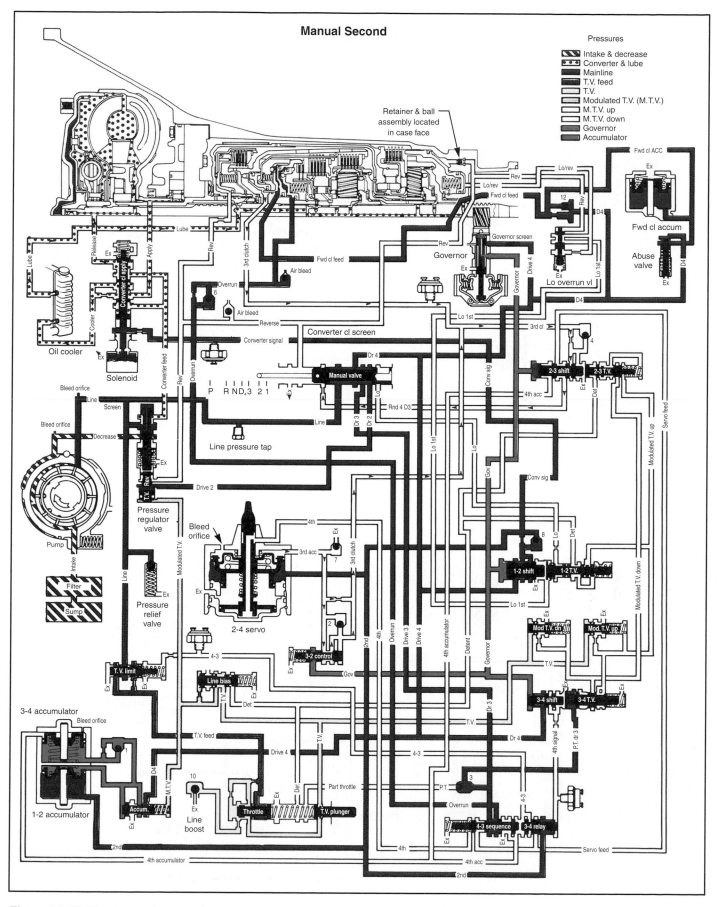

Figure 11-20. *The 4-speed automatic transmission in the manual second range. Which extra holding members applied in this gear? (General Motors)*

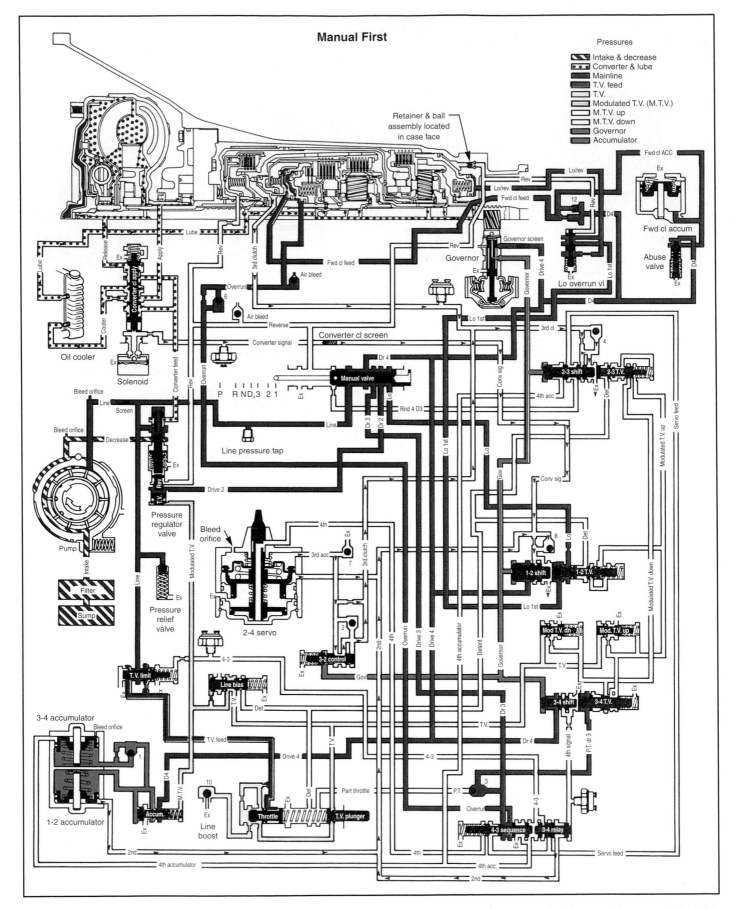

Figure 11-21. *The 4-speed automatic transmission in the manual first (or lo) range. Which extra holding members applied in this gear? Is this different from manual second? (General Motors)*

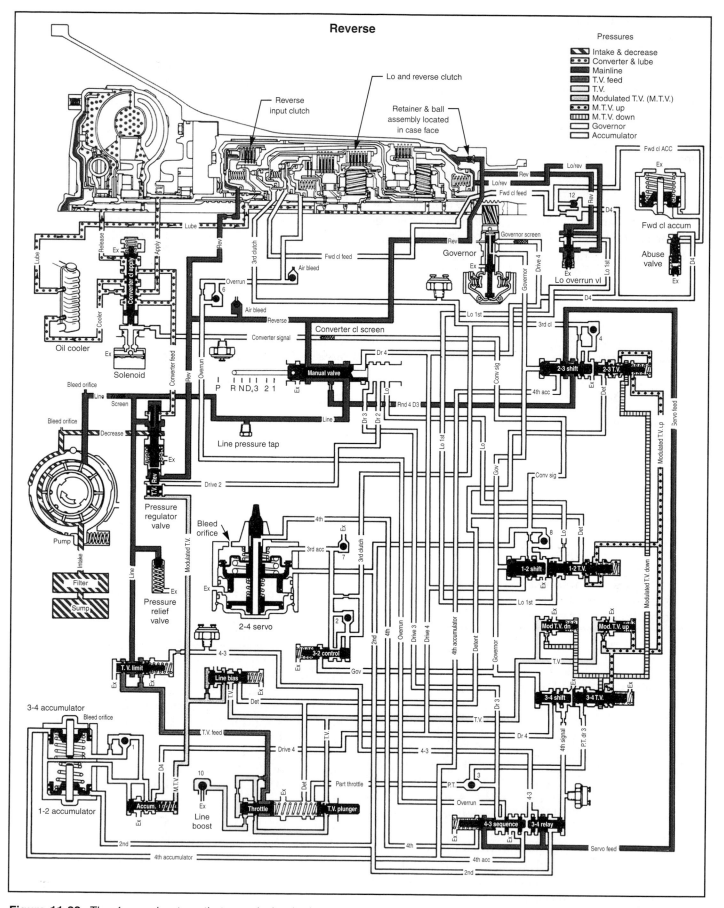

Figure 11-22. *The 4-speed automatic transmission in the reverse range. Why is pressure directed from the manual valve to the spring side of the pressure regulator? (General Motors)*

3-4 relay and 4-3 sequence valves. However, this happens only because of the valve body design, and it has no effect on reverse operation.

Transaxle Hydraulic System Operation

The following section explains how the hydraulic control system and holding members operate in a four-speed automatic transaxle. You will notice that hydraulic operation of a transaxle is similar to that of the rear-wheel drive transmission described earlier. **Figures 11-23** through **11-28** show the hydraulic system operation in drive and reverse, as well as during the application of the converter lockup clutch. This transaxle was used on a popular series of front-wheel drive cars from 1986 to 1992 and in mini-vans using the same drive train. An electronic version of this transaxle is widely used on 1994 and up versions of these same vehicles. The electronic version is discussed in Chapter 12.

Trace the oil circuits as you read the following paragraphs. Note how the governor, throttle, and shift valves operate to precisely time the shifts. Also note how the other valves, as well as accumulators, calibrated orifices, and check balls, affect the shifts for maximum smoothness, as well as band and clutch life.

> **Note: The operation of some valves and check balls is not described individually. This is done to help clarify the operation of the holding members and the main components of the shift system. Operation of these parts is similar to the operation of related parts in the rear-wheel drive circuits discussed earlier.**

Overdrive—First Gear

When the selector lever is placed in the overdrive position, the transaxle can be in any one of four gears, depending on the vehicle speed and throttle opening. Overdrive first gear is shown in **Figure 11-23**. In this gear, the pump is turning and producing pressure. The pump also keeps the torque converter filled. At this time, the converter bypass solenoid is not energized and the lockup clutch is not applied.

The main pressure regulator valve and spring control pump output by directing oil behind the valve. This produces regulated line pressure, as explained in Chapter 9. Regulated pressure is sent to the manual valve. With the manual valve in the overdrive position, pressure is sent to the forward clutch through the 3-4 shift valve. Oil pressure also passes through the 2-3 servo regulator valve into the

low/intermediate band servo. The pressure between the 2-3 servo regulator valve and low/intermediate servo is called low/intermediate servo apply pressure, or LISA pressure.

Pressure also goes to the shift valves, the throttle valve (called the TV limit valve), and the governor valve. If the vehicle is not moving, there is no governor pressure. Throttle pressure holds the shift valves in the downshifted position. With all valves in the downshifted position, the transaxle is in first gear.

Overdrive—Second Gear

As vehicle speed increases, the governor pressure increases until it can move the 1-2 shift valve to the upshifted position. When the 1-2 shift valve moves, it allows oil from the manual valve to apply the intermediate clutch, **Figure 11-24**. When the intermediate clutch applies, the transaxle shifts into second gear. Oil also goes to the accumulator capacity modulator and 1-2 accumulator. These absorb some of the oil going to the intermediate clutch and help to cushion the 1-2 shift.

Overdrive—Third Gear

As the vehicle continues to move faster, governor pressure continues to increase. At a certain speed, the governor pressure overcomes throttle pressure on the 2-3 shift valve. The 2-3 shift valve moves to the upshifted position. Oil flows through the 2-3 shift valve to the direct clutch. Oil pressure also flows to the low and intermediate band release side and the 2-3 servo regulator. The release oil pushes the band servo to the released position, while the 2-3 servo regulator causes the servo apply side oil to exhaust. Oil pressure movement is shown in **Figure 11-25**. This action applies the direct clutch while releasing the low and intermediate band. The transaxle shifts into third gear.

> **Note: Some pressure is always present in the low/intermediate servo apply (LISA) circuit. LISA pressure varies with line pressure. Varying the LISA pressure allows proper shift feel when the transaxle downshifts to second.**

Converter Clutch Apply

The converter lockup clutch can be applied in third or fourth gears. **Figure 11-26** shows the converter clutch being applied with the vehicle in third gear. When de-energized, the converter bypass solenoid is open. Oil pressure in the passage to the converter clutch control valve is exhausted to the sump. When the solenoid is energized, it closes to seal the passage. Oil pressure builds up in the passage to the converter clutch control valve. This pressure moves the converter clutch control valve, exhausting oil

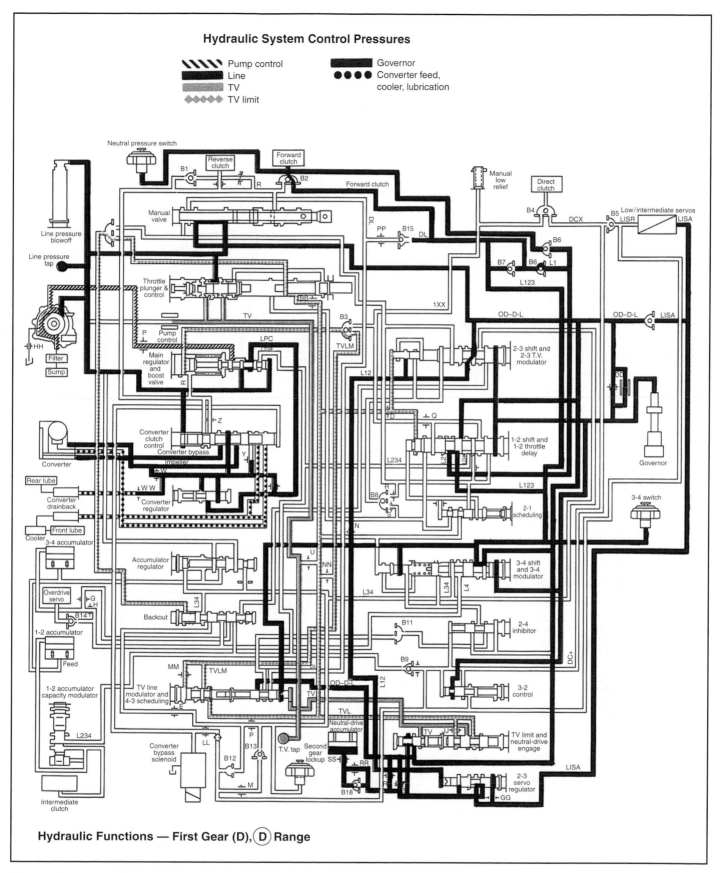

Figure 11-23. *This figure shows the hydraulic circuit of a 4-speed automatic transaxle in first gear with the selector lever in the overdrive range. (Ford)*

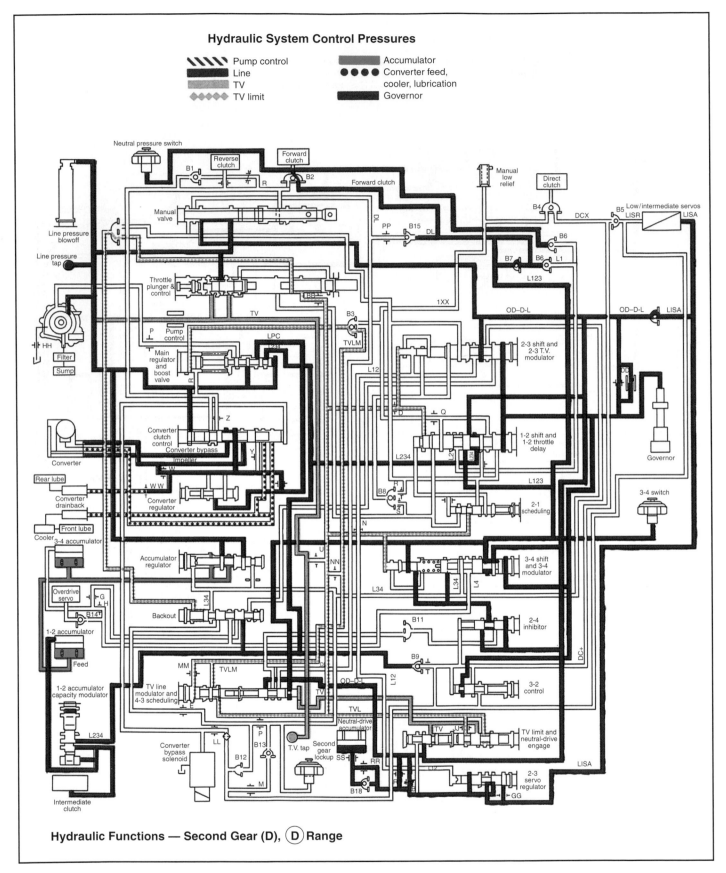

Figure 11-24. *In this figure, the 4-speed transaxle is in second gear of the overdrive range. Which extra holding members have been applied? (Ford)*

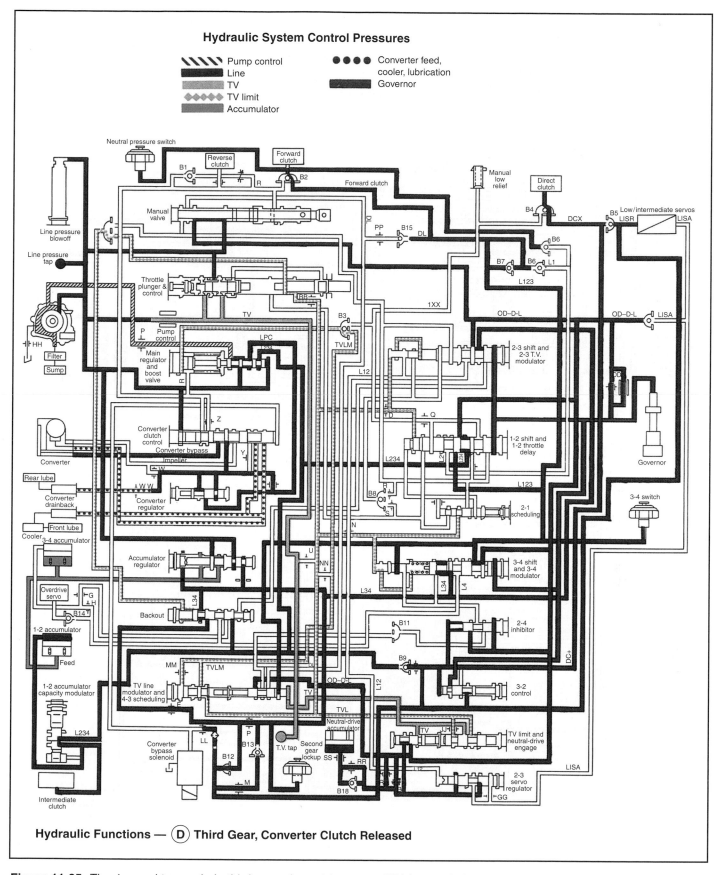

Figure 11-25. *The 4-speed transaxle in third gear of overdrive range. Which extra holding members have been applied? Which holding members have been released? (Ford)*

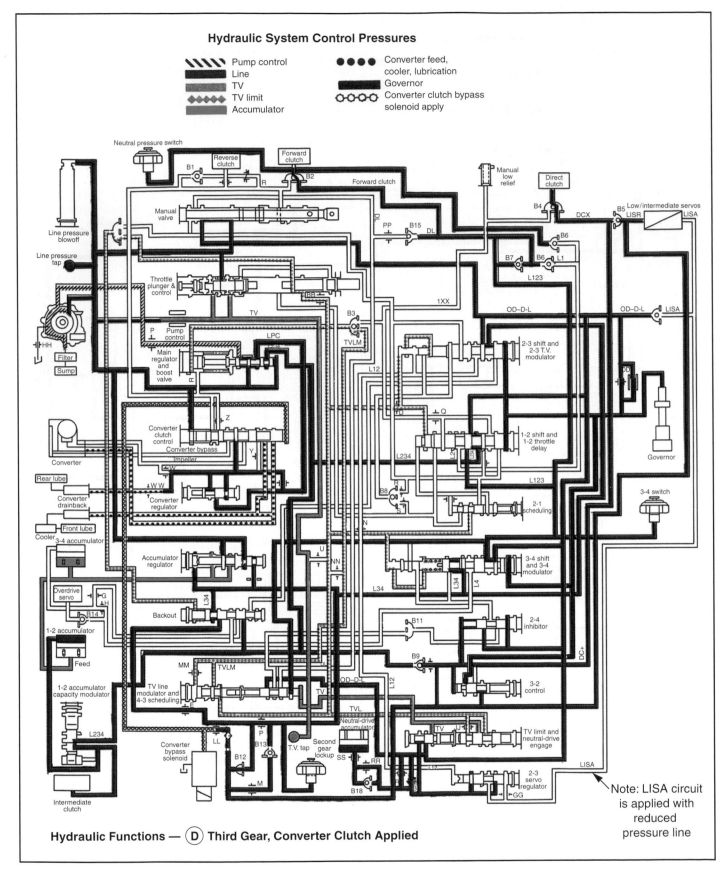

Figure 11-26. *In this figure, the converter clutch is being applied with the transaxle in third gear. Which valves move to apply the converter clutch? (Ford)*

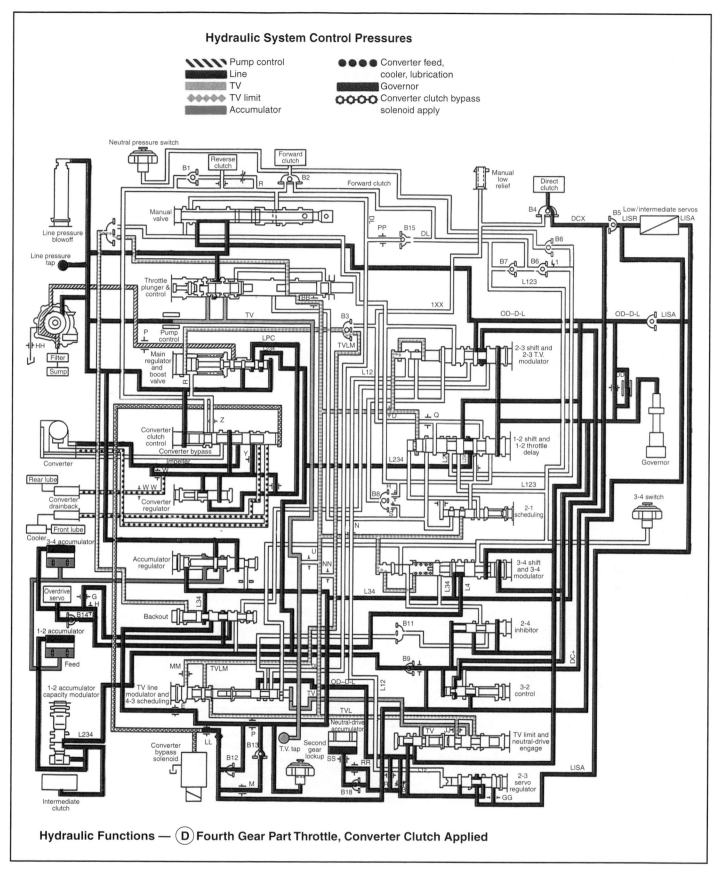

Figure 11-27. *The 4-speed transaxle in fourth gear of overdrive range. Which additional holding members have been applied? Which holding members have been released? (Ford)*

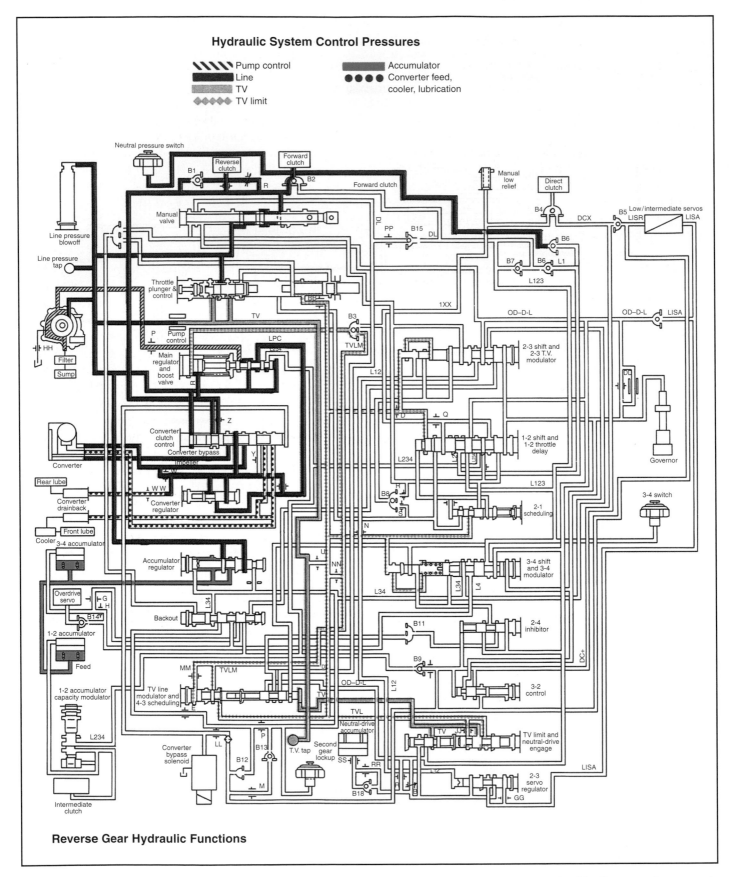

Figure 11-28. *The 4-speed transaxle in reverse. What similarities are there between reverse in this figure and reverse in Figure 11-22? (Ford)*

pressure from the release side of the converter clutch. Pressure remains on the converter clutch apply side, and the clutch is moved into engagement with the converter cover. The converter regulator pressure control valve varies the feel of the clutch apply.

Overdrive—Fourth Gear

Increasing vehicle speed eventually develops enough governor pressure to overcome throttle pressure on the 3-4 shift valve. The 3-4 shift valve moves to the upshifted position and applies the overdrive servo. Some overdrive servo oil is diverted to the 3-4 shift accumulator to cushion the shift. The 3-4 shift valve also cuts off oil pressure to the forward clutch. Leftover clutch pressure exhausts through the B2 check ball and through the manual valve to the sump.

The action of the 3-4 valve causes the overdrive servo to be applied and the forward clutch to be released. This places the transaxle in fourth gear. See **Figure 11-27**. The direct and intermediate clutches remain applied, but the direct clutch is ineffective since the direct one-way clutch is overrunning. The converter clutch can be applied or released, depending on whether or not the converter bypass solenoid is energized.

Reverse

When the manual valve is moved to the reverse position, oil is directed through the valve to the forward and reverse clutches. This causes the transaxle to shift into reverse. To raise pressures in reverse, reverse oil is directed behind the main pressure regulator valve, assisting the spring and raising line pressures. Reverse hydraulic system operation is shown in **Figure 11-28**.

downshifted position, moving the transmission or transaxle into a lower gear, depending on vehicle speed.

In a manually selected lower gear range, the transmission or transaxle is prevented from upshifting. Some manually selected low gear ranges will cause extra holding members to be applied by hydraulic pressure. These extra holding members keep the transmission from overrunning when a one-way clutch unlocks. Moving the selector lever into a lower gear position while the vehicle is moving will cause a downshift into the selected range. If vehicle speed is too great to allow a shift into the lowest gear, the next highest gear is applied. Reverse is obtained by direct hydraulic flow through the manual valve.

Transmission pressures vary between gears. Manually selected reduction ranges (manual low, manual second, and so on) use higher pressures than those used in normally selected drive or overdrive ranges. This is done to provide greater holding power when the vehicle is being used for towing or other heavy-duty work, and to provide sufficient oil to the extra holding members that are sometimes used in these ranges. Reverse range also requires higher gears, since reverse ratios require more holding power. Pressure is increased over normal ranges by directing oil pressure from the manual valve to the spring side of the main pressure regulator. This extra pressure assists the spring, causing the regulator valve to control pressure at a higher value.

Electronic transmission and transaxle oil circuits are similar in many respects. The major differences between various electronic transmissions and transaxles are the number, type, and arrangement of the holding members and the number of solenoids and sensors used. When attempting to diagnose a transmission or transaxle problem, always obtain the proper oil circuit diagram.

Summary

Transmission and transaxle hydraulic circuits are controlled by hydraulic valves, based on inputs from the selector lever (manual valve), road speed (governor), and throttle opening (throttle valve). The number of valves and their functions vary from one type of transmission to another.

The typical 3-speed hydraulically controlled transmission or transaxle has two shift valves, a throttle valve, and a governor valve. A 4-speed unit will have three shift valves. Throttle and governor pressures act on opposite ends of the shift valves. When governor pressure exceeds throttle pressure (plus spring pressure), an upshift occurs. The shift valves are designed so it takes less governor pressure to cause an upshift into lower gears.

A detent (passing gear) downshift is provided when the throttle pedal is completely depressed, pushing the throttle valve completely into its bore. Governor pressure will be overcome to force one or both shift valves into the

Review Questions—Chapter 11

Please do not write in this text. Place your answers on a separate sheet of paper.

1. In a transmission or transaxle hydraulic circuit, the _____ creates system pressure and the _____ _____ _____ regulates system pressure.

2. The vehicle speed at which an upshift occurs is governed by _____.
   (A) line pressure
   (B) throttle pressure
   (C) governor pressure
   (D) None of the above.

3. What causes a forced downshift?
   (A) A closed throttle.
   (B) A wide open throttle.
   (C) High governor pressure.
   (D) Low governor pressure.

4. A 3-speed automatic transmission would have _____.
   (A) one shift valve
   (B) two shift valves
   (C) four shift valves
   (D) It depends on the transmission design.

5. What causes the 1-2 shift valve to move before the 2-3 shift valve?

6. What is the purpose of directing line pressure through the TV limit valve?

7. During detent operation, line pressure is _____.
   (A) increased
   (B) decreased
   (C) zero
   (D) None of the above.

8. What happens if a driver tries to shift into manual first when the vehicle is traveling at high speeds?

9. Comparing the transmission and transaxle oil circuit diagrams in Figures 11-11 and 11-23 shows that the transmission has a _____ valve while the transaxle does not.
   (A) pressure regulator
   (B) governor
   (C) throttle
   (D) None of the above.

10. When looking at the oil circuit diagrams in Figures 11-11 and 11-23, how many shift valves do the transmission and transaxle have?
    Transmission _____
    Transaxle _____

ASE-Type Questions—Chapter 11

1. Each of the following is a direct function of the complete hydraulic circuit *except:*
   (A) to apply or release a band.
   (B) to route fluid for cooling and lubrication.
   (C) to engage or release a multiple-disc clutch.
   (D) to engage or disengage an overrunning clutch.

2. Which of the following components controls line pressure?
   (A) Oil pump.
   (B) Valve body.
   (C) Clutch pack.
   (D) Pressure regulator valve.

3. Throttle pressure will vary depending on:
   (A) engine load.
   (B) vehicle speed.
   (C) governor speed.
   (D) pump speed.

4. Each of the following inputs is used in the control of hydraulic system operation *except:*
   (A) road speed.
   (B) engine temperature.
   (C) engine load.
   (D) selector lever position.

5. Technician A says that an automatic transmission will always upshift at some predetermined road speed. Technician B says that hard acceleration will cause a vehicle to shift at a lower road speed. Who is right?
   (A) A only.
   (B) B only.
   (C) Both A and B.
   (D) Neither A nor B.

6. All of the following valves are found in the typical 3-speed transmission *except:*
   (A) throttle valve.
   (B) governor valve.
   (C) two shift valves.
   (D) overdrive valve.

7. The 1-2 shift valve on a particular transmission moves to the upshifted position. Which of the following is *most* likely to receive pressure from the valve?
   (A) Intermediate servo.
   (B) Reverse-high clutch.
   (C) Forward clutch.
   (D) Governor valve.

8. A cut back valve uses governor pressure to achieve:
   (A) reduced throttle pressure.
   (B) increased throttle pressure.
   (C) stabilized throttle pressure.
   (D) a cushioned shift.

9. Technician A says that shifting into second gear at cruising speed causes the transmission to downshift. Technician B says that third gear cannot be achieved in the manual second. Who is right?
   (A) A only.
   (B) B only.
   (C) Both A and B.
   (D) Neither A nor B.

10. With the engine running, the manual valve on a 3-speed transmission is moved to the reverse position. Which of the following is *most* likely to receive pressure from the valve?
    (A) High clutch.
    (B) Overrun clutch.
    (C) Forward clutch.
    (D) Intermediate servo.

Chapter 12

Electronic Control Systems

After studying this chapter, you will be able to:
- ❏ Explain how electronic controls differ from hydraulic controls.
- ❏ Identify input sensors and explain their operation.
- ❏ Identify output devices and explain their operation.
- ❏ Describe internal ECM components.
- ❏ Define control loops and explain their purpose.
- ❏ Explain how line pressure is controlled by solenoids.
- ❏ Explain how transmission/transaxle shifts are controlled by solenoids.
- ❏ Explain the shift sequence of an electronic automatic transmission.
- ❏ Explain the operation of ECM-operated lockup clutch controls.
- ❏ Explain the shift sequence of an electronic automatic transaxle.
- ❏ Explain the operation of a transmission with partial electronic shift controls.

Technical Terms

ECM	Engine coolant temperature sensor	Maintenance indicator light
OBD I	Air temperature sensor	Data link connector
OBD II	Exhaust temperature sensors	Trouble code
Input sensors	Manifold absolute pressure	Scan tool
Manual valve position sensor	Barometric pressure sensor	Solenoids
Throttle position sensor	Engine speed sensor	Duty cycle
Vehicle speed sensor	Mass airflow sensor	Dedicated solenoids
Pressure sensors	Oxygen sensor	Multi-shift solenoids
Temperature sensors	Chip	Limp-in mode
Input speed sensor	Microprocessor	Ratio control motor
Turbine speed sensor	Control loop	Variable ratio control valve
Engine temperature sensor	Diagnostic output	Manual shift program

Introduction

All modern transmissions and transaxles are fully or partially controlled by an on-board computer. To service these units, the technician must understand the electronic parts and circuits involved in their operation. This chapter covers the components and operating principles of electronic transmissions and transaxles. Studying this chapter will give you the knowledge needed to understand the electronic transmission and transaxle troubleshooting and service chapters.

Development of Electronic Controls

Electronic control of transmissions and transaxles began in the early 1980s. Transmissions and transaxles manufactured during this period used an on-board computer to control the converter lockup clutch. Later, some transmission shifts were made by electrical solenoids controlled by the computer. Pressure and temperature switches, input and output speed sensors, and fluid level sensors were also installed to monitor transmission operation.

Since the 1990s, almost all transmissions and transaxles have been computer controlled. In addition to shifts and converter clutch apply, the computer sometimes controls line pressures and detent operation. The computer can also use solenoids to control the initial application pressures of some clutches and bands to improve shift feel.

Electronic versus Hydraulic Controls

Electronic controls are more precise than hydraulic controls. They can take engine, vehicle, and atmospheric conditions into account. They can also precisely control the pressures to the holding members, resulting in smooth shifts that reduce holding member wear.

Precise control of shifts reduces transmission fluid overheating, lengthens transmission/transaxle life, and eliminates the need to adjust linkage.

Overall vehicle benefits from precisely controlled shifts are better mileage and performance, and decreased emissions. A vehicle with electronic transmission/transaxle controls also operates more smoothly, with shifts that are barely perceptible.

All automatic transmissions and transaxles, whether electronically or hydraulically controlled, have similarities. They all use a torque converter, hydraulic pump, planetary gears, clutches and bands, and manual linkage.

The main difference is the hydraulic control system. On older transmissions, as you have learned, the shift valves were operated by governor pressure and throttle pressure. When governor pressure rose above throttle pressure, the shift valve moved to the upshifted position. When throttle pressure overcame governor pressure, the shift valve moved to the downshifted position.

A completely electronic transmission does not have governor and throttle valves. In their place are electrical solenoids that control shifts and pressures, and sensors that monitor transmission, engine, and vehicle conditions. An on-board computer processes the inputs from the sensors and controls the operation of the output solenoids to produce the proper pressures and shift points.

Electronic Control System Components

To fully understand how electronic transmissions and transaxles operate, you must be familiar with the parts that make up the electronic control system. The computer receives inputs from sensors and issues output commands, which are based on these inputs, to the solenoids. The sequence of operation of the computer and related components is referred to as a control loop. Control loop operation was discussed briefly in Chapter 7. The following sections discuss the construction and operation of the input sensors, computer, and output solenoids used in electronic transmission control. Later in this chapter, we will explain how these parts work together to form a control loop that operates the transmission.

 Note: Vehicle computers are called by many names, depending on the manufacturer. In this chapter, as well as the chapters that follow, the transmission control computer will be referred to as the *ECM*. Older computer control systems are often referred to as *OBD I* systems. This is an abbreviation for on-board diagnostics, generation 1. Computer control systems in 1996 and later vehicles are called *OBD II systems*, standing for on-board diagnostics, generation 2. OBD II systems can be identified by their use of a standardized 16-pin diagnostic connector.

Input Sensors

Input sensors provide information to the ECM. The ECM will make decisions based on this information and issue output commands to the transmission. Some input sensors are located in the transmission and monitor transmission operation. Other sensors are primarily used to monitor engine operation. Information from these sensors is used by the computer to control engine operating systems and to control the transmission. Both types of sensors are discussed in the following sections.

Transmission Shift Control Sensors

On older transmissions, the governor and throttle valve operated the shift valves to change gear ratios. In an electronic transmission, throttle position sensors and vehicle speed sensors are used in place of these valves. The following sections explain the operation of these and other sensors used to help control shifts.

Manual Valve Position Sensor

The **manual valve position sensor** provides the simplest input to the ECM. This sensor, which is sometimes called the shift lever position sensor, is a set of on-off switches. Each gear shift lever position selected by the driver closes a particular switch in the manual valve position sensor, telling the ECM which transmission operating position has been selected. Some manual valve position sensors also operate the vehicle's backup lights and serve as neutral safety switches. This eliminates the need for separate switches. The sensor may be installed on the valve body or the external manual valve linkage. **Figure 12-1** shows a typical manual valve position sensor installed in the valve body.

Throttle Position Sensor

The **throttle position sensor,** or **TPS,** is mounted on the throttle body and is connected to the throttle linkage or the throttle shaft. Throttle position sensor input tells the ECM how hard the driver is accelerating. There are two types of throttle position sensors. The most common is the **variable resistor** type, **Figure 12-2.** This sensor consists of a sliding contact that moves along a resistor wire. Current

flows through the wire and the siding contact connection. Moving the throttle plate moves the contact. As the contact moves, the amount of the resistor wire and, therefore, the amount of resistance the current must flow through changes. This change in the amount of resistance in the circuit changes the current flow and voltage through the circuit. The ECM uses this voltage change to calculate the throttle position.

Some throttle position sensors are **transducers.** A transducer is a coil of wire wrapped around a movable iron core. See **Figure 12-3.** A small current flows through the coil, producing a magnetic field. The iron core is connected to the throttle linkage. Moving the throttle linkage causes the iron core to move inside the coil, affecting the magnetic field. Affecting the magnetic field causes the amount of current and voltage in the coil circuit to change. The ECM uses this change to compute throttle position.

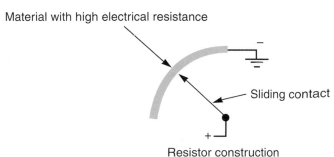

Variable Resistor Operation

Material with high electrical resistance

Sliding contact

Resistor construction

Current must travel a long distance through the high resistance material

High resistance

Current travels a short distance through the high resistance material

Low resistance

Figure 12-2. *Most throttle position sensors are variable resistors. Changes in the position of the sliding contact as the throttle is pressed vary the amount of voltage reaching the ECM. The throttle position sensor is always located at the throttle valve.*

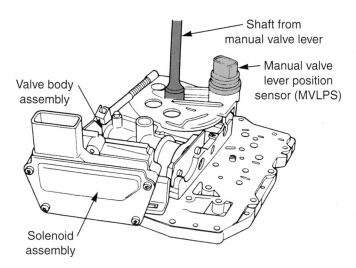

Shaft from manual valve lever

Manual valve lever position sensor (MVLPS)

Valve body assembly

Solenoid assembly

Figure 12-1. *The manual valve position sensor is installed on some electronically controlled transmissions and transaxles. This illustration shows the position of the sensor in the valve body. (DaimlerChrysler)*

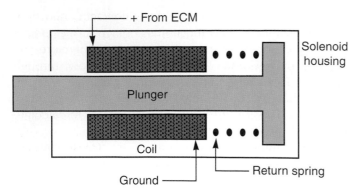

Figure 12-3. *A transducer consists of a metal plunger surrounded by a metal coil. Current flows through the wire whenever the ignition is on. Movement of the plunger changes the amount of current flowing in the wire. This affects the voltage sent to the ECM.*

Vehicle Speed Sensor

The **vehicle speed sensor** is mounted on the output shaft of the transmission or transaxle and tells the ECM how fast the vehicle is traveling. Speed sensors consist of a toothed wheel and a magnetic pickup. The wheel is attached to the output shaft and rotates with it, or it is driven by a gear on the output shaft. The pickup is mounted to the transmission case. **Figure 12-4** illustrates the relationship of the wheel and pickup when the wheel is installed directly on the output shaft. As the toothed wheel turns, it induces an alternating current (ac) in the magnetic pickup. Since alternating current changes direction, current flow in the pickup is reversed every time one of the wheel's teeth passes it. The faster the wheel is moving, the more often the current changes direction. The speed, or rate, of the change in current flow is called the *frequency.* The ECM measures this frequency and uses it to calculate input shaft speed.

Other Input Sensors: Transmission

In addition to the sensors discussed above, other sensors are used throughout the vehicle. Three of these sensors can be found inside the transmission or transaxle.

Pressure Sensors

Every electronically controlled transmission or transaxle has several **pressure sensors.** Most pressure sensors are simple on-off switches. They are connected to pressure passages in the transmission that feed certain clutches and bands. When the clutch piston or band servo is pressurized, the pressure also operates the pressure switch. Whether the pressure switches are on (pressure at the switch port) or off (no pressure at the switch port) tells the ECM what gear is engaged. The design of these switches is similar to that of the oil pressure switches used on engines with oil pressure lights. When no oil pressure is

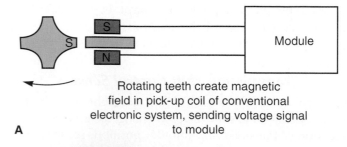

Rotating teeth create magnetic field in pick-up coil of conventional electronic system, sending voltage signal to module

A

B

Figure 12-4. *A—The relationship of the speed sensor and the shaft-mounted rotor. Rotation of the rotor induces a magnetic field in the sensor. This creates a voltage signal that is sent to the ECM. B—This photo shows the sensor and rotor from one automatic transmission.*

present, the diaphragm cannot close the contacts. When pressure reaches the switch, the diaphragm is pushed up, causing the contacts to close. **Figure 12-5A** shows a pressure sensor that completes a circuit between two electrical components. Note that it has two terminals. **Figure 12-5B** shows a single-terminal pressure sensor. This sensor grounds through the transmission case to complete the circuit.

Note: Switches that are held closed by spring pressure until hydraulic pressure is present are called *normally closed*, or *NC*, switches. Switches that are held open until hydraulic pressure is present are called *normally open*, or *NO*, switches.

Transmission Temperature Sensors

Temperature has a large effect on the operation of the transmission fluid. Fluid temperature can affect the viscosity (thickness) of the fluid and the holding ability of the clutches and bands. Therefore, many transmissions have **temperature sensors.** There are two types of temperature sensors. The simplest type of temperature sensor is the on-off switch, **Figure 12-6.** This switch can be normally open or normally closed. When the transmission fluid reaches a certain temperature, the switch position changes and signals the ECM. These switches were widely used on older transmissions to warn of overheating. Other temperature switches prevented converter lockup clutch application until the transmission reached normal operating temperature.

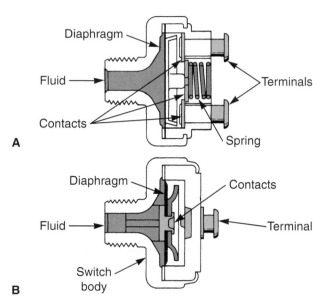

A

B

Figure 12-5. *Pressure switches transmit information to the ECM, usually telling it what gear has been selected by the driver or the hydraulic control system. A—This switch is a pass-through type that connects the battery with another electrical device. B—This switch grounds against the case to complete the circuit. (General Motors)*

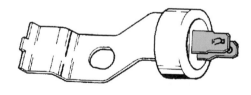

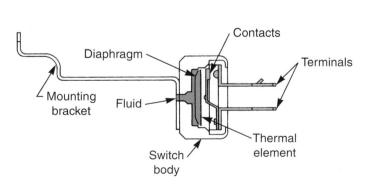

Figure 12-6. *This temperature sensor is an on-off switch used to signal the ECM when the transmission fluid reaches a certain temperature. A bimetallic strip completes the circuit at a set temperature. (General Motors)*

Some temperature sensors are resistor-type units that send a varying electrical signal based on transmission fluid temperature changes to the ECM. **Figure 12-7** shows a *resistor temperature sensor.* In some designs, current flows through the ECM, through the sensor, and then to ground. In other designs, a reference voltage is sent to the sensor and grounds through the ECM. In either case, any change in fluid temperature causes a change in the resistance of the sensor. Changes in sensor resistance affect the amount of current flow and, therefore, the voltage in the sensor circuit. The ECM reads this change in voltage as a temperature signal.

A resistor sensor allows the ECM to monitor transmission temperature and, therefore, fluid viscosity. Monitoring changes in viscosity helps the ECM modify shifts for maximum smoothness and durability. In addition, the sensor input allows the ECM to compensate for transmission overheating. If the transmission fluid begins to overheat, the ECM can increase pressures to prevent holding member damage.

Either type of temperature sensor may be used to signal the ECM that the transmission fluid is overheating. If the ECM receives an excessive temperature signal over a certain period of time, it will illuminate the dashboard warning light. **Figure 12-8** shows an assembly that includes both the pressure switches and a temperature sensor.

Input Speed Sensors

Some transmissions and transaxles have an *input speed sensor,* sometimes called a *turbine speed sensor.* This sensor tells the ECM how fast the transmission input shaft is turning. Input speed sensors operate in the same manner as vehicle speed sensors. The toothed wheel is mounted on the input shaft and the magnetic pickup is mounted to the case. **Figure 12-9** shows a typical input speed sensor.

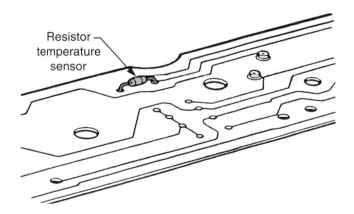

Figure 12-7. *This temperature sensor changes its resistance in response to temperature changes. Resistance changes affect the amount of current flowing in the wire, which in turn affects the voltage sent to the ECM as a signal. (Ford)*

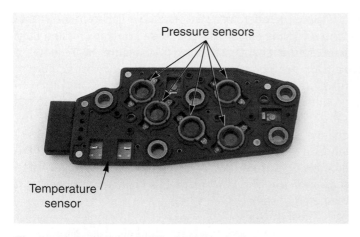

Figure 12-8. *This assembly contains both pressure sensors and temperature sensors.*

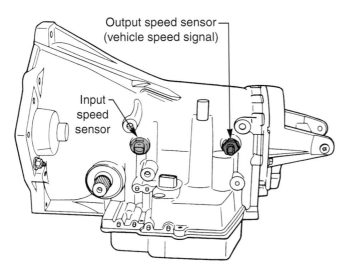

Figure 12-9. *Two speed sensors installed on a transaxle case. The front sensor measures input shaft speed and the rear sensor measures output shaft speed. The ECM compares the two signals to determine the transaxle gear and detect any slippage. (DaimlerChrysler)*

Other Input Sensors: Engine

Engine and body sensors also provide input to the ECM. Some of the more common sensors that affect the control of the transmission are discussed below.

Engine Temperature Sensors

Every electronic control system receives input from an **engine temperature sensor.** Engine coolant temperature is a good indicator of overall engine temperature. Therefore, every computer-control system has at least one **engine coolant temperature sensor,** which is threaded into an engine coolant passage. Most late-model engine controls also have an input to the ECM from an **air temperature sensor,** and a few vehicles have **exhaust temperature sensors.** These sensors are similar to those used to monitor transmission temperature. Resistance changes affect current

and voltage in the sensor circuit. The ECM interprets changes in current and voltage as changes in temperature.

> **Note: Some electronically controlled transmissions monitor the ambient (surrounding) air temperature. A few systems also monitor battery temperature. These sensors operate in the same manner as engine and transmission temperature sensors.**

Manifold Absolute Pressure Sensors

To monitor the load placed on the engine, a **manifold absolute pressure,** or **MAP,** sensor is attached to intake manifold vacuum. This sensor measures the intake manifold vacuum and converts it to an electrical signal, which is sent to the ECM. The vacuum developed in the intake manifold can be thought of as a pressure difference between the air inside the manifold and the outside air. Therefore, these sensors are called pressure sensors, even though they measure vacuum. MAP sensors contain a flexible diaphragm and a pressure-sensitive material called a **piezoelectric crystal.** The crystal produces an electrical current that is proportional to the pressure exerted upon it.

Figure 12-10 shows a common MAP sensor. Changes in manifold vacuum cause a change in the pressure on the piezoelectric crystal. The pressure change creates a small voltage pulse in the crystal. The ECM reads this voltage pulse as manifold vacuum.

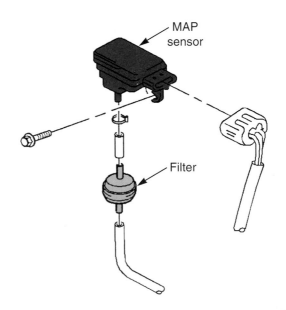

Figure 12-10. *Typical manifold absolute pressure, or MAP, sensor installation. Note the vacuum line and electrical connector. Most MAP sensors do not have a filter. Many MAP sensors are combined with the BARO sensor. (General Motors)*

Barometric Pressure Sensors

A **barometric pressure sensor,** or **BARO sensor,** measures the pressure of the surrounding air and converts it to an electrical signal. This sensor is very similar to and operates in the same manner as a MAP sensor. Some vehicles have both a MAP sensor and a BARO sensor.

Engine Speed Sensors

Some engines make use of an **engine speed sensor.** Most late-model systems use sensors mounted on the camshaft or crankshaft, while the distributor provides the engine speed signal for some systems. The operating principle of the engine speed sensor is identical to that of the vehicle speed sensor. The toothed wheel may be turned by the crankshaft or camshaft, **Figure 12-11,** or it may be part of the distributor pickup assembly. Many transmission control designs use the engine speed sensor input instead of having a separate transmission input shaft sensor.

Mass Airflow Sensors

Some vehicles use a **mass airflow sensor,** or **MAF sensor,** installed in the air intake system. See **Figure 12-12.** This sensor measures the amount of air entering the engine. Most MAF sensors use a **heated wire** located in the intake air stream. The ECM sends a reference voltage through the wire, causing it to become hot and increasing its resistance to current flow. Airflow into the engine cools the wire. This reduces the wire's resistance and allows more current to flow. Changes in current flow change the voltage in the circuit. The ECM interprets this voltage change as a change in airflow.

Other MAF sensors use a mechanical valve called a **vane,** which is moved by airflow. The valve is connected to a variable resistor using a sliding contact. Some vehicles

use a MAF sensor that operates from turbulence created by a calibrated restriction in the air intake passage. These are called **Karmann vortex** sensors.

Oxygen Sensors

The **oxygen sensor,** or **O_2 sensor,** is a vital part of all modern engine emission controls. Input from the oxygen sensor may also be used to control some transmission functions, especially when a rich condition occurs. In most common designs, the oxygen sensor produces electricity

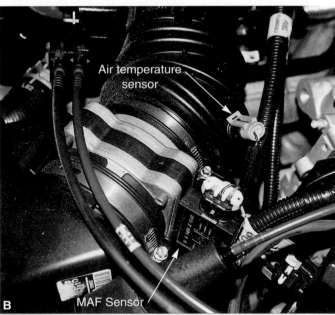

Figure 12-12. *The mass airflow sensor, or MAF, is usually installed on the air duct leading to the throttle plate. A—This sensor is attached to the air duct and can be removed without removing the ductwork. B—This sensor is part of an assembly that can only be removed by removing the duct. Note the air temperature sensor in the duct immediately after the MAF sensor.*

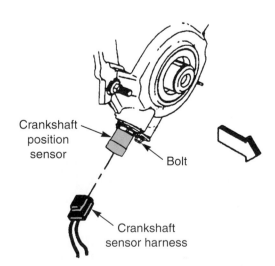

Figure 12-11. *This speed and crankshaft position sensor is installed on the front timing cover. Every modern vehicle has at least one of these sensors. They may be installed in several places on the engine, depending on the vehicle manufacturer and the type of engine. (General Motors)*

through a chemical reaction. The materials in the oxygen sensor generate a small voltage because of the difference in the amount of oxygen in the exhaust gases and the outside air. The greater the difference in the oxygen content between the exhaust and the outside air, the greater the voltage produced. This voltage corresponds to the air-fuel ratio of the engine—the richer the mixture, the greater the voltage. The ECM reads this voltage as a rich or lean mixture.

A few vehicles use a resistor-type oxygen sensor. This type of sensor changes resistance as the oxygen content of the exhaust gases changes. Most modern engines use several oxygen sensors, **Figure 12-13.**

Other Vehicle Inputs

Other vehicle inputs to the ECM include the charging system voltage, which the ECM reads directly from a wire connection at the alternator. On some vehicles, the ECM monitors the operation of the ABS or traction control system, as well as the air conditioning system. These inputs are taken directly from the unit as it operates. There is a hard-wired connection to the ECM. The presence of voltage at this wire tells the ECM that the device is operating.

ECM

The ECM is composed of various electronic devices, such as diodes, transistors, capacitors, and resistors. These devices are combined onto a complex electronic circuit by etching the circuitry on small pieces of semiconductor material. A circuit made in this manner is called a **chip,** or a **microprocessor.** These devices can precisely control transmission operation through solenoids, replacing the hydraulic and mechanical components previously used.

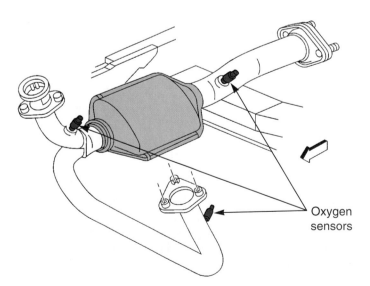

Figure 12-13. *Oxygen sensors may be located on the exhaust manifold or in the exhaust pipes. This illustration shows an OBD II system with sensors installed both before and after the catalytic converter. (General Motors)*

Main ECM Sections

As discussed in Chapter 7, all ECMs contain two main sections: the central processing unit, or CPU, and the memory. The CPU and memory are built into the ECM, and the entire ECM must be replaced if either is defective. The CPU receives the input sensor information, compares this information with the information stored in memory, performs calculations, and makes output decisions. There are two basic types of memory: read only memory (ROM) and random access memory (RAM). ROM contains the permanently installed programs and operating standards that tell the ECM what to do under various operating conditions. Disconnecting the battery or removing the system fuse will not erase ROM. Portions of the ROM are permanent, while other parts can be reprogrammed or changed. On late-model vehicles, special tools can be used to access ROM for reprogramming. In older vehicles, a separate chip called a PROM (programmable ROM) can be replaced.

RAM is a temporary storage place for data from the input sensors. Information received from the input sensors is temporarily stored in RAM, overwriting old information. When a trouble code (discussed later) is generated by the system, it is stored in RAM. The RAM has a temporary, or volatile, memory. On some older systems, RAM memory will be lost if the system fuse is removed or the battery is disconnected. Newer systems retain memory even when voltage is removed from the ECM.

Control Loops

The microprocessors that make up a modern ECM control transmission operation by processing information from the input sensors and issuing output commands. This ability allows them to operate as part of a **control loop,** **Figure 12-14.** A control loop can be thought of as an endless circle of causes and effects that is used to operate part of the vehicle, such as the automatic transmission. When the control loop is operating, the input sensors furnish information to the computer, which makes output decisions and sends commands to the output devices. The operation of the output devices affects the operation of the vehicle system, causing changes in the readings furnished by the input sensors. There is a continuous loop of information from the input sensors, to the computer, to the

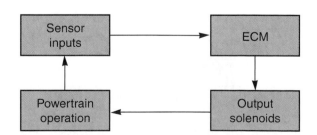

Figure 12-14. *The input sensors, computer, and output actuators form a continuous loop of inputs, ECM operation, outputs and power train operation. Note that the operation of the engine and transmission or transaxle is part of the loop.*

output devices, to the vehicle system, and back to the input sensors.

For example, the speed sensor tells the ECM that a certain speed has been reached. The throttle position sensor informs the ECM that throttle opening is not changing. The ECM will then decide to energize or de-energize the necessary solenoid(s) to shift into the next higher gear. If the throttle position sensor tells the ECM that the throttle is being opened quickly, the ECM will decide to operate the necessary solenoid(s) to shift into a lower gear. These two basic inputs are affected by inputs from the other sensors. The ECM operates many control loops at the same time.

The ECM can also compensate for a failed sensor by substituting other sensor inputs for those from the faulty component. For example, if the throttle position sensor fails, some ECMs can temporarily operate the transmission or transaxle by substituting the input of the MAP or MAF sensor for the input of the TPS. The ECM will also store information about the problem in memory, so the technician can access it when diagnosing system malfunctions.

ECM Diagnostic Outputs

In addition to operating the output devices, the ECM provides a **diagnostic output** to the technician. Internal ECM processes store diagnostic information. The external parts of the diagnostic output are:

❑ A dashboard-mounted warning light that illuminates when a problem is detected. This light is generally referred to as a **Maintenance Indicator Light,** or **MIL.** Other terms for this warning light were used on older (OBD I) vehicles.

❑ A **data link connector,** or **DLC,** which is used to retrieve information about system problems. This connector can be located in several places, depending on the vehicle year and model. It allows the technician to directly access the information stored in ECM memory. All OBD II systems use standardized 16-pin connectors. The connectors used in OBD I systems may vary from one manufacturer to another.

When a system defect occurs, the ECM stores information about the defect in memory. On older (OBD I) systems, the trouble codes identify components or systems that have failed completely. On OBD II systems, the ECM can store information about developing problems, out of range sensors, and systems that seem to be losing their effectiveness.

Information about defects is stored in the form of a **trouble code.** On vehicles up to 1995, trouble codes are 2- or 3-digit numbers that correspond to a specific defect. On vehicles with OBD II systems, the trouble code consists of a 5-character, alphanumeric code.

To determine whether trouble codes are present in the ECM's memory, the technician must use an electronic test device called a **scan tool.** Attaching the scan tool to the data link connector, **Figure 12-15,** causes the ECM to enter

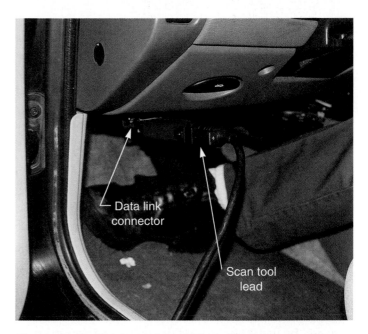

Figure 12-15. *In late-model vehicles, data link connectors are always located under the dashboard. In older vehicles, the data link connectors were located in various places, depending on the manufacturer.*

the self-diagnostic mode. The ECM will then release stored trouble codes to the scan tool. Use of scan tools will be discussed in detail in Chapter 14.

Output Solenoids

Solenoids are the primary output devices used in electronic transmissions and transaxles. Solenoids are installed in the valve body. Solenoids can be on or off for long periods of time, or they can be **pulsed** (turned on and off rapidly) to provide a **duty cycle** (a ratio of on to off). The ECM **drivers** operate the output solenoids. Drivers are special ECM circuits that consist of power transistors and related electronic devices. The power transistors in the driver circuits can handle the high current flows needed to operate the solenoids.

The following sections explain the various functions of the solenoids used in late-model automatic transmissions and transaxles. Note that not every electronic transmission or transaxle has all these components.

Line Pressure Control

As discussed in Chapter 9, a valve and spring assembly controls the line pressure. This assembly is usually called the main pressure regulator. Spring-operated main pressure regulator valves are used on all older automatic transmissions and on some newer models. Pressure from the throttle valve enters the spring side of the valve and assists pressure developed by spring tension. This is a simple and generally effective way of controlling pressure in response to load. However, it does not provide the precise control needed for maximum fuel economy and shift feel.

Many modern transmissions control the line pressure through an *electronically controlled pressure regulator.* Electronically controlled pressure regulators are operated by a solenoid controlled by the ECM. The solenoid is pulsed to create a duty cycle. The duty cycle creates a precisely controlled pressure leak at the main pressure regulator. This controlled leak accurately modifies pressure. See **Figure 12-16.** In other systems, the solenoid controls pressure by controlling the output of the vane pump. Operation of a pump output control solenoid is shown in **Figure 12-17.**

Converter Lockup Clutch Control

The basic operation of torque converter lockup clutches was covered in Chapter 8. Remember that on hydraulically operated units, the converter clutch was engaged by the ECM. Some older units used a pressure or speed switch to engage the converter clutch. On an electronically controlled transmission or transaxle, the ECM operates the lockup clutch control solenoid. The ECM makes the decision to apply the clutch based on inputs from the engine and drive train sensors. On some vehicles, an additional solenoid controls *clutch apply pressure.* This reduces roughness when the clutch is applied. The ECM varies the solenoid duty cycle based on sensor inputs. See **Figure 12-18.** On other vehicles, the main apply solenoid itself can provide cushioning based on an ECM-supplied duty cycle.

Shift Control

The basic principle of electronic shift control is the replacement of certain hydraulic components with ECM-operated solenoids that control fluid flow. On some transmissions and transaxles, energizing the solenoid opens and closes passages to the holding members, **Figure 12-19.** On other systems, the solenoid affects hydraulic pressure on one side of a shift valve, causing it to move. See **Figure12-20.**

In some transmissions and transaxles, the solenoids are directly connected to check balls that act as shift valves, **Figure 12-21.** Energizing the solenoids causes the check balls to move and either open or close pressure passages leading to the holding members. On a few transmissions, the solenoids are mechanically connected to the shift valves. Energizing the solenoids directly operates the shift valves. Refer to **Figure 12-22.**

On other transmissions, the valve body contains conventional shift valves. The solenoids indirectly move the shift valves by closing or opening pressure passages. Opening a passage causes a pressure drop, while closing the passage causes pressure to increase. Any pressure change on one side of a valve causes it to move toward the side with the lowest pressure, **Figure 12-23A.** When the solenoid closes the same passage, pressure equalizes on both sides of the valve. A spring then moves the valve back to its original position, **Figure 12-23B.** As with hydraulically controlled transmissions, pressure from another valve may be used to assist the spring.

Once moved by the solenoids, the shift valves perform the same job that they do in a hydraulically operated transmission. The shift valves direct oil pressure to the various holding members through passages in the valve body and case.

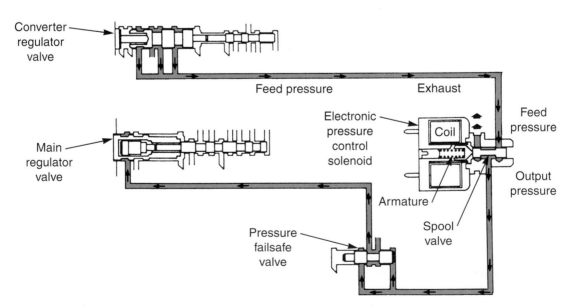

Figure 12-16. *This solenoid controls line pressure by controlling the assist pressure behind the main pressure regulator. On this particular transaxle, pump pressure reaches the solenoid through the converter pressure regulator. (Ford)*

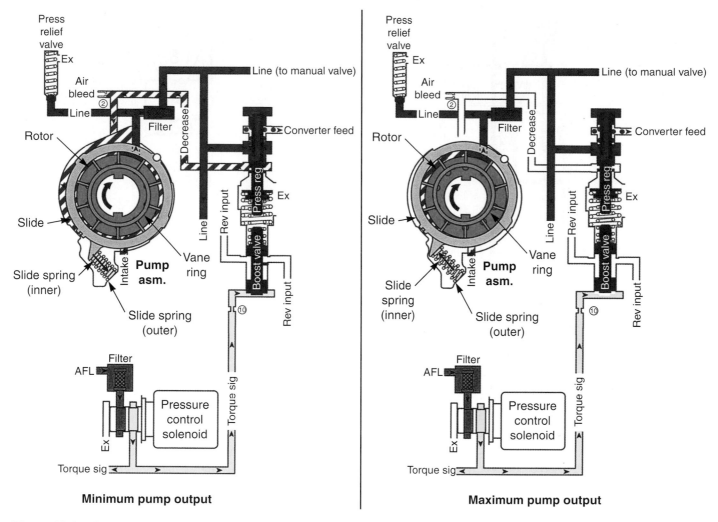

Figure 12-17. *Pressure behind the pressure regulator valve controls pump output. The regulator valve then controls pump output. (General Motors)*

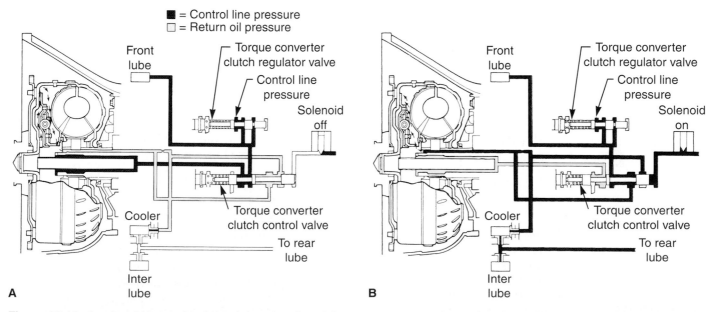

Figure 12-18. *A solenoid is used to control the operation of the torque converter clutch. Clutch apply feel is controlled by line pressure. A—The solenoid is off and fluid cannot flow to the apply side of the converter clutch control valve. The converter clutch is released. B—The solenoid is on and pressure flows to the apply side of the converter clutch control valve. (Ford)*

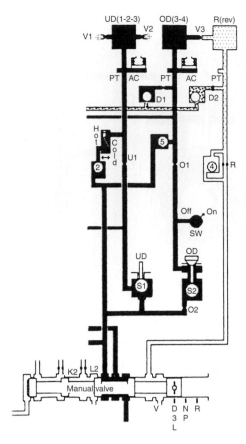

Figure 12-19. *Solenoid-controlled check balls can be used to operate the clutch packs by controlling oil flow through the passage. No valves are needed. (DaimlerChrysler)*

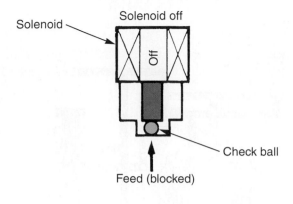

Figure 12-21. *These illustrations show a solenoid with an internal check ball. Energizing the solenoid causes the check ball to either seal or open a passage in the solenoid assembly. (Ford)*

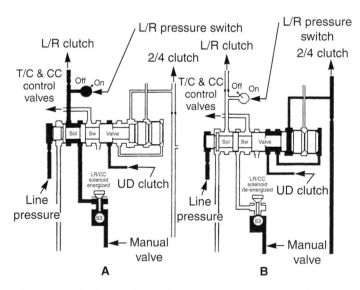

Figure 12-20. *These illustrations show how the operation of a solenoid can be used to deliver oil pressure to one side of a valve. A—The pressure increase causes the valve to move. B—Removing the pressure allows pressure on the other side of the valve to return the valve to its original position. (DaimlerChrysler)*

There are two ways to install solenoids so that they apply the needed clutches and bands:

❑ *Dedicated solenoids*—each solenoid is dedicated to operating only its associated valve. There is one solenoid for each shift valve. In this arrangement, a four-speed transmission with three shift valves would have three shift solenoids.

❑ *Multi-shift solenoids*—used in combination with pressure passages in the valve body. The pressure passages are arranged so that multi-shift solenoids affect some valves directly and others indirectly. In this arrangement, a four-speed transmission with three shift valves might have only two shift solenoids.

Dedicated Solenoids

On systems with a dedicated solenoid for each shift valve, each solenoid directly controls the operation of its related valve by removing or restoring pressure.

Trace the oil circuit pathways in **Figure 12-24** and you will notice that each solenoid controls the related shift valve by controlling the pressure delivered to that valve. Each of these solenoids sends pressure to the related valve to cause it to move. Other dedicated solenoids move the shift valve by exhausting (removing) pressure already

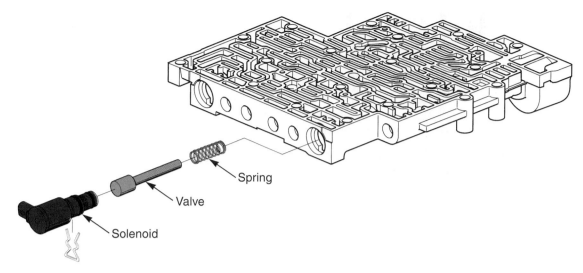

Figure 12-22. *This solenoid is directly connected to the valve. When the solenoid opens, it moves the valve against spring pressure, opening the valve. When the solenoid is de-energized, the spring returns the valve to its original position. (General Motors)*

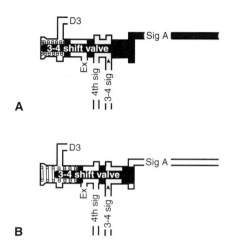

Figure 12-23. *Study the effect of pressure on this 3-4 shift valve. A—Pressure from the solenoid moves the 3-4 shift valve against spring pressure, allowing pressure to flow through other passages. B—When pressure is removed, the spring moves the valve to its original position. (General Motors)*

present on one side of the valve. These solenoids usually exhaust pressure directly to the oil pan or sump. A few dedicated solenoids operate by moving the valve through linkage or by directly moving the valve.

Multi-Shift Solenoids

In transmissions with multi-shift solenoids, the solenoids work directly on some shift valves and work through the valves and related oil passages to control other shift valves. The solenoids are arranged with the shift valves and oil passages so that they move the shift valves, depending on whether the solenoids are energized or de-energized. **Figure 12-25** shows a schematic of a four-speed transmission with two multi-shift solenoids. Shift solenoids have two possible positions: energized (on) and de-energized (off). When two multi-shift solenoids are used together, the possible sequences are shown in **Figure 12-26.**

The on-off sequence of the solenoids controls the application of the clutches and bands to produce various gears. An explanation of how the solenoids operate to provide four forward speeds on a common rear-wheel drive automatic transmission and front-wheel drive automatic transaxle will be presented later in this chapter.

Shift Cushioning

On hydraulically controlled transmissions, shifts are cushioned (shift feel improved) through the use of accumulators, calibrated orifices, and check balls. On many electronic transmissions and transaxles, shift feel is controlled by a pulsed solenoid. The solenoid restricts fluid flow according to ECM commands that are based on sensor inputs. The solenoid is pulsed to cause a soft apply under light throttle. When the throttle opening is large or the vehicle is heavily loaded, the solenoid is pulsed to allow maximum pressure to quickly reach the holding members. **Figure 12-27** shows a typical pressure control solenoid. Note that it is installed between the shift valve and the holding member. Most electronically controlled transmissions and transaxles still rely to some extent on accumulators, orifices, and check balls.

Governor Pressure Solenoids

On some partial electronic transmissions, governor and throttle pressure move the shift valves. The governor valve, however, has been replaced by a solenoid controlled by the power train ECM. The ECM receives inputs from speed sensors and a governor pressure sensor. Based on these inputs, it varies the solenoid duty cycle to send the correct governor pressure to the shift valves. **Figure 12-28** shows the solenoid and pressure sensor installed on the valve body.

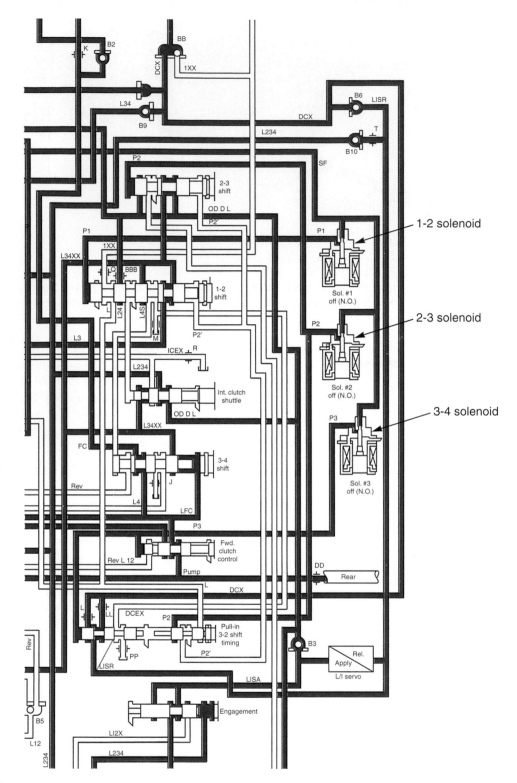

Figure 12-24. *Three dedicated solenoids are shown with their associated shift valves. Each solenoid operates one shift valve. (Ford)*

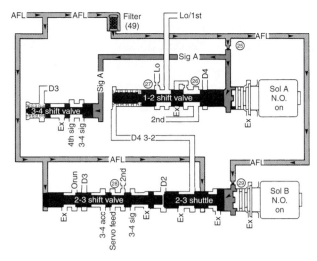

Example A: Park/reverse/neutral/& first gear

Solenoid A	Solenoid B
on	on
off	off
on	off
off	on

Figure 12-26. *Possible solenoid position sequences for a transmission with two multi-shift solenoids.*

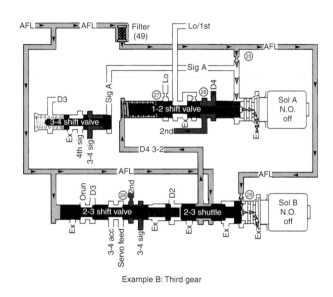

Example B: Third gear

Figure 12-25. *Multi-shift solenoids. Although there are only two solenoids, their sequence of operation causes three shifts in drive. (General Motors)*

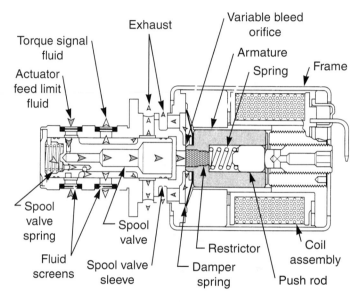

Figure 12-27. *Pressure control solenoids pulse on and off many times per second. The length of the pulses is called the duty cycle. The length of the duty cycle determines the pressure output of the valve. (General Motors)*

Electronic Transmission and Transaxle Operation

The following sections cover the operation of electronic transmissions and transaxles. Emphasis is on the operation of the solenoids and related hydraulic system parts. Remember that the holding members and many hydraulic control system parts are similar to those used on hydraulically controlled transmissions.

Sensor Inputs and Computer Operation

The transmission solenoids cannot operate unless the ECM tells them what to do. In turn, the ECM is helpless unless it receives input from the sensors. In its most basic operation, the ECM compares the vehicle speed with the throttle opening and tells the solenoids when to shift the transmission into the next gear. For example, assume that while accelerating from stop, the vehicle speed sensor indicates that the vehicle is traveling 12 mph (kph) and the throttle position sensor indicates a small throttle opening (light acceleration). The ECM will command the appropriate solenoid to shift the transmission from first to second. However, if the speed sensor indicates that the vehicle is traveling 12 mph (kph) and the throttle position sensor indicates a large throttle opening (heavy acceleration), the ECM will not energize the solenoid until a higher speed is reached. **Figure 12-29** shows how this simple arrangement works. Note how this compares to the action

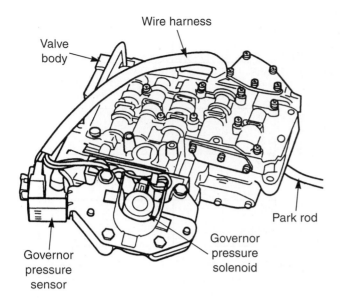

Figure 12-28. *The governor pressure solenoid is placed on the valve body and takes the place of the shaft-mounted governor valve. The assembly also contains a governor pressure sensor. (DaimlerChrysler).*

of the throttle and governor valves in a hydraulically operated transmission.

Other inputs to the ECM also affect shift speeds. For instance, the MAF or MAP sensors may indicate that the engine is under a heavy load, and therefore, the ECM may decide to shift the transmission into a lower gear, even though throttle position and speed sensor inputs call for a higher gear. The ECM may also select a lower gear if the temperature sensor indicates that the engine is beginning to overheat. Other sensor inputs may cause the ECM to raise line pressure or delay the application of the converter lockup clutch.

If one of the sensors fails, the ECM is able to make shift decisions based on inputs from other sensors. If, for instance, the throttle position sensor fails, the ECM may make shift decisions based on input from the MAP sensor. In other instances, failure of a sensor may cause the ECM to shift into *limp-in mode.* In limp-in mode, the ECM can make shifts as soon as possible, eliminate some gears, raise line pressure, or prevent lockup clutch apply. The ECM will also turn on the malfunction indicator light (MIL) on the dashboard to let the driver know there is a problem with the system.

Electronic Transmission with Two Shift Solenoids

The following section explains how a widely used electronically controlled transmission uses two multi-shift solenoids to provide four forward gears and reverse. Operation in Neutral and Park is also covered, as well as the application of the converter lockup clutch. This transmission is found on many late-model cars and light trucks.

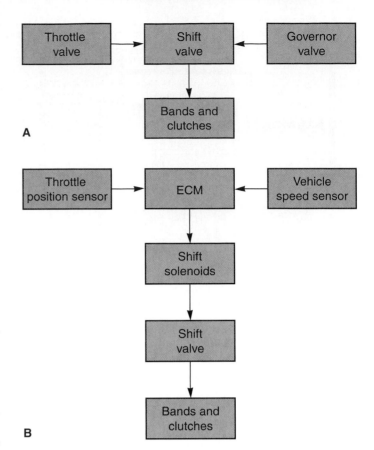

Figure 12-29. *Compare the interaction of the governor, throttle valve, and shift valve in a hydraulically controlled transmission (A) with the interaction of the throttle position sensor, vehicle speed sensor, ECM, and solenoids on an electronically controlled transmission (B).*

Refer to **Figures 12-30** through **12-41** as you read the following paragraphs. You may also compare the operation of this hydraulic system with the operation of the hydraulic system on an older version of this transmission shown in Figures 11-11 through 11-22. When you make the comparison, note the similarity between the two types of transmissions. They use the same number and type of holding members for each gear. Some of the hydraulic valves are the same or similar.

Park, Engine Running

The shift lever is in the Park position with the engine running, **Figure 12-30.** The pump is turning and supplying pressure to the main pressure regulator. Pump pressure is also sent to the pressure relief valve. The main pressure regulator consists of the regulator valve and a boost valve. The pressure regulator spring is installed between the two valves.

Line pressure from the pump also passes through the manual valve and the low overrun valve, and applies the low and reverse clutch. The low and reverse clutch has no effect on operation in Park and is only applied because of the valve body and manual valve design.

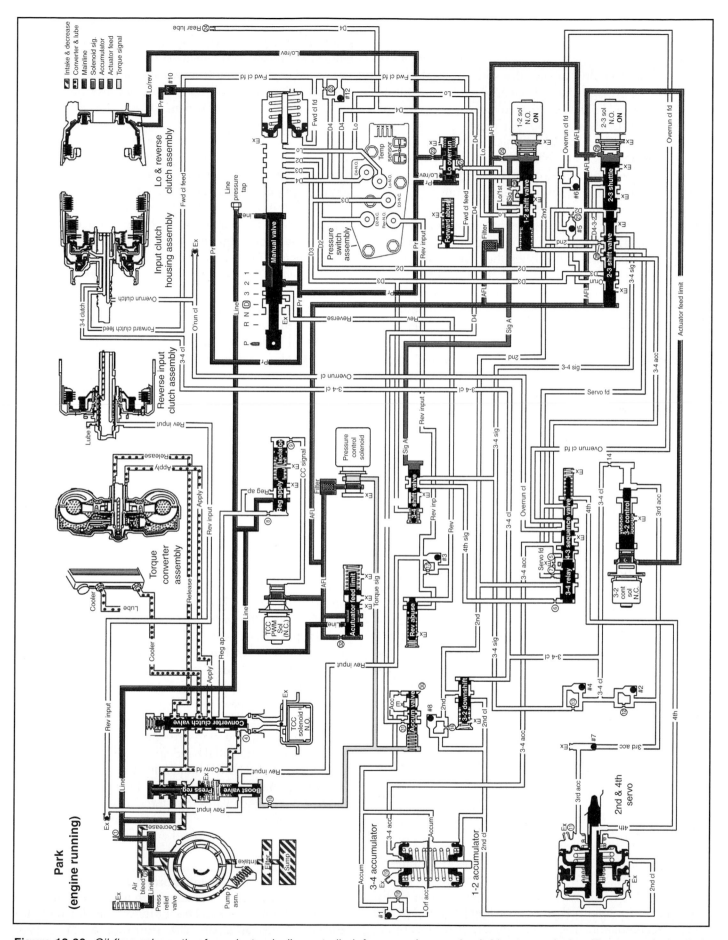

Figure 12-30. *Oil flow schematic of an electronically controlled, four-speed, rear-wheel drive transmission. The transmission is in Park and the engine is running. Study the components in this schematic before proceeding. (General Motors)*

Line pressure also travels to the actuator feel limit (AFL) valve. The AFL valve limits the amount of pressure in the pressure control and shift solenoid feed circuits, since excess pressure could damage the solenoids. AFL oil is directed to the pressure control solenoid, which is a pulsed solenoid operated by the ECM based on sensor inputs. The pressure control solenoid sends oil pressure to the main pressure regulator boost valve. This pressure moves the boost valve and increases the pressure on the pressure regulator spring. The amount of pressure on the spring varies the pressure output of the main pressure regulator.

AFL oil pressure is also delivered to the shift valves and solenoids. The ECM energizes both shift solenoids in Park. AFL oil from the accumulator limit valve cannot exit through the solenoids and pushes the shift valves to the left. However, the solenoids are not supplied with drive oil and, therefore, cannot apply any holding members.

The main pressure regulator also directs some oil to the torque converter clutch valve. The torque converter clutch valve is in the released position at this time. Oil passes through the upper part of the valve to the torque converter release circuit, through the torque converter, and out through the torque converter apply circuit. The converter oil then passes through the bottom part of the torque converter valve to the oil cooler. In *Park*, the operation of the solenoid has no effect, since line oil pressure is stopped at the converter clutch valve until the transmission shifts into second gear. Oil flows through the release and apply sides of the converter clutch plate, and the converter is unlocked.

Neutral, Engine Running

In Neutral with the engine running, **Figure 12-31,** the pump is turning and supplying pressure to the main pressure regulator just as it was in Park. AFL oil through the pressure control solenoid continues to regulate line pressure through the main pressure regulator boost valve. AFL pressure is also sent to the shift valves but has no effect. The main pressure regulator also continues to send oil to the torque converter clutch valve, which remains inoperative.

In *Neutral*, the manual valve is positioned to cut off oil to the low reverse clutch piston. Therefore, no holding members are applied in *Neutral*.

Drive, First Gear

When the driver places the vehicle in Drive (sometimes called Overdrive) with the vehicle not moving, **Figure 12-32,** line oil is sent to the forward clutch piston, engaging the forward clutch. This oil is called D4 oil. Under normal conditions, the check ball (#12) seals the oil passage, forcing the D4 oil to travel through the smaller orifice next to it. This ensures that the shift is made gently. Some forward clutch oil is also diverted to the forward clutch accumulator. The accumulator uses this oil to compress a spring, reducing the force of the oil sent to the clutch. When the spring is fully compressed, full oil pressure is available at the clutch piston. Notice that D4 oil is called FWD clutch feed oil when it reaches the forward clutch and the forward clutch accumulator piston.

If the engine speed is above idle, the pressure control solenoid will raise line pressure. This will cause the D4 oil to be higher than normal, opening the forward abuse valve. When the forward abuse valve is open, D4 oil is sent directly to the forward clutch piston, bypassing the orifice and check ball assembly. The forward abuse valve prevents clutch plate damage by quickly and firmly applying the clutch when the vehicle is placed in *Drive* at high engine RPM.

D4 oil is also directed to the accumulator valve, which fills and pressurizes the 1-2 and 3-4 accumulators for later use. Notice that the accumulator valve is affected by oil pressure from the pressure control solenoid. D4 oil is also delivered to one of the pressure switches in the pressure switch assembly. This normally open (N.O.) switch closes when it is pressurized to tell the ECM that the transmission is in *Drive*.

D4 oil is also sent to the 1-2 shift valve for use in later gears. The shift solenoids are both energized. The 1-2 shift solenoid sends oil pressure to the 3-4 shift valve, but this has no effect on transmission operation since no other oil pressure is present at the valve. Mechanically, the operation of the transmission is the same as in the older version discussed in Chapter 11.

Drive, Second Gear

When the ECM decides the transmission should be shifted into second gear, it de-energizes (turns off) 1-2 shift solenoid. Since the 1-2 shift solenoid is a normally open (N.O.) type, oil exits through the solenoid when it is de-energized. Spring pressure pushes the 1-2 shift valve to the right, allowing D4 oil to pass through the valve and apply the 2-4 band servo, **Figure 12-33.** This oil pressure, called 2nd oil, applies the band and shifts the transmission into second gear. The check ball and orifice, shown directly above the 3-2 downshift valve (#8), are used to restrict 2nd oil pressure to the servo and cushion the shift. Pressure also becomes available at the 3-2 downshift valve.

To further cushion the shift, some 2nd oil is sent to the 1-2 accumulator. This oil moves the 1-2 accumulator piston against spring pressure, diverting some pressure from the servo. The initial band apply will be soft, followed by full pressure to hold the band tightly.

The 2nd oil is also sent to the 2-3 shift valve, 3-4 relay valve, 4-3 sequence valve, and the converter clutch valve for use in other gears. The converter clutch may apply in second gear, as explained below.

 Note: In the left corner of Figure 12-33, a previously unseen valve and solenoid have been added. The solenoid and valve are supplied with AFL oil. The purpose of this valve and solenoid will be explained in the section on downshifts.

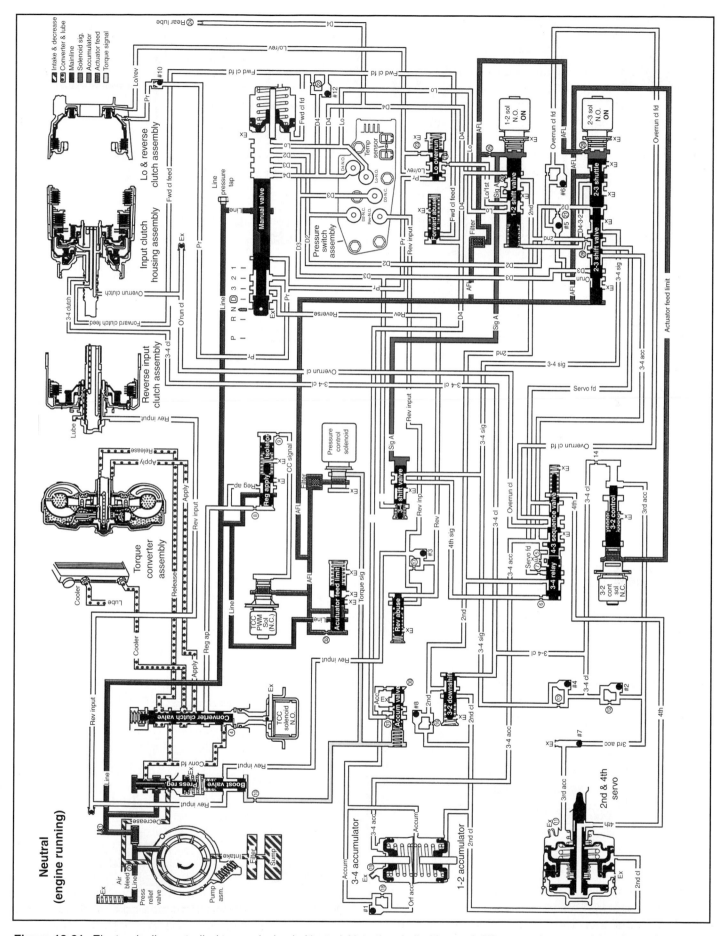

Figure 12-31. *Electronically controlled transmission in Neutral. Note the similarities and differences between this schematic and that of Figure 12-30. (General Motors)*

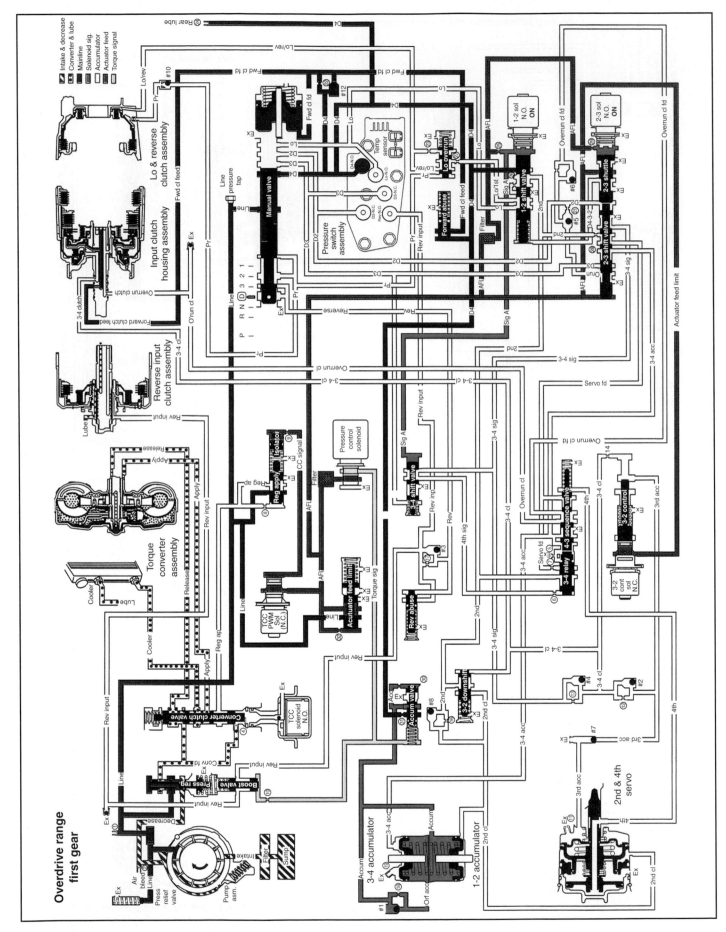

Figure 12-32. *Electronically controlled transmission in Overdrive range, first gear. (General Motors)*

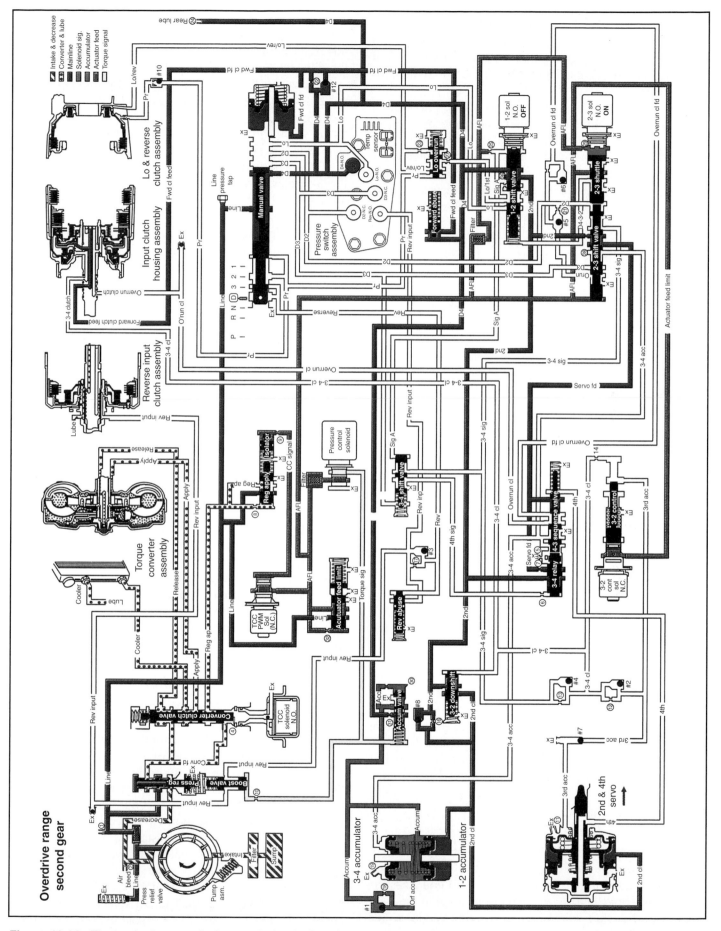

Figure 12-33. *Electronically controlled transmission in Overdrive range, second gear. Compare solenoid operation with first gear.* (General Motors)

Drive, Third Gear

When the ECM decides that the transmission should be shifted into third gear, it de-energizes the 2-3 shift solenoid, **Figure 12-34.** The 1-2 shift solenoid remains de-energized.

Trace the oil circuit pathways and you will notice that each solenoid controls the related shift valve by controlling the pressure delivered to that valve. Each one of these solenoids operates by sending pressure to the related valve to cause it to move. Other dedicated solenoids move the shift valve by exhausting (removing) pressure already present on one side of the valve. These solenoids usually exhaust pressure directly to the oil pan or sump. A few dedicated solenoids operate by directly moving the valve through linkage or by directly moving the valve.

The 2-3 shift solenoid is a normally open solenoid, and oil exits through the valve when it is de-energized. AFL oil pushes the 2-3 shift valve and 2-3 shuttle valve to the right. 2nd oil can now pass through the 2-3 shift valve. This oil is renamed 3-4 SIG oil after it passes through the 2-3 shift valve. In addition to applying and releasing holding members, 3-4 SIG oil goes to the 3-2 downshift valve, 3-2 control valve, and 3-4 shift valve for later use.

To apply the clutch, the 3-4 SIG oil passes through a check ball and orifice (#4). The #4 check ball and orifice restricts oil flow to the clutch piston to cushion the shift. Notice that this oil is renamed 3-4 CL (CL for clutch) oil after it passes through the #4 check ball and orifice. 3-4 CL oil is also sent to the 3-2 downshift valve and the 3-2 control valve for use during 3-2 downshifts, explained later.

3-4 CL oil also goes to the #2 check ball and orifice, then to the 2-4 servo. The oil, now called 3RD ACC oil, pressurizes the servo release passage, causing the servo to release the band. The action of the #2 check ball and orifice, and the accumulation action of the servo as it releases, cushions the shift. A separate accumulator is not used. 3RD ACC oil also goes to the 3-2 control valve for later use.

With the 3-4 clutch applied by 3-4 CL oil and the band released by 3RD ACC oil, the transmission is now in third gear.

Since the 2-3 shuttle valve also moves to the right when shift solenoid B is de-energized, this opens a passage for AFL oil to the spring side of the 1-2 shift valve. How this affects 1-2 shift valve operation is explained in the next section.

Drive, Fourth Gear

When enough speed is attained and other sensor inputs are correct, the ECM decides that the transmission should be shifted into fourth gear. The ECM energizes the 1-2 shift solenoid, sealing the exhaust passage in the solenoid. The 2-3 shift solenoid remains de-energized. See **Figure 12-35.**

With the 1-2 shift solenoid back on, AFL oil pressure can build up on the right side of the 1-2 shift valve through the orifice above the valve (#25). Once through the orifice, the oil is renamed SIG A oil.

However, the extra oil supplied by the 2-3 shuttle valve assists the spring and keeps the 1-2 shift valve from moving. The AFL oil, renamed the SIG A oil, builds up to its full value and moves the 3-4 shift valve.

With the 3-4 shift valve in the upshifted position, 3-4 SIG oil, renamed 4th SIG oil in this circuit only, pushes the 3-4 relay and 4-3 sequence valves to the right. This allows 2nd oil to charge the 3-4 circuit that sends oil to the outer apply side of the servo. The outer side is an apply area that helps the 2nd clutch oil overcome release oil and spring pressure and reapply the band. Some 3-4 SIG oil is diverted to the 3-4 accumulator to improve shift feel. This is renamed 3-4 ACC oil.

Converter Clutch Application

The converter clutch apply is shown in **Figure 12-35.** It is almost identical to the apply used on the nonelectronic version of this transmission. The converter clutch can be applied in any forward gear except first. When the converter clutch signal valve is closed by 2nd oil, line pressure can reach the converter clutch valve. As long as the torque converter clutch solenoid is de-energized, this oil pressure leaks out past the internal check ball and seat. With no pressure applied to it, the valve remains in the unapplied position, and oil flows through apply and release circuits.

When the torque converter clutch solenoid is energized, the check ball seats and closes the passage. Oil pressure builds up and becomes high enough to move the converter clutch valve. When the valve moves, oil pressure on the lockup clutch release side is exhausted. Oil pressure remains on the clutch apply side and pushes the clutch plate into contact with the converter cover, locking the converter.

Downshifts

As with upshifts, the ECM determines when the transmission should downshift and performs the shift by energizing or de-energizing the 1-2 and 2-3 solenoid.

The ECM bases these decisions on engine operating conditions, vehicle speeds, and other inputs supplied by the sensors. Based on these inputs, the ECM also engages and disengages the torque converter clutch. The term **downshifts** refers to the normal downshifts that occur when the vehicle is brought to a stop, as well as the detent (passing gear) downshifts that occur when the throttle is suddenly depressed.

To shift from fourth to third, the ECM de-energizes the 1-2 solenoid, **Figure 12-36.** The 2-3 solenoid remains de-energized. De-energizing the 1-2 solenoid removes pressure from the SIG A circuit and causes the 3-4 shift valve

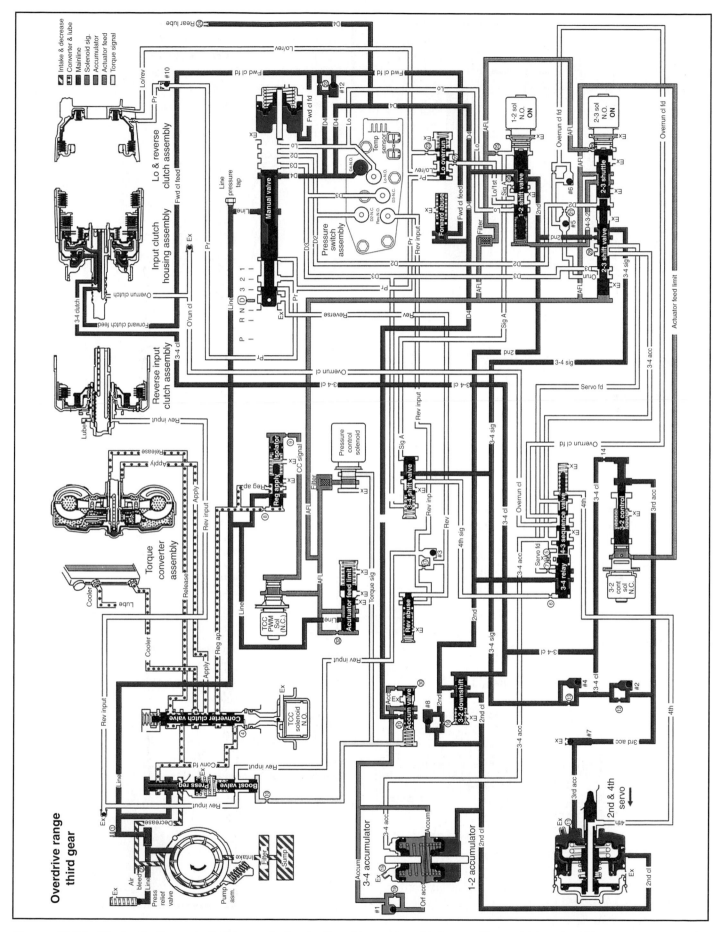

Figure 12-34. *Electronically controlled transmission in Overdrive range, third gear. Note which solenoids are energized and de-energized compared with earlier gears. (General Motors)*

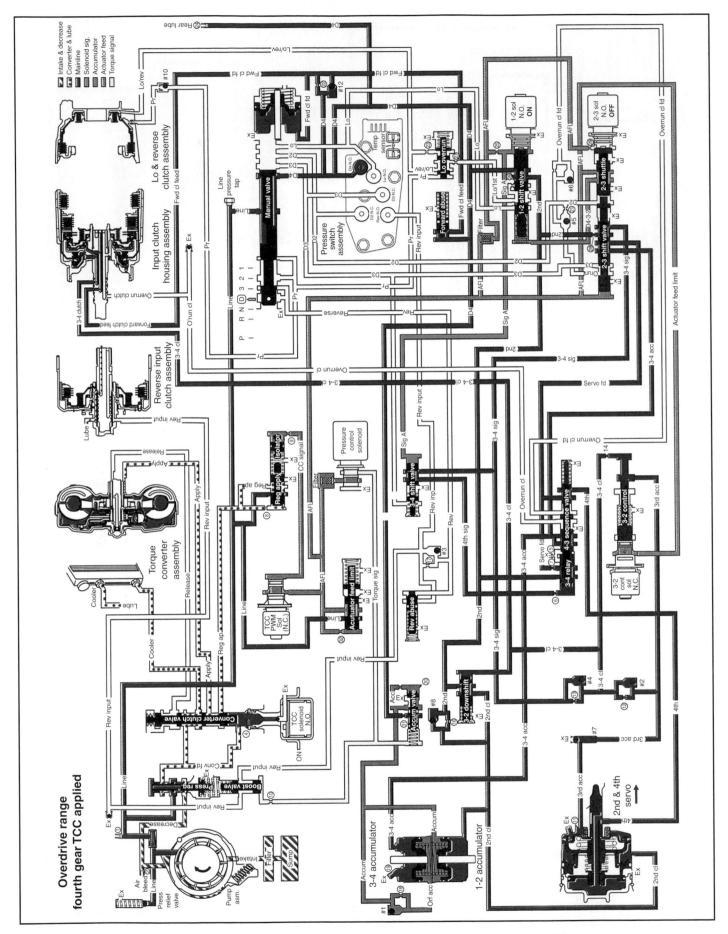

Figure 12-35. *Electronically controlled transmission in Overdrive range fourth gear. In this gear the torque converter clutch is applied. Note which solenoids are energized and de-energized and compare with earlier gears. (General Motors)*

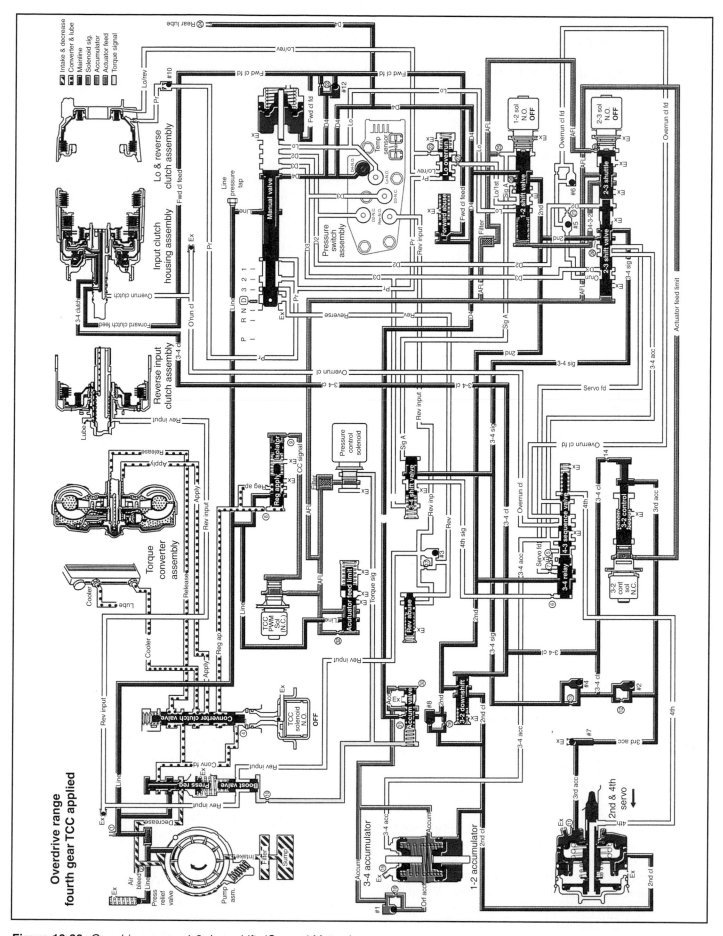

Figure 12-36. *Overdrive range, 4-3 downshift. (General Motors)*

to move to the downshifted position. This cuts off the oil to 3-4 relay and 4-3 sequence valves. Spring pressure moves the 3-4 relay and 4-3 sequence valves to their original positions, exhausting the oil in the servo apply chamber. The transmission returns to third gear.

To shift from third to second, the ECM re-energizes the 2-3 solenoid, **Figure 12-37.** The 1-2 solenoid remains de-energized. With the 2-3 solenoid back on, AFL pressure dead ends at the solenoid and pushes the 2-3 shuttle and 2-3 shift valves to the downshifted position. Moving the 2-3 valves cuts off the oil supply to the 3-4 clutch and the 2-4 servo release chamber. This places the transmission in second gear.

To more accurately match the 3-2 shift timing and feel to engine and road conditions, the 3-2 downshift control solenoid (CONT SOL in the illustration) is pulsed to vary the pressure to the 3-2 control valve. This varies the speed and timing of the 3-4 clutch release and the 2-4 band apply. The 3-2 downshift valve and #8 check ball and orifice also help create a downshift that matches vehicle speed and power output.

Manual Third

When the shift lever is moved to the manual third gear position, **Figure 12-38,** the manual valve directs oil, called D3 oil, to the back of the 3-4 shift valve to keep it from upshifting to third, and also opens a normally closed (NC) pressure switch on the pressure switch assembly. The pressure switch tells the ECM that manual third has been selected, and the ECM ensures that the solenoid sequence that is necessary to obtain fourth gear (1-2 solenoid on, 2-3 solenoid off) cannot be obtained. Therefore, fourth gear is prevented both hydraulically and electrically.

Operation in manual third is similar to operation in Drive. Starting from rest, the transmission will shift through first, second, and third. However, the transmission will not shift into fourth. As with the nonelectronic version of this transmission, oil pressure is delivered through the 2-3 shift and 4-3 sequence valves to the overrun clutch.

The overrun clutch is applied to keep the input roller clutch from overrunning. This provides engine braking for descending steep grades. Oil passes through an orifice and check ball assembly (#6) to ensure a smooth overrun clutch application. The converter clutch can be on or off, depending on the inputs to the ECM.

Manual Second

Selecting manual second causes the transmission to operate in second gear only, no matter what the road speed or throttle opening. When the manual valve is moved to the manual second gear position, **Figure 12-39,** the valve directs oil, called D2 oil, to the back of the 2-3 shift valve to keep it from upshifting to third. Oil from the manual valve also opens a normally closed (NC) pressure switch on the pressure switch assembly. The pressure

switch tells the ECM that manual second has been selected, and the ECM ensures that the 1-2 shift solenoid will not be energized. The transmission will stay in second gear under all conditions. The overrun clutch remains applied for engine braking. The converter clutch may be applied in this gear.

Manual First

In manual first, **Figure 12-40,** the manual valve sends pressure to a normally open switch in the pressure switch assembly. This switch is closed by oil pressure and tells the ECM that manual first has been selected. The ECM uses this input to ensure that both the 1-2 and 2-3 solenoids are on, holding the transmission in first gear. The manual valve also sends pressure to the center of the 1-2 shift valve, where it is directed to the low-overrun valve and into the low-reverse clutch apply piston. The low-reverse and low-overrun clutches prevent roller clutch overrunning and provide for engine braking. If manual first is selected at high vehicle speed, the transmission will not downshift to first until speed is reduced.

Reverse

When the transmission is placed in reverse, **Figure 12-41,** oil is directed to the reverse input clutch piston and both pressure areas of the low and reverse clutch piston. To raise line pressures in reverse, reverse oil is directed to the reverse input clutch piston through the boost valve of the main pressure regulator. This oil assists spring pressure in pushing on the main pressure regulator, raising line pressure. Reverse oil is also directed to a normally open switch on the pressure switch assembly. This switch tells the ECM that the transmission is in reverse.

Note that AFL pressure is still being produced. AFL pressure may be used by the pressure control solenoid to further raise line pressures if the vehicle is heavily accelerated in reverse.

For smooth reverse engagement, oil sent to the reverse input clutch piston must pass through an orifice and check ball (#3). If the engine speed is above idle, reverse pressure will be higher than normal, opening the reverse abuse valve. When the reverse abuse valve is open, oil pressure bypasses the orifice and check ball assembly, applying the clutch much more quickly and firmly.

Electronic Transaxle with Three Shift Solenoids

The following sections explain how a common electronically controlled transaxle uses three multi-shift solenoids to operate through four forward gears and the application of the converter lockup clutch. The action of the solenoids and hydraulic system is similar to that of the electronically controlled transmission described previously.

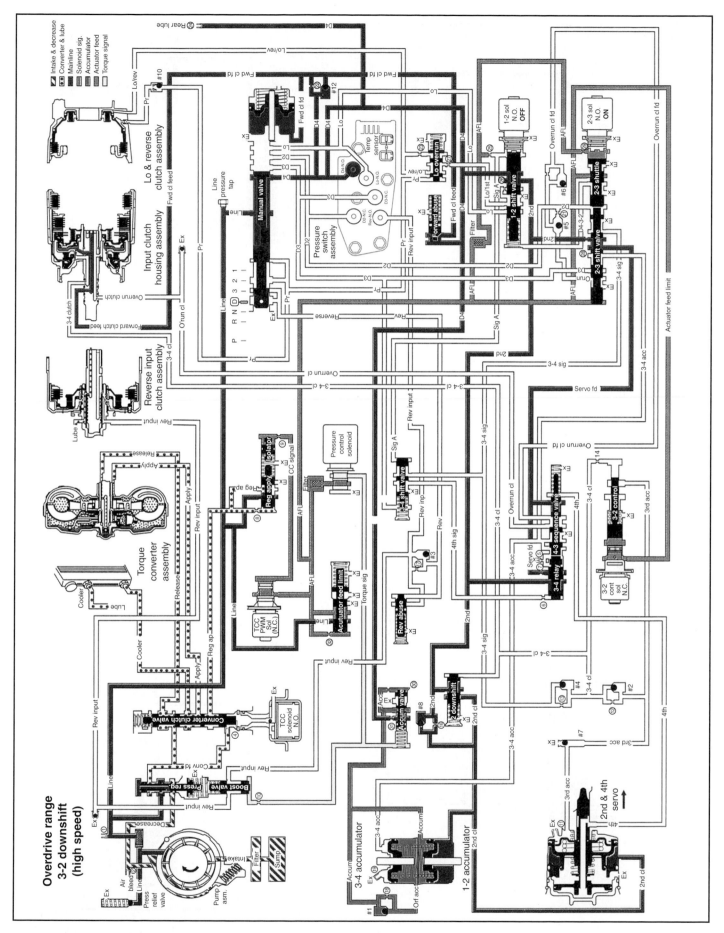

Figure 12-37. *Overdrive range, high speed 3-2 downshift. (General Motors)*

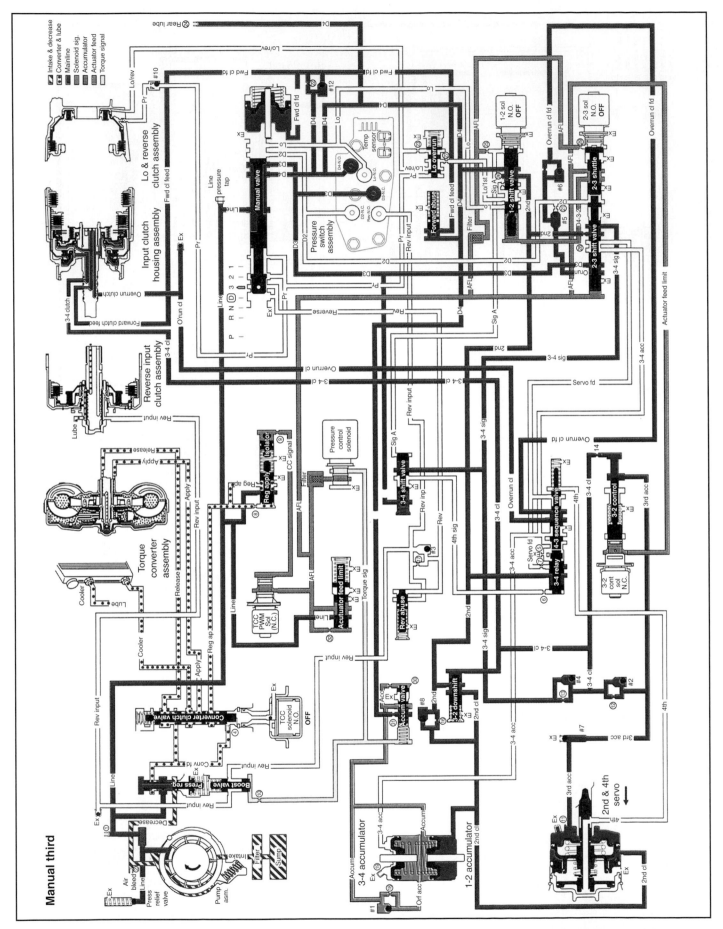

Figure 12-38. *Electronically controlled transmission in manual third range. Compare this range with third gear in overdrive range.* (General Motors)

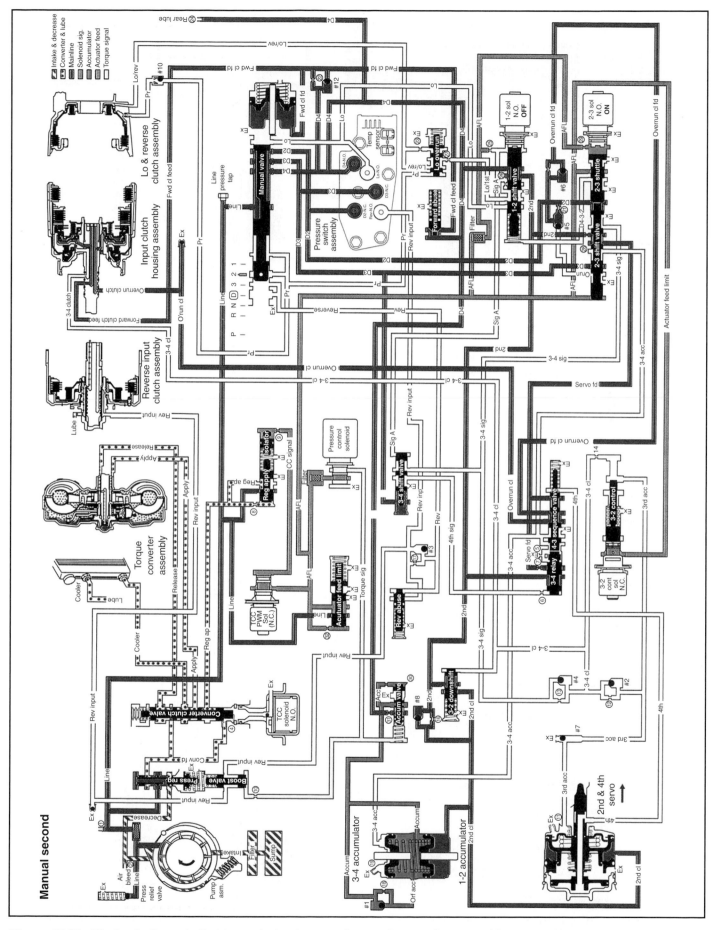

Figure 12-39. *Electronically controlled transmission in manual second range. Compare this range with second gear in overdrive range. (General Motors)*

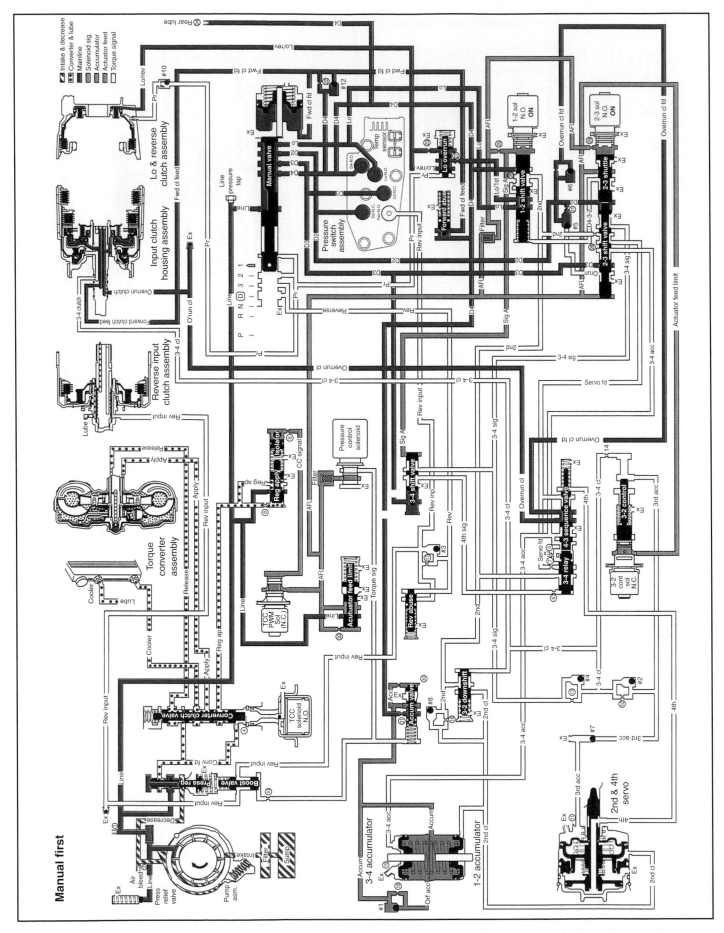

Figure 12-40. *Electronically controlled transmission in manual first range. Compare this range with first gear in overdrive range.* (General Motors)

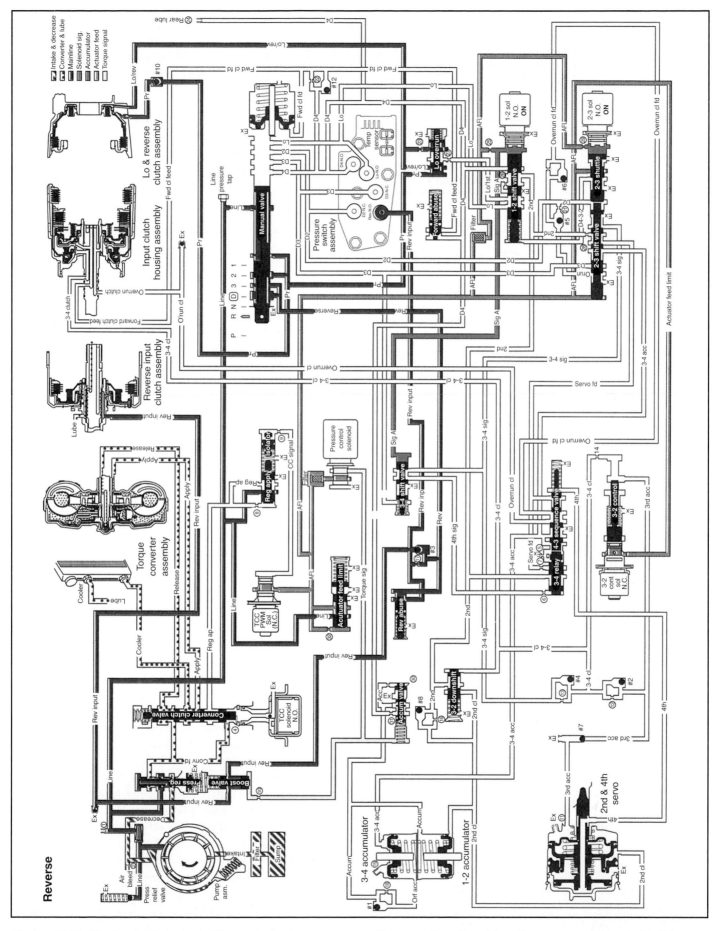

Figure 12-41. *Electronically controlled transmission in reverse range. Take special notice of how line pressure is boosted in this gear.* (General Motors)

Refer to **Figures 12-42** through **12-45** as you read the following paragraphs. You may want to compare the operation of this hydraulic system with the operation of the hydraulic system on an older version of this transaxle shown in Figures 11-23 through 11-28. For easier understanding of the solenoid action, the operation of some valves is not covered.

Note: The solenoid marked VFS TV SOL is the main pressure regulator control solenoid. It adjusts line pressure to match throttle opening and engine load in all gears.

Overdrive, First Gear

In first gear, **Figure 12-42,** solenoid #1 is de-energized. Oil flows through the solenoid to the 1-2 shift valve, holding it in the downshifted position. Solenoid #2 is energized, and no oil can flow through it. Solenoid #3 is de-energized, and oil flows through the solenoid to the forward clutch control valve and the 3-2 shift timing valve. This pressure moves the forward clutch control valve to the right. Oil pressure flows through the forward clutch control valve, through the 3-4 shift valve, and into the forward clutch apply piston. Some fluid also travels to the neutral-drive (ND) accumulator to cushion the application of the forward clutch. With all shift valves in the downshifted position and the forward clutch applied, the low roller clutch locks up and the transaxle is in first gear.

Overdrive, Second Gear

To shift the transaxle into second gear, the ECM energizes solenoid #1. Solenoid #1 opens, removing the pressure from the upshift passage to the 1-2 shift valve. The 1-2 shift valve moves to the left, opening a line oil pressure passage to the intermediate clutch shuttle valve.

The intermediate clutch shuttle valve sends oil pressure to the intermediate clutch and 1-2 accumulator. This action is illustrated in **Figure 12-43.** Applying the intermediate clutch causes the low roller clutch to unlock and shifts the transaxle into second gear. Note that the pressure first travels through the 1-2 capacity modulator (1-2 cap mod). This device further controls shift feel.

Overdrive, Third Gear

To shift into third gear, the ECM de-energizes solenoids #1 and #2, and energizes solenoid #3, **Figure 12-44.** When solenoid #3 is energized, the 2-3 shift valve moves back to the first gear position. This will allow oil to flow to the direct clutch.

When solenoid #2 is de-energized, oil flows to the 2-3 shift valve, moving it to the upshifted position. Line oil passes through the 1-2 shift valve to the direct clutch. This oil is also sent to the intermediate clutch shuttle valve to ensure that the intermediate clutch stays applied even though the 1-2 shift valve has returned to the first gear position.

When solenoid #3 is energized, there is no longer any pressure directed to the forward clutch control valve. The forward clutch control valve moves to its original position,

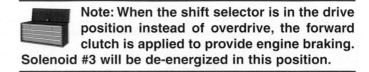

Note: When the shift selector is in the drive position instead of overdrive, the forward clutch is applied to provide engine braking. Solenoid #3 will be de-energized in this position.

and oil pressure is removed from the forward clutch piston. This disengages the forward clutch. With the direct and intermediate clutches applied, and the forward clutch released, the transmission is in third gear.

Overdrive, Fourth Gear

In fourth gear, the ECM energizes solenoid #1. Solenoids #2 and #3 remain in the same positions as in third gear. See **Figure 12-45.** Energizing solenoid #1 moves the 1-2 shift valve back to the upshifted position. Oil from the 1-2 shift valve is delivered to the upshift side of the 3-4 shift valve, causing it to move to the upshifted position. The 3-4 shift valve directs oil to the overdrive servo. Pressure to the overdrive servo causes it to apply the overdrive band, placing the transaxle in fourth gear. Some overdrive servo oil is diverted to the 3-4 accumulator to cushion the shift.

In this circuit diagram, the converter clutch has been applied. Oil from the bypass clutch control solenoid moves the bypass clutch control valve to the left, cutting off converter release pressure. This allows the pressure on the apply side of the converter clutch plate to push it into engagement with the converter cover.

Partial Electronic Shift Controls

Some transmissions and transaxles use a combination of electronic and hydraulic shift controls. In most cases, these are older three-speed designs that have been redesigned to upgrade them to four speeds. A conventional shift valve acted on by the throttle and governor valves controls the 1-2 and 2-3 shifts. The 3-4 shift is controlled by a solenoid operated by the vehicle's computer.

The transmission oil pressure diagrams illustrated in the next two figures are those of an updated 4-speed version of a once-common 3-speed transmission. The major differences are on the right side of the illustration. Note the presence of the overdrive clutch, 3-4 accumulator, and three valves related to the 3-4 shift. The 3-4 shift valve is held in the downshifted position by a spring.

Note that the first two shifts (1-2 and 2-3) are completed hydraulically by the action of the governor and throttle valves on the shift valves. On later versions of this transmission, the governor valve has been replaced by a governor pressure solenoid like the one shown in **Figure 12-28.** The final shift (3-4) is performed by the solenoid. This solenoid operates conventional shift valves by opening or closing pressure passages. Also, note that the transmission contains an additional solenoid. This solenoid is used to

First Gear Hydraulic Flow – D, Ⓓ Range

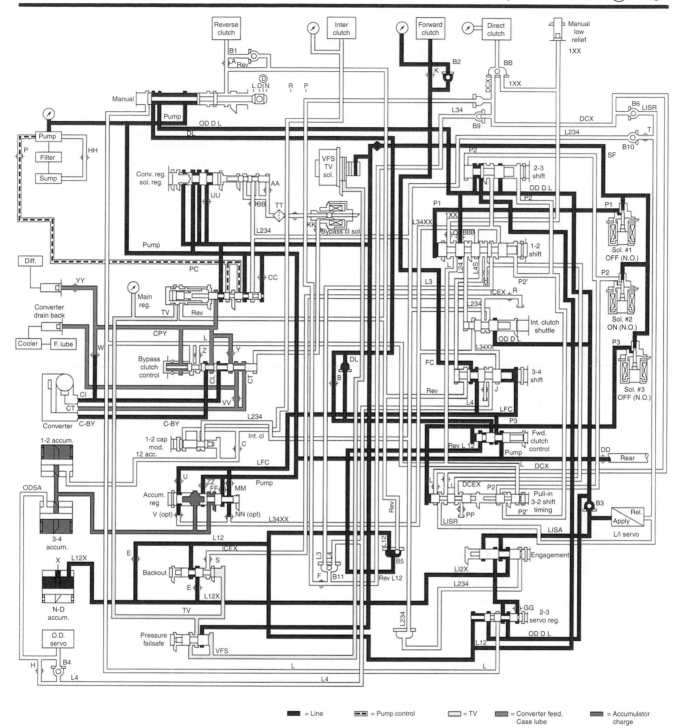

Figure 12-42. *The oil flow schematic of an electronically controlled, four-speed, front-wheel drive transaxle. The transaxle is in the overdrive range, first gear, and the engine is running. Study the components in this schematic before proceeding. (Ford)*

apply the converter lockup clutch, and operates independently of the overdrive solenoid. The lockup clutch can be applied in third or fourth gears.

In **Figure 12-46,** governor pressure on one end of the 2-3 shift valve has overcome throttle pressure on the other end and moved the 2-3 shift valve to the upshifted position. The transmission is now in third gear, which is direct

drive (1:1). Movement of the shift valve allows fluid into the direct (front) clutch apply piston and intermediate (front) servo release chamber.

This same oil pressure is also delivered to the 3-4 shift and 3-4 timing valves. However, the oil going to the 3-4 shift valve must pass through a restricting orifice, and leaks out through the check ball as long as the overdrive sole-

Second Gear Hydraulic Flow – D, (D) Range

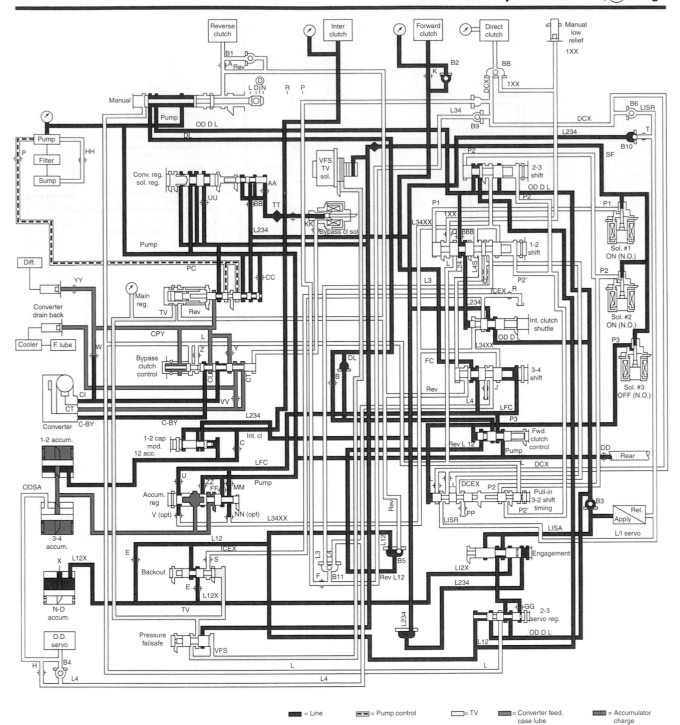

Figure 12-43. *Overdrive range, second gear. Note which solenoids are energized and de-energized. Compare this with first gear.* (Ford)

noid is de-energized. There is not enough pressure to move the 3-4 shift valve against spring pressure.

When the vehicle reaches a certain speed, the ECM will decide that a shift to overdrive is desirable, and it will energize the overdrive solenoid. When the overdrive solenoid is energized, the check ball seals the exhaust passage.

Oil pressure can no longer leak through the check ball. Pressure rises and moves the 3-4 shift valve against spring pressure. This action sends line oil pressure to the overdrive clutch and shifts the transmission into fourth gear. Some oil is also diverted to the 3-4 accumulator to cushion the shift. See **Figure 12-47.**

Third Gear Hydraulic Flow – (D) Range, Converter clutch released

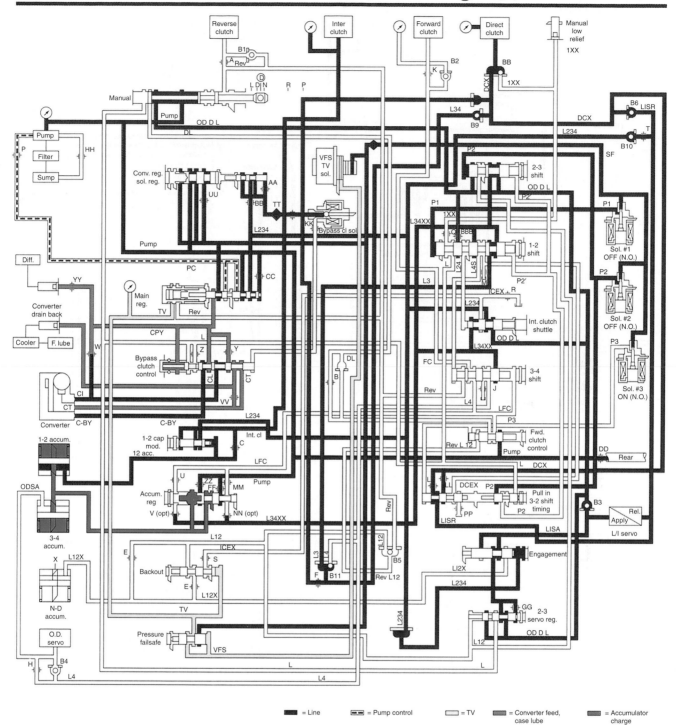

Figure 12-44. *Overdrive range, third gear. Again, note which solenoids are energized and de-energized. (Ford)*

The overdrive section has no control over the governor and throttle valves. The governor and throttle valves could move the 2-3 shift valve and shift the transmission into second while the overdrive unit is engaged. To prevent this, the same pressure that moved the 3-4 shift valve also acts on the 3-4 timing valve, moving it against spring pressure. The 3-4 timing valve sends oil pressure to the upshift side of the 2-3 shift valve. This extra pressure keeps the 2-3 shift valve from moving to the downshift position as long as overdrive is engaged.

The 3-4 shuttle valve is used to vary the firmness of the up and downshifts of the overdrive clutch, based on the pressure supplied to it by the governor valve. More governor pressure means a higher speed and results in a firmer apply. Low governor pressure is the result of lower speeds and causes a softer apply.

Fourth Gear Hydraulic Flow – ⒟ Range, Converter clutch released

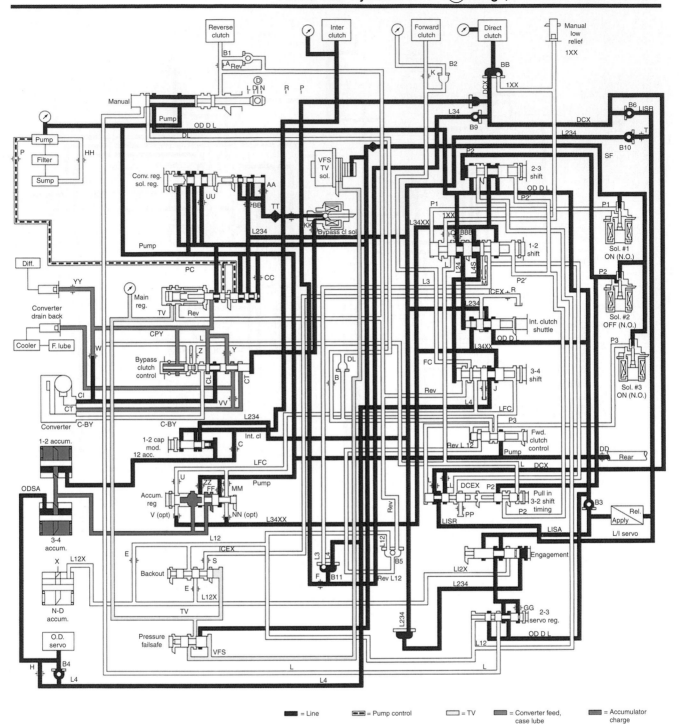

Figure 12-45. *Overdrive range, fourth gear. The torque converter clutch is applied. Compare solenoid operation with that in earlier gears. (Ford)*

To downshift back to third gear, the ECM de-energizes the solenoid. Pressure drops at the 3-4 shift valves, and it returns to the downshifted position. Oil pressure is removed from the overdrive clutch, causing it to disengage. The transmission returns to third gear.

CVT Pulley Controls

All CVTs are electronically controlled. **Figure 12-48** illustrates the electronic control system for a typical CVT. The *ratio control motor* or ratio control solenoid receives output commands from the ECM and moves the *variable ratio control valve.* Moving the control valve varies

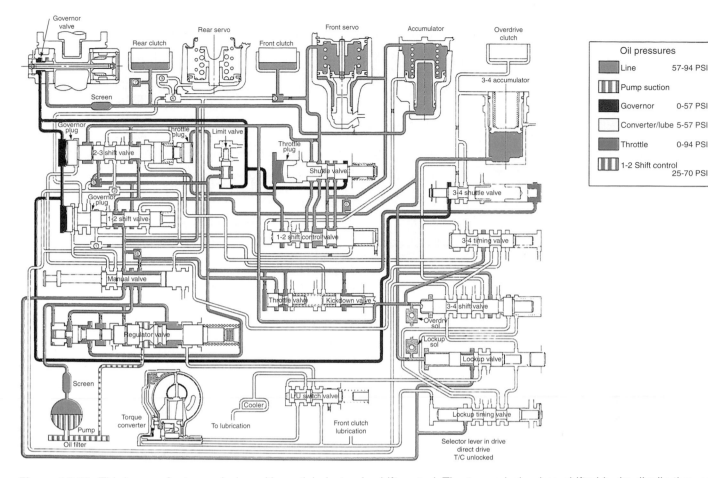

Figure 12-46. *Third gear of a transmission with partial electronic shift control. The transmission has shifted hydraulically through first, second, and third gears. Study the position of the solenoids and shift valves. (DaimlerChrysler)*

hydraulic pressure to the drive pulley piston. Note that the piston moves one pulley side. Fluid pressure, called primary feed oil, moves from the variable ratio control valve through a calibrated orifice to the drive pulley piston. The calibrated orifice restricts oil flow to prevent overly rapid movement of the pulley in the inward direction. When fluid pressure is reduced, the check ball unseats and allows the primary feed oil to exhaust rapidly.

When the drive pulley moves, force is transmitted through the metal belt to overcome hydraulic pressure in the driven pulley piston. The drive pulley piston has a greater surface area than the driven pulley piston, and will always force the driven pulley to respond to size changes in the drive pulley. Line pressure is fed to the driven pulley piston to match pulley clamping force to engine output.

The primary limit valve is a pressure regulator valve that keeps excessive pressure from damaging the drive pulley piston and other parts. Primary feed oil enters a chamber on the side opposite the spring. When feed oil pressure exceeds spring pressure, the valve moves toward the spring side, opening an exhaust passage.

Manual Shifting Programs

A *manual shift program* is a method of manually changing gears on electronically controlled transmissions and transaxles. The manual shift program allows the driver to change between gear ratios by moving the shift lever, **Figure 12-49.** The transmission's electronic control module performs the actual gear changes based on shift lever movement, while continuing to monitor engine and vehicle sensor inputs. The transmission/transaxle unit itself is identical to one without a manual shift program.

Moving the shift lever to a special position switches the control module to the manual shift program. See **Figure 12-50.** Once the manual shift program is activated, the shift lever is momentarily pushed up (the + direction in Figure 12-50) to shift the transmission or transaxle to the next higher gear. Momentarily pushing the shift lever down (the – direction in Figure 12-50) shifts to the next lower gear. On some manual shift programs, the shift lever is moved to the left or right to change gears. Notice that in addition to the manual shift program (M program), the shift lever in **Figure 12-50** has a provision for fully automatic shifts (XE program). It also has a provision for performance

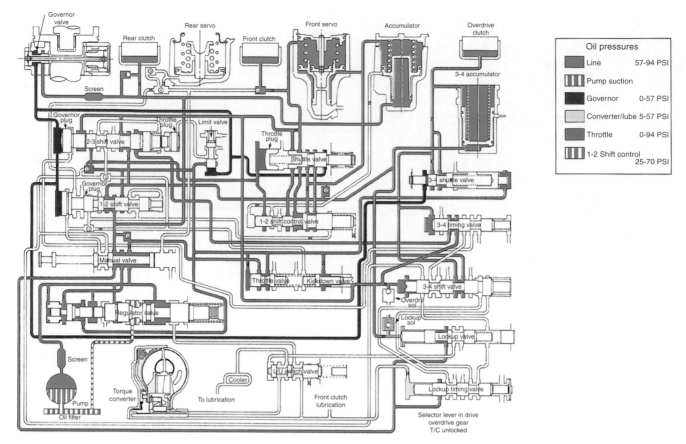

Figure 12-47. *Fourth gear of a transmission with partial electronic shift control is achieved when the overdrive solenoid is de-energized. Compare the solenoids, shift valves, and oil circuitry with those in Figure 12-45. (DaimlerChrysler)*

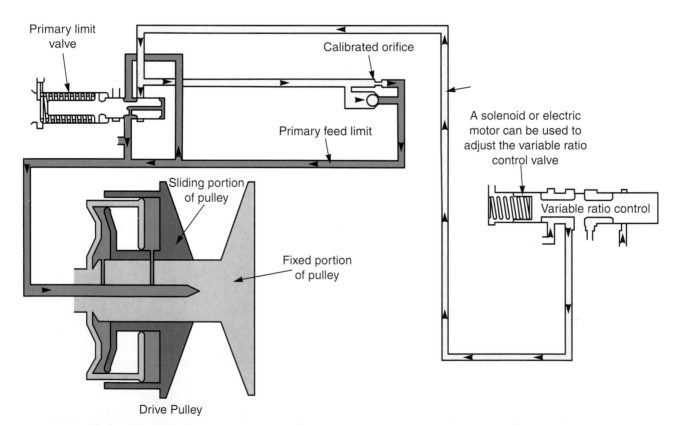

Figure 12-48. *Moving the control valve varies hydraulic pressure to the drive pulley piston. Changes in the hydraulic pressure applied to the drive pulley piston causes one side of the drive pulley to move toward or away from the fixed side, changing the effective diameter of the drive pulley.*

Figure 12-49. *A typical shift lever used with a manual shift program. Most manual shift programs are used with floor shifters. (Volkswagen)*

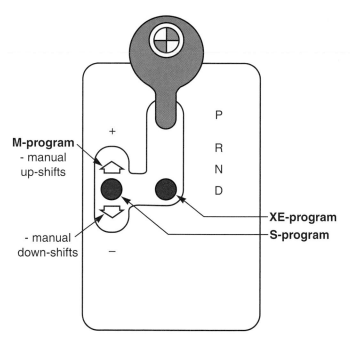

Figure 12-50. *This illustration shows how a typical manual shift program shift lever operates. Note that three different shift programs are available: normal shifting (XE), performance shifting (S), and manual shifting (M). (ZF)*

shifts (S program), providing fully automatic shifts that occur at higher engine speeds.

The electronic shift control system will ignore a manual shift command that would result in engine or transmission damage. For instance, the control system will neither allow the driver to shift into first gear when the vehicle is traveling at a high rate of speed, nor will it allow a shift into overdrive until a certain speed is reached.

Manual shifting programs have many names, depending on the manufacturer. However, the general design of all manual shift programs is similar. Almost all modern manual shift programs make use of a floor- or console-mounted shift lever.

Summary

On all vehicles made during the last decade, some or all transmission shifts and other functions are made by electrical solenoids controlled by the computer, called the ECM. Various sensors monitor engine speed and torque requirements, other engine and vehicle condition, and transmission pressures and temperatures.

Using these inputs, the ECM operates electrical solenoids to control transmission operation. Solenoids control gearshifts and converter clutch apply, and sometimes line pressures, shift feel, and detent operation. The sensors, ECM, and solenoids replace the governor and throttle valves used to move the shift valves on older transmissions.

Some input sensors are located in the transmission to monitor transmission operation. These sensors monitor pressures to determine the transmission gear, and fluid temperature sensors. Main shift control sensors are the sensors for throttle position and vehicle speed. Some transmissions and transaxles have an input speed sensor that tells the ECM how fast the input shaft is turning.

Other input sensors inform the ECM of engine temperature, manifold air pressure, barometric pressure, and engine speed. Some ECMs use input from a mass air-flow sensor and one or more oxygen sensors. Other vehicle inputs to the ECM include the charging system voltage, the operation of the ABS or traction control system, and whether the air conditioner is on.

The ECM is composed of many small electronic components. The entire ECM must be replaced if any electronic part fails. The CPU makes the output decisions based on inputs. The memory is the storage place for information. The two main types of memory are read-only memory (ROM) and random access memory (RAM). ROM contains permanent information installed at the factory. RAM is nonpermanent, or volatile, memory. It changes as operating conditions change.

Output devices consist of solenoids installed on the transmission or transaxle valve body. The solenoids operate check balls that act as shift valves, or indirectly move conventional shift valves by opening or closing pressure passages. Solenoids may also control line pressure and shift feel. Some transmissions have a solenoid that provides governor pressure. Depending on their use, solenoids can be on or off for long periods, or may be pulsed many times per second.

The input sensors, ECM, and solenoids create a control loop that is constantly reacting to changes in operating conditions to provide the proper pressures and gear ratio.

The computer also provides diagnostic outputs in the form of dashboard warning lights and a diagnostic connector. The control module saves information about the system defects in the form of trouble codes. Trouble codes are a series of numbers and, on OBD II systems, letters that correspond to a specific system defect. The technician must use a scan tool to retrieve trouble codes

Transmission and transaxle solenoid functions include line pressure control, lockup clutch control, shift control, and shift cushioning. Most shift control solenoids are multi-shift types that control some valves directly and others indirectly. Some transmissions have dedicated solenoids that directly control each shift valve.

The pressure development, shifting, and other functions of modern transmissions and transaxles are fully controlled by the ECM. Interaction of the solenoids and the hydraulic components of the transmission selects the correct gear and applies the converter clutch. Many ECMs control overall pressures and sometimes shift feel. Some transmissions have partial electronic shift controls. These are usually found on older three-speed transmissions that have been updated to add a fourth gear. CVT ratios are controlled by the ECM through valves that control the movement of the drive pulley.

A manual shift program allows the driver to manually change gear ratios on an electronically controlled transmission or transaxle.

Review Questions—Chapter 12

Please do not write in this text. Place your answers on a separate sheet of paper.

1. What two valves are not used on a completely electronic transmission?

Matching

Match the input sensor with its description.

2. Sliding contact that moves along a resistor wire. _____
   (A) Speed sensor.
   (B) Oxygen sensor.
3. Creates an AC signal. _____
   (C) MAP sensor.
4. Pruduces electricity through a chemical reaction. _____
   (D) Throttle position sensor.

5. Uses piezoelectric crystal to produce voltage. _____

6. What are the three main types of MAF sensor?

7. Which of the following sensors uses chemical action to produce electricity?
   (A) Speed sensor.
   (B) Oxygen sensor.
   (C) MAP sensor.
   (D) Throttle position sensor.

8. What are the two main external components of the ECM self-diagnostic system?

9. A solenoid's ratio of on to off is called its _____ _____.

10. How many shift valves are operated by a dedicated solenoid?

 Note: Refer to Figures 12-29 through 12-40 to answer questions 11–20.

For the following gears, state whether the solenoids are on (energized) or off (de-energized).

	Solenoid A	Solenoid B
11. Neutral	_____	_____
12. Drive first gear	_____	_____
13. Drive second gear	_____	_____
14. Drive third gear	_____	_____
15. Drive fourth gear	_____	_____
16. Manual second	_____	_____
17. Reverse	_____	_____

18. Why is the low reverse clutch piston applied in Park?

19. Downshifts on this transmission are controlled by the _____.

20. What gears are available in Manual Second gear?

 Note: Refer to Figures 12-41 through 12-44 to answer the following questions 21–25.

For the following gears, state whether the solenoids are on (energized) or off (de-energized).

	Solenoid 1	Solenoid 2	Solenoid 3
21. Overdrive first gear	_____	_____	_____
22. Overdrive second gear	_____	_____	_____
23. Overdrive third gear	_____	_____	_____
24. Overdrive fourth gear	_____	_____	_____

25. In the above transaxle, when the shift selector is in the drive position instead of overdrive, the forward clutch is applied to provide engine braking. What is the difference in the status of the solenoids (on or off)?

ASE-Type Questions—Chapter 12

1. All of the following statements about electronic transmissions are true *except:*
   (A) sensors for throttle position and vehicle speed are used in place of the governor and throttle valve.
   (B) electronic solenoids operate the shift valves.
   (C) no hydraulic parts are used.
   (D) holding members are identical to those used in non-electronic versions.

2. Whether a transmission pressure switch is on or off tells the ECM _____.
   (A) the input shaft speed
   (B) the output shaft speed
   (C) what gear is engaged
   (D) whether any slippage is present

3. Technician A says that the viscosity of the fluid affects transmission operation. Technician B says that all transmission temperature sensors are on-off types. Who is right?
   (A) A only.
   (B) B only.
   (C) Both A and B.
   (D) Neither A nor B.

4. Which of the following pairs of sensors uses the same type of measuring device?
   (A) MAP and BARO.
   (B) Speed and throttle position.
   (C) MAF and MAP.
   (D) BARO and O_2.

5. Speed sensors can be installed in all of the following locations on the vehicle *except:*
   (A) transmission/transaxle input shaft.
   (B) engine crankshaft or camshaft.
   (C) ignition distributor.
   (D) throttle shaft.

6. Technician A says that the ECM can read alternator charging voltage without using a sensor. Technician B says that the ECM may monitor air conditioner operation. Who is right?
   (A) A only.
   (B) B only.
   (C) Both A and B.
   (D) Neither A nor B.

7. All of the following statements about governor pressure are true *except:*
   (A) the ECM controls the operation of the governor solenoid.
   (B) a pressure sensor monitors governor solenoid pressure.
   (C) the solenoid and pressure sensor are installed on the valve body.
   (D) the shaft-mounted governor is used to control solenoid pressure.

8. All of the following statements about shift solenoids are true *except:*
   (A) some solenoids are directly connected to check balls that act as shift valves.
   (B) some solenoids indirectly move the shift valves.
   (C) when the solenoid opens a pressure passage, pressure rises.
   (D) when the shift valve is moved by the solenoid, it performs the same job that it does in a hydraulically operated transmission.

9. Technician A says that shift solenoids are often pulsed (duty cycle) types. Technician B says that pressure regulator solenoids must be pulsed types. Who is right?
   (A) A only.
   (B) B only.
   (C) Both A and B.
   (D) Neither A nor B.

10. Continually Variable Transmissions (CVTs) are being discussed. Technician A says that movement of the drive pulley will always force the driven pulley to move. Technician B says that the driven pulley has a larger diameter apply piston than the drive pulley. Who is right?
    (A) A only.
    (B) B only.
    (C) Both A and B.
    (D) Neither A nor B.

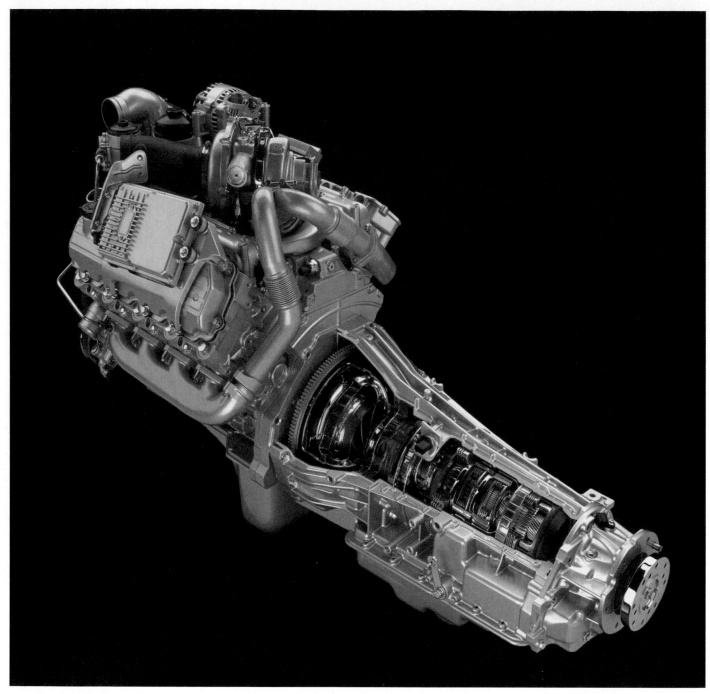

This five-speed automatic transmission is mated to a diesel engine. (Ford)

Chapter 13

Troubleshooting Mechanical, Hydraulic, and Electrical Problems

After studying this chapter, you will be able to:
- ❑ Describe typical automatic transmission and transaxle mechanical and hydraulic problems.
- ❑ State possible causes of transmission and transaxle problems.
- ❑ Identify logical troubleshooting procedures.
- ❑ Describe typical diagnostic charts.
- ❑ Name diagnostic tests used for troubleshooting.
- ❑ Explain how a road test should be conducted.
- ❑ Explain how to perform a stall test and discuss the significance of the results.
- ❑ Explain how air and hydraulic pressure tests are performed and discuss their usefulness.
- ❑ Discuss the use of hydraulic circuit diagrams to analyze automatic transmission and transaxle problems.

Technical Terms

Root cause

Slippage

No-drive condition

Shift problems

Abnormal noise

Overheating

Lubricant leakage

Noise, vibration, and harshness

Diagnostic charts

Band and clutch application chart

Automatic shift diagram

Shift points

Road test

Hydraulic pressure test

Stall test

Air pressure tests

Hydraulic circuit diagrams

Oil pressure diagrams

Introduction

Diagnosing automatic transmission and transaxle problems is as important as fixing them. You should always have at least a general idea of what is wrong before you begin repairs. This chapter will describe typical transmission/transaxle problems and troubleshooting techniques. Some of the diagnostic procedures in this chapter can be performed with the unit installed in the vehicle. Other procedures can be performed only after disassembly. Disassembly will be covered in Chapter 17.

To diagnose transmission and transaxle problems, you must have a thorough understanding of automatic transmissions. If you are not familiar with their operation, you should study Chapters 8 and 9 until you feel comfortable with the basics. As shown in **Figures 13-1** and **13-2,** an automatic transmission or transaxle can have many types of problems.

 Note: Troubleshooting of electronically controlled automatic transmissions is covered in Chapter 14, Electronic Control System Problems and Troubleshooting.

The Seven-Step Troubleshooting Process

The seven-step process of troubleshooting enables the technician to quickly isolate problems. This process uses seven logical steps to reach a conclusion about a malfunction. This logical process works to troubleshoot any transmission or transaxle and can be used to diagnose any vehicle system. **Figure 13-3** shows the seven-step process as it is described below. Refer to **Figure 13-3** as you read the following sections.

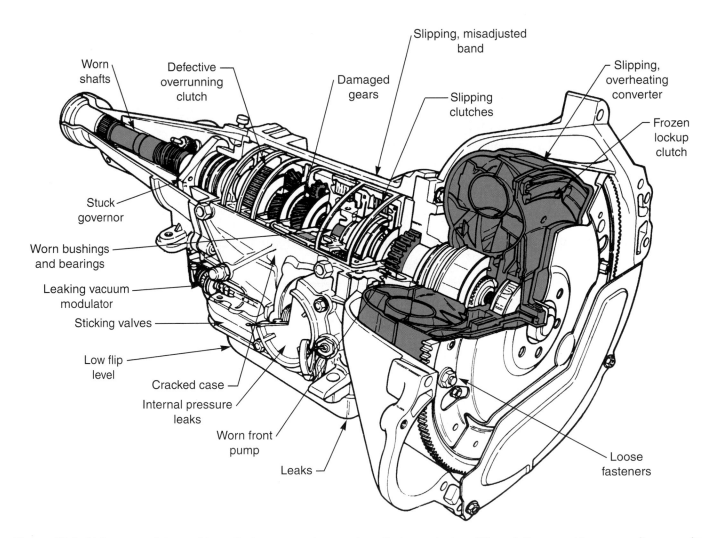

Figure 13-1. *Note some of the problems that can occur in an automatic transmission. Although these problems are often complex, they can be easily diagnosed by using logical troubleshooting procedures and by having a thorough knowledge of transmission operating principles. (Ford)*

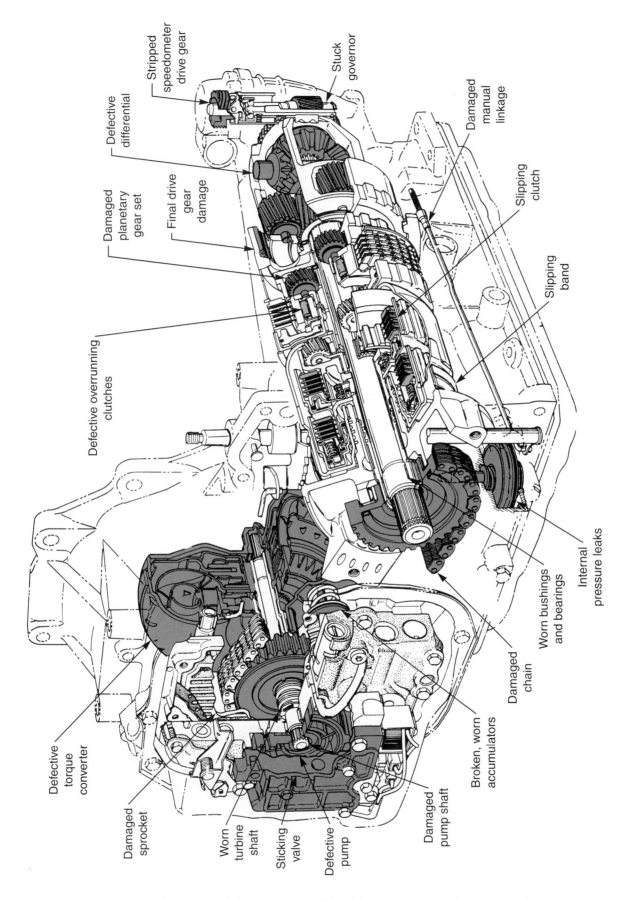

Figure 13-2. *An automatic transaxle has many of the same potential problems as a transmission, as well as some problems that can only occur in transaxles. (General Motors)*

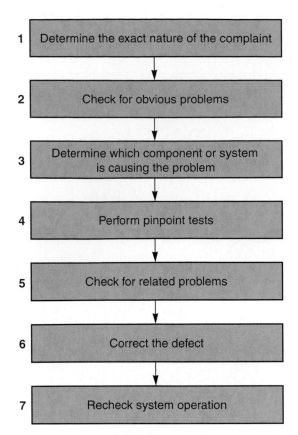

Figure 13-3. *The seven-step process shown here will be used in this chapter. Following this process will save time when diagnosing a transmission or transaxle problem.*

You may think that seven steps are too many to deal with. However, the secret to successful diagnosis is *checking, rechecking,* and *checking again* until you are certain that the cause of the problem has been found and corrected.

Step 1—Determine the Exact Nature of the Complaint

The first step when troubleshooting any problem is to verify the problem. Always determine the exact nature of the complaint. Begin by asking the vehicle's driver to describe what the vehicle is doing. Listen carefully to the driver's description, as this may give you significant clues as to the exact problem. Sometimes the driver's description must be interpreted. For instance, the driver may say, "It takes off OK, but all of a sudden it starts slipping." This may mean the transmission or transaxle is pulling well in first gear but slipping when it shifts into second. Drivers may try to describe the problem with gestures or by imitating noises.

After discussing the problem with the driver, you are ready to perform a road test. Do not be surprised if the actual symptom does not correspond to the driver's description. Remember that the average driver is not a technician, and he or she may misinterpret the symptoms.

In the example in the last paragraph, the driver described the problem as slipping. When you drive the vehicle, you find that the problem is excessively low upshifts, causing the engine to bog down. Without a road test, you would have been looking for causes of slipping, instead of concentrating on the shift controls.

 Note: Always check the fluid level before performing the road test, Figure 13-4. Checking fluid level is covered in Chapter 15. Never attempt to road test a vehicle with an extremely low fluid level.

Sometimes the driver may mistake a normal condition for a problem. A driver may trade in an older car with three speeds and no lockup torque converter for a newer vehicle with four speeds and a lockup converter. After driving the new vehicle, the driver may misinterpret the extra shifts and firm feel of the new vehicle's transmission or transaxle as a problem.

After taking into consideration all the information you have gathered from the driver and the road test, you can determine the exact nature of the complaint. Once you know what the exact problem is, proceed to the next step.

Step 2—Check for Obvious Causes

If the road test indicates that there is a problem, always check for obvious causes. Checking the simple things first can save a lot of time. The most obvious thing to check is the fluid level and condition. You should have already checked the level before performing the road test. Nevertheless, recheck the level at this time. Also, study the fluid for signs of overheating and check it for a burned smell. Chapter 15 has more information about fluid diagnosis.

If the road test indicates a shift problem, check the linkage adjustment. This is also covered in Chapter 15. Also, carefully inspect the linkage to ensure that it is not broken or disconnected, **Figure 13-5.** A broken throttle

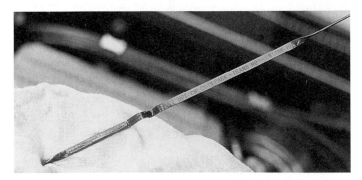

Figure 13-4. *Whenever you investigate any automatic transmission/transaxle problem, check the fluid level first. Many problems are caused by low fluid.*

cable may be hard to locate if it has snapped inside its sheath.

Check for loose electrical connectors at the transmission or transaxle case, **Figure 13-6.** If the vehicle has a vacuum modulator, make sure that the vacuum lines are connected and that there are no engine problems that could lower engine vacuum. Also check that the engine has no driveability-related problems, such as a miss caused by a defective plug or fuel injector. Engine problems are often misdiagnosed as transmission/transaxle defects.

 Note: If the vehicle's drive train is electronically controlled, check for and retrieve trouble codes before proceeding. Checking trouble codes and related diagnostic procedures are covered in Chapter 14. If you do locate obvious problems, correct them and recheck system operation. If you do not locate obvious problems, proceed to the next step.

Step 3—Determine Which Component or System Is the Most Probable Cause of the Problem

In this step, check the components and systems that could cause the problem. Combine the information gained in Step 1 with your knowledge of how transmissions or transaxles work and what happens when they malfunction. Using this information and knowledge in a logical manner will lead you to the components or systems that are *most likely* causing the problem. You can classify the problem as a failure in one of three basic systems: *the mechanical system, the hydraulic system, or the electrical system.*

Note: Problems in the electronic control system will be covered in Chapter 14.

Mechanical System Problems

Transmission and transaxle mechanical problems are those that develop in the mechanical system. This includes problems with the clutch packs (except for apply piston), bands, overrunning clutches, planetary gears, transmission shafts, chains and sprockets, and linkage. It also includes problems with the case and related components.

The most common mechanical system malfunction, aside from out-of-adjustment linkage, is burned and glazed clutch and band linings. Linings in this condition can prevent the holding members from functioning properly, which in turn can prevent the flow of power through the unit. As a result, the vehicle will not move. Defective overrunning clutches, broken or stripped planetary gears, and broken shafts are other typical problems. Among stationary mechanical parts, the most common problem is a leaking seal or gasket. A cracked case or extension housing is also a source of problems. A crack in either of these parts will cause an oil leak.

Hydraulic System Problems

Automatic transmission and transaxle problems often occur due to malfunctions in the hydraulic system. Hydraulic system problems are those that develop in the fluid, filter, pump, valves, apply pistons, accumulators, or hydraulic passageways.

The most common hydraulic system problem is a low fluid level. When the fluid level is low, the transmission may shift erratically, slip, or fail to move the vehicle. A

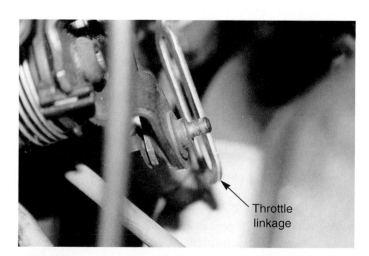

Figure 13-5. *If the throttle linkage is disconnected or broken, shifts will always occur at the same speed. Disconnected linkage is easy to spot, as shown here. Cables usually break inside the sheath. Therefore, broken cables are harder to spot.*

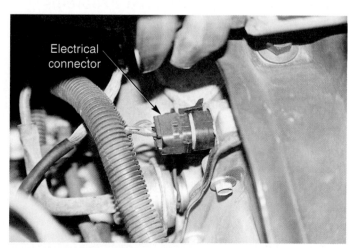

Figure 13-6. *Vacuum modulator lines and electrical connectors should be checked to determine whether they are properly attached. Simple observation will often uncover the source of a transmission or transaxle problem.*

fluid level that is too high causes similar problems but for different reasons. Specific high fluid level problems are discussed in Chapter 15.

Other common hydraulic system problems include plugged filters and sticking valves. A plugged filter prevents oil flow. Sticking valves cause pressure or shift problems. Defective oil pumps, torque converters, servo and clutch pistons, and accumulators can all be sources of hydraulic system problems.

Electrical System Problems

Electrical system problems are usually related to the neutral start switch or the detent solenoid. Many electrical problems are not directly related to the transmission and can be fixed by replacing fuses or by simple electrical repairs.

Computer control problems should be diagnosed by checking the entire computer system using suitable test equipment and troubleshooting guides. Refer to Chapter 14 for more information on this topic.

Once you have determined where the problem is most likely to be, you can proceed to the next step. Step 4 will either confirm your diagnosis or indicate that you must begin again.

Step 4—Perform Pinpoint Tests

In this step, you must check the components or systems that were identified in Step 3 as the most likely sources of the problem. In most cases, this process involves testing the major parts and systems. Start by making the easiest tests first, and then go on to tests that are more involved. This step takes the most time and involves using various testing methods described in this text. After the probable source of the problem is found, proceed to the next step.

Step 5—Check for Related Problems

In this step, you should review the various test procedures performed in the last step and determine whether the suspect component is the **root cause,** or actual cause, of the problem. Be sure to check out all other possible causes of the problem. If you do not locate the underlying problem, you will have to deal with a comeback. Always recheck the suspected part using the proper test equipment and procedures. If your testing causes you to reach the same conclusions as in the last step, you have probably found the problem.

Step 6—Correct the Defect

In this step, you make the necessary repairs to restore the transmission or transaxle to proper operation. Defect correction ranges from something as simple as adding fluid or adjusting linkage to the complete overhaul of the transmission or transaxle. Procedures for correcting various defects are covered later in this text.

Step 7—Recheck System Operation

In this step, you recheck the vehicle to make sure the problem has been corrected. This usually involves performing a road test to ensure that the transmission or transaxle shifts properly and does not slip or engage harshly. Always recheck the unit, even when the problem is as simple as a low fluid level. If, for example, the vehicle was driven with an extremely low fluid level, the holding members may have been damaged. Always recheck operation, no matter how simple the fix appears to be.

The Importance of Follow-Up

When performing troubleshooting procedures, always try to think about what you have discovered and how it will affect the operation of the transmission (and the entire vehicle) in the future. For instance, if a low fluid level caused the original problem, try to determine why the fluid is low. In this case, you will want to find out if the fluid is leaking out or if it was simply not checked for a long time. If the fluid is leaking out, you must tell the customer about it and make sure he or she knows that the original problem will reoccur if the leak is not fixed.

Automatic Transmission and Transaxle Problems

Like any other mechanical components, automatic transmissions and transaxles are subject to certain malfunctions. Typical operational problems experienced by the various automatic transmissions and transaxles are outlined in the following sections.

Slippage

Slippage occurs when the transmission or transaxle does not transmit all the engine power to the rear wheels. When slippage occurs, the engine will speed up without a corresponding increase in vehicle speed. In some cases, slippage will be constant. In other instances, the transmission will slip badly when starting off or while shifting and will then suddenly engage completely. Your first diagnostic step (after talking to the driver and checking fluid level) will be to determine if the vehicle slips in all gears or only in one gear. This is determined during a road test.

If the vehicle slips in every gear, recheck the fluid level. Fluid level rechecking is especially important on transaxles, where the fluid may be high when the transaxle is cold but drops to an abnormally low level as the transaxle heats up. Also, check for a plugged oil filter. A plugged filter will sometimes cause a whining noise that varies with engine speed. In some instances, a transmission with a plugged filter will operate satisfactorily at low speeds but will begin to slip at higher speeds. Changing the filter will often solve the problem.

If the vehicle has TV linkage, check its adjustment as described in Chapter 15. Since throttle valve adjustment affects overall transmission pressure, adjusting the linkage to raise the shift points will raise transmission pressures. This increased pressure may eliminate slippage.

If the transmission slips in one gear only, carefully note the gear in which slippage occurs. You can then refer to a *band and clutch application chart* to determine which clutches or bands are applied in this gear. This will tell you where the trouble lies. The use of band and clutch application charts is covered later in this chapter.

Slippage can be caused by too much clearance between a band and drum. Adjusting the band may solve the problem. Bands can be adjusted with the unit installed in the vehicle.

Slippage can also be caused by burned, glazed, and otherwise worn clutch pack and band linings. Slippage caused by such defective holding members can sometimes be seen as failure to retard vehicle speed in a manual reduction gear (engine braking). Adjusting a burned or glazed band will usually not eliminate slippage.

Checking for worn clutch packs or bands usually requires that the transmission be removed from the vehicle and at least partially disassembled so the parts can be visually inspected. Burned clutch packs and bands can often be confirmed by examining the transmission fluid. A burnt odor or sludge in the transmission oil pan is normally caused by failure of the lining material, typically brought on by overheating.

In rare instances, slippage may be caused by a defective torque converter stator or leakage within transmission hydraulic system. These problems will also be discussed later in this chapter.

No-Drive Condition

When the vehicle has a **no-drive condition,** the transmission fails to transmit power to the rear wheels. The vehicle may fail to move forward in one or all forward gears, or it may fail to move backward. In some cases, it will not move at all. There are two causes for a no-drive condition. One is a binding (locked-up) condition in the drive train or in the brake system. The other is severe slippage within the transmission.

To find out if the no-drive condition is caused by binding, raise the drive wheels off the ground. Then, race the engine with the transmission in gear and the brakes applied. If the engine strains and continues to strain when the brake pedal is released, the drive train or brakes are binding. A quick check for binding is to place the shifter in neutral (engine off) and to release the parking brake. You should now be able to push the vehicle. If the vehicle cannot be pushed, something is binding.

To locate the source of the binding condition, raise the vehicle so the drive wheels are off the ground and remove the drive shaft or CV axles. If the engine does not strain with the drive shaft or axles removed, the problem is in the rear axle assembly or the brake system. If the engine strains after drive shaft removal, the problem is in the transmission. If the parking pawl is not stuck, a clutch pack is welded together, a one-way clutch is jammed, or another internal part is locked.

If binding has been ruled out, the transmission or transaxle probably will not drive because of severe slippage. If the fluid level is low, add fluid and recheck operation. If, on the other hand, the fluid level is high, fluid may have drained back from the torque converter. Running the engine for a few minutes may allow the oil pump to refill the converter so the vehicle will move. Place the shifter in Neutral, since many units have difficulty filling the converter while running in Park.

If the fluid level is normal, make sure the shift linkage is connected externally to the outer manual lever at the transmission case and that the inner manual lever is connected to the manual valve at the valve body. If the linkage checks out okay, the problem is inside the transmission. The transmission must be disassembled to check for internal problems.

Shift Problems

Shift problems are those problems that result in improper gear changes. Examples are late or early upshifts, no upshifts or no downshifts, harsh or soft shifts, and no passing gear. Many shift conditions can be corrected easily.

Always check the transmission fluid level first. If the fluid level is okay, adjust the TV linkage or check the vacuum modulator, as applicable. Lengthening or shortening the linkage rod or *TV cable* will raise or lower shift points. If the transmission does not upshift, make sure the kickdown linkage, if applicable, is not jammed in the wide-open throttle position.

If the vehicle is equipped with a vacuum modulator, make sure that the modulator receives full intake manifold vacuum and that the engine is in good condition, since engine condition can affect vacuum development. A simple test is to disconnect the vacuum hose at the modulator with the engine idling. This should cause the engine to speed up and idle roughly. If the engine idle is unchanged, check for a plugged or a leaking vacuum hose. Also, make sure the hose is properly connected to intake manifold vacuum. If the modulator is visibly damaged, or if there is transmission fluid in the vacuum line, the modulator should be replaced.

If servicing the linkage and vacuum modulator does not correct the problem, check the operation of the governor. Sticking governors can cause erratic shifting or *no shifts.* Many case-mounted governors can be removed and checked, **Figure 13-7.** Output shaft–mounted

Figure 13-7. *Case-mounted governors can be removed and checked for sticking valves and damage to the drive gear. If the governor is mounted on the output shaft, the tail shaft housing must be removed to inspect it. A few shaft-mounted governors can be removed by removing the oil pan and valve body.*

governors can be checked after the extension housing is removed.

Next, check the valve body. Sticking accumulators or valves in the valve body can cause late or early shifts, erratic shifts, hard shifts, or excessively soft shifts. Sludge or metal particles from other transmission parts will often cause these components to stick. In most cases, checking the valve body will require removing the oil pan.

Abnormal Noise

An *abnormal noise* coming from the automatic transmission or transaxle may be a whine, rumble, scrape, or other sound. Some noises can be heard in every gear, including neutral and park. Other noises occur in only one gear.

If the noise seems to occur in all gears, check the transmission fluid level and oil filter. Noises caused by a plugged filter will be heard in every gear and will become louder and higher pitched as engine speed is increased. This noise may be accompanied by slippage. The noise will stop as soon as the engine is turned off, even if the vehicle is still moving.

Other possible causes of noise occurring in every gear are the torque converter and the oil pump. Torque converters can make a whining or scraping noise when the vehicle is starting off or accelerating. This noise will not be heard when the vehicle is at cruising speed, since all the converter components are turning at the same speed. Oil pumps will usually run quietly unless they were improperly assembled. Most oil pump noises occur just after the transmission has been overhauled. These noises resemble the noise caused by a plugged filter.

If the noise seems to occur in only one gear, the most common cause is the planetary gearsets. Planetary gearset noises can occur only when the gear train is in a gear that causes the planetary gears to turn in relation to each other—reduction, overdrive, neutral, or reverse. Planetary gearsets are quiet in direct drive. Unless they are badly damaged, noisy planetary gearsets rarely show visible signs of problems.

Overheating

Overheating may occur when transmission fluid reaches temperatures above 300ºF (150ºC). Transmission overheating is not always obvious; it is usually noticed as soft shifts, gear slippage, or delayed gear engagement. Often, the unit works well when cold but begins to slip when it warms up. When the transmission or transaxle overheats, the fluid *and* the transmission components are heated excessively.

As previously mentioned, fluid temperature should generally be kept below 275ºF (135ºC). Occasional operation above 275ºF or even 300ºF will not permanently damage the fluid or the clutches and bands. However, a constant pattern of fluid temperatures exceeding 275ºF leads to broken-down fluid, causing the loss of its lubricating properties. This will show up in the form of plugged valves, burned clutch and band linings, and plugged filters. Extremely overheated oil will cause the exterior of the converter metal to turn blue.

If you suspect overheating, check the oil cooler. Make sure fluid is flowing through it. Make sure the cooler is not clogged or covered with sludge. The engine cooling system should be checked to ensure that it is not operating at excessive temperatures. If the vehicle is being used for towing, the extra load may produce too much heat for the original-equipment oil cooler to handle. In this case, an auxiliary oil cooler should be installed.

Lubricant Leakage

Lubricant leakage is an unwanted loss of fluid from inside the transmission or transaxle. Before attempting to correct a suspected leak, make sure the leaking fluid is actually coming from the transmission or transaxle. Automotive fluids are usually carried rearward by the airflow under the vehicle. Thus, finding fluid on the transmission or transaxle does not necessarily mean the unit is leaking. Also, while fluid is typically red, fluid color is not a reliable indicator for the type of fluid and, therefore, the source of the leak. Engine oil, power steering fluid, and even antifreeze can be mistaken for automatic transmission fluid. After determining that the fluid is coming from the transmission or transaxle, you must pinpoint the exact location of the leak.

If the transmission or transaxle leaks when the engine is running, a usual cause is the failure of a seal or gasket that is under pressure when the unit is operating. Start the diagnostic procedure to determine where the leak is occurring by thoroughly cleaning the transmission or transaxle case. Also, clean the torque converter, pump face, and nearby surfaces with solvent. Removing the inspection cover or dust cover will give you access to these components.

Start the engine and allow it to run for 10 minutes at high idle. Then, examine the case for fresh fluid. Check the rear seal and the manual lever/shaft seal. Check other parts on the side of the case, such as the covers over the servos, governors, or accumulators; oil cooler lines and fittings; electrical connectors; and pressure tap plugs. The area around the torque converter may show traces of transmission fluid streaking outward from the center of the converter, where it enters the front pump seal. This area is easily seen from under the vehicle.

If the transmission or transaxle leaks primarily when the engine is not running, the cause is usually a seal or gasket that is below the level of fluid in the pan. Check the transmission oil pan for a defective gasket or warped sealing surfaces. Also, check the pan for loose bolts or for a possible puncture caused by road debris.

If the pan is okay, clean the case so you can look for fresh fluid. Then, check seals at the manual lever/shaft assembly, TV cable, band adjustment screw, and electrical connectors. Also, check the extension housing gasket, rear seal, and front pump seal.

Some leak sources are difficult to spot, especially when they are under other vehicle components. Other leaks are caused by an unusual problem, such as case porosity or a hairline crack. Many technicians use a black light to locate hard-to-find leaks. To use the black-light method, pour a small amount of a fluorescent dye into the transmission through the filler tube. Then operate the vehicle for about ten minutes to distribute the dye and allow some fluid to leak out. Raise the vehicle on a lift and point the black light at suspected leak areas, **Figure 13-8.** The black light will cause the dye in the fluid to glow, identifying the source of the leak. The dye does not affect the fluid and can be left in the transmission or transaxle after leak detection is finished.

Most gaskets and seals can be replaced without removing the transmission or transaxle. An exception is the front pump seal. When changing a rear seal, make sure the extension bushing is good. Even a new seal cannot adequately protect against leakage if the bushing is defective.

A vacuum modulator can also be a source of leakage. Check this component if the transmission fluid level seems to drop without being accompanied by leakage. A ruptured vacuum modulator diaphragm will allow transmission fluid to be drawn into the engine and burned.

Water Damage

Water is an often-overlooked cause of transmission and transaxle failure. Water (and antifreeze) can enter the unit through leaking oil cooler in the vehicle's radiator. A

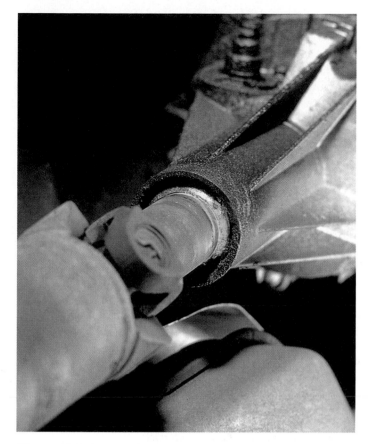

Figure 13-8. *Hard-to-find leaks can be isolated with a special dye and a black light. Add the dye to the transmission or transaxle and allow it to circulate. Then raise the vehicle and shine the black light on suspected leak areas. The dye will glow at the site of the leak. It may be necessary to drive the vehicle for a while to allow enough of the dye to leak out. (Spectronics)*

small pinhole or a hairline crack is enough to allow leakage. With the vehicle running, radiator pressure and oil cooler pressure are both about the same, and only small amounts of leakage will occur through the cooler. When the engine is turned off, however, the cooling system remains pressurized, while oil cooler pressure drops to zero. The water and antifreeze mixture will enter the oil cooler through the leak. Another cause of water entry is driving through deep water. If the vehicle has reached normal operating temperature, driving through deep water will cause the transmission or transaxle to suddenly cool off. This cooling causes contraction of the fluid in the case, creating an internal vacuum that draws water through the case vent and into the interior of the transmission or transaxle. Excessive engine washing or steam cleaning can also force water into the transmission or transaxle.

Water can rust internal components. It can also cause excessive wear, since it is a poor lubricant. Water can cause damage to the friction materials used on the clutches and bands. Since friction materials are partially composed of paper, they are *hygroscopic* (they absorb water). Water entering the transmission or transaxle will be absorbed by the clutch friction plates and the band lining. Applying the bands and clutches during heavy acceleration can momentarily raise the temperature of the friction materials above the boiling point of water. This causes the "popcorn effect," turning the water to steam and blowing pieces of the clutch or band friction material off the steel backing. Sometimes the entire lining will be forced from the backing. This results in severe slippage. If antifreeze has entered with the water, it will soften seals and friction materials, dissolve the friction material bonding agent, and damage plastic parts.

Signs of water in the transmission or transaxle are milky pink transmission fluid, transmission fluid in the engine coolant, and rust on the transmission or transaxle dipstick. When the water turns to steam inside the case, it may cause transmission fluid to be blown out of the vent, **Figure 13-9.** The only sure cure for water contamination is to completely overhaul the unit, thoroughly flush the oil cooler lines, and replace the torque converter. If there is any chance that the water contamination was caused by a leaking oil cooler, the cooler must be checked. If it is defective, it must be repaired or bypassed using an auxiliary air cooler.

Automatic Transmission and Transaxle Component Malfunctions

As you are probably aware, automatic transmission and transaxles have many parts that can fail. It is helpful to learn how these parts become defective, what problems they can cause, and what is needed to repair or replace them.

Flywheel Malfunction

The flywheel is a steel stamping connecting the crankshaft to the torque converter. On most vehicles, the

Figure 13-9. *Most transmission or transaxle vents are located at the top of the case. Fluid blowing out of a vent may be difficult to trace, since it will completely coat the case.*

flywheel contains the ring gear that engages the starter teeth for cranking. The flywheel usually lasts the life of the vehicle. Occasionally, it may develop a crack, causing knocking noises and, possibly, a vibration that varies with engine speed. The ring gear teeth may also become stripped. The only cure for a defective flywheel is replacement. The transmission must be removed to replace the flywheel. Some vehicles use a flexplate to connect the engine and converter. When a flexplate is used, the ring gear is installed on the converter.

The bolts holding the flywheel or flexplate to the crankshaft or the torque converter cover may become loose. Loose bolts will cause a knocking sound, especially at lower speeds, and will possibly cause vibration. Loose attaching bolts may also make the starter operate noisily or jam.

Torque Converter Malfunction

Torque converter defects are relatively rare. The torque converter may slip or make noises. A failed stator one-way clutch may cause the converter to overheat the fluid. If the converter elements begin to break up, metal particles will be carried throughout the transmission. This is usually caused by excessive clearances between the internal elements.

Torque converter noise is a scraping sound that can be heard only when starting off or accelerating. It tapers off and stops at cruising speeds, when all the internal converter elements are turning at the same speed.

If the stator overrunning clutch locks up and then fails to release, the transmission or transaxle will overheat at high speeds. If the clutch fails to lock the stator, the converter will slip at low speeds.

If the transmission or transaxle has heavy sludge deposits, the converter will be full of sludge. The converter may continue to operate properly, but it will contaminate the transmission if the converter is reused.

The torque converter is probably not defective if the malfunction occurs in only one gear. Sometimes, transmission fluid will drain from the converter into the oil pan overnight. This is caused by a sticking valve, a worn connecter bushing, or an oil pump malfunction. It does not mean the converter is defective.

If the outside of the converter has turned blue, this is a sign of severe transmission overheating. The converter should be replaced, and the cause of overheating should be determined.

The only correction for a problem inside a modern welded converter is to replace the converter. The only exception to this is for sludge buildup. Sometimes, sludge can be removed with a converter flusher, but often the only sure cure is to replace the converter. Before replacing the converter, make sure all other transmission or transaxle components have been thoroughly checked and eliminated as possible sources of the problem.

Shaft Defects

The input and output shafts tend to be relatively trouble free. Some shafts wear at bushing contact points. Occasionally a shaft will break, but this is rare in normal use. Seal rings and seal ring grooves can wear, causing pressure loss and slippage. The transmission or transaxle must be removed and disassembled to service the shafts. The fit of the shaft in its bushings and the fit of the seal rings should always be checked when a transmission or transaxle is overhauled.

Planetary Gearset Defects

Planetary gearset noises can be heard when the gear train is in reduction, overdrive, neutral, or reverse. The gearsets do not make noise in direct drive. Gearset noise may vary with gear changes. This noise is usually loudest in low gear and decreases as gear reduction decreases. Often, noise is heard in only one gear. The gears may be badly damaged, or they may seem to have no damage at all. Excessive end play between the planet gears and the planet carrier can also cause noise. Some planetary gearset noise is caused by worn bushings, which allow the gears to move out of alignment. The transmission or transaxle must be disassembled to replace faulty planetary gearsets.

Clutch Pack Malfunction

Worn clutch packs almost always cause slippage. In a few cases, a clutch pack may become so hot that it welds itself together, causing the transmission to stick in gear. If the transmission or transaxle slips in only one gear, the problem is one particular clutch pack. For example, a transmission or transaxle that drives well in forward but slips in reverse most likely has a malfunction of its reverse clutch.

Slightly glazed friction discs may work well when the transmission or transaxle is cool but slip or chatter once it warms up. Slippage may also be caused by leaking clutch piston seals, a missing air bleed or pressure relief check ball, or tight clutch return spring(s). Weak clutch return springs can cause harsh clutch engagement.

A worn clutch pack can often be confirmed by transmission fluid that has a burnt odor or by sludge in the oil pan. If the sludge is more than 1/16" (1.59 mm) deep, or if it is built up in spots, the unit should be disassembled and overhauled. It is always necessary to remove the transmission or transaxle to overhaul clutch packs.

Band and Servo Malfunctions

Worn bands will slip, especially when they are first applied. A common example is an intermediate (second gear) band that causes slippage during a shift into second. A band that is slipping severely can rarely be corrected by adjustment. The lining will be too damaged to hold as it should. If the band ear or linkage is broken, the transmission or transaxle may skip the forward gear or gears that utilize the band.

Most bands are replaced by disassembling the transmission and installing a new band. A few bands can be replaced without removing the transmission from the vehicle. Servos can lose pressure due to leaking seals. Sometimes, the wrong return spring may be installed, causing a soft or harsh shift.

Overrunning Clutch Malfunction

Overrunning clutches will usually work well or not at all. If the one-way planetary holding member fails to lock, the automatic transmission or transaxle will not drive in that gear. The vehicle will not move if the overrunning clutch is used for first gear. If an overrun band is used in manual low, the unit will move when manual low is selected. If an overrunning clutch is used in any other gear, the transmission or transaxle will skip that gear. If the one-way clutch fails to unlock, the transmission will lock up except in the gear that normally uses the one-way clutch.

Oil Pump Malfunction

The front pump can make noises, or it can wear out. Noisy pumps usually make a whining sound. This sound, which is caused by tight pump clearances, will increase as engine speed increases. Pumps do not tighten up with use, so a whining pump is usually noticed just after overhaul. Noise from a rear pump on an older transmission will vary with road speed.

Sometimes, the oil pump or oil pump bushings will wear and develop excessive clearances. This can cause oil to drain out of the torque converter when the vehicle is not operating. As a sign of this problem, the transmission fluid level will be above the full mark on the dipstick after the vehicle has been standing for several hours. The torque converter will usually refill after the engine is operated for a few minutes.

 Note: On some transaxles, the oil level will be higher when the transaxle is cold. Do not confuse this with a pump bushing defect.

Another result of excessive pump wear is low oil pressure. Most pumps are designed to generate much more system oil pressure than needed. The pump will be very worn if a reduction in pressure is evident.

The rear pump is found on many older vehicles. A defective rear pump may make it impossible to push start the vehicle. Rear pumps provide pressure to the governor valve. A defective rear pump may also prevent the transmission from upshifting.

Most worn or otherwise defective oil pumps are replaced as a unit. Occasionally, the pump bushing can be replaced in an otherwise good pump. The front pump seal should be replaced as part of any pump service. The transmission or transaxle must be removed from the vehicle to repair or replace the oil pump.

Oil Filter Malfunction

The better a filter does its job, the more likely it is to plug up. Dirt and other debris in the fluid are trapped in the filter. Eventually, it may become plugged. Filters can also plug up because of excess sludge in the transmission or transaxle. Sludge is caused by the failure of other parts. A plugged filter will cause a whining noise that varies with engine speed. The transmission or transaxle may begin to slip at higher engine rpm. As the restriction gets worse, the transmission or transaxle will begin to slip badly, and eventually, it will not move at all. The oil filter can be replaced by removing the oil pan and removing the screws or clips holding the filter in place. The filter should be checked for debris buildup or obvious damage, **Figure 13-10.** It is often helpful to break open the filter to check for metal particles or friction material. A plugged filter is usually an indication that something else is wrong.

Oil Cooler Malfunction

The oil cooler removes heat from the transmission or transaxle. Oil cooler failures cause the fluid and the unit to overheat. Overheating leads to slipping, delayed gear engagement, oil sludging, and eventual holding member failure.

The oil cooler can fail due to internal plugging of the cooler lines, pinched or kinked lines, sludge or debris on the cooler surfaces, or a problem in the engine's cooling system. The parts of the cooler are largely external. These can develop leaks and rattles. **Figure 13-11** shows some typical oil cooler system malfunctions. Cooler malfunctions can usually be solved by flushing out the cooler and radiator. The cooler can also allow coolant to leak into the transmission passages, leading to severe damage. Also, be sure to check the engine's cooling system temperature. Engine cooling system problems are often the source of the transmission or transaxle cooling problem.

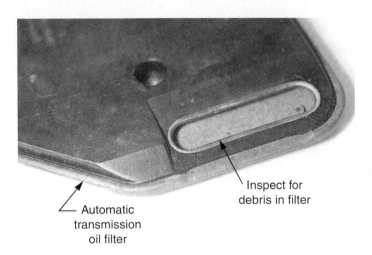

Inspect for debris in filter

Automatic transmission oil filter

Figure 13-10. *A plugged filter can cause many problems, including slippage, noise, and buildup of dirt or metal particles in the transmission and converter. The filter should be replaced at regular intervals. It should also be replaced whenever the transmission or transaxle is disassembled for service. (BBU, Inc.)*

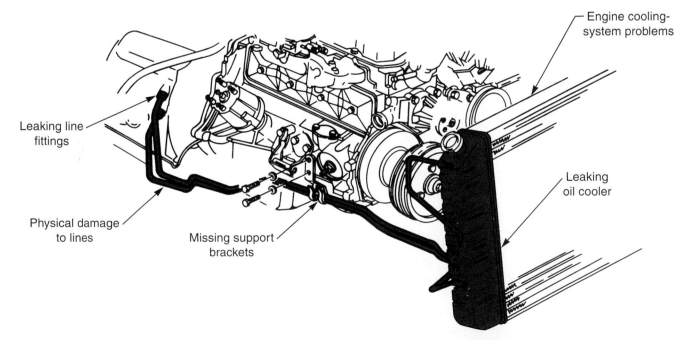

Engine cooling-system problems

Leaking line fittings

Physical damage to lines

Missing support brackets

Leaking oil cooler

Figure 13-11. *Typical oil cooler problems. The cooler can plug up and cause overheating. It can also allow water and antifreeze to enter the transmission, causing friction lining damage. The cooler lines can leak or become kinked or crushed. (General Motors)*

Valve Body Malfunction

Defective valve bodies can cause many hydraulic system problems. Valves can stick or they may have been improperly assembled. The valve body can become warped, leading to cross leaks (unwanted pressure leaks between passages) and erratic operation. Most valve body problems will show up as shift problems, although sticking torque converter or pressure regulator valves can cause pressure problems and slippage. Wear of moving parts may result in the deposit of metal particles in the valve body, causing valves to stick. Note that aluminum valve bodies seem to develop sticking valves more often than cast iron valve bodies.

Accumulators in the valve body may leak or stick, affecting shift quality. Accumulator pistons can break or wear out, causing slippage. Check balls may have been left out during overhaul, leading to slippage or erratic shifting.

A valve body can sometimes be repaired by a thorough cleaning. Most valve bodies can be removed for inspection and cleaning by first removing the oil pan. If any valves cannot be freed by cleaning, the valve body should be replaced.

Vacuum Modulator Malfunction

Many vacuum modulator malfunctions are not in the modulator itself, but in related parts or systems. In many instances, the modulator is not receiving full engine vacuum because of plugged, disconnected, misconnected, or leaking vacuum hoses. See **Figure 13-12.** A worn or improperly tuned engine can also cause the vacuum to be lower than it should be. In rare instances, the throttle valve may be sticking in its bore.

Of course, the vacuum modulator itself can be faulty. If the modulator is leaking or bent, low or high line and throttle pressures result. This, in turn, will cause low or high shift points. Low line pressures, specifically, can cause soft shifts, leading to slippage and burning of clutch and band friction linings. In addition to causing pressure problems, a split or hole in the modulator diaphragm can cause fluid loss, as transmission fluid is drawn into the intake manifold and burned in the engine.

Some vacuum modulators are adjustable. In this case, a transmission or transaxle may be experiencing pressure problems because the modulator is out of adjustment. A simple adjustment may correct the problem.

If the vacuum modulator is bent or leaking, or if adjusting an adjustable type does not fix a pressure problem, the modulator must be replaced. Modulators can be removed from the case without removing other parts.

Governor Valve Malfunction

The governor valve is often the cause of shift problems. It tends to stick more readily than other valves. Metal particles caused by internal component wear can make the governor stick—even a few miles after the valve has been

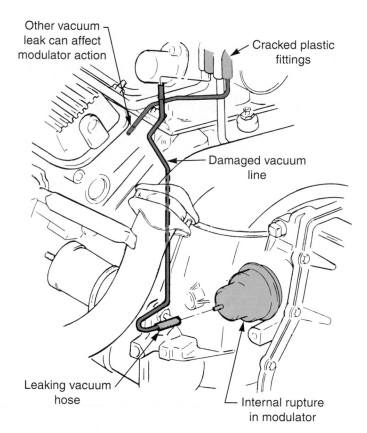

Figure 13-12. *Vacuum modulator problems can be caused by a defective modulator or by failure of the modulator to receive full manifold vacuum. Always check the vacuum at the modulator before replacing it. (General Motors)*

cleaned. Most governor circuits have small filters in the governor pressure inlet passage. These filters can plug up and reduce governor pressure. In addition, the valve, valve bore, and seal rings for the governor can wear, causing pressure loss. A common problem with gear-driven governors is apple coring of the driven gear, **Figure 13-13.** Apple coring occurs when the speedometer driven gear is worn in such a way that it resembles an apple core.

Governors can usually be restored to service by cleaning. Many governors are mounted on the case and can be easily removed. However, some governors are installed on the output shaft. To service these, the extension housing must be removed.

Automatic Transmission Linkage Malfunctions

Many vehicles use TV linkage, which consists of levers connected by a series of rods or a cable, to control shift timing and pressures, as well as to provide a detent for passing gear. Some vehicles have a kickdown linkage. The TV linkage or kickdown linkage is connected to the engine's throttle, which in turn is connected through the throttle linkage to the accelerator pedal. These linkages are often out of adjustment because of wear or damage. In some cases, linkage is misadjusted at the factory and never noticed by the owner.

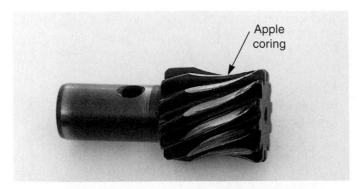

Figure 13-13. *Governor gear damage like this is often called apple coring. Although this is a steel governor gear, plastic gears are much more likely to exhibit this type of damage.*

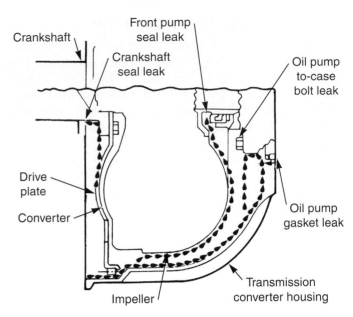

Figure 13-14. *The area around the torque converter is often a source of leaks. The front pump seal can leak when the engine is running. Other seals and gaskets can leak when the engine is off (when fluid level is high). Do not confuse transmission leakage with leakage from the rear crankshaft seal. (DaimlerChrysler)*

The shift selector lever operates the shift linkage, which in turn operates the manual valve. The inability to select gears is a common problem caused by a shift linkage malfunction.

Most linkages can be adjusted to eliminate problems. Adjustment is covered in Chapter 15. Replacement of linkage can generally be performed without removing other parts. In some instances, however, the oil pan must be removed to reconnect the internal linkage to the manual valve.

Parking Pawl Malfunction

The parking pawl is operated by the driver. A common problem is failure of the parking pawl to hold. Most problems can be corrected by adjustment. In some instances, the oil pan must be removed to reconnect the parking pawl linkage.

Bearing and Thrust Washer Malfunctions

Bearing or thrust washer failures can cause noise and wear of other parts, such as shafts, clutch drums, and gearsets. Defective bushings or thrust washers can allow moving parts to shift enough to cause internal hydraulic pressure loss or external leaks. A worn extension bushing or output shaft bearing, which is a large bearing that supports the output shaft, may cause vibration when the vehicle is accelerated.

Bearings and thrust washers are located throughout the transmission. Bearing and thrust washer replacement usually requires transmission or transaxle disassembly.

Front Pump Seal and Output Shaft Seal Malfunctions

The front pump seal and the output shaft seal can develop leaks as they age. The front pump seal is under pressure when the engine is running, and a faulty seal will leak during vehicle operation. See **Figure 13-14.** A front pump leak may appear as traces (streaks) of fluid directed radially outward from the pump seal. The converter will

usually be streaked also. The transmission or transaxle must be removed to service the front pump seal.

The output shaft seal (there are two seals on a transaxle) will leak when the vehicle is moving. Even though there is no pressure behind the output shaft seal, fluid will be thrown against the seal by internal part movement. This splash will make a bad seal leak profusely. In addition, the seal of a rear-wheel drive transmission will sometimes leak when the vehicle is parked facing uphill.

The output shaft seal can be replaced by removing the drive shaft or CV axle shaft and prying out the seal. A new seal can then be installed. Always check the bushing located near the seal for wear. A worn extension bushing, as it is called, may result in excessive movement of the driveline slip yoke or CV axle shaft. This will cause a replacement seal to leak as the shaft vibrates in use.

Other Seal and Gasket Malfunctions

There are many case or component seals used on automatic transmissions and transaxles, including manual lever and shaft assembly seals, TV cable seals, band adjustment screw seals, and electrical connector seals. The typical transmission or transaxle also uses a number of gaskets, including oil pan and extension housing gaskets.

Some seals and gaskets are under pressure when the transmission or transaxle is operating and will leak during vehicle use. Many seals and gaskets will not leak when the engine is running. This is because fluid level is lower when the oil pump is drawing fluid from the hydraulic reservoir for the hydraulic system. Leakage will usually occur when the vehicle sits idle for an extended period, such as overnight, allowing drain back to raise the oil level above the defective seal or gasket.

Seals and gaskets must be replaced if they leak. Most can be replaced without removing the transmission or transaxle from the vehicle.

Noise, Vibration, and Harshness (NVH) Diagnosis

Many transmission and transaxle complaints involve **noise, vibration, and harshness,** usually referred to as **NVH.** Every engine, transmission, drive axle, or other collection of moving parts produces some vibration and noise. This is reduced on modern vehicles by careful balancing and fitting of parts, dampening vibration and noise with weights on the rotating parts, and insulating the vehicle's body from its moving parts. However, some noise and vibration transfer is unavoidable. If a defect occurs, this noise and vibration it causes may become severe enough to cause a driver complaint.

An automatic transmission or transaxle may cause NVH complaints. However, there are many other possible, more likely causes of an NVH complaint, and these should be investigated. Complaints of harshness, such as severe shocks over bumps, are usually caused by the suspension. Sources of vibration often occur in the engine, drive shaft, CV axles, rear axle, brakes, and tires. Before deciding that the transmission or transaxle is the source of the problem, check for the following:

❑ Engine cylinder defect (fouled plug, clogged injector, burned valve).
❑ Engine balance defect (vibration damper slipped, crankshaft bent or unbalanced).
❑ Broken engine or transmission/transaxle mounts or loose mount fasteners.
❑ Exhaust component contacting the vehicle's body or frame.
❑ Bent drive shaft or CV axle shaft(s).
❑ Worn U-joints or CV joints.
❑ Loose wheel bearings.
❑ Worn or unbalanced tires.
❑ Bent or cracked wheel rims.
❑ Dragging brakes or warped rotors/drums.
❑ Loose suspension or steering parts.
❑ Loose transmission/transaxle to engine mounting bolts.
❑ Loose or cracked flywheel (ring gear) or missing balance weights.
❑ Cracked transmission/transaxle case.
❑ Loose or missing torque converter bolts or missing balance weights.

If you decide that the transmission or transaxle is the source of the NVH complaint, the most likely culprit is the torque converter. To test the converter, allow the engine to idle with the transmission/transaxle in *Neutral* or *Park.* If the vibration is excessive (and all engine problems have been eliminated), the torque converter should be replaced.

Sometimes the torque converter and flywheel positioning can contribute to vibration. It may be possible to reduce vibration by repositioning the converter in relation to the flywheel. Begin by removing the dust cover and locating the converter-to-flywheel bolts or nuts. Then remove the bolts or nuts and turn the converter either 1/4 turn (if the converter has 4 fasteners) or 1/3 turn (if the converter has 3 fasteners) in relation to the flywheel. See **Figure 13-15.** The converter can be turned in any direction. Reinstall the fasteners and recheck the vibration with the engine idling. Repeat this procedure, being sure to move the converter in the same direction each time, until the least vibration is obtained. If this method does not reduce vibration to an acceptable level, replace the torque converter.

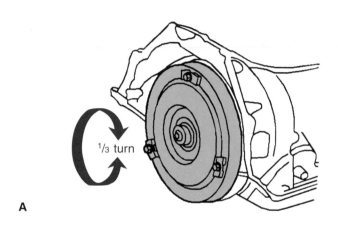

A

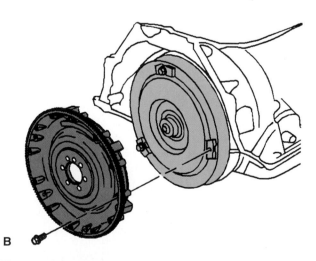

B

Figure 13-15. *Torque converter and flywheel positioning can sometimes contribute to vibration. Repositioning the converter can sometimes cure this type of vibration. A—After removing the flywheel, turn the converter in relation to the flywheel. In this case, the converter should be moved 1/3 turn. B—Reinstall the flywheel and recheck the vibration. (General Motors)*

Automatic Transmission Diagnostic Charts

Diagnostic charts will help you to determine the exact cause of a problem. These charts are prepared by the transmission or transaxle manufacturer. Each chart applies to one specific type of transmission or transaxle and is designed to match the transmission problem to the specific part malfunction.

There are numerous types of diagnostic charts. A portion of a typical diagnostic chart is shown in **Figure 13-16.** It lists various problems, or conditions, and possible causes

Condition	Inspect Component	For Cause
Oil leak	• Oil pan	– Bolts not correctly torqued. – Improperly installed or damaged pan gasket. – Oil pan gasket mounting face not flat.
	• Filler pipe	– Multi-lip seal damaged or missing.
	• Filler pipe bracket	– Mispositioned.
	• Thottle valve cable	– Multi-lip seal missing, damaged, or improperly installed.
	• Rear seal assembly	– Damaged or improperly installed.
	• Speedometer driven gear	– O-ring damaged.
	• Manual shaft	– Lip seal damaged or improperly installed.
	• Case	– Line pressure tap plug. – Fourth clutch pressure tap plug. – Porous.
	• Intermediate servo	– O-rings damaged.
	• Oil pump assembly	– Front pump seal leaks: Seal lip cut–check converter hub for nicks, etc.; bushing moved forward and damaged; garter spring missing from seal. – Front pump attaching bolts loose or bolt seal damaged or missing. – Front pump housing O-ring damaged or cut. – Porous casting. – Inspect converter weld area.
	• Vent pipe	– Transmission overfilled. – Water in oil. – Foreign matter between pump and case or between pump cover and body. – Case porous; front pump cover mounting face shy of stock near breather. – Pump to case gasket mispositioned. – Incorrect dipstick. – Pump shy of stock on mounting faces, porous casting, breather hole plugged in pump cover.

Figure 13-16. *The transmission diagnostic chart is a convenient guide to the common problems of a particular transmission. (General Motors)*

in tabular form. Such charts usually take up several pages of a service manual. Another common type takes the form of a flowchart.

A more condensed diagnostic chart is shown in **Figure 13-17.** The *conditions,* which appear down the side of the chart, list many of the problems that are common to this kind of transmission or transaxle. The *possible causes,* which appear across the top of the chart, list common malfunctions in this unit. An "X" is placed on the line wherever a certain part could be the cause of the problem. Studying the chart will reveal, for example, that slipping in reverse can be caused by ten different conditions. Projecting up from the "X" to the top of the chart, you can learn what might be causing the problem. The lowest numbered items are the most likely causes; the highest numbered items are the least likely causes.

As an example, suppose the vehicle had a problem of *no upshift.* Reading to the right, the first "X" encountered is in the fourth column—*hydraulic pressures too low.* This would be the most likely cause of the problem. The cause that would be the least likely lies in column 34—*overrunning clutch inner race damaged.*

Another type of diagnostic chart is the **band and clutch application chart.** This chart shows which holding members are applied in a particular gear. This type of chart can be very useful if trying to isolate a defective holding member when the transmission or transaxle is slipping in a particular gear.

A typical band and clutch application chart is shown in **Figure 13-18.** This particular chart is used by finding the gear (first, second, etc.) in the two leftmost columns and reading to the right to find the clutches and bands that are used in that particular gear. Note that the same holding members are used in different combinations to obtain the different gears. If, for instance, a transmission or transaxle takes off well from stop but slips while trying to shift into second, the clutch or band that is applied to obtain second gear is probably at fault.

Sometimes a band or clutch will slip in one gear but not in another. For instance, one clutch may slip when applied in high gear but not when used in reverse. The reason for this is that the pressures used in reverse are higher than those used in the forward gears. Even though the friction discs are faulty, the higher reverse pressures apply them tightly, and they do not slip. Lower pressures are used for high gear, allowing the discs to lose their grip and slip. When hydraulic principles are understood, diagnosing this condition is not too difficult.

Still another type of diagnostic chart is the **automatic shift diagram, Figure 13-19.** This chart shows the vehicle's approximate **shift points,** or points at which the particular unit should shift. The chart can be compared with actual shift points, and transmission or transaxle operation can be determined accordingly.

Diagnostic Testing

Automatic transmissions and transaxles can be tested by many methods, including *road tests, hydraulic pressure tests, stall tests,* and *air pressure tests.* The technician must be guided by experience to determine which of these tests should be performed on a particular unit. The road test is an exception. It is of critical importance and should *always* be performed if the vehicle is driveable. The procedures for performing these diagnostic tests are explained on these next few pages.

Road Test

A **road test** involves driving the vehicle to observe transmission or transaxle operation under actual road conditions. Road tests should be performed whenever the vehicle is driveable. Always check the fluid level before beginning the road test and observe all traffic laws during the test. If possible, take the vehicle's owner along on the road test.

Start the road test by accelerating normally and comparing actual shift points to the published shift points, such as those shown in **Figure 13-19.** Decelerate gradually, taking note of downshift speeds. Make sure the detent and part-throttle downshifts occur at the proper times. During these tests, also take note of the shift feel, missed shifts, slipping, noises, and whatever other problems may be present.

Hydraulic Pressure Tests

A **hydraulic pressure test** involves connecting one or more pressure gauges to the transmission or transaxle pressure fittings and observing the hydraulic fluid pressures. Actual pressures are then compared to published values to decide if there is a hydraulic system problem. Hydraulic pressure tests are useful in pinpointing suspected problems or in cases where the problem cannot be determined by any other method.

To perform a hydraulic pressure test, first locate the pressure tap(s) you will need. **Figure 13-20** shows some typical pressure tap locations on one make of transmission. **Figure 13-21** shows the pressure taps for a transaxle. Remove the pressure tap plug(s) and install the pressure gauge. A typical hookup is shown in **Figure 13-22.** Make sure the range of the pressure gauge exceeds the maximum pressure the system can produce. Install a tachometer if the instrument panel does not indicate engine rpm. Start the engine and observe the readings on the pressure gauge, checking pressures in neutral, all drive ranges, and reverse. Make the checks at idle speeds, noting pressures at specified engine rpm. Then, raise the engine speed as specified in the service manual and again note pressure readings.

Possible Cause

No.	Possible Cause
35	Faulty lockup clutch.
34	Overrunning clutch inner race damaged.
33	Overrunning clutch worn, broken, or seized.
32	Planetary gear sets broken or seized.
31	Rear clutch dragging.
30	Worn or faulty rear clutch.
29	Insufficient clutch plate clearance.
28	Faulty cooling system.
27	Kickdown band adjustment too tight.
26	Hydraulic pressure too high.
25	Breather clogged.
24	High fluid level.
23	Worn or faulty front clutch.
22	Kickdown servo band or linkage malfunction.
21	Governor malfunction.
20	Worn or broken reaction shaft support seal rings.
19	Governor support seal rings broken or worn.
18	Output shaft bearing and/or bushing damaged.
17	Overrunning clutch not holding.
16	Kickdown band out of adjustment.
15	Incorrect throttle linkage adjustment.
14	Engine idle speed too low.
13	Aerated fluid.
12	Worn or broken input shaft seal rings.
11	Faulty oil pump.
10	Oil filter clogged.
9	Incorrect gearshift control linkage adjustment.
8	Low fluid level.
7	Low-reverse servo, band or linkage malfunction.
6	Valve body malfunction or leakage.
5	Low-reverse ban out of adjustment.
4	Hydraulic pressures too low.
3	Engine idle speed too high.
2	Stuck lockup valve.
1	Stuck switch valve.

Condition — cross-reference of possible cause numbers (X)

Condition	1	2	3	4	5	6	7	8	9	10	11	12	13	14	15	16	17	18	19	20	21	22	23	24	25	26	27	28	29	30	31	32	33	34	35
Harsh engagement from neutral to D or R			X			X																				X									X
Delayed engagement from neutral to D or R				X		X		X	X	X																				X					
Runaway upshift				X		X		X	X	X			X		X				X	X		X	X												
No upshift				X		X		X	X		X				X	X			X	X		X	X											X	
3-2 kickdown runaway				X		X		X	X				X		X	X			X	X		X	X												
No kickdown or normal downshift						X								X	X						X	X	X												
Shifts erratic				X		X		X	X	X	X		X		X				X	X	X		X												
Slips in forward drive positions				X		X	X	X	X	X	X	X	X		X				X	X	X	X	X							X			X		
Slips in reverse only				X	X	X	X	X	X	X	X	X	X		X		X			X			X										X		
Slips in all positions				X		X	X	X	X	X	X	X	X																						
No drive in any position				X		X		X		X	X					X																X			
No drive in forward drive positions				X		X		X	X			X	X				X			X															
No drive in reverse	X			X	X	X	X	X		X	X												X							X			X		
Drives in neutral				X					X																				X	X	X				
Drags or locks		X			X						X																X					X	X		
Grating, scraping growling noise					X													X														X	X		
Buzzing noise						X		X			X		X																					X	
Hard to fill, oil blows out filler hole													X											X	X										
Transmission overheats			X	X					X																		X	X	X						X
Harsh upshift															X	X										X									
Delayed upshift															X	X			X	X	X	X	X												
Slips in reverse of manual low					X						X																								

Figure 13-17. *The problems and the possible causes are cross-referenced by an X or other mark at the appropriate spot on this type of chart. (DaimlerChrysler)*

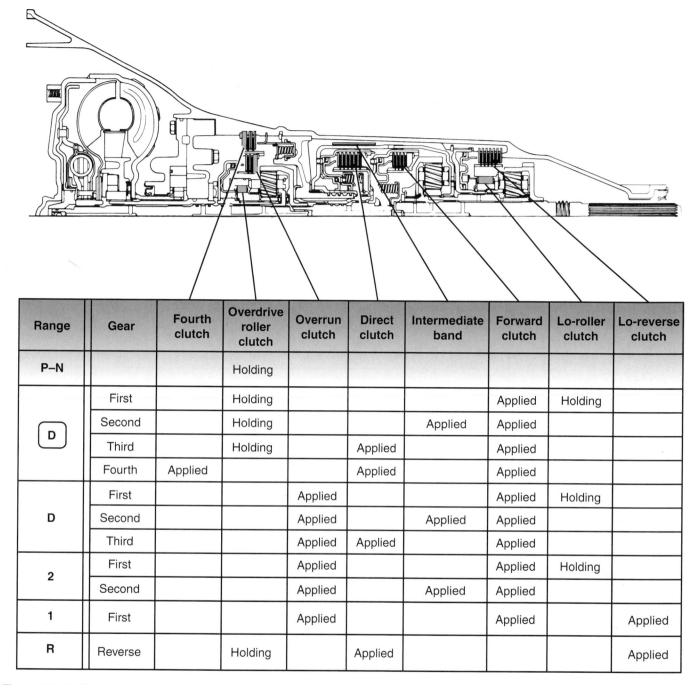

Range	Gear	Fourth clutch	Overdrive roller clutch	Overrun clutch	Direct clutch	Intermediate band	Forward clutch	Lo-roller clutch	Lo-reverse clutch
P–N			Holding						
D	First		Holding				Applied	Holding	
	Second		Holding			Applied	Applied		
	Third		Holding		Applied		Applied		
	Fourth	Applied			Applied		Applied		
D	First			Applied			Applied	Holding	
	Second			Applied		Applied	Applied		
	Third			Applied	Applied		Applied		
2	First			Applied			Applied	Holding	
	Second			Applied		Applied	Applied		
1	First			Applied			Applied		Applied
R	Reverse		Holding		Applied				Applied

Figure 13-18. *The band and clutch application chart will help the technician determine which holding members are applied in each gear. This will help isolate problems occurring in a particular gear. (General Motors)*

A hydraulic pressure test procedure is depicted in **Figure 13-23.** When performing this test on the ground, apply the parking brake. In addition, hold one foot on the brake pedal while depressing the accelerator. If the vehicle is being pressure tested off the ground, be careful of moving wheels and driveline parts.

Compare the gauge readings with pressure figures published in the factory service manual. Excessive pressure readings indicate that a pressure regulator or pressure relief valve is stuck closed. Low pressure indicates a worn front pump, a pressure regulator or pressure relief valve that is stuck open, internal leaks, or a plugged filter. If pressure is low in only one gear, the apply piston seals or seal rings are probably leaking.

If the transmission or transaxle has a TV linkage, disconnect the linkage rod or cable at the intake manifold. Start the engine, apply the parking brake, and place the selector in drive. Operate the linkage by hand while observing the pressure gauge. Line pressure should increase as the cable or rod is moved toward the full throttle position and decrease as the cable or rod is moved to the idle position. This indicates that the throttle and pressure regulator valves are able to respond to the changes in throttle position.

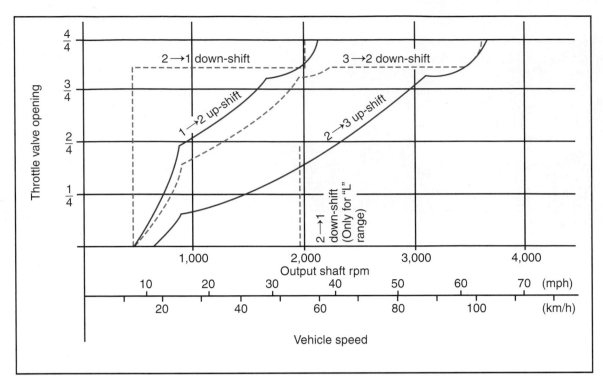

Figure 13-19. *The automatic shift diagram shows the approximate points for transmission shifts. These shift points are compared to the transmission's actual shift points. Some manufacturers simply list shift points in their service manuals. (Toyota)*

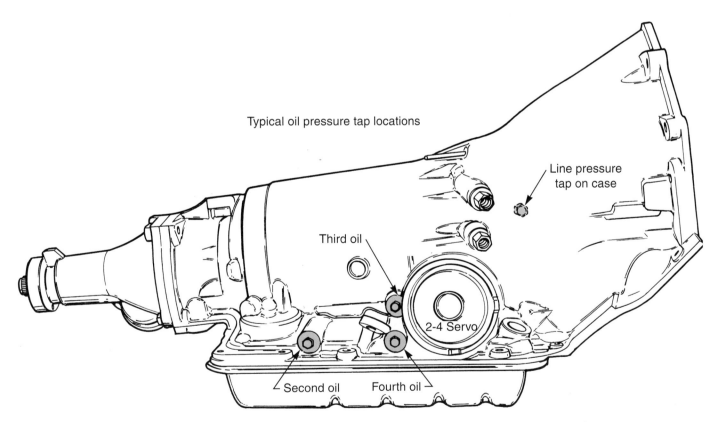

Figure 13-20. *Pressure taps for performing control pressure tests are located on various parts of the transmission case. All automatic transmissions have at least one pressure tap; most have several. (General Motors)*

If the transmission or transaxle has a vacuum modulator, disconnect the vacuum line with the engine running. The transmission or transaxle should be in drive and the parking brake applied. Pressure should increase, indicating that the vacuum modulator and regulator valve are responding to the loss of manifold vacuum.

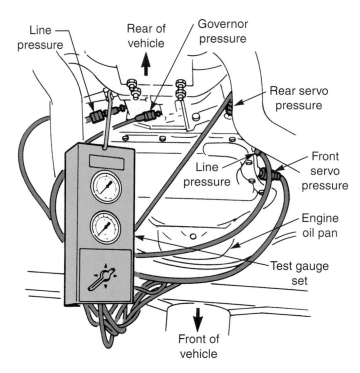

Figure 13-22. *The hookup for a hydraulic pressure test is shown here. In this illustration, the test is performed with the vehicle on a hoist. More than one gauge can be installed to test several pressures at the same time. (DaimlerChrysler)*

Figure 13-21. *This photograph shows the pressure taps on a common transaxle. Most transmissions and transaxles do not have this many pressure taps.*

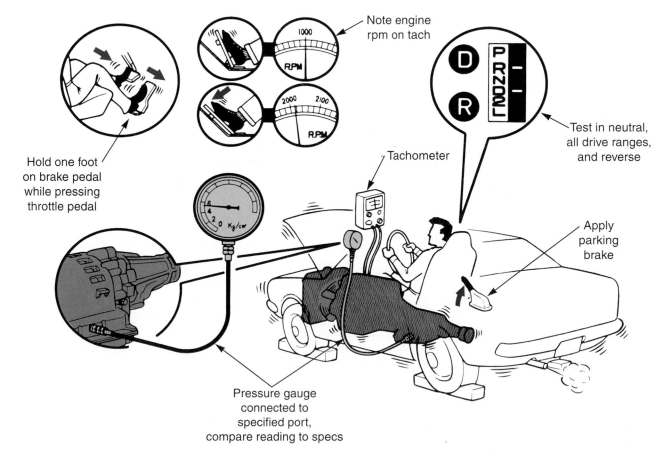

Figure 13-23. *Transmission pressures should be tested in all transmission gears, both in the shop and during a road test. The latter allows readings to be obtained while the transmission shifts through all gears at various throttle openings. Pressure readings can then be compared with the factory specifications. (Toyota)*

Note: Some manufacturers advise against removing the vacuum modulator line with the engine running.

Perform a road test with a pressure gauge installed. The gauge must have a hose that is long enough to reach the pressure tap from the vehicle's passenger compartment. If possible, have someone else drive as you watch the gauge. Observe the gauge needle as each shift occurs. The needle should drop during each shift. A slow pressure rise or low pressure in any one gear is caused by internal leakage from defective piston seals or seal rings in that particular gear. Consult the band and clutch application chart to pinpoint hydraulic leakage to a specific clutch or servo.

The pressure gauge can also be used to check throttle and governor pressure on many vehicles. Throttle pressure should increase with throttle opening. Governor pressure should increase with road speed. Consult the service manual for the proper pressure tap locations and pressure specifications.

Stall Test

A *stall test* is a way of loading the engine through the transmission or transaxle to isolate a problem in the torque converter or transmission. A stall test can be used to check the holding ability of the clutch packs and bands, as well as the operation of the stator one-way clutch. The test is sometimes useful to diagnose slipping holding members, low hydraulic pressures, or problems with the torque converter stator.

To conduct a stall test, install a tachometer on the engine. Then, start the engine and place the selector in drive. With the parking brake applied and the brake pedal fully depressed, press the accelerator until the engine speed will not increase. The tachometer reading at this point indicates maximum stall speed. This speed should not exceed the published figures. If it does, there is a problem in the transmission or transaxle. Release the accelerator immediately to prevent further damage. A higher-than-specified stall speed indicates defective holding members or low hydraulic pressures. If the engine will not reach the specified stall speed, the engine requires service or the torque converter stator is not locking.

Stall tests should be performed in each driving range. Allow a cooling period of at least two minutes between each test with the selector in neutral. Do not run the engine at stall speed for more than 5 seconds because of the rapid heat buildup that results from this test.

Some manufacturers do not recommend stall tests because of the stress they put on the engine and drive train. Therefore, these manufacturers do not publish stall speed specifications. There is no purpose in conducting a stall test on a vehicle when stall speed specifications are not available.

 Warning: While conducting a stall test, apply both the parking brake and the service brakes. As an added precaution, block both front wheels. Make sure no one is standing in front of or behind the vehicle. The vehicle could lurch forward or backward as the engine reaches maximum rpm, even with the brakes applied and the wheels blocked.

Testing Oil Cooler Flow

Low flow through the oil cooler can cause transmission/transaxle overheating and failure. To test for insufficient oil cooler flow, determine which of the oil cooler lines is the outlet line. Then place a drip pan under the oil cooler line fittings. Next, detach the outlet oil cooler line fitting from the radiator oil cooler. Install a fitting and a length of hose on the cooler outlet. Place the other end of the hose in a container. An empty gallon milk jug is often used for this purpose. Then start the vehicle and observe the oil flow from the outlet line. Flow should be at least one quart (0.9 liters) every 20 seconds. A greater flow rate is acceptable, but less than one quart every 20 seconds indicates a problem. Stop the engine as soon as the flow rate has been established.

If the flow rate is below normal, check for a clogged oil cooler in the radiator by repeating the flow rate test at the cooler inlet line. If the flow rate is now above one quart every 20 seconds, the cooler is restricted. If the flow rate remains low, check for a kinked line or an internal pressure or flow problem in the transmission or transaxle.

Air Pressure Tests

Air pressure tests involve applying compressed air to the hydraulic pressure ports that supply hydraulic fluid to the holding members, and sometimes, to the governor valve. See **Figure 13-24.** When air pressure is applied, operation of these components can be seen or heard to determine if they are working correctly. Air pressure tests may reveal that a component is hung up, or that there is a leak or obstruction within the hydraulic passageway. Air pressure tests may help determine whether the clutch and servo apply piston seals, seal rings, or gaskets are sealing as they should.

The oil pan and valve body must be removed to perform an air pressure test. Before removing these components, the fluid must be drained from the transmission or transaxle. Once the valve body is removed, apply compressed air to the proper hydraulic passage, **Figure 13-25.** Air pressure should be regulated to about 30 psi–40 psi (207 kPa–276 kPa). Air supply must be free of all moisture and dirt.

If compressed air is delivered to a clutch passage, the clutch should make a sharp *thunk* as it applies. The application can sometimes be felt by placing your fingers on the clutch drum as air is applied. Servos can be seen operating the band as air is applied. A defective clutch or servo apply piston will not apply strongly and may leak air. If the air pressure test indicates that a leak exists somewhere, the transmission must be disassembled to check the apply piston seals and other parts. Some governor valves can also be checked with air pressure. If the governor buzzes when air is applied to the passage, it is free in its bore (it is not sticking).

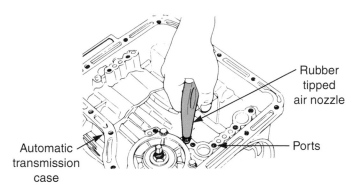

Figure 13-24. *An air pressure test is performed to determine if holding members—and, sometimes, the governor valve and accumulators—are operable. This test can be used to check for hydraulic system leaks and obstructions. It is performed by applying air pressure to the ports leading to the clutch pack, band, or other component. The transmission fluid must first be drained and the oil pan and valve body removed. (Nissan)*

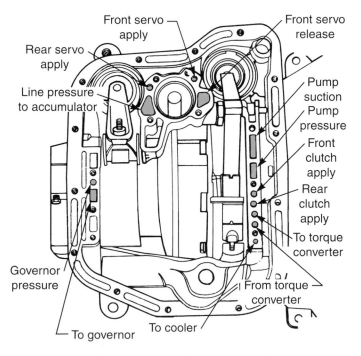

Figure 13-25. *Bottom view of a transmission case with the oil pan and valve body removed. The factory service manual will usually have a diagram like this, showing where the holding member apply ports are located on the case. Other hydraulic system devices, such as the governor, can also be checked with air pressure. (DaimlerChrysler)*

Using Hydraulic Circuit Diagrams to Troubleshoot

Hydraulic circuit diagrams, or *oil pressure diagrams,* can be used to trace circuits in the hydraulic system. Knowing exactly where the different hydraulic passageways go and what they are for will enable you to understand many seemingly obscure problems. Most service manuals contain at least one hydraulic diagram for a specific make of transmission, and many contain a diagram for every gear. The hydraulic circuits can be traced to isolate problems and determine the cause of hydraulic malfunctions.

Figure 13-26 shows the hydraulic flow in a common 3-speed transmission. The flow can be traced through the passageways to the manual and shift valves and on to the clutches and bands.

Suppose one of the check balls (shown in **Figure 13-26** as numbered circles inside of hydraulic lines) were missing. Checking the transmission line pressure would allow you to determine that a missing check ball was causing a transmission malfunction in a certain gear.

If, for instance, the #3 check ball were missing, fluid would be dumped out through the passageway to the manual valve whenever the transmission shifted into third gear. Refer to **Figure 13-26.** This flow would cause a drop in line pressure, as well as prevent full application of the front and rear clutches and full release of the front servo. Either the transmission would slip badly in third, or it would not upshift out of second. By connecting a pressure gauge to the transmission and observing a line pressure drop in third gear, you could determine that the #3 check ball was missing.

> **Note: Check balls do not suddenly disappear. A missing check ball would be the result of carelessness during valve body overhaul and would be evident when the vehicle was driven after an overhaul.**

Notice on this diagram that the fluid used by the 2-3 shift valve to apply the front clutch and release the front band comes from the 1-2 shift valve. Therefore, if the 1-2 shift valve sticks in the upshifted position, the transmission will start in second gear, but the 2-3 shift will be normal. If the 1-2 shift valve sticks in the downshifted position, however, the transmission will stay in first gear, even though the 2-3 shift valve is not sticking. On some transmissions, however, the 2-3 shift valve is fed line pressure, and the unit will shift from first to third, even though the 1-2 shift valve was stuck. This illustrates the importance of studying the specific hydraulic circuit diagrams for the transmission you are working on to avoid spending much time looking for a nonexistent problem. In this case, misunderstanding the hydraulic system could mean wasting time checking the governor or throttle valves.

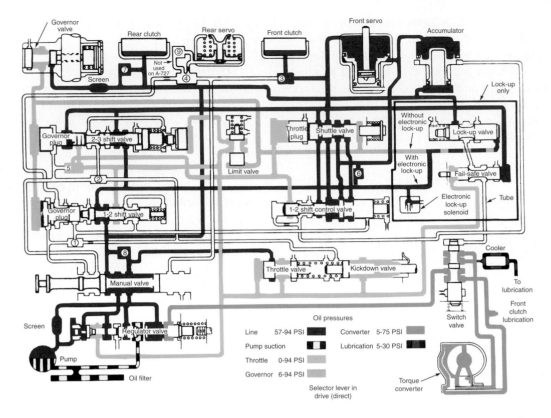

Figure 13-26. *Hydraulic diagrams such as this can be used to troubleshoot problems in the hydraulic system. The diagram shows the hydraulic flow of a rear-wheel drive transmission in third gear (direct drive). (DaimlerChrysler)*

Figure 13-27 shows the hydraulic flow in a common 4-speed automatic transmission. Extra holding members are required for overdrive. Notice that pressure from the *vacuum throttle valve* is used, in part, to modify the action of the pressure regulator valve. Throttle pressure acts on the regulator valve to assist the spring. Changes in throttle pressure will affect the action of the pressure regulator valve. This helps explain why problems with the vacuum modulator or throttle valve may cause rough downshifts or slipping.

Valves in this hydraulic system that modify the line pressure and can affect upshift or downshift quality are the pressure regulator valve, vacuum throttle valve, pressure modifier valve, throttle back-up valve, governor valves, and throttle relief valve. Shift timing can be affected by the shift valves, governor valves, throttle valves, 3rd-2nd timing valve, solenoid downshift valve, and throttle back-up valve. Valves that can affect the quality of the shift by modifying pressure during the shift are the pressure modifier valve, any one of the check valves, the accumulator, and the throttle back-up valve. Tracing the fluid flow through these valves can often lead you to a sticking valve that is the source of a slippage problem or a shift timing or quality problem.

Figure 13-28 shows a common 4-speed transaxle. The throttle valve is controlled by a TV cable, but line pressures are controlled by a vacuum modulator. Knowing this would help with the diagnosis of a shifting problem. If the

TV cable breaks, shifts will be excessively low but not soft. Knowing that line pressures are not controlled by the TV cable will make this situation less of a mystery.

Another feature of this transmission is the torque converter clutch control system. The solenoid is designed to direct pressurized fluid to the converter clutch valve when energized. Therefore, if the converter clutch will not apply, it may be because the solenoid has failed.

Electrical Problem Diagnosis

Some transmissions have electrically operated detent systems to cause a forced downshift. On other transmissions and transaxles, the converter lockup clutch is energized through pressure or vacuum switches. These systems can fail and cause transmission problems.

Detent System Diagnosis

If the detent system fails to operate, the transmission will not downshift at wide-open throttle. Begin by checking that debris is not keeping the detent switch from operating. Then remove the electrical connector and check for power. One of the connector terminals should illuminate a test light with the ignition switch in the *on* position. If the electrical connector has no power, check the system fuse and wiring. If the electrical connector has power, reattach it and press the accelerator pedal to the floor. Touch

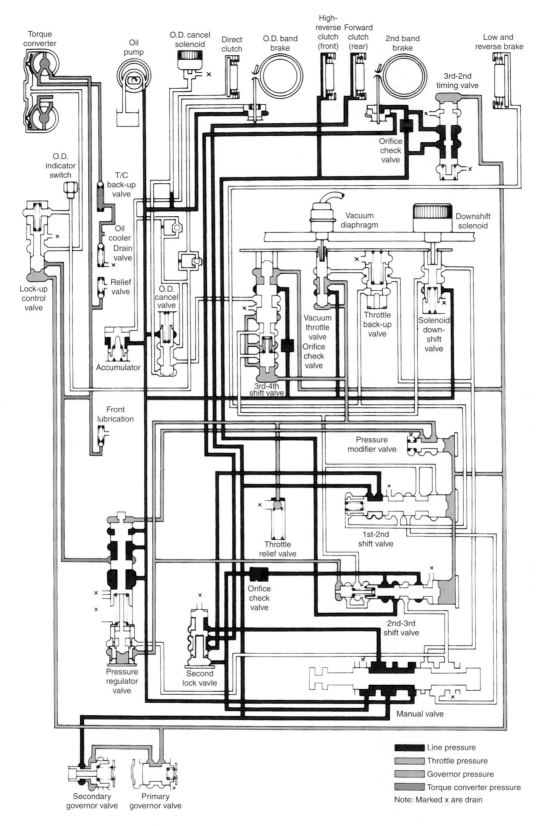

Figure 13-27. *This diagram shows the fluid flow in a typical 4-speed automatic transmission. The hydraulic system controls four clutch packs, two bands, and a lockup torque converter. The transmission hydraulic system is acted upon externally by a vacuum modulator and two electric solenoids.*

Park – engine running

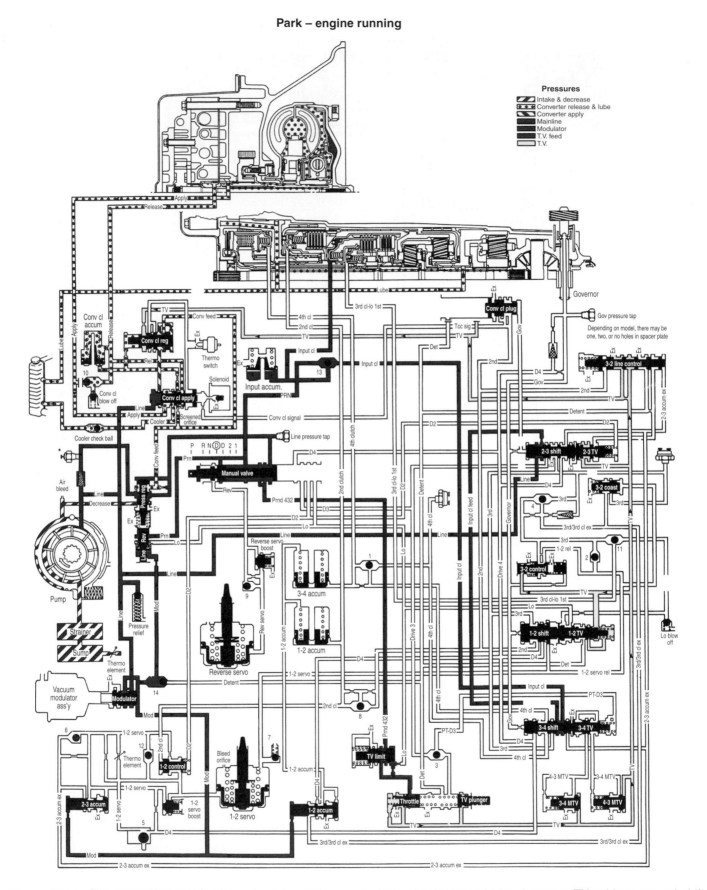

Figure 13-28. *The 4-speed transaxle shown here is used on many front-wheel drive vehicles. It uses a TV cable to control shift speeds and a modulator to control pressure. (General Motors)*

the other connector terminal with the test light probe. If the test light does not light, the switch is defective or out of adjustment. If power is leaving the switch, check that it is reaching the detent solenoid connector on the transmission. If power is reaching the detent solenoid connector, the solenoid is defective or the transmission has an internal hydraulic problem.

If the detent solenoid is always energized, the transmission will only upshift at high speeds. To test for this condition, turn the ignition switch to the *on* position and make sure that the accelerator is not touching the detent switch. Then probe the solenoid connector with a test light. If the light comes on, the detent switch is stuck or misadjusted, or there is a short in the detent system wiring.

Converter Clutch System Diagnosis

A converter clutch control system can fail to apply, remain applied at all times, or apply at the wrong time. If the converter clutch does not apply, trace the power supply using a test light. If no power is available to the system, check the fuse and look for wiring problems. When a vacuum switch is used, check the switch with an ohmmeter. The switch should show zero resistance when the engine is running at idle. If the switch is not operating, check for a leaking or clogged vacuum hose. Some vacuum switches can be adjusted. Pressure switches should show zero resistance when pressurized. Check the service manual to determine when the switch is supplied with pressure. Some systems pressurize the switch in a certain gear, while others use governor pressure to close the switch at a certain speed. If power is reaching the apply solenoid, suspect a defective solenoid or an internal hydraulic problem.

If the converter clutch is applied at all times, the engine will die when the vehicle is shifted into gear. Look for a switch that is stuck in the closed position. Also look for places in the wiring harness where the wiring has melted together. If the clutch applies at the wrong speeds or gears, check the vacuum switch and hose. Also check the pressure switch. A common cause of a pressure switch malfunction is an internal hydraulic problem or sludge in the pressure passage.

Summary

While automatic transmissions and transaxles are complex, they can be diagnosed and repaired if the technician understands their operation, and uses logical troubleshooting procedures. Before beginning repairs, you must first try to find out what the problem is.

A general troubleshooting procedure is to first have the owner describe the problem. Road test the car with the driver, if possible. Check the transmission fluid level prior to the road test. Check the linkage if the road test points to a possible linkage malfunction. Adjust the linkage as necessary. Check vacuum at vacuum modulator, as applicable. Inspect electrical connections. Perform such diagnostic tests as hydraulic and air pressure tests, if necessary, and consult diagnostic charts. Make sure the problem is not caused by another vehicle system. Sometimes, transmission disassembly is required if the foregoing procedures are inconclusive.

Causes of automatic transmission problems are often identified as belonging to one of three systems. These are the hydraulic system, the mechanical system, and the electrical system. Hydraulic system problems affect the devices that produce and control hydraulic pressure in the transmission. Mechanical system problems affect the shafts, gearsets, holding members, case, etc. Electrical system problems affect components such as solenoids, neutral start switch, and computer sensors.

Typical transmission problems include slippage, no-drive condition, shift problems, noise, overheating, and leakage. There can be many causes of these problems, and the transmission should be thoroughly checked to determine the exact problem.

Typical transmission problems can be attributed to a variety of component malfunctions. For example, a noisy transmission can be attributed to a broken flexplate, a torque converter malfunction, a damaged planetary gearset, and other defects.

Diagnostic charts can make troubleshooting easier. One type of diagnostic chart lists common transmission problems and typical causes. The band and clutch application chart allows the technician to determine which holding members are applied in which gears. This makes diagnosis of a problem in a particular gear easier. The automatic shift diagram allows the technician to compare transmission shift points against published specifications.

The automatic transmission should be tested before it is disassembled. One such test is the road test, in which the vehicle is driven and transmission action observed. Pressure tests require a pressure gauge. They are conducted to determine whether problems exist in the hydraulic system. Stall tests can be used to determine whether the holding members and torque converter are functioning properly. Stall tests are very hard on the engine and drive train components, and some manufacturers do not recommend them. Air pressure tests are used to determine if the band and clutch apply pistons are operating properly and to spot hydraulic system leaks or obstructions. Air pressure tests are made after the oil pan and valve body are removed from the transmission case.

Automatic transmission hydraulic diagrams can be used to diagnose many problems. They can help the technician spot many unusual problems and locate sources of pressure leaks or obstructions. Manufacturers provide hydraulic diagrams for every type of transmission that they use.

Review Questions—Chapter 13

Please do not write in this text. Place your answers on a separate sheet of paper.

1. During the first step of the seven-step troubleshooting process, the technician should:
   (A) perform pinpoint tests.
   (B) determine which component is the most probable cause of the problem.
   (C) determine the exact nature of the complaint.
   (D) verify the root cause of the problem.

2. Which problem is evident when a transmission or transaxle does not transfer enough power to the drive wheels?
   (A) Erratic shifts.
   (B) Slippage.
   (C) Early upshift.
   (D) No passing gear.

3. A burnt odor or sludge in the oil pan is normally caused by failure of the _____ material.

4. How would you analyze a noisy transmission?

5. An automatic transmission makes a whining noise in first gear only. The problem is probably in the _____ _____.

6. Transmission overheating may occur when the temperature of the fluid reaches roughly _____°F, or _____°C.

7. Which of the following service procedures requires transmission removal?
   (A) Adjusting transmission band.
   (B) Replacing front pump seal.
   (C) Replacing rear seal.
   (D) Cleaning valve body.

8. An automatic transmission slips in second gear only. What could be the cause?
   (A) Defective forward clutch.
   (B) Broken pump gear.
   (C) Defective intermediate band.
   (D) Stuck 1-2 shift valve.

9. When checking for a possible vacuum modulator malfunction, you should check for a ruptured diaphragm and a blocked, leaking, or otherwise damaged _____ line.

10. A 4-speed automatic transmission shifts late into second and does not shift into third or fourth. The most likely cause is a stuck _____.

ASE-Type Questions—Chapter 13

1. Technician A says that automatic transmission fluid level should be checked before road testing the vehicle. Technician B says that a road test may indicate that the vehicle driver is mistaking a normal condition for a problem. Who is right?
   (A) A only.
   (B) B only.
   (C) Both A and B.
   (D) Neither A nor B.

2. All the following statements about automatic transmission diagnosis are true *except:*
   (A) the vehicle's driver can accurately describe any problem.
   (B) in many cases, the problem can be determined without removing the transmission from the vehicle.
   (C) automatic transmissions can have many types of problems.
   (D) shift linkage adjustment is a common source of problems.

3. Technician A says that burned and glazed clutch and band linings are a common source of automatic transmission problems. Technician B says that many automatic transmission electrical problems require that the transmission be completely disassembled. Who is right?
   (A) A only.
   (B) B only.
   (C) Both A and B.
   (D) Neither A nor B.

4. All the following conditions can cause automatic transmission slippage *except:*
   (A) a plugged transmission filter.
   (B) high pump output.
   (C) low fluid level.
   (D) worn holding members.

5. Technician A says that some engine and exhaust system problems can be mistakenly identified as a transmission problem. Technician B says that a no-drive condition can be caused by locked brakes or jammed differential gears. Who is right?
   (A) A only.
   (B) B only.
   (C) Both A and B.
   (D) Neither A nor B.

6. Which of these would be the least likely cause of erratic transmission shifts?

   (A) Misadjusted throttle valve linkage.

   (B) Sticking governor.

   (C) Stuck converter stator.

   (D) Metal particles in the valve body.

7. Technician A says that automatic transmission noise in only one gear is a sign of a plugged oil filter. Technician B says that sticking valves are a common cause of automatic transmission shift problems. Who is right?

   (A) A only.

   (B) B only.

   (C) Both A and B.

   (D) Neither A nor B.

8. All of the following automatic transmission components can cause noises *except:*

   (A) torque converter.

   (B) oil pump.

   (C) clutch packs.

   (D) planetary gearsets.

9. Technician A says that a ruptured vacuum modulator diaphragm will cause the transmission fluid level to be low. Technician B says that a ruptured vacuum modulator will cause incorrect transmission shift points. Who is right?

   (A) A only.

   (B) B only.

   (C) Both A and B.

   (D) Neither A nor B.

10. A sure sign of an excessively worn pump is:

    (A) low oil pressure.

    (B) pump noise.

    (C) low fluid level.

    (D) transmission overheating.

11. A band and clutch application chart, combined with a road test, can be used to locate a defective _____.

    (A) oil pump

    (B) holding member

    (C) governor

    (D) oil cooler

12. An automatic transmission slips badly when shifting into second gear. All other gears work properly. Technician A says the problem could be caused by a leaking second gear servo or improper band adjustment. Technician B says the problem could be caused by a badly worn second gear band. Who is right?

    (A) A only.

    (B) B only.

    (C) Both A and B.

    (D) Neither A nor B.

13. A pressure gauge is attached to an automatic transmission pressure tap. The engine is started and the transmission is shifted through all gears. The transmission has low oil pressure in only one gear. Technician A says that low pressure in only one gear indicates a leaking governor circuit. Technician B says that low pressure in only one gear indicates a worn oil pump. Who is right?

    (A) A only.

    (B) B only.

    (C) Both A and B.

    (D) Neither A nor B.

14. Technician A says that the transmission oil pan and valve body must be removed to make air pressure tests. Technician B says that air pressure testing requires clean and dry air at 120 psi (800 kPa) minimum. Who is right?

    (A) A only.

    (B) B only.

    (C) Both A and B.

    (D) Neither A nor B.

15. Technician A says that check balls can dissolve inside the transmission valve body. Technician B says that hydraulic circuit diagrams can be used to locate sticking valves. Who is right?

    (A) A only.

    (B) B only.

    (C) Both A and B.

    (D) Neither A nor B.

Digital multimeters, such as the one shown above, can be used to pinpoint problems in an electronic control system. (OTC)

Chapter 14

Troubleshooting Electronic Control System Problems

After studying this chapter, you will be able to:
- ❏ Determine the exact nature of an electronic transmission/transaxle complaint.
- ❏ Determine whether the problem is in the electronic control system.
- ❏ Make visual checks for problems.
- ❏ Retrieve trouble codes and match codes to problem areas.
- ❏ Use scan tools and other test equipment to locate problem areas.
- ❏ Use test equipment to test individual components.
- ❏ Determine repair steps necessary to correct electronic transmission/transaxle defects.

Technical Terms

Pattern failures

Limp-in mode

Gear skipping

OBD I

OBD II

Scan tool

Trouble codes

Snapshots

Freeze frames

Testing by substitution

Zirconia-type oxygen sensor

Titania-type oxygen sensor

Introduction

To diagnose problems on electronically controlled transmissions and transaxles, the technician must be able to proceed logically, without randomly replacing expensive parts. This chapter covers the logical processes that must be used to diagnose electronic control systems. It also covers diagnosis of electronically controlled automatic transmission and transaxle components. Information on trouble code retrieval and other uses of scan tools is also presented.

Electronic Transmission and Transaxle Problems

Many of the problems that occur in an electronically controlled transmission or transaxle are the same as those that occur in hydraulically controlled transmissions. Other problems are unique to the electronic control system. The following section covers common electronic control system problems. **Figure 14-1** shows the possible sources of electronic control system problems in a modern transaxle.

Many computer control system problems are called **pattern failures.** A pattern failure is a problem that is common to a certain type of vehicle. The experienced technician will learn to spot pattern failures and quickly determine the defective part. The chart in **Figure 14-2** shows some common electronic control system failures that occur in one manufacturer's automatic transmissions and transaxles.

Limp-In Mode

If a major problem occurs in the computer control system, the ECM will place the system in **limp-in mode.** When the system is in the limp-in mode, the ECM ignores most input sensor readings and operates the engine and drive train output devices based on internal settings. The transmission or transaxle solenoids are energized in a way that gives the unit only one or two forward gears. Whenever the system goes into limp-in mode, the ECM it will illuminate the dashboard maintenance indicator light, or MIL. On older vehicles, the MIL is called the *service engine soon* light. An illuminated MIL is always an indication that the computer control system has a problem and that trouble codes should be retrieved.

Input Sensor Problems

In many cases, the transmission or transaxle reacts to problems in various vehicle sensors. Some sensors are located on the valve body, **Figure 14-3.** Other sensors are primarily monitor engine operation and are installed on the engine.

Faulty input sensors are the most common cause of electronic transmission or transaxle problems. A defective throttle position sensor (TPS) can cause improper shifting or erratic application of the converter lockup clutch. A failed TPS can also cause rough upshifts or a bump on closed throttle downshifts.

Speed sensors can cause problems by sending incorrect engine or output shaft information to the ECM. This usually affects shift speeds. A speed sensor located inside the case can collect metal filings that affect the signal. Some metal particles are produced as part of normal

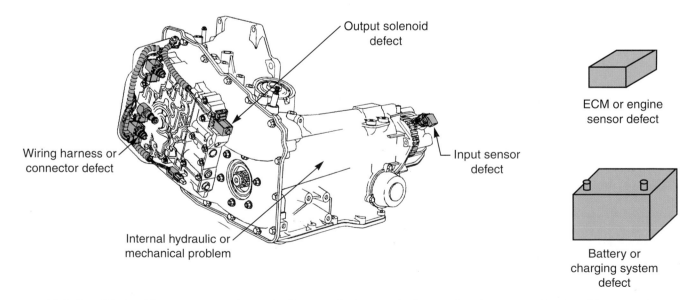

Figure 14-1. *Possible trouble spots on an electronically controlled transaxle. Remember that conventional problems, such as worn clutches or internal leaks, are just as likely as electronic component failure. (General Motors)*

Electronic Component Malfunctions	
Component/system	**Can affect**
Throttle position (TP) sensor	• Shift pattern (erratic) • Shift quality (firm or soft) • Engine (rough)
Automatic transmission output shaft speed (AT OSS) sensor	• Shift pattern (erratic) • TCC solenoid valve apply (at wrong time) • Shift quality (harsh or soft)
Transmission fluid pressure (TFP) manual valve position switch	• TCC solenoid valve apply (no apply if diagnostic code is set) • Shift pattern (no fourth gear in hot mode) • Shift quality (harsh) • Line pressure (high) • Manual downshift (erratic)
Automatic transmission fluid temperature (TFT) sensor	• TCC solenoid valve control (on or off) • Shift quality (harsh or soft)
Engine coolant temperature (ECT) sensor	• TCC solenoid valve control (no apply) • Shift quality (harsh)
Shift solenoid valves	• Gear application (wrong gear, only two gears, no shift)
Brake switch	• TCC apply (no apply) • No 4th gear if in hot mode
System voltage	• Line pressure (high) • Gear application (third gear only) • TCC control (no apply) • No 4th gear if in hot mode
3-2 control shift solenoid valve assembly	• Gear application (third gear only) • 3-2 downshifts (flare or tie-up)
Pressure control solenoid	• Line pressure (high or low) • Shift quality (harsh or soft)
TCC solenoid valve	• TCC solenoid valve apply (no apply) • No 4th gear if in hot mode
Cruise control	• Delays 3-4 upshift and TCC apply during heavy throttle
Acceleration slip regulation (ASR)	• Downshifts

Figure 14-2. *A manufacturer's list of possible problems for one electronic transmission. Symptoms will vary from one manufacturer to another, as well as from one type of transmission or transaxle to another. (General Motors)*

operation, and they often become stuck on the sensor magnet, affecting the production of the magnetic field. Wheel-mounted speed sensors can be damaged by road debris, or they can become coated with road tar.

Temperature sensors that are defective or produce out-of-range signals can cause hard or soft shifts. Some defective temperature sensors can keep the transmission from upshifting or prevent converter clutch apply. A common cause of engine temperature sensor problems is cooling system deposits that coat the sensor element. This causes the sensor to respond slowly or not at all to temperature changes.

Other sensors can indirectly affect shift speeds and shift quality. Most defects in engine sensors will show up first as engine-related problems. Defective pressure or temperature switches in the transmission or transaxle will usually not cause engine problems.

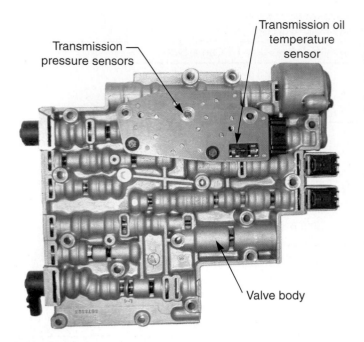

Figure 14-3. *Most transmission sensors are located inside the case, usually on the valve body. On this particular valve body, all the pressure sensors and the temperature sensor are installed in a plate. The plate is then attached to the valve body. Also note the pressure control solenoids on the outside edges of the valve body. (Sonnax)*

A loss of voltage to a sensor (usually called reference voltage) will result in an inoperative sensor. Reference voltage problems are usually caused by a defective ECM or a ground problem.

Solenoid Problems

Solenoid problems will vary depending on whether the solenoid is an on-off type or a pulsed type. On-off solenoids can stick open or closed. The usual result is the loss of some gears. In many cases, the transmission or transaxle will take off from a stop in a gear other than first. Other solenoid failures can cause *gear skipping* (shifting from first to third, for example), failure to shift into higher gears, or incorrect application of the converter lockup clutch. Occasionally, a solenoid will stick at times and work properly at other times. Typical causes of intermittent sticking include a high-resistance solenoid winding, a bad electrical connection, or buildup of sludge. Erratic shifting is a common symptom of an intermittently sticking solenoid. The transmission or transaxle may work well most of the time, with only occasional shift problems. In some cases, the transmission will shift improperly only when cold or only when hot.

Pulsed solenoids are generally used to control pressures. Therefore, instead of skipping gears, a failed pulsed solenoid can cause slipping or excessively hard or soft

shifts. A defective line pressure solenoid will cause problems in any or all gears. Many pulsed solenoids control pressures of a specific operation, such as converter clutch apply feel or part throttle downshift. Defects in the related solenoid will cause problems during that process only.

Almost every solenoid has one or more small filters, **Figure 14-4.** If the filter becomes plugged, oil pressure will not be able to pass through the solenoid valve to the rest of the hydraulic system. Additionally, a torn filter can cause the solenoid to stick.

ECM Problems

Since it contains many complex circuits, a faulty ECM can cause a variety of problems, depending on which internal part or circuit has failed. Sometimes, the ECM will keep the transmission or transaxle from shifting. A defective ECM may cause the unit to stick in one gear. Another common ECM problem is erratic shifting. Examples include downshifting at cruising speeds or occasionally failing to upshift. The operation of the ECM is often affected by heat. A cold ECM may work well when the vehicle is first driven but cause problems as it heats up.

A defective ECM can hang up, or stick in one mode. Turning the ignition switch off and then back on may temporarily correct the problem. If the ECM controls pressures through a pulsed solenoid, failure may cause slipping, hard shifting, shudder during shifts, and other problems. In many cases, a failed ECM will also cause engine drivability problems.

A defective ECM can set false trouble codes and may sometimes set codes that do not exist. A failed ECM may illuminate the MIL when there is no problem, or it may fail to illuminate the MIL when a problem is present.

Figure 14-4. *The valves of a solenoid are machined to extremely small clearances and cannot tolerate the presence of even the smallest particle of dirt or metal. Most transmission solenoids are equipped with small filters to reduce the possibility of contamination. These filters can plug up, especially if other transmission or transaxle parts have failed.*

Wiring and Connector Problems

The voltages used to operate the sensors are usually much lower than battery voltage. Therefore, any wire damage or corrosion at the connectors greatly affects the sensor inputs to the ECM. Note that even slight resistance can cause incorrect sensor inputs, leading to improper or erratic shift points. A commonly overlooked wiring problem is a corroded ground connection or disconnected ground wires. Remember that the return path for the current is as important as the input path. Resistance through a ground circuit can affect several transmission control loops at once. Therefore, when a transmission/transaxle control system seems to have several unrelated or intermittent problems, look for a poor ground.

Another common wiring problem is a wire that has been chafed or broken by movement. This commonly occurs where wires must pass through confined spaces or small openings in the body or other sheet metal parts. Manufacturers often place electrical connectors near or under the vehicle's battery. The battery acid and hydrogen gas often cause corrosion inside the connector. Another common problem is a wire that is allowed to touch an exhaust system part. The insulation melts and the wire grounds against the exhaust component.

Battery Voltage and Computer System Operation

The voltage input to the computer control system must be at or near battery voltage (12.6 volts). Defects in the charging system or battery can pull voltage below 12 volts. Low voltage can confuse the ECM, causing computer control system problems. In many cases, the voltage will be only slightly below normal. Slightly low voltage will cause computer control system problems but not starting problems. Typical results of low voltage include intermittent engine and/or drive train problems, or an engine or drive train problem that cannot be isolated to a specific system or component. If any of the above situations occur, check the battery and charging system before performing further troubleshooting procedures. Begin by checking the tension and condition of the alternator drive belt, **Figure 14-5.** Also, check the belt pulley for wear or damage. If the drive belt is tight and appears to be in good condition, check the charging rate and battery condition using an electrical system tester, **Figure 14-6.**

The Seven-Step Troubleshooting Process

The seven-step troubleshooting process was discussed in Chapter 13. This process can be used to

Figure 14-5. *Worn and slipping alternator drive belts are the source of many computer control system complaints. Inspect the belt for a glazed or cracked surface and for proper tension. Also check the alternator drive pulley for wear.*

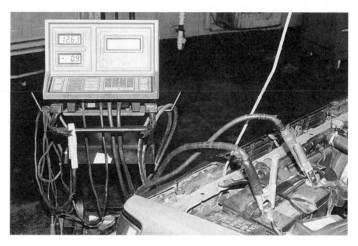

Figure 14-6. *Using an electrical system tester to conduct a battery load test. This unit will perform the test automatically.*

diagnose electronic transmissions and transaxles, as well as hydraulic models. The seven steps are as follows:

Step 1. Verify the problem—Interview the driver and make a road test to determine the exact problem.

Step 2. Check for obvious causes—Check for obvious faults, such as disconnected wires or misadjusted linkage.

Step 3. Determine which component or system is the most probable cause of the problem—Combine your knowledge and what you have learned in the previous steps to make a preliminary diagnosis of the problem.

Step 4. Perform pinpoint tests—Check all systems and components, eliminating them as possibilities.

Step 5. Check for related problems—Repeat test procedures, or perform new tests, to make sure that the problem identified in step 4 is the root cause of the problem.

Step 6. Correct the defect—Repair or replace the defective components as necessary.

Step 7. Recheck system operation—Road test the vehicle to make sure that the problem has been corrected and that no trouble codes have reappeared.

If necessary, review the detailed explanation of the seven-step process in Chapter 13. Remember that on modern vehicles, almost all systems are interconnected and affect each other.

Is the Trouble in the Electronic Control System?

Electronically controlled transmissions and transaxles may have problems caused by the mechanical and hydraulic components. Slipping and erratic shifting are as likely to be caused by hardened or damaged seals or burned holding members as by defects in the electronic control system components.

To isolate the problem to the mechanical, hydraulic, or electronic components, you must conduct some of the same tests used when checking hydraulic systems and components. Always begin any test procedure by removing the dipstick and checking fluid level and condition. As with a hydraulically controlled transmission or transaxle, this may give you all the information you need to diagnose a problem. Also check the manual linkage for looseness and proper positioning. On a few electronic transmissions, some of the shifts are hydraulically controlled. On these transmissions, check the adjustment and condition of the throttle linkage, and check governor operation.

When you perform a road test, note the condition of the engine and other vehicle systems, as well as the operation of the transmission or transaxle. A miss, surge, hesitation, or other problem may be mistaken for a defect in the transmission or transaxle. Engine missing and lockup torque converter shudder are often confused. Occasionally noises or roughness in the suspension may be blamed on the drive train.

If the preliminary checks lead you to believe that the problem is in the electronic control system, proceed with the diagnostic procedures outlined below.

Basic Electrical Checks

Before performing complex electronic tests, make some basic electrical checks. Begin by ensuring that the system fuses are not blown. Note that there may be more than one fuse protecting the system.

 Note: Do not remove any fuses until you have retrieved trouble codes from the ECM.

Next, make a careful inspection for burned, chafed, and disconnected wires. In some cases, the wiring harness wrapping may have to be pulled back to expose damaged wires. Be sure to look carefully for burned insulation at any fusible link, **Figure 14-7.** Pull on the link to determine whether the wire has broken internally. Fusible links are usually located at the battery positive terminal, a nearby power relay, or the starter solenoid.

Check all vehicle ground wires. See **Figure 14-8.** On many vehicles, several ground wires are attached to one of the bolts on the engine thermostat housing. Slight coolant leaks can cause these wires to corrode. Often, the ground wires are removed during cooling system service and never reattached.

OBD II Computer Control Systems

Early computer control systems were designed to monitor the operation of computer system parts, such as

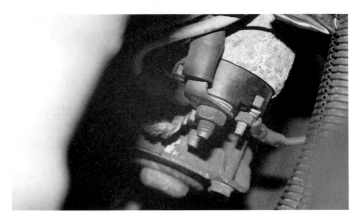

Figure 14-7. *Fusible links may be installed at the starter's positive terminal. A melted fusible link is sometimes hard to spot, since fusible links often melt open without damaging the insulation. Always tug at the fusible link to make sure it has not opened internally.*

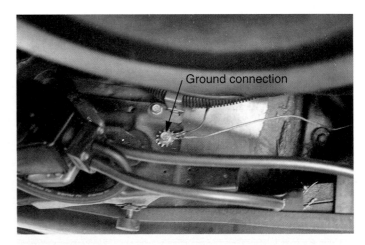

Ground connection

Figure 14-8. *Bad grounds are a common source of trouble. Always check grounds by visual inspection and by pulling on the connections. Also try to tighten the attaching bolt when applicable.*

sensors and output devices. If one of these parts failed, the ECM received an out-of-range electrical reading. The ECM would then set one or more trouble codes and turn on the dashboard-mounted warning light. These early systems are called on-board diagnostics, generation I, or **OBD I,** systems. OBD I was replaced in 1995-96 by an updated system called **OBD II** (on-board diagnostics, generation II) systems. The ECM in OBD II systems still detects failed parts, but it also monitors air-fuel ratio changes, engine misfires, temperature changes, and other operating conditions. From these inputs, the ECM can determine that a part is about to fail, or that excessive emissions will be produced in the near future. The ECM will then set one or more trouble codes and turn on the maintenance indicator light (MIL) in the dashboard. OBD I systems react to a problem that is already causing high emission or loss of performance. OBD II systems have the ability to catch potential problems *before* they begin to cause high emission or loss of performance. To accomplish this, the OBD II system has a misfire monitor, an extra oxygen sensor after the converter, and an evaporative emissions monitor.

Using Scan Tools

If the basic electrical tests do not reveal a problem, the next step is to check the unit using a **scan tool**. The scan tool is a hand-held electronic device used to retrieve **trouble codes** and perform other diagnostic procedures. **Figure 14-9** shows a typical scan tool. The scan tool can be thought of as a portable computer that can communicate with the vehicle's on-board computer. While they are expensive, scan tools are becoming common because they save diagnostic time and reduce the chance of misdiagnosis. The scan tool also allows the technician to obtain information directly from the ECM. This information would not be available by any other means. Additionally, many scan tools can be used to reprogram the vehicle's computer.

Figure 14-9. *Scan tools can be used to retrieve trouble codes and obtain other information about the operation of an electronic transmission or transaxle.*

 Caution: On OBD II–equipped vehicles, the proper scan tool *must* be used to retrieve trouble codes. Never attempt to retrieve trouble codes from an OBD II system by grounding one of the diagnostic connector terminals. Grounding any terminal will damage the ECM.

The scan tool must be connected to the system through the diagnostic connector. Always locate the proper diagnostic connector. On OBD II vehicles, a 16-pin connector called the data link connector (DLC) is used to access all vehicle computer systems. On older (non-OBD II) vehicles, there may be separate diagnostic connectors for the engine, drive train, anti-lock brakes, air conditioning, and suspension.

The power of the modern scan tool increases the amount of information available to the technician. Unfortunately, it also makes it easier to misinterpret the information. For instance, with hundreds of potential trouble codes, it is easy to look up the wrong code. Sometimes the amount of available scan tool information can cause the technician to miss the actual cause of a problem. For example, if the technician concentrates on differences in input and output shaft speeds in all gears (information that is available on modern scan tools), he or she may not notice that the torque converter clutch is not engaging. Much time could be wasted looking for slipping holding members when the actual problem is a defective converter control system.

For this reason, you must carefully interpret all trouble codes and other scan tool information before proceeding with diagnosis and repair.

Trouble Codes

OBD I computer systems have a two-digit trouble code system. The two-digit system limits the number of trouble codes to 100. OBD II systems use 5-character alphanumeric codes. Each code contains a letter and a four-digit number. See **Figure 14-10.** The letter identifies the general system causing the problem. The letter codes cover the three major vehicle subdivisions. They are B (body), C (chassis), P (power train), and U (internal computer communications network). The first number indicates whether the code is a standard code, which is assigned by SAE, or a non-uniform code, which is assigned by the vehicle manufacturer. The second number indicates the specific system in which the problem is occurring, and the final two numbers indicate which devices or circuits are causing the problem.

There are 8,000 possible OBD II codes. While all this capacity is not presently being used, a modern vehicle ECM may be programmed with several hundred trouble codes. **Figure 14-11** is a list of some current OBD II transmission trouble codes.

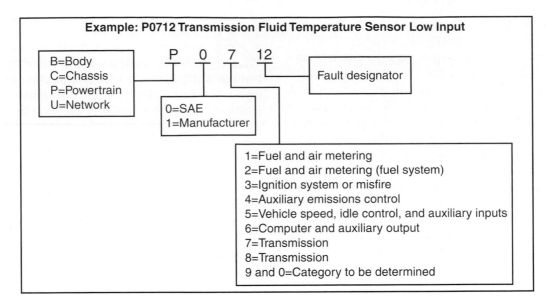

Figure 14-10. *Typical OBD II trouble code. This particular code indicates a low input signal to the transmission fluid temperature sensor.*

Other Scan Tool Functions

In addition to retrieving trouble codes, the scan tool can be used to retrieve other information from the vehicle computer. This information may include speeds at which shifts occur, converter clutch application speed, current gear, and transmission/transaxle temperatures and pressures. Some scan tools also display the desired gear for the particular engine and vehicle speed. This allows the technician to instantly determine whether the transmission is in the proper gear. Other scan tool readings give input and output shaft rpm, which can be used by the technician to diagnose slippage in a particular gear. **Figure 14-12** shows transmission-related data that can be obtained using a scan tool.

Other conditions that can be monitored by the scan tool include engine RPM and temperature, sensor inputs, vehicle speed, and outside air temperature. In OBD II systems, scan tools can also tell the technician whether or not there is an engine miss, and if so, which cylinder is causing the problem.

Some scan tools can interface with the ECU to provide **snapshots,** or **freeze frames,** of engine and transmission operation. The snapshot is a scan tool feature that records vehicle operating conditions immediately before and after a malfunction occurs. This information helps the technician determine exactly what happened to cause the problem. To record the snapshot information, the technician must drive the vehicle with the scan tool attached until the malfunction occurs. The technician can then access the readings through the scan tool. Some vehicle ECMs save snapshot information when a malfunction occurs during normal operation. The technician can retrieve this information using the scan tool. **Figure 14-13** shows typical snapshot information captured during one particular malfunction.

Scan tools can also be used to test output solenoids by bypassing the ECM and commanding the solenoids to operate. One of the most useful ways a scan tool can be used is during a road test. Attach the scan tool to the vehicle's diagnostic connector and observe readings as you drive the vehicle. Information on shift speeds, shift quality, slippage, and other data can be observed under actual driving conditions. The scan tool can be connected and then placed on the dashboard or seat for road testing.

The safest method of performing a road test while using a scan tool is to have someone else drive while you check the scan tool readings. Always use lightly traveled roads during the road test, and be alert for other drivers, traffic signals, and road conditions. The scan tool will display readings similar to those shown in **Figure 14-14.** These readings can be compared with actual transmission or transaxle operation as the malfunction occurs. This helps determine whether the problem is in the computer control system or another system. Actual scan tool displays vary widely. Scan tool instructions should be followed exactly to ensure that you obtain the proper readings.

Checking System and Component Operation

Once you have isolated the problem using the scan tool, you can proceed to test individual systems and components as outlined in the following sections.

Test Equipment

To test electronic components, you will need some or all of the following testers. This test equipment was

PO7XX OBDII Transmission/Transaxle Trouble Codes

PO700 Transmission Control System Malfunction	PO742 Torque Converter Clutch Circuit Stuck On
PO701 Transmission Control System Range/Performance	PO743 Torque Converter Clutch Circuit Electrical
PO702 Transmission Control System Electrical	PO744 Torque Converter Clutch Circuit Intermittent
PO703 Torque Converter/Brake Switch B Circuit Malfunction	PO745 Pressure Control Solenoid Malfunction
PO704 Clutch Switch Input Circuit Malfunction	PO746 Pressure Control Solenoid Performance or Stuck Off
PO705 Transmission Range Sensor Circuit Malfunction (PRNDL Input)	PO747 Pressure Control Solenoid Stuck On
	PO748 Pressure Control Solenoid Electrical
PO706 Transmission Range Sensor Circuit Range/Performance	PO749 Pressure Control Solenoid Intermittent
PO707 Transmission Range Sensor Circuit Low Input	PO750 Shift Solenoid A Malfunction
PO708 Transmission Range Sensor Circuit High Input	PO751 Shift Solenoid A Performance or Stuck Off
PO709 Transmission Range Sensor Circuit Intermittent	PO752 Shift Solenoid A Stuck On
PO710 Transmission Fluid Temperature Sensor Circuit Malfunction	PO753 Shift Solenoid A Electrical
	PO754 Shift Solenoid A Intermittent
PO711 Transmission Fluid Temperature Sensor Circuit Range/Performance	PO755 Shift Solenoid B Malfunction
	PO756 Shift Solenoid B Performance or Stuck Off
PO712 Transmission Fluid Temperature Sensor Low Input	PO757 Shift Solenoid B Stuck On
PO713 Transmission Fluid Temperature Sensor Circuit High Input	PO758 Shift Solenoid B Electrical
	PO759 Shift Solenoid B Intermittent
PO714 Transmission Fluid Temperature Sensor Circuit Intermittent	PO760 Shift Solenoid C Malfunction
	PO761 Shift Solenoid C Performance or Stuck Off
PO715 Input/Turbine Speed Sensor Circuit Malfunction	PO762 Shift Solenoid C Stuck On
PO716 Input/Turbine Speed Sensor Circuit Range/Performance	PO763 Shift Solenoid C Electrical
	PO764 Shift Solenoid C Intermittent
PO717 Input/Turbine Speed Sensor Circuit No Signal	PO765 Shift Solenoid D Malfunction
PO718 Input/Turbine Speed Sensor Circuit Intermittent	PO766 Shift Solenoid D Performance or Stuck Off
PO719 Torque Converter/Brake Switch B Circuit Low	PO767 Shift Solenoid D Stuck On
PO720 Output Speed Sensor Circuit Malfunction	PO768 Shift Solenoid D Electrical
PO721 Output Speed Sensor Circuit Range/Performance	PO769 Shift Solenoid D Intermittent
PO722 Output Speed Sensor Circuit No Signal	PO770 Shift Solenoid E Malfunction
PO723 Output Speed Sensor Circuit Intermittent	PO771 Shift Solenoid E Performance or Stuck Off
PO724 Torque Converter/Brake Switch B Circuit High	PO772 Shift Solenoid E Stuck On
PO725 Engine Speed Circuit Malfunction	PO773 Shift Solenoid E Electrical
PO726 Engine Speed Circuit Range/Performance	PO774 Shift Solenoid E Intermittent
PO727 Engine Speed Circuit No Signal	PO780 Shift Malfunction
PO728 Engine Speed Circuit Intermittent	PO781 1-2 Shift Malfunction
PO730 Incorrect Gear Ratio	PO782 2-3 Shift Malfunction
PO731 Gear 1 Incorrect Ratio	PO783 3-4 Shift Malfunction
PO732 Gear 2 Incorrect Ratio	PO784 4-5 Shift Malfunction
PO733 Gear 3 Incorrect Ratio	PO785 Shift/Timing Solenoid Malfunction
PO734 Gear 4 Incorrect Ratio	PO786 Shift/Timing Solenoid Range/Performance
PO735 Gear 5 Incorrect Ratio	PO787 Shift/Timing Solenoid Low
PO736 Reverse Incorrect Ratio	PO788 Shift/Timing Solenoid High
PO740 Torque Converter Clutch Circuit Malfunction	PO789 Shift/Timing Solenoid Intermittent
PO741 Torque Converter Clutch Circuit Performance or Stuck Off	PO790 Normal/Performance Switch Circuit Malfunction

Figure 14-11. *OBD II trouble codes for transmissions and transaxles. Older systems may also have transmission trouble codes.*

discussed in Chapter 2 and will be briefly reviewed here. Scan tools were discussed previously.

Voltmeters can be connected to read the voltage available at an electrical connection. The voltmeter in **Figure 14-15** is being used to measure voltage at a wire connector. The positive lead is placed on one of the connector terminals, and the negative lead is connected to a good ground. In **Figure 14-16,** the voltmeter is reading the voltage across a connection as current flows through it. If the connection has excessive resistance, current will try to flow through the meter, creating a voltage reading. The connection must be cleaned or replaced if the voltage is higher than the maximum specified.

Typical Scan Tool Data Values

Tech 1 Parameter	Units Displayed	Typical Scan Values
MAP	kPa	20-48 kPa, depending upon altitude
BARO	kPa	70-100kPa, depending upon altitude
TP sensor	Volts	0.3-0.9V
TP angle	Percent	0%
Engine speed	RPM	+/− 100 RPM from desired idle
Desired idle	RPM	+/− 100 RPM from engine speed
Current DTC set	Yes/no	No
ECT sensor	Volts	Varies
TFT	C° (F°)	Varies
TFT sensor	Volts	Varies
Trans. OSS	RPM	0
TCC duty cycle	Percentage	0%
TCC enable	Yes/no	No
TCC slip speed	RPM	+/− 50 RPM from the engine speed
Brake switch	Closed/open	Closed
TCC ramp	Seconds	0.00
TCC apply	Seconds	0.00
TCC out of range	Yes/no	No
TCC min. TP	Yes/no	No
TCC delta TP	Yes/no	No
TCC duty open	Yes/no	No
TCC duty shorted	Yes/no	No
TCC enable open	Yes/no	No
TCC enable short	Yes/no	No
Mph (km/h)	0-158 (0-255)	0
MAP sensor	Volts	1-2V, depending on altitude
Engine run time	Hr/min/sec	Varies
Ignition 1	Volts	Varies
ECT	C° (F°)	Varies
PC act. current	Amps	Varies (0.1-1.1 amps)
PC ref. current	Amps	Varies (0.1-1.1 amps)
PC duty cycle	Percentage	Varies
PC sol. low volts	Normal/low volts	Normal
TFP switch A/B/C	On/off	Off/on/off
Power enrichment	Yes/no	No
Cruise enables	Yes/no	No
A/C clutch	Yes/no	No
Kickdown enabled	Yes/no	No
4WD low	Yes/no	No

Figure 14-12. *Common data provided by the ECM through a scan tool. These readings can be used to quickly isolate a problem area. (General Motors)* *(Continued)*

Tech 1 Parameter	Units Displayed	Typical Scan Values
4WD	Yes/no	No
Current TAP	0-16	0
Hot mode	Yes/no	No
Adaptable shift	Yes/no	No
Long shift delay	Yes/no	No
Long shift time	Yes/no	No
TP range	Yes/no	Yes
TP delta	Yes/no	No
VSS delta	Yes/no	No
Max TAP	Yes/no	No
Trans. range	Error, rev, overdrive, drive 3 drive 2, low 1, P/N	P/N
Turbine speed	N/A	—
Input speed	N/A	—
Current gear	1,2,3,4	1
1-2 sol. 2-3 sol.	On/off	On/on
1-2 shift time	Seconds	Varies
2-3 shift time	Seconds	Varies
1-2 shift error	Seconds	Varies
2-3 shift error	Seconds	Varies
3-4 shift time	Seconds	Varies
3-4 shift error	Seconds	Varies
3-2 DS sol.	Yes/no	Yes
1-2 sol. open	Yes/no	No
1-2 sol. short	Yes/no	No
2-3 sol. open	Yes/no	No
2-3 sol. short	Yes/no	No
Shift delay	Seconds	0.00
Speed ratio	Ratio	8.00:1
Shift pattern	Number	8
Start of shift	Yes/no	Yes
End of shift	Yes/no	Yes
Upshift	Yes/no	No
Shift completed	Yes/no	Yes
Gear ratio	N/A	—

Figure 14-12. *Continued.*

TECH 1 DATA ANALYSIS

| Next Sample | Prev Sample | Select Sample | Scrl Dn | Scrl Up | Plot | Print | Exit |

M/Y: 1990 RPO Codes: LHO		**Data File: LUMINA. SMP 04/03/90 08:45 am**	
PROM ID	6521	Vehicle Speed	0 MPH
Trouble Codes		Torque Converter Clutch	OFF
Open/Closed Loop	OPEN	2nd Gear	NO
Engine Speed	1112 RPM	3rd Gear	NO
Desired Idle	1050 RPM	4th Gear	NO
IAC Position	60	EGR 1	OFF
Coolant Temperature	25°C	EGR 2	OFF
Manifold Air Temperature	19°C	EGR 3	OFF
Oxygen Sensor	306 mU	Spark Advance	15 deg
Injector Pulse Width	2.7 mS	Knock Signal	NO
Fuel Integrator	128	Knock Retard	0 Deg
Block Learn	126	Canister Purge Duty Cycle	0 %
Block Learn Cell	0	Fan 1	OFF
Manifold Absolute Pressure	1.64 V	Fan 2	OFF
Barometric Pressure	4.82 V	A/C Request	NO
Throttle Position	0.64 V	A/C Clutch	OFF
Throttle Angle	0 %	A/C Pressure (H-Car)	0.00 V
Time From Start	00:00:59	Power Steering	NORMAL
Park/Neutral	P-N	Battery	14.4 V

Choose Scrl Dn or Scrl Up to view the next or previous page

First Sample: –40	**Current Sample: +0**	**Last Sample: +39**

Figure 14-13. *Typical snapshot data. This information is saved at the time a malfunction occurs. (General Motors)*

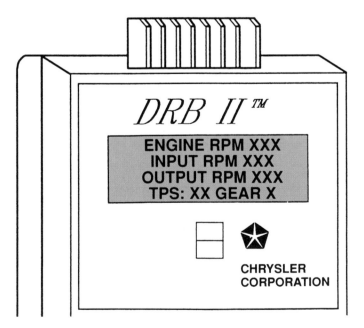

Figure 14-14. *Data displayed on the screen of a typical scan tool during a road test. The display shows engine, input shaft, and output shaft speeds, as well as the transmission gear and the throttle position (TPS reading). From this information, the technician can determine whether the transmission is slipping in a particular gear. (DaimlerChrysler)*

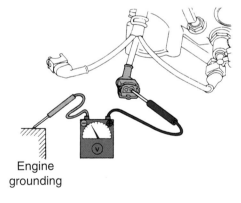

Engine grounding

Figure 14-15. *A voltmeter will check for voltage at a terminal. The voltmeter can also be used to check for current flow between two terminals of a connector. (Subaru)*

Caution: Never use a test light to check any electronic circuit unless specifically recommended by the manufacturer. The current flow through a test light is great enough to damage most electronic circuits.

Ammeters, such as the one in **Figure 14-17,** can measure amperage flows up to 10 amps. More amperage will damage the ammeter, so most modern ammeters are equipped with an inductive pickup, **Figure 14-18.** This pickup is clamped over the current-carrying wire. The pickup reads the magnetic field created by current flowing through the wire and converts it to an amperage reading.

Ohmmeters, **Figure 14-19,** are used to check for continuity or the presence of resistance. Wire resistance should be at or near zero. The resistance of an electronic part should be as specified in the service literature.

Modern voltmeters, ammeters, and ohmmeters are generally combined into a single unit called a multimeter. A multimeter used on any electronic system must have high impedance, or resistance to current flow. Many solid-state components, such as ignition modules and ECMs, can be severely damaged by careless use of test lights and multimeters. Specialized testers are often needed to check the operation of specific electronic devices or systems.

Waveform meters can be used to detect problems in sensors and solenoids. The technician connects the meter and observes the waveform produced by the operation of the suspected device. Then the technician compares the actual waveform with the standard waveform. The standard waveform is the waveform the device is supposed to produce when operating properly. If the waveforms do not

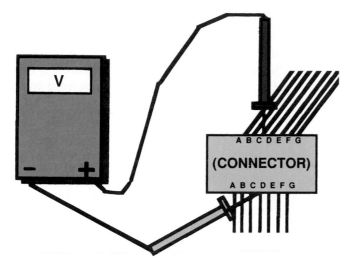

Figure 14-16. *Checking the voltage drop across a connector. More than a small voltage drop indicates a high-resistance connection. Actual maximum voltage drop varies from one circuit to another and depends on the amount of current flowing in the circuit. (General Motors)*

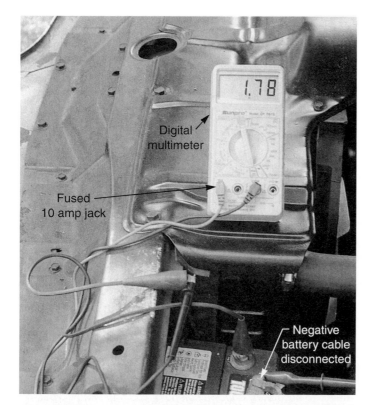

Figure 14-17. *Checking amperage draw. The ammeter will register the amount of current flowing in a circuit. Modern multimeters can read up to about 10 amps. (Fluke)*

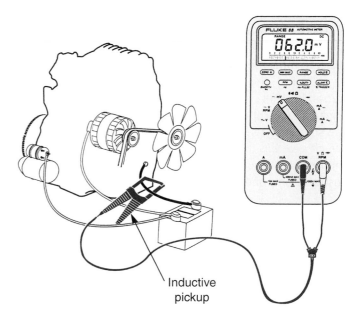

Figure 14-18. *An inductive pickup allows the meter to read a greater amount of amperage without subjecting the meter itself to high current. (Fluke)*

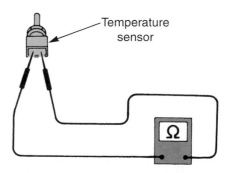

Figure 14-19. *Ohmmeters are commonly used to check temperature sensors. (Nissan)*

match, the device is usually defective. Occasionally, an incorrect waveform is caused by a related electrical problem instead of by the device itself. Therefore, even if the waveform is incorrect, it is important to make further checks before condemning the device. Many shops have lab scopes, which are similar to engine ignition oscilloscopes. Lab scopes perform the same functions as waveform meters. If you are unclear about the operation of waveform meters and lab scopes, refer to Chapter 3 for more information.

Temperature sensing tools can be used to determine the exact operating temperature of the engine or transmission. The actual temperature can be compared with the sensor readings recorded by the scan tool to determine whether the sensor is sending the correct signals to the ECM.

System Operation Testing

Once a suspect system has been isolated, the technician should determine whether that system and its components are working correctly. The easiest way to diagnose an electronic control system is by using the proper scan tool. If the transmission is operable, drive the vehicle and check the applicable scan tool readings. At the same time, observe transmission and transaxle operation.

During the test drive, note the transmission or transaxle shift pattern. Also note any slipping, harsh engagements, or noises. Try to match the malfunction with the scan tool readings. This will provide an indication of which gear has the problem and possible causes. With this information, you should be able to proceed to test the suspect components and systems.

Component Testing

This following section explains how to test individual electronic transmission and transaxle components. Before going on the individual component testing, be sure to read the general test information presented in the following section.

Types of Component Tests

There are three major ways to test electronic components: visual checks, electrical tests, and testing by substitution. General information on performing these tests is given in the following sections.

Making Visual Checks

You have already performed some visual checks as part of your preliminary diagnosis. If a problem is suspected in a specific system, check the system for disconnected or damaged wiring, obvious component defects, or problems in related areas. For instance, if your preliminary investigation indicates a problem with the MAP sensor, do not assume the sensor itself is bad. First, check the vacuum hoses to the sensor and all nearby

vacuum connections for damage (splits, cracks), kinking, and clogging. Make sure that hoses are not disconnected or connected to the wrong parts. Next, look at the wiring connector and check the sensor for physical damage.

Making Electrical Tests

 Caution: Some procedures call for energizing a solenoid or other device with battery power. Always consult the proper factory service manual to be sure that the device is designed to operate on full battery voltage. Some electrical devices will be destroyed if battery voltage is applied to them. When using jumper wires to operate a solenoid or other device, be sure that the wiring to the ECM is disconnected. Full battery voltage to some ECM connectors can destroy the ECM.

 Note: Before making ohmmeter checks, be sure that the device or wiring is disconnected from all sources of electrical power. Ohmmeter readings taken on an energized device or wire are useless. Electrical power can also damage the ohmmeter.

Solenoids can be tested by making a continuity check, **Figure 14-20**. Additionally, jumper wires can be used to apply power to the solenoid while observing its

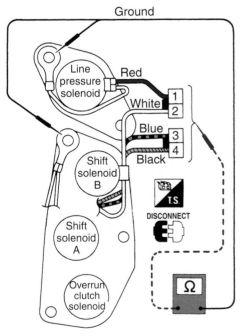

Figure 14-20. *Many solenoids can be checked with an ohmmeter. To accurately check a solenoid, you must know what the correct resistance reading should be. If the solenoid resistance does not match the specifications, the solenoid should be replaced. (Nissan)*

operation. Most solenoids do not have a visible plunger, and the technician must listen carefully for a click to determine whether the solenoid is operating. Some solenoids and other components can be checked through the wiring harness without removing the part from the transmission. Sometimes, the part can be checked though the transmission or transaxle wiring connector, and the oil pan does not need to be removed. However, you must refer to the service manual to determine which connector pin energizes which component, **Figure 14-21.**

Although most solid-state components require the use of specialized test equipment, basic checks can be made to some solid-state components. Common checks include using an ohmmeter to check a terminal for proper grounding. This check must be made very carefully to avoid damage to the unit. A few solid-state components can be checked for proper operation.

When checking any electrical component for proper voltage or resistance, consult the manufacturer's service literature to determine what the readings should be.

On some vehicles, the solenoid amperage draw can be read and compared to the manufacturer's specifications. Some pressure control solenoids can be checked by comparing amperage draw to the corresponding line pressure. A typical amperage-pressure chart is shown in **Figure 14-22.**

Testing by Substitution

In many cases, the only way to determine whether a component is defective is to replace it with a unit that is known to be in good working condition. When **testing by substitution,** however, the technician must keep the following things in mind:

❏ Obtaining a known good part may be expensive and time-consuming. In many cases, especially when substituting an electrical or electronic part, the part will not be returnable.

❏ If the actual problem is a short or high amperage draw somewhere else in the related circuit, the heavy current flow may ruin the replacement part.

❏ If the part requires a great deal of labor (time) to replace, this time will be wasted if the suspect part is not the problem.

Be sure to eliminate all other sources of a problem before substituting parts. Always consult the manufacturer's manual to determine whether a part can be tested before replacing it. Technicians can partially offset the problem of obtaining substitute parts by acquiring a stock of commonly needed replacement parts.

Testing Manual Valve Position Sensors

 Note: Some manual valve position sensors are called range sensors, PRNDL switches, or position switches.

An ohmmeter can be used to test the manual valve position sensor. Check the resistance at each switch position. Resistance should change as the shift lever is moved. See **Figure 14-23.** If resistance is not within specifications or does not change when the shifter is moved, the sensor should be replaced.

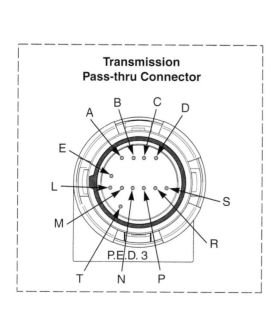

Cavity	Function
A	1-2 shift solenoid (low)
B	2-3 shift solenoid (low)
C	Pressure control solenoid (high)
D	Pressure control solenoid (low)
E	Both shift solenoids, TCC solenoid, and 3-2 control solenoid (high)
L	Transmission fluid temperature (high)
M	Transmission fluid temperature (low)
N	Range signal "A"
P	Range signal "C"
R	Range signal "B"
S	3-2 control solenoid (low)
T	TCC solenoid (low)

Figure 14-21. *To check an electronic transmission component through the case connector, you must know which connector pin goes to which component. A connector diagram, such as the one shown here, can be very helpful. (General Motors)*

Pressure control solenoid current (amp)	Approximate line pressure (psi)
0.02	170–190
0.10	165–185
0.20	160–180
0.30	155–175
0.40	148–168
0.50	140–160
0.60	130–145
0.70	110–130
0.80	90–115
0.90	65–90
0.98	55–65

Transmission manual lever position	Resistance (ohms)	
	Min	Max
P	3770	4607
R	1304	1593
N	660	807
Ⓓ	361	442
2	190	232
1	78	95

Figure 14-23. *Manual lever position sensors can usually be checked with an ohmmeter. One manufacturer's resistance specifications for various manual lever position sensor.*

Figure 14-22. *To make an amperage check of a solenoid on the vehicle, you must know what the current draw is supposed to be. Compare the correct reading in a table like this with actual current draw to determine whether the solenoid is good. (General Motors)*

Some manual valve position sensors can be checked with a scan tool. The sensor should be connected for this test. Follow the scan tool menu directions to test the sensor.

Testing Throttle Position Sensors

Before testing the throttle position sensor, determine what type of sensor it is. Most throttle position sensors are resistor types and can be tested with an ohmmeter. To make an ohmmeter check, disconnect the sensor wiring harness connector. Then attach an ohmmeter to the proper sensor leads. Operate the throttle and observe the ohmmeter readings. Many manufacturers specify checking the resistance at idle and wide open throttle positions. If the readings are incorrect, or if the ohmmeter readings do not change steadily as the throttle is opened, the sensor is defective.

Some throttle position sensors are transducers and must be checked with a voltmeter. A special adapter must be used to take readings with the wiring harness connected and the ignition switch in the *on* position. Testing procedures are similar to those for checking a resistor-type sensor.

> **Note: It is often easier to detect erratic throttle position sensor readings with an analog (needle-type) multimeter than with a digital meter. See Figure 14-24. The needle should move smoothly as the throttle is moved.**

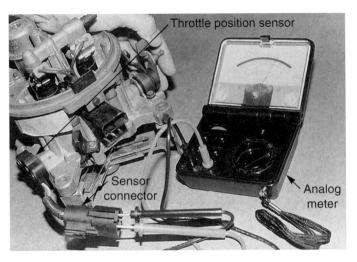

Figure 14-24. *While outdated for most automotive uses, an analog meter is useful for checking some throttle position sensors. Move the throttle while monitoring the meter.*

Some throttle position sensors can be adjusted if they are out of range. However, if the throttle position sensor gives erratic readings, it is defective and should be replaced.

Checking Speed Sensors

The simplest way to check a speed sensor is to use the proper scan tool. If the sensor is part of the distributor pickup assembly, the scan tool will be able to detect the ignition speed signal. Scan tools can also check the operation of speed sensors installed at the engine crankshaft or camshaft, as well as those located on the transmission or transaxle input shaft. Modern scan tools can read the speed sensor output as miles per hour. Comparing this to the actual speed of the vehicle will immediately tell the technician whether a vehicle speed sensor is operating properly. Some scan tools are designed to display speed sensor readings as drive shaft revolutions.

As was discussed in Chapter 12, speed sensors produce an alternating current (ac) output. Therefore, the operation of most vehicle speed sensors can be checked by measuring the output in ac volts. To make this test, raise the vehicle and disconnect the speed sensor at the transmission or transaxle.

> **Note: Vehicles with anti-lock brakes (ABS) or traction control may have speed sensors at each wheel. Consult the service manual to determine which speed sensors provide input to the transmission or transaxle.**

Next, set the multimeter to the proper ac voltage range. With the drive wheels off the ground, shift the transmission or transaxle into drive and accelerate. The multimeter should read increasing ac voltage as the engine speed is increased. Some manufacturers call for checking the resistance of the speed sensor winding as shown in **Figure 14-25.** This should generally be done only after other tests have indicated a sensor problem.

Before condemning the speed sensor, check the tip of the sensor for a buildup of metal shavings. Often the sensor will operate properly when the shavings are removed. Because they are magnetized, both internal and external sensors can collect metal particles.

Testing Pressure Sensors

Since most pressure sensors are on-off devices, they can be easily tested with an ohmmeter. A typical procedure for testing a pressure sensor is shown in **Figure 14-26.** With no pressure supplied, the sensor will be in its normal position, either normally open (NO) or normally closed (NC). If a normally closed sensor reads infinite resistance,

it is defective. If a normally open sensor has a low resistance reading, it too is defective.

However, even if a pressure sensor has the proper ohmmeter readings with no pressure applied, it may still be defective. It is always possible for a pressure sensor to fail and remain stuck in its normal (no pressure) position. If you suspect that a pressure sensor is not operating properly, the best way to check it is with a scan tool. Use the scan tool to access the pressure sensor signal. For example,

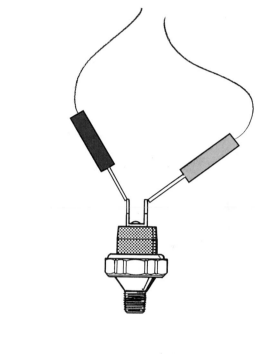

A

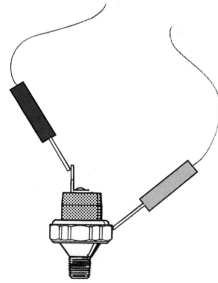

B

Figure 14-26. *An ohmmeter check of a pressure sensor. A normally open pressure sensor should have infinite resistance. A normally closed pressure sensor should have very low resistance. A—Connect the ohmmeter leads to the two terminals to obtain a reading. B—Connect the ohmmeter leads to the terminal and the sensor body.*

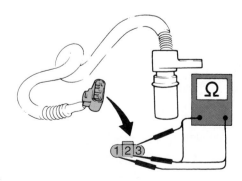

Figure 14-25. *Some speed sensors can be tested with an ohmmeter. You must have access to the proper resistance specifications. Other speed sensors can only be checked with a scan tool or an oscilloscope. (Nissan)*

if a pressure sensor should be closed in a particular gear but the scan tool does not indicate a closed signal when the unit is in that gear, the sensor is probably defective.

Testing Temperature Sensors

Temperature sensors make use of resistance changes to send a temperature signal to the ECU. Transmission temperature sensors are normally installed on the valve body or on the case inside the oil pan. Engine coolant temperature sensors are installed in such a way that they contact the engine coolant. They are located near the thermostat on many engines. Incoming air temperature sensors may be installed in the intake manifold, the plenum, or the ductwork connecting the air cleaner to the engine.

Checking temperature sensors with an ohmmeter is relatively simple. Unlike other common resistors, temperature sensors show a decrease in resistance as temperature increases. Therefore, both the resistance and temperature of this type of sensor must be monitored. **Figure 14-27** shows one method of checking a temperature sensor with an ohmmeter and a thermometer. The sensor has been removed from the vehicle and is being heated in a container of water. Sensor resistance should be checked at various temperatures. Another method involves using an infrared temperature gauge and observing the temperature of the sensor as the engine warms up. Check the resistance reading at various temperatures. Then compare the resistance and temperature readings to the manufacturer's specifications. If the readings do not closely match the specifications, the sensor is defective.

Another test can be made to check the temperature signal being sent to the ECM by the sensor. To make this test you must have a temperature tester and a scan tool. Use the temperature tester to measure the exact temperature at the sensor. Then use the scan tool to determine the temperature reading the sensor is sending to the ECM. If the readings do not match, the sensor is defective.

Note: Be sure to make sensor tests over a range of temperatures. Many sensors will read incorrectly only in certain temperature ranges.

Checking Manifold Air Pressure (MAP) Sensors

To check a MAP sensor, attach a voltmeter to the MAP sensor connections or to the proper MAP sensor wires at the ECM. Turn the ignition switch to the *on* position and apply vacuum to the sensor vacuum port as shown in **Figure 14-28.** Observe the voltage reading as the vacuum is developed. If the voltage reading increases with increases in vacuum, the MAP sensor is probably working properly. Many manufacturers publish figures that specify the correct MAP sensor voltage at various vacuum levels. If the MAP sensor does not attain these voltages, it should be replaced.

A similar test can be made to the barometric pressure (BARO) sensor on some vehicles. Check the output voltage at the proper ECU terminals (ignition on, engine not running) and compare it to the specifications for the altitude in your area. See **Figure 14-29.**

Checking Mass Airflow (MAF) Sensors

Regardless of their design, all mass airflow, or MAF, sensors produce one of three types of outputs, depending

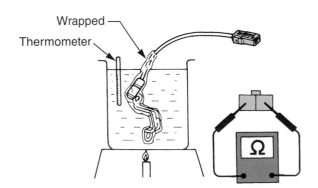

Figure 14-27. *Always test a temperature sensor at different temperatures. Place the sensor in water and gradually heat the water. Measure the resistance at different temperatures and compare the readings with factory specifications. (Nissan)*

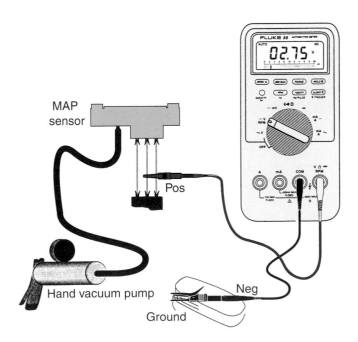

Figure 14-28. *One type of meter and vacuum pump setup for checking the output of a MAP sensor. Many MAP sensors can be checked with a scan tool. (Fluke)*

| Altitude | | Voltage range |
Meters	Feet	
Below 305	Below 1000	3.8–5.5v
305–610	1000–2000	3.6–5.3v
610–914	2000–3000	3.5–5.1v
914–1219	3000–4000	3.3–5.0v
1219–1524	4000–5000	3.2–4.8v
1524–1829	5000–6000	3.0–4.6v
1829–2133	6000–7000	2.9–4.5v
2133–2438	7000–8000	2.8–4.3v
2438–2743	8000–9000	2.6–4.2v
2743–3048	9000–10,000	2.5–4.0v

Low altitude = high pressure = high voltage

Figure 14-29. *An altitude compensation chart is needed to ensure that MAP and BARO sensor readings are correct for the altitude in your area. (General Motors)*

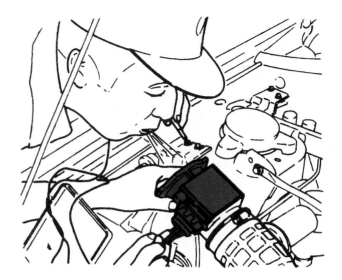

Figure 14-30. *Blowing through a MAF sensor will simulate the air that enters while the engine is running. (Nissan)*

on the manufacturer. These are analog dc voltage, low-frequency pulse, or high-frequency pulse. Older MAF sensors generally produce a dc voltage output, while many newer designs produce frequency pulse outputs.

It is sometimes possible to check the output of a dc voltage MAF sensor, but this must be done carefully to avoid damaging the electronic circuits. The safest way to do this is to use a scan tool to measure voltage changes while varying engine speed. Most MAF sensors have a voltage range of 0–5 volts. There are exceptions to this rule, however, and the technician should obtain the proper specifications before condemning the MAF sensor.

To avoid misdiagnosis and prevent damage to the MAF sensor and ECM, refer to the appropriate service manual for testing procedures. Always use a high imped-ance multimeter to avoid damaging the MAF or ECM.

The output of a frequency-type MAF sensor can be measured by most scan tools or by multimeters capable of reading RPM or duty cycles. Consult the multimeter manual to determine the exact meter settings. The frequency will be at a set value with the ignition on and the engine off. When the engine is started, increasing airflow through the MAF sensor will cause an increase in the frequency reading.

A variation of this procedure is shown in **Figure 14-30.** Disconnect the MAF sensor from the ducts. Then blow through the MAF sensor with the ignition on. Frequency should rise, indicating that the sensor is responding to air movement.

 Note: Always determine the type of MAF sensor you are dealing with before performing tests, since frequency-type sensors closely resemble dc output types.

Testing Oxygen Sensors

 Caution: Oxygen sensors are sensitive to excess current. Improper grounding or the use of test lights and low-impedance multi-meters can destroy them. Some manufacturers do not recommend meter tests of the oxygen sensor.

There are two basic types of oxygen sensors, the elec-trically heated and the non-heated. Oxygen sensors with one or two lead wires are non-heated types. If an oxygen sensor has three or more leads, it is a heated type.

Testing Oxygen Sensor Output

To test an oxygen sensor, it must be at a temperature of approximately 650° F (361° C). The sensor must be heated either by its heating element (when applicable) or by exhaust heat.

Begin by obtaining the correct specifications for the oxygen sensor that you are testing. Oxygen sensors use either Zirconia or Titania elements. A *Zirconia-type oxygen sensor* will produce a voltage reading, while a *Titania-type oxygen sensor* will produce a resistance reading.

After determining the type of sensor you are testing, set the multimeter to the proper range (usually 2V for a Zirconia sensor and 200KΩ for a Titania sensor). After the range is set, connect the positive (red) sensor lead to the sensor's signal wire and the negative (black) lead to ground. One- and three-wire sensors are grounded through the sensor housing. On two- and four-wire sensors, one of the wires is a ground wire. Be very careful to make the proper connections, as even slight voltage surges can ruin an oxygen sensor.

Make sure the sensor is at the correct operating temperature. Non-heated oxygen sensors should be heated by running the engine at high idle for at least 5 minutes. Then observe the multimeter readings. A rich mixture will cause a Zirconia sensor's voltage to increase, while causing a Titania sensor's resistance to decrease. A lean mixture, on the other hand, will cause a Zirconia sensor's voltage to decrease, while causing a Titania sensor's resistance to increase. If the sensor output responds quickly to changes in mixture ratio (within 1 to 3 seconds, depending on the manufacturer), the sensor is good.

Testing Oxygen Sensor Heater Circuit

The heater circuit of a heated oxygen sensor must be checked for proper resistance to ensure that the heater resistor is not burned out or shorted. To test the heater, set the multimeter to the 200KΩ range. Then connect the multimeter leads to the oxygen sensor heater terminals. Polarity is not important, but the leads must not contact the sensor signal terminal. If the resistance at the heater terminals is within specifications, the heater circuit is good.

 Note: If the above tests are made while the oxygen sensor is installed on the vehicle, running the engine with the sensor disconnected will probably set false trouble codes. After all tests are complete, be sure to clear the ECM memory.

Testing Output Solenoids

Once the scan tool has isolated a solenoid as a potential problem, the solenoid can often be tested by one of three methods. One method is to use an ohmmeter to measure the resistance of the solenoid windings. Obtain a high-impedance ohmmeter and set it on the required range. Then remove the solenoid wiring harness and measure the resistance. The resistance should be within the manufacturer's specified range. A few solenoids will read very low resistance. Resistance should never be zero, as this indicates a shorted winding. Infinite resistance indicates an open winding.

Using jumper wires to apply power to solenoids is another way to test their operation. An operating solenoid will make at least one click when it is energized.

Caution: Some solenoids are not designed to operate on 12 volts. Full battery voltage may destroy the solenoid. Some solenoids can be tested if a voltage-dropping resistor is inserted in the jumper wire circuit. When testing a solenoid, attach the jumper cables just long enough to listen for a click.

A third method of testing a solenoid is to use an ammeter to check for excessive current draw. Most solenoids will draw one amp or less. Always check the manufacturer's specifications for exact current draw.

Checking Electric Motors

Most electric motors used on modern vehicles can be checked with the appropriate scan tool. Scan tool use was discussed earlier in this chapter. The windings of some motors can be checked with an ohmmeter. Connect the ohmmeter to the input and output connectors, or to the connector and the motor body. As a general rule:
- ❏ If the resistance reading is zero or very low, the winding is shorted.
- ❏ If the ohmmeter reads infinite resistance, the winding is open.

If winding specifications are available, check that the motor winding has the proper resistance. Specifications are usually given as a range of resistance, for instance 500 to 800 ohms.

Checking Wiring

When troubleshooting a wiring problem, remember that for any electrical device to operate properly there must be a complete circuit. This means that voltage must be available to the device, the device must be in operating condition, and the electrical circuit must be completed through a good ground.

Always obtain the proper wiring diagram when checking for a wiring problem. The correct wiring diagram will lead you to the most obvious connectors and trouble spots. See **Figure 14-31.**

Never assume that a connector is good just because it looks good on the outside. Separate any suspicious connectors and make a careful visual inspection for bent or corroded pins. Also, check for pins that have been pushed out of their holders. Make sure there are no signs of overheating on the connector. Also make ohmmeter checks for continuity. When using a diagnostic chart, such as the one in **Figure 14-32,** always follow procedures exactly.

Carefully inspect wire splices, **Figure 14-33.** Also, look for wires that are rubbing on moving parts or insulation that has melted on hot exhaust system components. Either of these conditions can cause a short that affects system operation and may destroy the ECM. The wiring harness in **Figure 14-34** must pass through a confined area between the body and the transmission. In this situation, even slight movement or incorrect positioning can cause a problem.

Testing the ECM

Some manufacturers have ECM testing procedures. Most of these procedures involve checking ECM voltage outputs. **Figure 14-35** shows an ECM being tested for proper voltage at the wiring harness. This procedure must be done very carefully to prevent ECM damage. It is usually safer to check for correct voltages at the sensors rather than at the ECM.

Most ECMs can only be tested by substitution. As was discussed earlier, this should be done after careful checking has determined that the ECM is the only possible source of the problem. Some manufacturers have special testers for checking the ECM. On most vehicles, however, the scan tool can be used to test the ECM.

Most modern ECMs have a provision for changing the information in the programmable read only memory

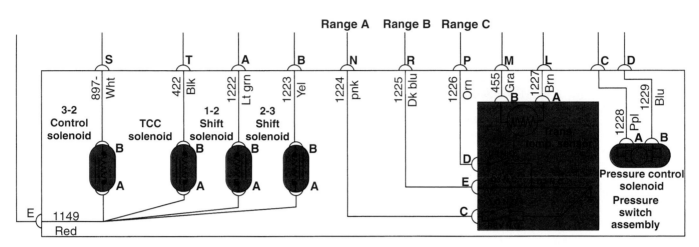

N.C. = Normally closed switch
N.O. = Normally open switch

Figure 14-31. *To properly check the electrical system of an electronic transmission or transaxle, the technician must have access to the proper electrical schematic. (General Motors)*

C8 Check internal axode harness (continuity)	Results	Action to take
• Disconnect the internal harness from the soleniod (MCCC/CCC wire connector) **NOTE:** Do not probe into connector terminals. • Connect the positive lead from an ohmmeter to the tester MCCC/CCC jack and negative lead at the Black (91/92 MY) or Brown (93 MY) wire at the MCCC/CCC connector. • Record resistance. Should be less than .5 ohms. • Next, connect the positive lead from an ohmmeter to the tester VPWR jack and the negative lead at the Red (91/92 MY) or Green (93 MY) wire of the MCCC/CCC connector. • Record resistance. Is the resistance less than .5 ohms?	Yes No	Go to C9. Replace internal harness. Go to C10.
C9 Check internal axode harness • Check for continuity between BAT – (engine ground) and the appropriate wire with an ohmmeter or other low current tester (less than 200 milliamps). | Wire color		
Solenoid	91/92	93
MCCC/CCC	Black	Brown
VPWR	Red	Green | • Connection should not show continuity (infinite). • Is there continuity?	Yes No	Replace internal harness. Go to C10. Go to C10.

Figure 14-32. *This diagnostic chart is used when checking a wiring harness on a common automatic transaxle. Note that, even though the second procedure (C9) is a check for continuity (a complete circuit), an ohmmeter must be used. Using an ohmmeter instead of a test light ensures that the internal transaxle parts will not be damaged. (Ford)*

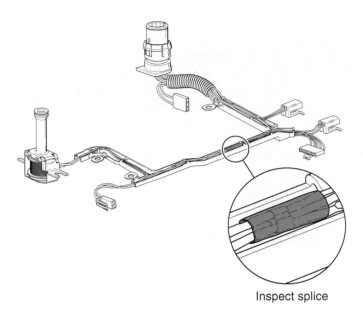

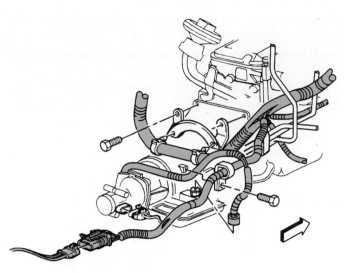

Figure 14-34. *A close-fitting wiring harness is subject to damage. Wiring harnesses should be inspected in areas where they pass through tight spots and near engine and exhaust system parts. (General Motors)*

Inspect splice

Figure 14-33. *Splices are a common source of problems. Carefully inspect the splice for looseness or signs of overheating. (General Motors)*

(PROM). Often, the original ECM programming was not perfectly calibrated to the vehicle. Sometimes actual operating conditions cause minor drivability problems with the original PROM programming. When this occurs, the manufacturer issues updated PROM information. The only way to determine whether updated information will cure a transmission or transaxle problem is to check the manufacturer's service bulletins or other update information.

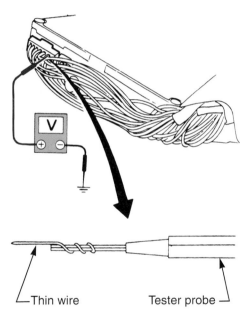

Thin wire   Tester probe

Figure 14-35. *One method of checking voltage output from the ECM. Be extremely careful when making checks like this, as the current flow through the meter could damage the ECM. (Nissan)*

Deciding on Work Needed

After locating a defective part, never assume that it is the root cause of the problem. Always try to determine what caused a part to fail. For example, if the ECM is defective, test all sensors and output solenoids for excessive current draw that could have caused the module to fail. Careless or incomplete diagnosis can allow further damage to the electronic components. Sometimes, a missed electronic defect can cause the transmission or transaxle to slip, destroying the holding members.

Summary

Many electronically controlled transmission or transaxle problems are the same as those encountered in hydraulically controlled transmissions. Other problems are unique to the electronic control system. Many computer control system problems are referred to as pattern failures, since they occur regularly on a particular system.

In many cases, the transmission or transaxle is reacting to problems in the engine. Sensors are the most common cause of electronic transmission/transaxle problems. Many sensor defects first show up as engine performance problems.

If a solenoid controls an entire system, failure may cause a transmission defect in all gears. If a defective solenoid controls only one gear, the problem will show up in that gear only.

A failed ECM can cause slipping, hard shifting, improper shift speeds, shudder during shifts, and other problems. A failed ECM may also cause engine drivability problems.

Wiring and connector problems are common sources of computer system problems. The voltages used to operate the sensors are usually much lower than battery voltage, and a slight increase in resistance can cause problems. Defects in the ground circuit are commonly overlooked.

The seven-step diagnosis process is very helpful in determining the exact cause of a problem. It is sometimes difficult whether the problem is in the electronic control system or in the basic hydraulic and mechanical components. Always make the same preliminary checks that you would with hydraulically controlled transmission and transaxles when troubleshooting an electronically controlled transmission or transaxle.

Before performing complex electronic tests, make some basic electrical checks. Check fuses, wiring, and ground wires. If these checks do not reveal a problem, use a scan tool to check the system. The scan tool allows the technician to access trouble codes and obtain information on shift speeds, converter clutch application speed, and transmission/transaxle temperatures and pressures. The scan tool can also provide information on engine RPM and temperature readings.

Multimeters are used to read electrical values such as voltage, resistance, and amperage. Many multimeters can read the waveforms produced by the system components as they operate. Lab scopes can also produce these readings.

The easiest way to diagnose the operation of an electronic control system is by using the proper scan tool. If possible, drive the vehicle and observe the scan tool readings. During the test drive, note the transmission or transaxle shift pattern, as well as any slippage or noises.

To test individual electronic transmission and transaxle components, you should make visual checks, perform electrical tests, and test by substitution. Begin by looking for disconnected or damaged wiring, obvious component defects, or problems in related areas. Then make electrical tests to sensors and output devices as needed. As a last resort, test by substituting a known good part.

Most throttle position sensors can be tested with an ohmmeter. Some throttle position sensors are transducers and, therefore, must be checked with a voltmeter.

To check a speed sensor, use a multimeter to check for the presence of an ac waveform. Speed sensors can also be checked with a scan tool. Some manufacturers recommend checking the resistance of the speed sensor winding.

Pressure sensors can usually be checked with an ohmmeter. If a normally closed sensor reads infinite resistance with no pressure applied, the sensor is defective. If a normally open sensor reads low resistance with no pressure applied, it is defective.

To test temperature sensors, check the resistance reading at various temperatures. If the readings are not within specifications, the sensor is defective. Many sensors read incorrectly only in certain temperature ranges.

To check a MAP sensor, attach a voltmeter and vacuum pump to the sensor. If MAP sensor voltage reading increases with increases in vacuum, the MAP sensor is probably good. There are several types of mass airflow (MAF) sensors. Test procedures vary with each type. Always disconnect the MAF wiring to make readings.

Testing oxygen sensors requires care, since oxygen sensors are sensitive to excess current flow. Some manufacturers do not recommend meter tests of the oxygen sensor. Oxygen sensors use either Zirconia or Titania elements. Oxygen sensors with one or two lead wires are non-heated types. If an oxygen sensor has three or more leads, it is a heated type. The sensor must be at its operating temperature before testing it.

To test an output solenoid, remove the solenoid wiring harness and measure solenoid resistance. Some solenoids can be operated with jumper wires. Full battery voltage may destroy other solenoids, so always check the manufacturer's specifications before trying this test.

Schematics are useful when looking for defective wires, connections, and devices. Always obtain the proper wiring diagram when checking for a wiring problem.

Some manufacturers have procedures or equipment for testing for proper ECM voltage outputs. On most vehicles, a scan tool can be used to test the ECM. The ECM may have to be tested by substituting a known good unit and rechecking system operation.

Always double-check your results before replacing a defective component.

Review Questions—Chapter 14

Please do not write in this text. Place your answers on a separate sheet of paper.

1. Electronically controlled transmission or transaxle problems often resemble those that occur in _____ transmissions.

2. A pattern failure is a _____ type of problem.

3. _____ are the most common cause of electronic transmission/transaxle problems.

4. On-off _____ _____ can stick open or closed.

5. On a few electronic transmissions, some shifts are _____ controlled.

6. Engine missing is often confused for lockup torque converter _____.

7. Do not remove any _____ until you have retrieved trouble codes from the ECM.

8. Where are fusible links usually located?

9. Scan tools can be thought of as a portable _____ that can communicate with the vehicle's ECM.

10. OBD I computer systems can have a maximum number of _____ trouble codes. The OBD II system can have as many as _____ potential codes.

11. The snapshot is a scan tool feature that records vehicle operating conditions just before and just after a _____ occurs.

12. The three ways of checking computer control systems are _____ checks, _____ tests, and testing by _____.

13. Speed sensors produce a(n) _____ current output.

14. If a normally closed (NC) sensor reads zero resistance, it is _____.

15. If the MAP voltage reading increases with increases in _____, the sensor is probably OK.

16. The three types of MAF sensor outputs are analog _____ voltage, low _____ pulse, or high _____ pulse.

Matching

Match the following test equipment or testing method with the device that it is best at testing.

17. Jumper wires _____   (A) Temperature sensor
18. Ohmmeter _____   (B) Speed sensor
                             (C) Output solenoid
19. Substitution _____   (D) ECM
20. Waveform meter _____   (E) Throttle position sensor

ASE-Type Questions—Chapter 14

1. Technician A says that electronically controlled transmission or transaxle problems are always caused by the electronic control system. Technician B says that an electronically controlled transmission or transaxle may have some of the same problems as a hydraulic unit. Who is right?
   (A) A only.
   (B) B only.
   (C) Both A and B.
   (D) Neither A nor B.

2. A defective throttle position sensor can cause all the following electronic transmission problems except:
   (A) harsh upshifts.
   (B) downshift bump.
   (C) converter vibration.
   (D) improper shift speeds.

3. Technician A says that a transaxle suffering from gear skipping may shift from first to third gear. Technician B says that erratic shifting is a symptom of an intermittently sticking solenoid. Who is right?
   (A) A only.
   (B) B only.
   (C) Both A and B.
   (D) Neither A nor B.

4. A problem thought to be in the transmission may actually be in the _____.
   (A) ignition system

(B) fuel system
(C) tires
(D) All of the above.

5. All the following are OBD II trouble code letters except:
   (A) B (body).
   (B) T (transmission/transaxle).
   (C) C (chassis).
   (D) P (power train).

6. Each of the following statements about temperature sensor testing is true except:
   (A) temperature sensors show a decrease in resistance as temperature increases.
   (B) transmission temperature sensors are normally installed on case, outside of the oil pan.
   (C) temperature sensors are tested with a voltmeter.
   (D) the resistance and temperature of a temperature sensor must be monitored.

7. Testing a barometric pressure (BARO) sensor is similar to testing a _____ sensor.
   (A) MAF
   (B) MAP
   (C) speed
   (D) oxygen

8. Most ECMs can only be checked by:
   (A) using jumper wires to energize the ECM drivers.
   (B) checking ECM input with a test light.
   (C) checking ECM terminals with an ohmmeter.
   (D) substituting with a known good ECM.

9. Updated _____ information may cure some driveability problems.
   (A) PROM
   (B) RAM
   (C) speed sensor
   (D) solenoid winding

10. Technician A says the careful technician will not replace a part without checking for the cause of its failure. Technician B says that the only way to get rid of the vehicle is to replace the first thing that appears to correct the problem. Who is right?
    (A) A only.
    (B) B only.
    (C) Both A and B.
    (D) Neither A nor B.

Chapter 15

Transmission and Transaxle In-Vehicle Service

After studying this chapter, you will be able to:
- ❏ Change transmission/transaxle oil and filter.
- ❏ Highlight important points regarding transmission fluid service.
- ❏ Adjust automatic transmission linkages and bands.
- ❏ Perform service operations on valve bodies.
- ❏ Perform governor service.
- ❏ Perform service operations on modulators.
- ❏ Fix external fluid leaks.
- ❏ Order needed transmission/transaxle parts.
- ❏ Repair bad servos.

Technical Terms

Varnish formation

Adjuster

Alignment pin

Production label

Vehicle identification number

Parts catalog

Part number

Shift kits

Servo-cover depressor

Servo release port

Introduction

Automatic transmission service is a specialized activity that requires extensive knowledge and careful work habits. This chapter will outline the automatic transmission/transaxle service procedures that can be performed without removing the unit from the vehicle. If you have studied the information in previous chapters, you will be familiar with the components and tools discussed in this chapter.

Remember that the information in this chapter is general in nature. Since there are many types of automatic transmissions and transaxles, you should always consult the appropriate factory service manual for the particular unit you are working on.

Quite often, an automatic transmission will not require a complete overhaul. The subassembly repair techniques in this chapter can be used to service many transmission or transaxle parts without the removal or complete teardown of the unit.

There are a number of service procedures that can be performed without removing the unit from the vehicle. Obviously, checking fluid level and condition is one; so is changing the fluid and filter. Adjustments to linkages, the neutral start switch, and the transmission bands are also in-car service procedures.

There are still other service procedures that can be performed while the transmission or transaxle is installed in the vehicle. The speedometer pinion gear, rear seal, vacuum modulator, oil cooler lines, extension bushing and output shaft bearing, governor, parking gear and parking lock components, valve body, and accumulator piston can all be serviced without removing the unit from the vehicle. See **Figure 15-1.** Many of these procedures are covered in Chapter 17, rather than in this chapter.

Checking Fluid Level and Condition

Like engine oil, automatic transmission fluid should be checked at specified intervals. Many manufacturers recommend *changing* the fluid at specified intervals. The fluid can become contaminated with metal, dirt, water, and friction material from internal parts. Contaminated fluid can cause rapid part wear and premature transmission or transaxle failure.

The most common hydraulic system problem is a low fluid level. When the fluid level is low, air from the oil pan can enter the hydraulic system through the pump intake. When the hydraulic system contains air, hydraulic power will be wasted compressing it. The air will compress instead of moving valves and holding members. The unit will shift erratically, or it may slip or fail to move the vehicle. The remaining fluid will become overheated. Proper lubrication will not be maintained, and parts will be damaged.

If a transmission is overfilled, the turning planetary gears will churn the transmission fluid into foam. The foaming fluid contains air, causing the same hydraulic system problems as a low fluid level. In addition, foaming can result in fluid escaping from the transmission vent, where it may be mistaken for a leak.

In addition to the problems previously mentioned, the introduction of air into the ATF results in severe and rapid oxidizing of the fluid, which seriously affects the fluid's properties. One result of fluid oxidation is **varnish formation.** This problem appears as a sticky, burned-on substance that resembles furniture-refinishing varnish.

To check the fluid level, locate the filler tube and dipstick. Remove the dipstick and wipe it clean. Then reinsert it into the filler tube until fully seated. Finally, pull the dipstick from the tube and observe the level reading.

Figure 15-2 shows typical markings on a dipstick. The fluid level should be checked as specified on the dipstick or in the owner's manual: generally at normal operating temperature with the engine running, the shift selector lever in park or neutral, and the vehicle on level ground. Some dipsticks are marked so that the level can be checked with the fluid hot or cold. Some transmission and transaxles will have different readings in neutral and park, so always check with the shift selector lever in the position recommended by the service manual.

Fluid levels in automatic transaxles are sometimes hard to read accurately. Many transaxle dipsticks will read high when the fluid is cold, but after the fluid heats up, the dipstick will read less than full or even low. On some transaxles, fluid level cannot be read accurately immediately after fluid is added. Before checking fluid level in these units, the vehicle must be driven to restore the normal fluid level.

The best procedure for checking fluid level in the automatic transaxle is to drive the vehicle several miles, park it on a level surface, and then check the fluid without turning the engine off.

 Note: If the transaxle has a separate reservoir for differential lubricant, this level should also be checked.

While checking automatic transmission fluid level, the condition of the fluid should also be examined. If the fluid smells burnt and is orange or dark brown, it has been overheated. The clutch and band friction linings are probably burned and glazed, impairing transmission or transaxle operation. Changing the fluid will not solve the problem.

If the fluid is milky looking, it is contaminated with water or coolant. Water enters when the vehicle is operated in very deep water. A leak in the oil cooler will allow coolant to leak in from the engine cooling-system radiator. Water or coolant will ruin the clutch and band friction linings.

Transmission Service Guide

Transmission component	Serviced with transmission in vehicle	Transmission must be removed from vehicle
Accumulators	X	
Bands		Most makes
Clutch pistons		X
Clutch plates		X
Converter		X
Electrical switches	X	
Extension housing	X	
Filter	X	
Governor	X	
Input shaft		X
Manual linkage	X	
Modulator	X	
Modulator valve	X	
Oil pan	X	
Output shaft		X
Parking gear linkage	X	
Planetary gears		X
Pump or pump seal		X
Seal rings		X
Servos	X	
Speedometer gear	X	
Throttle linkage	X	
Throttle valve	X	
Valve body	X	

Figure 15-1. *As shown in this chart, numerous service procedures can be performed on an automatic transmission without removing the transmission from the vehicle.*

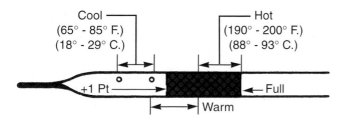

Cool
(65° - 85° F.)
(18° - 29° C.)

Hot
(190° - 200° F.)
(88° - 93° C.)

+1 Pt

Full

Warm

Note: Do not overfill. It takes only one pint
to raise level from "add" to "full"
with a hot transmission

Figure 15-2. *The fluid level is important to the proper operation of a transmission or transaxle. Both hot and cold levels are given on most dipsticks. On the dipstick shown, for example, fluid is at an acceptable level if it falls within the ranges shown at the respective temperatures. (General Motors)*

Automatic Transmission and Transaxle Adjustments

There are several adjustments that can be made with the transmission or transaxle installed in the vehicle. These relatively simple adjustments will solve many problems. The most common adjustments are made to the following components:

❑ Shift linkage.
❑ Throttle, TV, and kickdown linkages.
❑ Electric detent switches.
❑ Transmission bands.
❑ Neutral start switch.

Shift Linkage Adjustment

Shift linkage adjustments should be done whenever the shift selector lever position does not correspond to the actual transmission gear selection. The inner manual lever, which holds the manual valve in the proper position, is not adjustable. The only adjustment that can be made to the shift linkage is between the outer manual lever and the shift selector lever. If this adjustment is off, the transmission may not be in the gear position the driver thinks it is. **Figure 15-3** shows typical shift linkage in both column- and floor-shift versions. **Figure 15-4** illustrates the adjustment of a cable shift linkage used on some vehicles with floor-mounted shifters.

Before attempting to make adjustments, inspect the linkage for loose or worn parts. Tighten or replace worn parts before making the adjustment. Check shift and throttle cables for binding or damage to the sheath. If the cable appears to be damaged from overheating, check the engine ground straps. Disconnected ground straps can cause the vehicle's electrical equipment to ground through the cable, melting the sheath.

Exact procedures for adjusting the shift linkage of an automatic transmission vary. Generally, you must make sure the position of the outer manual lever is synchronized with that of the shift selector lever. For example, if the selector lever is set in park, the outer manual lever must also be in the park position.

If you are sure the shift linkage is in good shape, place the selector lever in park. Then, loosen the locknut holding the *adjuster*, which may be an adjustment swivel, a bracket, or a slot in the linkage rod. Some adjusters are under the vehicle, while others are located under the dash.

Once the locknut is loose, make sure the shift selector indicator is still in park. Then move the outer manual lever on the transmission case all the way to the park, or rear detent, position. With both the shift lever and the outer manual lever in park, tighten the locknut and recheck shift linkage operation. Although this procedure is typical for many transmissions, some manufacturers will specify that the selector lever and outer manual lever be placed in a position other than park during adjustment. See **Figure 15-5**.

Throttle, TV, and Kickdown Linkage Adjustments

Linkages should be adjusted whenever the transmission shift points are not correct. The problem could be high or low shift points. On transmissions without a vacuum modulator, the throttle valve (TV) linkage, which is operated by the throttle linkage, controls shift timing and pressures. In some cases, the TV linkage also provides a detent for passing gear. If a transmission has a separate detent valve for passing gear, it may be operated by a kickdown linkage through the throttle linkage. Adjustment procedures are similar for both the TV and kickdown linkage.

A typical TV linkage assembly is shown in **Figure 15-6**. Note that some assemblies have threaded swivels that are used for adjustment. Always check the condition of linkage parts before making any adjustments and replace defective parts.

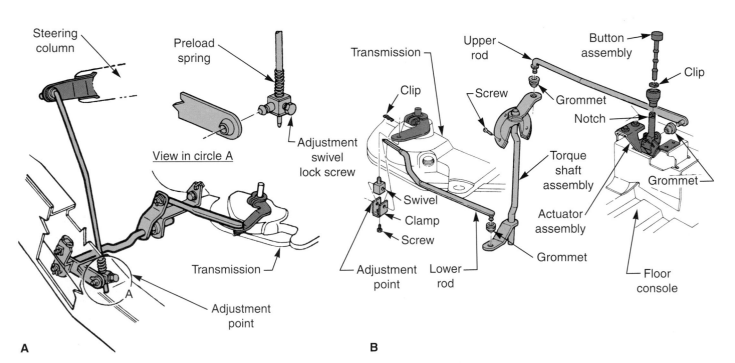

A   **B**

Figure 15-3. *Note how automatic transmission shift linkage is designed in a common vehicle. Other linkage systems are similar. Adjustment of these linkages is made by means of an adjustment swivel. A—Linkage with column-shift selector lever. B—Linkage with floor-shift selector lever. (DaimlerChrysler)*

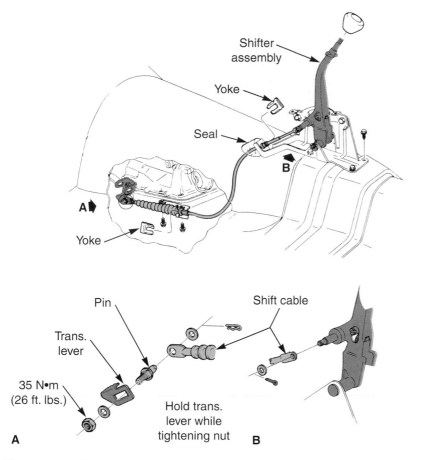

Shift cable adjustment
1. Loosen shift rod clamp screw. Loosen pin in transmission manual lever.
2. Place shift lever in "P" position. Place transmission manual lever in "P" position and ignition key in lock position.
3. Pull shift rod lightly against lock stop and tighten clamp screw.
4. **Move pin in manual transmission lever to give "free pin" fit and tighten attaching nut.**
5. Check operation:
   A. Move shift handle into each gear position and see that transmission manual lever is also in detent position.
   B. With key in "run" position and transmission in "reverse," be sure that key cannot be removed and that steering wheel is not locked.
   C. With key in "lock" position and transmission in "park", be sure that key can be removed and that steering wheel is locked.
   D. Engine must start in park and neutral.
   E. With key in the "lock" position, there is to be no rotation of the column shift bowl when the shift lever T-handle is released and the lever moved toward reverse.

Figure 15-4. *Many floor shifters and a few column shifters use a cable to connect the shift lever to the manual valve. Adjustment procedures vary. A common adjustment is shown here. (General Motors)*

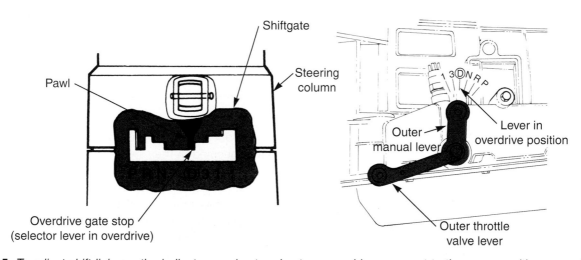

Figure 15-5. *To adjust shift linkage, the indicator quadrant and outer manual lever are set to the same position—overdrive in this particular transmission. The locknut is then tightened, and correct operation is verified in all selected positions. (Ford)*

If the linkage is in good condition, start linkage adjustment by having a helper press the throttle pedal completely to the floor (engine off). Next, check the throttle plate(s) in the throttle body (or carburetor). They should be completely open. If they are not, adjust the throttle linkage, which is the linkage between the throttle pedal and throttle plates, until the throttle plates open fully when the throttle pedal is fully depressed. If adjusting the throttle linkage fails to remedy the problem, the TV linkage may be the cause. Disconnect the TV linkage and recheck the plate opening. If the throttle plate still does not open completely, reconnect the TV linkage and recheck the adjustment. Also check for damaged throttle linkage between the throttle plate and the accelerator pedal, as well as for debris or bunched carpeting under the accelerator pedal.

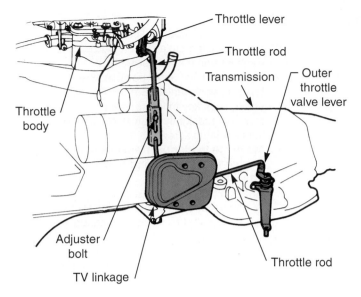

Figure 15-6. *This TV linkage is adjusted by first loosening the adjuster locknut and then lengthening or shortening the linkage as needed. Always road test the vehicle after any linkage adjustments are made. (DaimlerChrysler)*

After rechecking the throttle linkage, loosen the adjustment locknut on the TV linkage (or disconnect the swivel retaining clip). Do this while your helper continues to hold the throttle pedal completely to the floor. Then, pull the TV linkage at the transmission to the wide-open throttle position. (Pull in the direction that the linkage moves when the throttle pedal is depressed.) While holding the linkage in the wide-open throttle position, tighten the locknut (or reinstall the swivel and retaining clip). As a final step, road test the vehicle to check the adjustment.

TV cable adjustment is similar to rod-type linkage adjustment. However, most TV cable adjusters are located at the throttle lever of the throttle body or carburetor, and most adjustments can be accomplished by one person. Many adjusters consist of a cable holder that snaps up to release the cable. Many TV cables are self-adjusting—simply pressing the throttle pedal to the floor completes the adjustment. See **Figure 15-7.**

Electric Detent Switch Adjustment

If the vehicle uses an electrically operated detent switch, the switch must be checked and, if necessary, adjusted. Many switches can be checked using a test light. Turn the ignition switch to the *on* position, but do not start the engine. Then check that the switch has power to one terminal. Next, press the accelerator to the floor and check for the presence of voltage at the other terminal. If the switch has more than two terminals, consult the service manual to determine which terminals should have power. If the switch has no power, check the fuse and wiring. If the switch has power to both terminals at all times, check for shorted wires. A common cause of shorts is a fire or

exhaust heat that has melted the input and output wires together.

If the switch seems to be in good condition, check that it is properly adjusted. Some switches are adjusted by loosening the locknuts and moving the switch into the proper position. See **Figure 15-8.** If the switch is installed at the accelerator pedal, it can often be adjusted by bending its mounting bracket. Be sure that the switch is energized during approximately the last 10% of pedal travel only.

Transmission Band Adjustment

Although adjusting a band will not cure severe slippage, it is worth trying in cases where the shift is becoming soft. On some transmissions, band adjustment should be done as part of routine maintenance. There are many different types of band apply systems and many methods of adjustment. **Figure 15-9** shows a typical band adjusting screw. Other bands can only be adjusted by changing the length of the band apply pin. This is covered in more detail in Chapter 17. Some bands are not adjustable.

Begin band adjustment by loosening the locknut on the band adjustment screw. (If the adjuster is inside the oil pan, the pan must be removed first.) Next, tighten the band adjustment screw until the band is snug against the drum, counting the number of turns required. If the number of turns in is equal to the factory specified number of turns out, the original adjustment was correct.

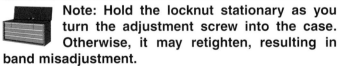

Note: Hold the locknut stationary as you turn the adjustment screw into the case. Otherwise, it may retighten, resulting in band misadjustment.

Torque the band adjustment screw to the exact torque value specified in the service manual. Afterward, back the screw off the exact number of turns specified in the manual. In some cases, it may be possible to back the screw off less than specified to get a tighter band adjustment and a firmer shift. However, never vary more than one quarter to one half turn from factory values. Once the band is adjusted, the locknut should be tightened to keep the band adjustment screw from turning. To avoid changing the adjustment, be sure to hold the adjustment screw in place while tightening the locknut.

Neutral Start Switch Adjustment

The neutral start switch, for floor-shift and a few column-shift designs, should be adjusted when the vehicle will *not* start in park or neutral or when it *will* start in any *other* gear. After making sure the shift linkage is properly

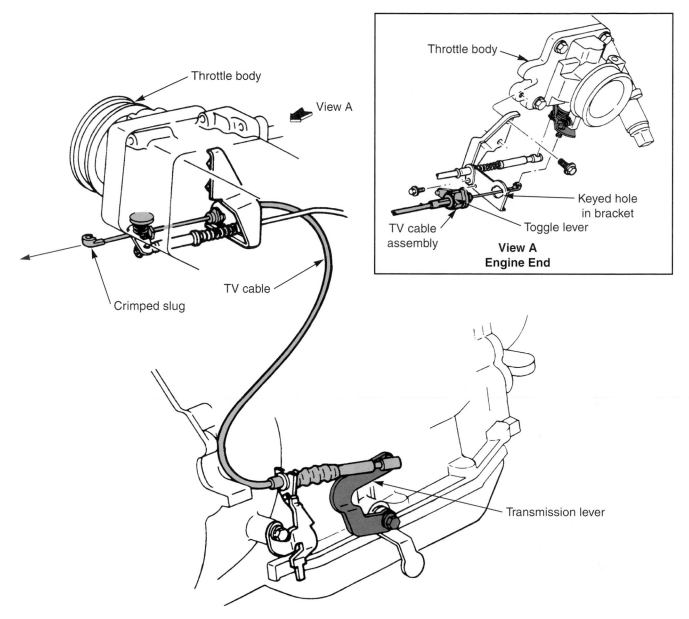

Figure 15-7. *Most modern vehicles without electronic shift controls use a cable to connect the throttle valve to the engine throttle body. The linkage is almost always adjusted at the throttle body, usually with a simple adjusting device. Many throttle cables are self adjusting. (Ford)*

adjusted, begin neutral start switch adjustment by locating the switch and loosening its holddown. Then, place the shift selector lever in neutral, apply the parking brake, and try to start the engine. If the starter will not operate, rotate the neutral start switch in small steps until the starter operates when the ignition switch is turned. Note that it may be easier to have a helper hold the ignition switch in the start position as you move the neutral start switch.

The switch is sometimes adjusted with an ***alignment pin,*** **Figure 15-10.** The procedure is rather simple. Loosen attaching bolts, place transmission in neutral, and then rotate the switch to align the adjustment holes. Insert the alignment pin in the holes to the specified depth. Tighten the switch holddown and remove the alignment pin.

Once the engine starts in neutral, check operation in all other shift quadrant positions. The engine should crank in neutral and park, but in no other positions. Readjust the switch as needed and tighten it down.

Ordering Transmission and Transaxle Parts

Faulty transmission or transaxle parts should be identified and replacement parts ordered as soon as possible to prevent unnecessary delays. One of the important factors in successful transmission/transaxle service is obtaining the

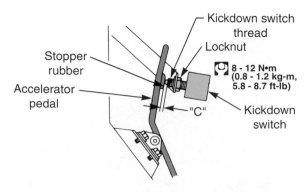

Figure 15-8. *Electric detent (passing gear) switches are usually installed on the firewall inside the passenger compartment. The switch may have a threaded adjustment as shown here, or it may have a slotted mounting bracket. The switch can be adjusted so that it closes to energize the detent solenoid when the accelerator pedal is opened to about 10% of wide-open throttle. (Nissan)*

Figure 15-9. *Band adjusting screws are usually installed on the outside of the case. On a few transmissions and transaxles, the oil pan must be removed to reach the adjusting screw.*

proper replacement parts. There are several reasons for this. To begin with, there are many types of transmissions and transaxles. Manufacturers use over 100 different transmissions and transaxles in their vehicles. In addition, the original design of a transmission or transaxle is often changed, and many manufacturers produce several variations of the same unit. Changes are often made at the beginning of a model year, and sometimes during production runs.

Modern transmissions and transaxles also tend to wear out hard parts, such as pumps, drums, planetary gears, shafts, and center supports. Hard parts for the same transmission or transaxle often vary between model years and engine sizes, even when the clutch plates and gaskets are the same.

In addition, late-model transmissions and transaxles use many electrical and electronic devices. Some electrical and electronic parts commonly fail after overhaul. Technicians usually replace these parts to prevent future

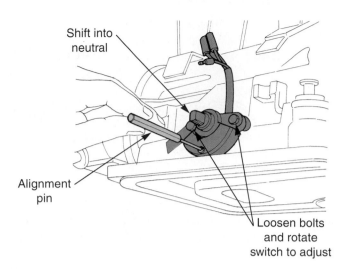

Figure 15-10. *An alignment pin makes neutral start switch adjustment easier. Some neutral start switches are adjusted without using an alignment tool. (Nissan)*

problems. Electrical and electronic devices vary from one vehicle to another. Identical models of one popular rear-wheel drive transmission can take one of 19 different torque converter lockup solenoids.

Torque converters are also ordered more than they were in the past. Many shops replace the torque converter instead of attempting to clean it, especially when the transmission or transaxle is very dirty.

Given all these factors, how does the technician obtain the proper parts? One way is through experience. Seasoned technicians instantly recognize a transmission or transaxle by such visual clues, such as case shape and size, and external features, such as reinforcing ribs or the placement of linkage, cooler lines, or modulators. Sometimes, however, a unit that looks like all the others but is slightly different internally will fool even an experienced technician. There are methods, however, of determining the exact transmission or transaxle being serviced. The following steps will enable you to obtain the proper transmission or transaxle parts.

Determining the Transmission Type

Before attempting to order parts, determine exactly what transmission or transaxle is being serviced. A good general guide to transmission or transaxle type is the shape of the bottom pan. Some manufacturers stamp the name or model number of the transmission or transaxle on the pan. A few transaxles have side and bottom pans. The pan shape is especially useful when you suspect that the original transmission or transaxle has been replaced with a different model. Pan shape, however, may not identify all the variations of a particular unit. The pan shape must sometimes be combined with information about the vehicle type, drive type, vehicle model year, and engine size to obtain the proper parts. For instance, the Ford A4LD varies internally, depending on whether it is used behind a 2.3-liter engine or a 4.0-liter engine.

Overall transmission/transaxle size and shape, as well as the placement of external parts, can also be used as identifiers. Some manufacturers install **builder plates** directly onto the transmission or transaxle, **Figure 15-11**. Parts catalogs often contain information that allows the type of transmission or transaxle to be determined from the make, year, and model of the vehicle.

Some manufacturers have a special **production label** with a code identifying the type of transmission used in a vehicle. Sometimes, information about the type of transmission or transaxle is part of the **vehicle identification number,** or VIN. Transmission/transaxle information is contained in the fourth through eighth digits of the VIN. To use the production label or VIN numbers, you must have specific decoding information from the manufacturer of the vehicle.

Identifying Defective Parts

Many experienced rebuilders can rely on previous experience to determine which parts will need to be replaced. Defective transmissions, however, often contain unsuspected defects. Once the transmission is completely disassembled, carefully inspect every part. Many transmission and transaxle hard parts commonly wear out. Other components are damaged by nearby part failures. As previously noted, many technicians prefer to replace major electrical components to eliminate potential problems.

Once you have determined which parts must be replaced, make a written list of all the needed parts. Part

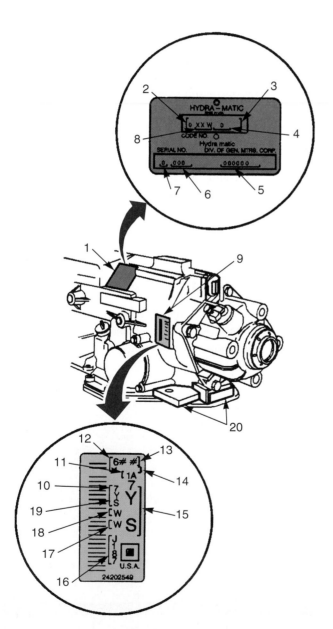

Legend
(1) Goodwrench
(2) Year
(3) Not used
(4) Remanufacturing site code
(5) Serial number
(6) Julian date
(7) Year remanufactured
(8) Model
(9) Transmission identification plate location
(10) Model year
(11) Line build
(12) GM production code
(13) Julian date
(14) Shift
(15) Model
(16) Serial number in base code 31
(17) W = Warren assembly plant
(18) W = 4T60E, B = 4T65E
(19) Model
(20) Vehicle identification number (VIN) Derivative stamping location

Transaxle identification nameplate
The transaxle ID nameplate contains all of the build information pertaining to that specific transaxle.

Figure 15-11. *Many transmission and transaxle manufacturers place a plate or label directly on the case. The plate or label contains the information needed to order the correct parts. (General Motors)*

lists are usually written on the shop ticket, but they may be written on a separate parts order form or a sheet of paper.

Using Parts Catalogs

After compiling a list of needed parts, look up the parts in the appropriate *parts catalog.* All catalogs are divided into sections based on transmission or transaxle type. Some sections may combine a newer four- and five-speed transmission with the older three-speed transmission that it is based on. For instance, the Ford C-6 three-speed and E40D four-speed are often combined into a single section. Similarly, electronic transmission or transaxle versions are often combined with earlier hydraulic versions. For example, the GM 4T60 (440-T4) and the 4T60E usually share a section.

An index at the front of the catalog will list the types of transmissions or transaxles, followed by the page numbers for each section. Some catalogs show the *group number* of the transmission. Group numbers make part stocking easier and allow technicians to be sure they have selected the proper part. The catalog may also contain directions for its use, **Figure 15-12.**

Once the correct catalog section has been located, you will notice that it is divided into subsections based on specific components, such as complete overhaul kits, clutches, bands, and hard parts. After turning to the desired subsection, you will notice that each component listed in the section will have a *part number* and, in most cases, a description. The part number may be correlated to the section and to a corresponding illustration, **Figure 15-13.** For example, if an illustrated part in section 56 has the number 127, the associated part in the part listings will be number 56127. Letters may be attached to the part number to indicate whether it is new or used, or whether it can only be obtained from the vehicle dealer. Typically N will stand for new, R for rebuilt, U for used, and D for dealer.

There may be many variations of a single part, depending on the type of transmission or transaxle. If the catalog lists several variations of a part, do not guess which one you need. Use the information given with each listing to determine the proper part. Build date, engine size, and other information is often used to determine which of several similar parts should be ordered. Sometimes, the part itself must be examined to determine such things as the number of vanes in a pump or whether a one-way clutch uses sprags or rollers.

Once you get all the part numbers together, you are ready to order the parts. Sometimes, you will have to order parts from different suppliers. It is common to order all the hard parts from one supplier, clutch plates and gaskets from another supplier, and the torque converter from a third supplier. Do not forget to order any special transmission fluid, if needed.

Ordering from a Parts Supplier

Call the appropriate suppliers and order the needed parts. The parts person may need vehicle information, such as that discussed previously, to accurately place the order. Sometimes, the VIN helps the parts person order the right parts. Ask for assistance if you are not sure about the needed parts.

If you are not ordering from a catalog, you must be ready to provide vehicle and transmission/transaxle information. Knowledgeable parts persons will be able to determine which part is needed based on transmission/transaxle type, engine size, vehicle year, and other information.

Check Parts when They Arrive

When the parts arrive, immediately check them to make sure that they are correct. In most cases, this can be done visually and by checking the description on parts

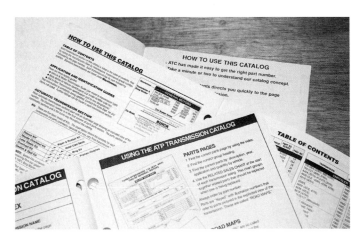

Figure 15-12. *Parts catalogs often contain directions for their use. Many catalogs also contain special sections that make it easy to locate commonly needed parts, and may have instructions for determining what kind of transmission or transaxle is being serviced.*

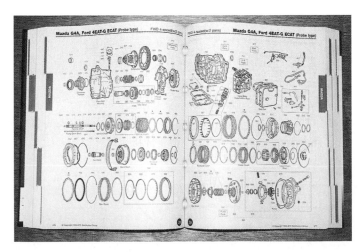

Figure 15-13. *Exploded views in many parts catalogs make it easy to find a particular part and then match it to a part number.*

packages. Be careful when checking hard parts. It is often necessary to compare the old and new parts to ensure that the new part is correct. Once all parts have been located, you can start rebuilding the unit with the assurance that the right parts are at hand.

Changing Fluid and Filter

To change the fluid and filter in an automatic transmission, warm the engine and transmission. Then raise the vehicle on a hoist and position a drain pan under the transmission oil pan.

⚠ **Warning: Be careful not to spill the hot transmission fluid. It can cause burns. If a drain plug is not provided, remove the transmission oil pan attaching bolts in the sequence described below to reduce spillage.**

Remove all but the front four bolts. Pry on the pan to loosen the gasket and allow the pan to drop at the rear. (Transmission fluid may begin to drain at this point.) Loosen the front four bolts, allowing the pan to drop further and the fluid to drain. When the fluid stops draining, remove the front bolts and the transmission oil pan. Examine the fluid. Take note of any debris. Then discard the fluid in an approved receptacle.

Remove the transmission filter and replace it with a new filter. Clean the oil pan. Scrape the old gasket from the transmission pan and housing. Position the new pan gasket using an approved sealer. Use sealer sparingly. Excess sealer could clog pressure passages.

Reposition the oil pan on the bottom of the transmission and start all the oil pan bolts with your fingers. Then, tighten them in a crisscross pattern to their specified torque requirement. Overtightening can split the gasket or distort the oil pan.

Refill the transmission with the correct type and amount of automatic transmission fluid. Using a funnel, pour the fluid into the transmission filler tube.

🔧 **Caution: Before adding fluid, determine the type of fluid to be used, Figure 15-14. If necessary, check the manufacturer's service manual for details. The wrong fluid could severely damage the transmission or transaxle.**

Start the engine, apply the parking brake, and move the selector lever to each position momentarily. Then, check under the car for leaks. Recheck the fluid level after the transmission reaches its normal operating temperature.

| Fluids other than Dextron® | |
Vehicle	Fluid
Accura	Honda Gen ATF
Audi	Audi/VW Synthetic ATF #G002000
BMW w/5-speed automatic only	BMW LA 2634
Chrysler '88-'98	ATF+3
Chrysler '99-'01	ATF+4
Dodge '88-'98	ATF+3
Dodge '99-'01	ATF+4
Eagle	ATF+3
Ford	Mercon V
Honda	Honda Gen ATF
Hundai	ATF+3
Isuzu	Honda Gen ATF
Jeep	ATF+3
Lexus	Toyota type T-IV ATF
Lincoln	Mercon V
Mazda	Mercon V
Mercury	Mercon V
Mistubishi	Mitsubishi Diamond ATF
Plymouth '88-'00	ATF+3
Saturn	Saturn Transaxle Fluid #21005966+
Toyota	Toyota type T-IV ATF
Volkswagen without filter tube	Audi/VW Synthetic ATF #G002000

Figure 15-14. *Many vehicles use Dexron® automatic transmission fluid in their transmission or transaxle. A considerable number of vehicles, however, use a fluid other than Dexron. The following is a list of vehicles that require special automatic transmission fluids.*

Governor Service

Most governors can be serviced without removing the transmission or transaxle from the vehicle. To remove a case-mounted governor, remove the governor cover and withdraw the governor from the case. See **Figure 15-15.** Some governor covers are bolted to the case. Other covers are held to the case by a snap ring. If the governor is mounted on the output shaft, there may be an access cover to allow the governor to be removed without removing the extension (tailshaft) housing. Many transaxles with shaft-mounted governors have an access cover. Some transaxle governors can be reached by removing the oil pan and valve body. On many older transmissions, the extension housing must be removed to reach the governor. Extension housing removal will be covered in Chapter 17.

Governors usually fail because of sticking weights and valves. To inspect the governor, check the weights and the valve, and make sure they move freely. Look for scored

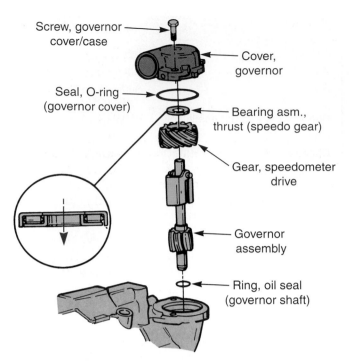

Screw, governor cover/case

Cover, governor

Seal, O-ring (governor cover)

Bearing asm., thrust (speedo gear)

Gear, speedometer drive

Governor assembly

Ring, oil seal (governor shaft)

Figure 15-15. *An exploded view of a gear-driven governor. Most gear-driven governors can be removed from the case without removing the transmission or transaxle from the vehicle. (General Motors)*

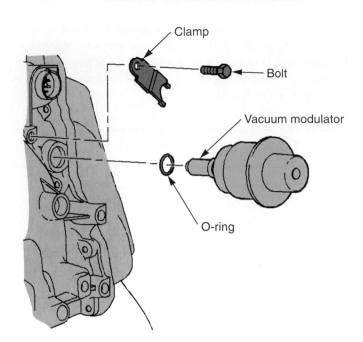

Clamp

Bolt

Vacuum modulator

O-ring

Figure 15-16. *Most modern modulators are held in place by a clamp and bolt. A few modulators screw into the case. (General Motors)*

surfaces and make sure mating surfaces are smooth and flat. The valve can be removed, cleaned, and polished in the same way as the valves in the valve body. If a sticking weight or valve continues to be a problem after cleaning and polishing, or if you have any doubts about the condition of these components, replace the governor.

Some governors contain small screens to filter out dirt or metal particles. These screens should be thoroughly cleaned. If a screen is torn or otherwise damaged, it should be replaced.

If the governor is driven from the output shaft, the driven gear may be defective. Most driven gears can be replaced by removing a retaining pin and sliding the gear from the governor assembly. After sliding the new gear into position, reinstall the pin and check the valve for free operation.

After servicing the governor, install it in reverse order of removal. If the governor cover uses seals or gaskets, replace them. In operation, the governor throws oil around and a defective gasket is sure to leak. If the governor is case mounted, slide the governor into position with a twisting motion to engage the teeth on the output shaft.

Vacuum Modulator Service

To remove a vacuum modulator, remove the vacuum hose. Next, remove the attaching bolt holding the modulator to the case. Then slide the modulator from the case. The exploded view in **Figure 15-16** illustrates the general layout of modulator components. A few modulators are

threaded into the case and can be loosened and removed. The modulator valve may come out with the modulator. If the valve stays in the case, it can be removed with a small magnet. Some oil may leak when the modulator is removed, so have a drip pan ready.

Check the vacuum modulator for a dented neck and for transmission fluid on the vacuum hose side. If the neck is bent or dented, or if there is fluid on the vacuum side, the modulator must be replaced. As a further check, use a hand-powered vacuum pump to place a vacuum of at least 15 inches on the modulator, **Figure 15-17.** If the modulator will not hold vacuum, it should be replaced.

Reinstall the modulator valve by carefully sliding it into position. Then install the modulator using a new O-ring. Install the attaching bolt and tighten it. If the modulator is threaded into the case, use a new gasket and tighten the modulator carefully. Then reinstall the vacuum line. Add fluid if necessary and road test the vehicle to ensure that the transmission or transaxle is operating properly.

 Note: The replacement modulator may look different from the original model. This is common with replacement modulators. It does not indicate that the part is incorrect.

Valve Body Service

Begin valve body service by removing the oil pan. After removing the oil pan, check it for metal particles, varnish buildup, and sludge, **Figure 15-18.** Excessive

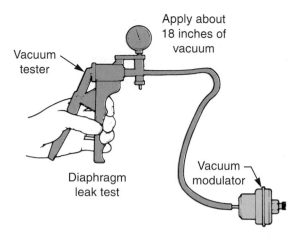

Figure 15-17. *A hand-powered vacuum pump can be connected to a vacuum modulator to check the diaphragm. (Ford)*

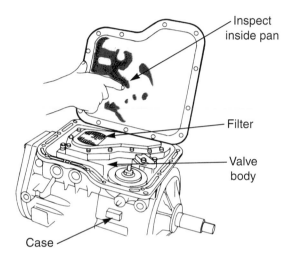

Figure 15-18. *Remove the oil pan and check for metal particles, varnish buildup, and sludge. Note that some oil pans have a small magnet to pick up metal pieces. Be careful not to lose the magnet. (Nissan)*

sludge indicates that the holding members are burnt and the fluid has been overheated. Metal particles usually indicate that the torque converter, oil pump, planetary gearsets, or other moving parts are wearing badly. If this is the first time the oil pan has been removed for service, a few aluminum particles may show up in the pan. These particles may have originated from case machining operations at the factory and are not a cause for concern.

Removing the oil pan gives you access to the filter and valve body. The filter may be pressed into a case passageway and held in place with a clip on the valve body, or it may be bolted directly to the valve body. Remove the filter at this time.

After the filter has been removed, the valve body bolts can be unfastened and the valve body can be removed from the case. As you remove the valve body, carefully note the position of all valve body parts, including springs, oil feed tubes, linkage attachments, and check balls, **Figure 15-19.** Place the assembled valve body

and related parts, as well as any modulator parts, together in a safe spot on the workbench.

The valve body should be thoroughly cleaned in solvent to remove dirt and metal particles. Before cleaning, be sure to remove any plastic parts or electrical devices that could be damaged by the solvent.

Note: Aluminum valve bodies should be handled very carefully. Do not use bottle brushes or any cleaning tools in aluminum valve bodies, as they can scratch the bores.

The valve body casting should be checked for porosity, cracks, and damaged machined surfaces, **Figure 15-19A.** Orifices and passageways must be clean and free from damage. Valves should be checked for burrs, nicks, and scores. Springs should be checked for distorted or otherwise damaged coils. **Figure 15-20** shows some of the parts that should be removed from the valve body for cleaning and inspection.

Check all valves for free movement in their bores. Any valve that does not move freely in its bore is sticking and can cause problems. Valves can be manually moved in their bores against spring pressure. If the spring does not return the valve to its original position, the valve is sticking. Some valves will not stick until they are removed and reinstalled. This is because the valve and bore have worn matching grooves as the valve moves in the bore. If the valve cannot be made to move freely, the valve body should be replaced.

To repair a sticking valve, remove the valve from the valve body. Most valves are held in the valve body by small retainer plates. To avoid mixing up the valves, remove one retainer plate at a time. Clean and reassemble the valves under that retainer plate before going on to the next one. If the retainer plate holds more than one valve, be careful to return the valves to their original positions.

Sticking valves can sometimes be loosened by lightly polishing the valve lands with crocus cloth. This is especially important when polishing aluminum valves. Some manufacturers recommend using a microfine lapping compound instead of crocus cloth. Always polish in a rotary motion, or *around* the surface of the valve lands. Be careful not to polish the sharp edges of the valve, as these help keep the valve from jamming on small pieces of dirt. Be sure to remove all traces of cleaning compound. Cleaning the valve *bore* may also restore a valve to proper operation. If proper operation cannot be restored, the valve or the valve body assembly must be replaced.

It is not necessary to completely disassemble the valve body if you just want to check the valves. However, if the transmission is very dirty or sludged up, the entire valve body must be disassembled. Accumulators must be removed to clean out sludge and metal particles. Always consult the service manual for a valve diagram.

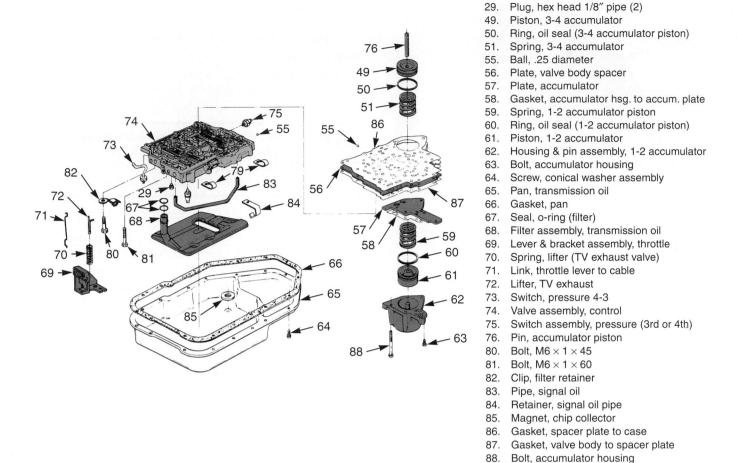

29. Plug, hex head 1/8″ pipe (2)
49. Piston, 3-4 accumulator
50. Ring, oil seal (3-4 accumulator piston)
51. Spring, 3-4 accumulator
55. Ball, .25 diameter
56. Plate, valve body spacer
57. Plate, accumulator
58. Gasket, accumulator hsg. to accum. plate
59. Spring, 1-2 accumulator piston
60. Ring, oil seal (1-2 accumulator piston)
61. Piston, 1-2 accumulator
62. Housing & pin assembly, 1-2 accumulator
63. Bolt, accumulator housing
64. Screw, conical washer assembly
65. Pan, transmission oil
66. Gasket, pan
67. Seal, o-ring (filter)
68. Filter assembly, transmission oil
69. Lever & bracket assembly, throttle
70. Spring, lifter (TV exhaust valve)
71. Link, throttle lever to cable
72. Lifter, TV exhaust
73. Switch, pressure 4-3
74. Valve assembly, control
75. Switch assembly, pressure (3rd or 4th)
76. Pin, accumulator piston
80. Bolt, M6 × 1 × 45
81. Bolt, M6 × 1 × 60
82. Clip, filter retainer
83. Pipe, signal oil
84. Retainer, signal oil pipe
85. Magnet, chip collector
86. Gasket, spacer plate to case
87. Gasket, valve body to spacer plate
88. Bolt, accumulator housing

Figure 15-19. *The valve body can be removed after the attaching bolts are removed. Many parts, such as spacer plates, accumulator pistons, oil supply tubes, and check balls, are held in place by the valve body. Make sure you do not lose any of these parts when removing the valve body. (General Motors)*

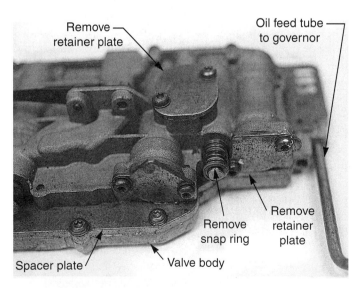

Figure 15-20. *Overhauling the valve body should be done carefully. Remove one retainer plate and valve at a time, if possible. Then, reinstall the valve exactly as it was originally. If you suspect that a valve is sticking, check it for free movement in its bore. Split valve bodies should be opened to replace the spacer plate gaskets and for thorough cleaning. (BBU, Inc.)*

Figure 15-21 shows the relative positioning of valves and springs. This illustrates how these components should be installed in the valve body. When disassembling the valve body, watch for check balls, springs, and other small parts that must be reinstalled.

Check the spacer plates for warping, baked-on gasket material, deep scratches, or pounded out check ball orifices. Damaged spacer plates should be replaced.

Also, carefully check the inner manual lever, which holds the manual valve in position. Check the teeth on the cockscomb to make sure they are not worn, **Figure 15-22.** Check the spring-loaded roller or ball to make sure it puts pressure on the teeth and holds the lever in position.

Shift Kits

Sometimes, the original transmission or transaxle pressure and oil flow designs are not adequate for older units. Age and wear often require that the original pressure and oil flow designs be modified to prevent problems. Many technicians install *shift kits* to improve the operation

No.	Description
301	Body, control valve
302	Ring, retainer (accumulator piston)
303	Piston, front accumulator
304	Ring, seal (accumulator piston)
305	Spring, front (accumulator piston)
308	Pin, grooved
309	Plug, valve bore (.56 O.D.)
310	Valve, 1-2 accumulator
311	Spring, 1-2 accumulator valve primary
312	Spring, 1-2 accumulator valve secondary
313	Pin, coiled spring
314	Plug, valve bore (.50)
315	Valve, detent
316	Valve, detent regulator

No.	Description
317	Pin, detent regulator valve
318	Spring, detent regulator
319	Valve, manual
320	Bushing, 1-2 modulator valve
321	Valve, 1-2 regulator
322	Spring, 1-2 regulator valve
323	Valve, 1-2 detent
324	Valve, 1-2
325	Spring, 1-2 modulator valve
326	Valve, 1-2 modulator
327	Pin, straight .12 dia. × 1.32

No.	Description
328	Bushing, 2-3 modulator valve
329	Spring, 2-3 valve (outer)
330	Valve, 2-3 modulator
331	Spring, 2-3 valve (inner)
332	Valve, 2-3
333	Pin, straight .12 dia. × .82
334	Plug, valve bore (.437)
335	Spring, 3-2 valve
336	Pin, 3-2 valve
337	Valve, 3-2
339	Bushing, 1-2 modulator valve
340	Spring, 1-2 accumulator primary
341	Valve, 1-2 accumulator
342	Bushing, 1-2 accumulator valve
343	Valve, 1-2 accumulator primary
344	Spring, 1-2 accumulator secondary

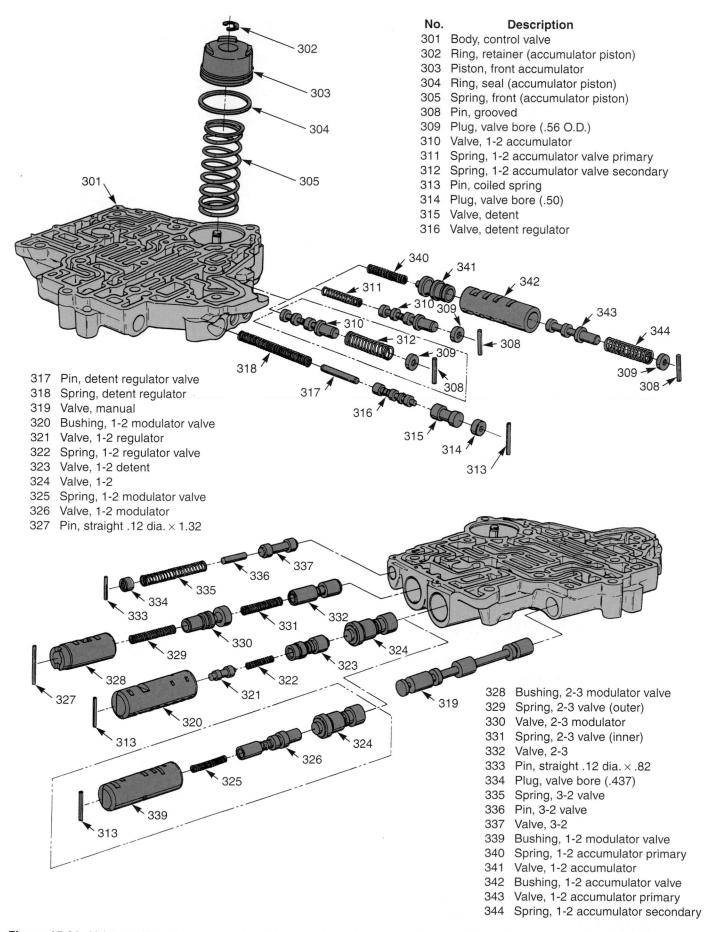

Figure 15-21. *Valve position diagrams, such as the one shown here, make reassembling the valve body an easier task. The diagrams often include variations of the basic valve arrangement. (General Motors)*

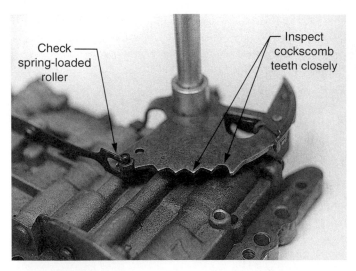

Figure 15-22. *Check the inner manual lever carefully to make sure that parts are not worn and that it operates properly to select shift positions. (BBU, Inc.)*

of a transmission or transaxle. Shift kits consist of upgraded springs, valves, seals, and other parts, most of which are installed in the valve body. These parts modify internal pressures and the fluid flow to improve shift feel and transmission/transaxle durability. Many shift kits contain drill bits for enlarging restrictor holes in the spacer plate. Shift kits always contain instructions for the installation of the new parts. Sometimes there are instructions for making modifications to other transmission parts to increase performance. Shift kits can be installed as a way to prevent transmission or transaxle problems, or during overhaul to improve the durability of the rebuilt unit.

Some shift kits, such as the one shown in **Figure 15-23,** are designed to cure all the potential problems in one transmission or transaxle model. Other shift kits contain only the parts and instructions to fix one common problem in a particular unit, **Figure 15-24.**

Reinstalling the Valve Body

To reinstall the valve body after cleaning, first remove all gasket material from the spacer plate and the transmission case. Be careful not to scratch the spacer plate or the case. Some technicians use solvents to dissolve the gasket material. Thoroughly clean the spacer plate and case to remove all particles of gasket material.

After everything has been cleaned, locate the position of all check balls and install them. Use assembly lube or petroleum jelly to hold the check balls in position. Also locate all feed tubes, electrical components, and other parts and reinstall them on the valve body.

Install the spacer plate over the valve body, using new gaskets. Do not use sealer on the gaskets. Install any check

Figure 15-23. *Many shift kits are designed to fix all weak areas of a particular transmission. Shift kits such as this one correct all problems caused by improper fluid flow and pressure. (Superior)*

balls that fit between the spacer plate and case, being sure to properly position them and hold them in place with assembly lube or petroleum jelly. If desired, install guide pins in the bottom of the case to make valve body positioning easier.

Carefully raise the valve body into position. Lightly install any feed tubes as the valve body is being raised. As the bolts are tightened, make sure that any feed tubes enter the case smoothly.

Caution: Do not overtighten the valve body attaching bolts. Overtightening and using an improper tightening sequence can cause the valve body to warp, leading to sticking valves.

Lightly tighten the attaching bolts, starting at the center of the valve body and working outward. Then return to the center of the valve body and finish tightening the bolts. If possible, use an inch-pound torque wrench and tighten the bolts to the specified torque. If torque specifications are not available, use a socket with a screwdriver handle to tighten the bolts.

Once the valve body is installed, lightly strike the feed tubes, if used, to ensure that they are firmly seated. Replace all electrical connectors and install the filter. Then reinstall the pan using a new gasket and refill the transmission or transaxle with fluid. Road test the vehicle to ensure that the transmission is operating properly.

Servo Service

Many band servos can be serviced without removing the transmission or transaxle from the vehicle. Servos seldom require service other than the replacement of the seals and, if used, the cover gaskets. Nevertheless, the servos should be disassembled and checked whenever the transmission is disassembled. Many servo covers are held to the transmission case with a snap ring. A special tool is required to compress the servo return spring so the snap ring can be removed. Other servo covers are held to the

case with bolts. Some servos are installed under the valve body, and the valve body must be removed to gain access to these units. In some designs, a link is used to connect the servo to the band. To keep the link from falling out of place, turn the adjusting screw all the way in (bottom the adjuster) before removing the servo.

To rebuild a servo, begin by removing the servo cover. To do this, remove the servo cover's retaining bolts or depress the servo cover and remove the snap ring. Never try to remove the snap ring until the cover has been pushed in and is out of the way. A *servo-cover depressor* can be used for this purpose. With the snap ring or bolts out of the way, remove the servo cover. See **Figure 15-25.**

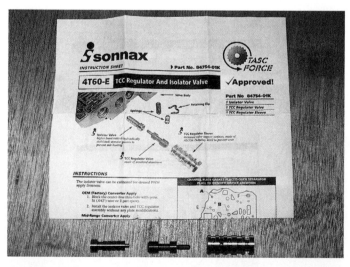

Figure 15-24. *Special shift kits are available to repair single problems in a particular transmission. This kit corrects problems with the torque converter apply on one type of automatic transaxle. These kits are a quick and relatively inexpensive way to fix a unit that has a single problem but otherwise operates well.*

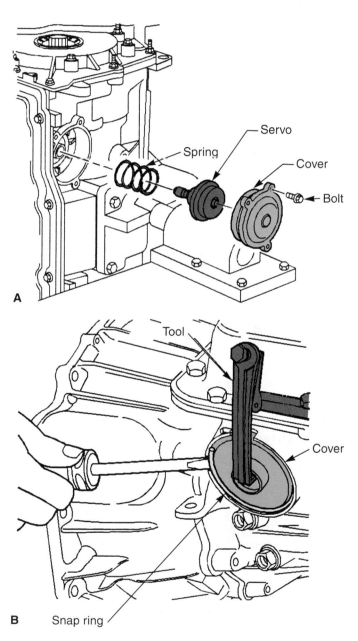

Figure 15-25. *Methods of servo cover removal. A—Removing the cover retaining bolts to allow the servo cover and internal servo parts to be removed. B—Using a special tool to depress the cover against spring tension so the snap ring can be removed. (General Motors)*

Pull the servo apply piston and the return spring from the servo cylinder. Some servos have more than one piston and more than one spring. Note the position of the parts as you remove them. A few pistons will be tight in their cylinders. These can be removed by *carefully* applying air pressure to the **servo release port**. See **Figure 15-26.** The valve body must be removed for this operation.

Remove the seals from the servo piston(s), noting the direction that the sealing lips on the lip seals face, if used. See **Figure 15-27.** Check the servo piston and piston pin for wear. Normally, servo parts do not wear very much. If a spring height measurement is given in the service manual, check the spring with a sliding caliper as shown in **Figure 15-28.** If the spring is not within specifications, replace it.

After cleaning all servo parts, select, lubricate, and install the new seals on the servo apply piston. Some servos have multiple pistons, and all seals should be replaced. There may be a seal on the servo apply piston pin. This seal should also be replaced.

Install the servo piston(s) in the cylinder. Make sure all servo pistons and springs are reinstalled in their original locations. Select the proper servo cover O-rings or gasket, and reinstall the servo cover.

Accumulator Service

 Note: Many servos contain internal accumulators or servo pistons that act as accumulators during certain shifts. For this reason, accumulator service is often part of servo service.

Accumulators visually resemble servos and should be checked and repaired in a similar manner. Access the accumulator by removing the accumulator cover, if one is used, or by removing the valve body. Many accumulators are installed under the valve body or are part of the valve body. Valve body removal was covered earlier in this chapter.

Withdraw the accumulator piston and spring(s). Some accumulators also have a removable pin, which will come out with the piston. Inspect all parts for wear, nicks, or burrs. Carefully inspect the pin and the internal accumulator bore where it rides on the pin. Wear in this area is

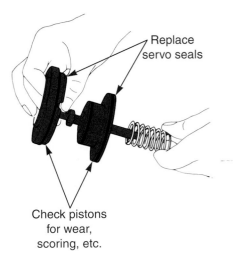

Replace servo seals

Check pistons for wear, scoring, etc.

Figure 15-27. *Most servos can be serviced by replacing their seals. The servo should be checked for any signs of wear. Return servo pistons and springs to their original positions.*

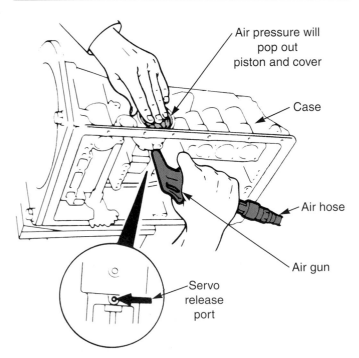

Air pressure will pop out piston and cover

Case

Air hose

Air gun

Servo release port

Figure 15-26. *Some servos must be removed with air pressure, as shown in this illustration. Always make sure that the servo does not fly out of the case when air pressure is applied. (DaimlerChrysler)*

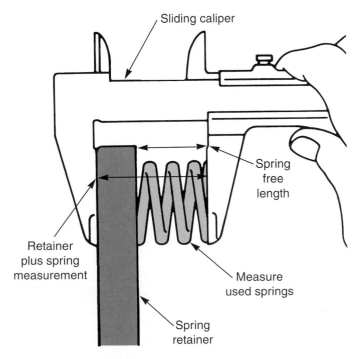

Sliding caliper

Spring free length

Retainer plus spring measurement

Measure used springs

Spring retainer

Figure 15-28. *Servo return spring length can be measured with a sliding caliper. Undersized springs should be replaced. (DaimlerChrysler)*

common, especially on plastic or aluminum accumulators that are used during more than one shift. You may find a broken accumulator piston, especially on an accumulator that is designed to cushion the application of the forward clutch. Also check the accumulator bore for scoring. Replace any defective parts. Always replace the accumulator seals and seal rings.

Replacing Leaking Seals, O-Rings, and Gaskets

Once a leaking seal, O-ring, or gasket has been located, it can be replaced by disassembling the mating parts and replacing the leaking component. The following general procedures cover the removal and installation of seals, O-rings, and gaskets that commonly leak. For additional gasket and seal replacement information, as well as information on adhesive compounds, RTV sealant, and gasket-making compounds, refer to Chapter 5.

> **Note: The replacement of some sealing devices, such as the front pump seal, O-ring, and pump-to-case gasket, requires that the transmission or transaxle be removed from the vehicle. Replacement of tailshaft housing seals and axle shaft seals requires that the drive shaft or CV axle be removed.**

Seal Replacement

Metal clad seals are used whenever a rotating part must be sealed in relation to a stationary part. Seals such as the manual shaft seal can be pried out of the case once the shaft has been removed, **Figure 15-29**. Clean the area where the seal was installed. Then install a new seal into the case. Most seals can be gently tapped into place using a socket and hammer. If necessary, use sealer on the outer part of the seal housing. Once the seal is in place, lightly lubricate the seal lip. Then reinstall the shaft and check for leaks.

O-Ring Replacement

Many case electrical connectors have O-rings that seal the connector to the case, as shown in **Figure 15-30**. To change the O-ring, remove the external electrical connector and depress the internal locking tabs to remove the case connector. In many cases, the oil pan must be removed to depress the locking tabs. After depressing the tabs, pull the case connector from the case. Remove the O-ring and clean the sealing area. Lightly lubricate the new O-ring and install it on the case connector. Then

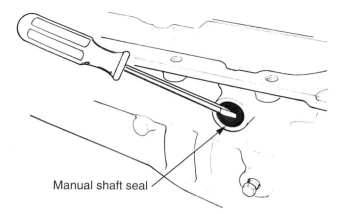

Manual shaft seal

Figure 15-29. *Manual shaft seals can be carefully pried from the case with a screwdriver. Special tools are available to remove the seal without removing the manual valve shaft. (General Motors)*

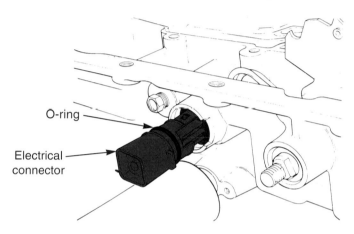

O-ring

Electrical connector

Figure 15-30. *An O-ring is commonly used to seal electrical connectors to the case. O-rings can be removed by prying them off with a pointed tool or by carefully pulling them from their grooves. Carefully clean the groove before installing the new O-rings. (General Motors)*

reinstall the case connector by pressing it into the case. Make sure the locking tabs engage to hold the connector in place. Then reinstall other parts as necessary, refill the transmission or transaxle, and check for leaks. Road test the vehicle if necessary.

Gasket Replacement

Gaskets are commonly used on transmission/transaxle oil pans, side covers, timing gear and differential covers, tailshaft housings, governor covers, and other places where two flat surfaces must be sealed. To replace a gasket, remove the pan or cover by removing the attaching bolts. Have a drain pan handy in case of oil leakage. Once the parts are separated, remove all old gasket material from the pan/cover and the case. Be careful not to scratch the sealing surfaces. Some technicians use solvents to dissolve the gasket material. Thoroughly clean the pan or cover and case to remove all particles of gasket material. Check the pan or cover for warping and damage at the sealing area. If the pan or cover cannot be straightened, it should be replaced.

Install the new gasket using the proper sealant. If the pan/cover uses a gasket-making compound, apply it according to manufacturer's directions. In most cases, the sealant is applied to only one of the mating surfaces. Then reassemble the parts, carefully tightening the attaching fasteners. Begin by slightly tightening the fasteners. Alternately tighten the fasteners on opposite sides of the part. If you are installing a part with many fasteners, such as an oil pan, begin tightening the center fasteners, slowly working outward. See **Figure 15-31.** Repeat the tightening sequence, gradually increasing the torque. Whenever possible, use a torque wrench to make the final tightening sequence. Then refill the transmission/transaxle if necessary and road test. Finally, recheck for leaks.

Adding an Auxiliary Oil Cooler

You may be asked to install an auxiliary, or external, oil cooler to a vehicle that will be used for towing or other severe service. Start by determining what size cooler is needed. For instance, towing a small boat would require a smaller cooler than towing a large travel trailer. Cooler manufacturers provide charts detailing what size cooler is appropriate for different situations. After obtaining the cooler, make sure that it is the correct size and that all hoses and attaching hardware are included.

Next, determine the best place to install the cooler. Coolers should be installed ahead of the radiator and air conditioner condenser, close enough so that the cooling fan or fans can pull air over the fins. Some technicians prefer to leave the cooler installation until last to ensure that the hoses can be connected to the cooler fittings. Attach the cooler to the vehicle using the straps or quick connect ties provided. If you are attaching the cooler directly to the radiator or condenser, make sure that there is at least .25″ air space between the two units. This reduces unwanted heat transfer into the cooler.

After the cooler is in place, determine which cooler line to the radiator is the outlet line. The cooler should always be installed in the outlet line from the radiator. The simplest way to do this is to disconnect the top line (which is usually the outlet line) from the radiator. Place a drip pan under the line and have an assistant briefly crank the engine. Note whether the fluid exits from the radiator or the line. If the fluid exits from the radiator, this is the outlet line. If the fluid exits from the line, the other line is the outlet.

After identifying the outlet line, connect the hoses. Most cooler installation packages have a fitting to be installed in the radiator outlet. After installing the fitting, install the hoses on the fitting and the cooler line. Some

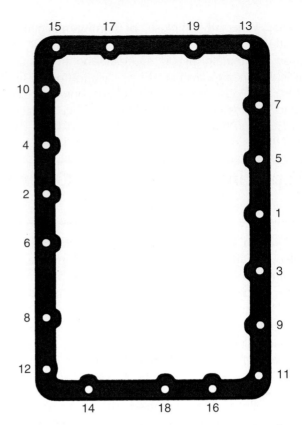

Figure 15-31. *Always start tightening the pan or cover from the center and work outward. This will allow the gasket to spread out and do a better job of sealing. Tightening from the edges inward may cause the gasket to become kinked. One manufacturer's recommended tightening sequence is shown here.*

technicians prefer to cut the original cooler line at a convenient point to install the hoses. Be sure to slip the clamps over the hoses before installation. After the hoses are in position, tighten the hose clamps.

Run the hoses through the radiator support, making sure that they will not contact moving parts or hot surfaces. Then connect the hoses to the cooler and tighten the clamps. Finally, start the vehicle and check for leaks. The cooler usually requires from .5 to 1 quart (.45 to .9 liters) of additional fluid. Check and add fluid before the vehicle is driven.

Summary

Many automatic transmission and transaxle parts can be serviced without removing the unit from the vehicle. Fluid and filters can be replaced by removing the bottom oil pan. Always obtain the right type of transmission fluid before beginning an oil or filter change, or any type of service work that will require fluid replacement.

Some transmission and transaxle problems can be corrected by adjusting manual valve and throttle valve linkage. Shift linkage should be adjusted whenever the shift indicator position and the actual transmission/transaxle gear do not agree. Throttle linkage can be adjusted to correct shift points and, sometimes, shift feel. Before adjusting throttle linkage, make sure that the linkage is not broken or disconnected.

A few transmission and transaxle bands have an adjustment screw. The screw is turned in after the lock nut is loosened. Once the screw is fully tightened, it is turned out a specified number of turns, and the lock nut is retightened.

Ordering replacement parts is often tricky. Before ordering parts, determine the exact type of transmission or transaxle being serviced. There are several methods for doing this, including observing the pan shape, checking transmission tags or labels, and checking the VIN or a body-mounted label. Once the unit has been identified, use the parts catalog to order parts.

Valve bodies, modulators, and governors can be serviced by removing them from the transmission or transaxle. To remove a valve body, drain the fluid and remove the oil pan. Then remove the fasteners holding the valve body to the case and remove the valve body. With the valve body on the bench, check each valve for free movement in its bore. If the valves must be removed, remove them one at a time. Valves can be cleaned and polished. If a valve body cannot be restored by light polishing and cleaning, it should be replaced.

To remove a case-mounted governor, remove the governor cover and slide the governor from the case. A shaft-mounted governor can be removed by first removing the tailshaft housing. Governors can be checked for sticking valves, missing return springs, and stripped driven gears. Most modulators can be removed by removing the attaching bolt and pulling the modulator from the case. Check the modulator for a leaking diaphragm or a bent neck.

Most fluid leaks can be corrected by replacing the defective part. Common transmission and transaxle sealing parts include seals, O-rings, and gaskets.

An external oil cooler can be added to reduce fluid temperatures. Generally, the parts included in the cooler package provide for easy installation.

Review Questions—Chapter 15

Please do not write in this text. Place your answers on a separate sheet of paper.
1. A valve body can be serviced without removing the
   _____.
   (A) oil pan
   (B) transmission
   (C) filter
   (D) Both A and B.

2. Problems resulting from an overfilled transmission are _____ (the same as/different than) those resulting from an under filled transmission.

3. What can be learned from studying the condition of the transmission/transaxle fluid?

4. When adjusting shift linkage of an automatic transmission or transaxle, you must make sure the position of the outer manual lever is synchronized with that of the _____ _____ lever.

5. Which of the following can be used to help identify the type of transmission or transaxle being serviced?
   (A) Shape of the bottom pan.
   (B) Builder plates or labels.
   (C) Location of external parts.
   (D) All of the above.

6. When used, the _____ may be held to the case with a bolt. Some versions screw into the _____.

7. Which of the following can be replaced after removing the oil pan?
   (A) Valve body.
   (B) Modulator.
   (C) Filter.
   (D) Both A and C.

8. When polishing valve lands, avoid rounding the _____.

9. Which of the following parts is sometimes installed under the valve body?
   (A) Modulator.
   (B) Servo.
   (C) Governor.
   (D) Filter.

10. When installing a pan or cover with many fasteners, start tightening the fasteners in the _____ and work _____.

ASE-Type Questions—Chapter 15

1. Technician A says that a malfunctioning automatic transmission always requires a complete overhaul. Technician B says that contaminated transmission fluid can cause rapid part wear and premature failure. Who is right?
   (A) A only.
   (B) B only.
   (C) Both A and B.
   (D) Neither A nor B.

2. Technician A says that the fluid level in an automatic transaxle is more sensitive to temperature than the fluid level in an automatic transmission. Technician B says that transaxle fluid level may be affected by whether the shift lever is in neutral or park. Who is right?

   (A) A only.
   (B) B only.
   (C) Both A and B.
   (D) Neither A nor B.

3. Technician A says that some automatic transaxles have a separate reservoir for the differential lubricant. Technician B says that the most accurate check of the transaxle fluid level is made just before starting a vehicle that has been parked overnight. Who is right?

   (A) A only.
   (B) B only.
   (C) Both A and B.
   (D) Neither A nor B.

4. Technician A says that some shift cables can be damaged if the engine ground straps are disconnected. Technician B says that the throttle valve linkage on vehicles without modulators is used only for passing gear. Who is right?

   (A) A only.
   (B) B only.
   (C) Both A and B.
   (D) Neither A nor B.

5. Technician A says that the band adjustment screw must be tightened against the drum and then backed off a certain number of turns to properly adjust the band. Technician B says that backing off the band adjustment screw less than specified will generally provide a firmer shift. Who is right?

   (A) A only.
   (B) B only.
   (C) Both A and B.
   (D) Neither A nor B.

6. Replacement parts are being ordered for an automatic transaxle. Technician A says that needed parts should be identified and ordered as soon as possible to reduce waiting time. Technician B says that the new parts should be compared to the old parts to ensure that they are correct. Who is right?

   (A) A only.
   (B) B only.
   (C) Both A and B.
   (D) Neither A nor B.

7. To drain a transmission pan that does not have a drain plug, the technician should remove all pan bolts *except:*

   (A) those at each corner.
   (B) the rear four.
   (C) the front four.
   (D) two on each side.

8. A case-mounted governor should be checked for all the following *except:*

   (A) a stripped gear.
   (B) missing springs.
   (C) leaking diaphragm.
   (D) a stuck valve.

9. Technician A says that a vacuum modulator, when used, should have a leak-free connection to the intake manifold. Technician B says that transmission fluid on the vacuum hose side of the modulator is a normal condition. Who is right?

   (A) A only.
   (B) B only.
   (C) Both A and B.
   (D) Neither A nor B.

10. Technician A says that all valves should be checked for free movement in the valve body. Technician B says that valves should be removed and reinstalled one at a time to reduce the possibility of mixing up valves or springs. Who is right?

    (A) A only.
    (B) B only.
    (C) Both A and B.
    (D) Neither A nor B.

Chapter 16

Transmission and Transaxle Removal and Installation

After studying this chapter, you will be able to:
- ❑ Remove an automatic transmission from a vehicle.
- ❑ Install an automatic transmission in a vehicle.
- ❑ Remove an automatic transaxle from a vehicle.
- ❑ Install an automatic transaxle in a vehicle.

Technical Terms

Converter housing attaching bolts

Engine holding fixture

Converter-to-flywheel attaching bolts

Transmission jack

Torque converter holding tool

Introduction

An automatic transmission or transaxle cannot be overhauled until it is removed from the vehicle. Transmission and transaxle removal and installation requires less knowledge than diagnosis or overhaul. Like any automotive job, however, removal and installation should not be started without an understanding of the tools and skills needed to perform the task safely and efficiently.

The general information in earlier chapters will help you remove and install transmissions and transaxles. Many of the operations will require little more than common sense. Other operations, however, will require knowledge, skill, and dexterity. The information presented in this chapter will enable you to remove and replace automatic transmissions and transaxles, and will aid you when overhauling a transmission or transaxle.

Commonly Overlooked Precautions

Removal and installation of an automatic transmission or transaxle should not be attempted without some basic preparation. Before reading further in this chapter, take the time to study the following basic precautions. While they are not the only precautions necessary for safe and successful removal and installation, they address the most common mistakes made during these operations

❏ Support the vehicle properly. If you are using a lift, make sure the safety catch is working. If you are using a floor jack and jack stands, make sure they are in good working order.

❏ Use a transmission jack to support the transmission or transaxle during removal and installation. Never rely on your strength or that of another person to hold a transmission or transaxle.

❏ Always have a drip pan and rags handy. Spilled transmission fluid is a potential accident.

❏ As you remove fasteners, place them in containers or store them in such a way that they will not become lost. Organize the fasteners so you can easily determine which fastener should be used on a particular transmission or transaxle part.

❏ Never force parts together. If parts will not fit together easily, remove them and determine the source of the problem.

❏ After installation is complete, recheck all fasteners for tightness. Also, recheck all fittings, fluid connections, vacuum lines, and electrical connectors.

❏ If you have any fasteners left over, find out where they go. Never assume that they are not important.

❏ Add transmission fluid to a newly installed transmission or transaxle *before* starting the engine.

❏ Before starting the engine, make sure any engine parts that were loosened or removed to gain access to the transmission or transaxle have been reconnected.

❏ If unusual noises are obvious when the engine is started, immediately stop the engine and determine the cause.

❏ Remember that the procedures given in this chapter are general in nature and do not cover every possible type of transmission or transaxle. Always refer to the manufacturer's service manual before starting a removal or installation job. Remember that your instructor is a great source of information. When in doubt, ask your instructor.

Automatic Transmission Removal

The first step in overhauling an automatic transmission is removing the unit from the vehicle. The general procedure for removing an automatic transmission is as follows:

1. Disconnect the negative battery cable. This will prevent accidental shorts.

2. While under the hood, also remove the upper *converter housing attaching bolts,* if possible. See **Figure 16-1.** In some cases, it may be necessary to install an *engine holding fixture.* The holding fixture will support the weight of the engine once the rear engine mount has been removed.

3. Raise the vehicle on a hydraulic lift to get enough clearance to remove the transmission. If you do not have access to a hydraulic lift, use a quality hydraulic floor jack and approved jackstands.

 Warning: Never support the vehicle with wood or cement blocks. Be sure the vehicle is secure before working underneath it.

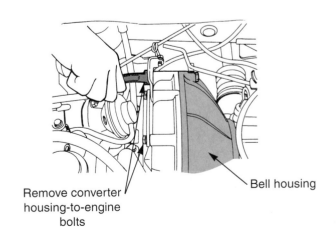

Remove converter housing-to-engine bolts

Bell housing

Figure 16-1. *Remove the upper converter housing attaching bolts while the vehicle is still on the ground. Also, remove anything else that can be reached from under the hood. (Subaru)*

4. Check the engine mounts to ensure that they are not broken.

 Warning: Broken engine mounts can allow the engine to fall backwards once the transmission cross member has been removed.

5. Drain the transmission fluid. Draining the fluid *now* will prevent spillage when the driveline and other components are removed. If the oil pan does not have a drain plug, the pan must be loosened to allow the fluid to drain out. If you remove the pan to drain the fluid, reinstall it after the fluid has been drained.

6. Mark the driveline at the rear axle assembly so that it can be reinstalled in the same position. See **Figure 16-2.** Then remove the bolts holding the rear U-joint to the rear axle pinion yoke.

7. Using a screwdriver or pry bar, pry the drive shaft away from the pinion yoke, **Figure 16-3;** then pull the drive shaft yoke from the transmission and remove the drive shaft.

Note: Many vehicles, especially trucks and SUVs, have two-piece drive shafts. These drive shafts always use a center support. In addition to performing the above steps, the technician must remove the fasteners holding the center support to the frame. Once the center support has been unfastened, the drive shaft and center support can be removed as a unit.

8. Remove miscellaneous items connected to the transmission, such as the speedometer cable assembly, the oil filler tube, and the electrical connectors. Label the connectors using masking tape to simplify reconnection.

9. Disconnect the TV, kickdown, and shift linkages at the transmission. Remove the vacuum line from the vacuum modulator, if applicable. Do not remove the modulator.

10. Disconnect the oil cooler lines at the transmission. Use two wrenches to avoid twisting the lines, **Figure 16-4.** The lines will usually leak when they are disconnected, and therefore, a pan should be positioned to catch any spilled fluid. Position the lines, linkages, and other parts out of the way to keep them from snagging on the transmission as it is lowered.

11. Remove the torque converter inspection cover or dust cover. Then, remove the **converter-to-flywheel attaching bolts** or nuts. See **Figure 16-5.** The converter

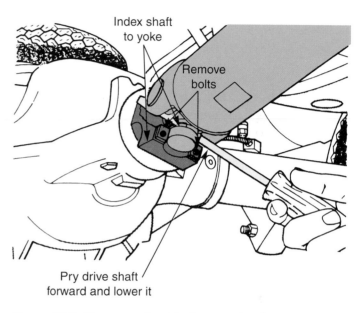

Figure 16-3. *To remove the driveline, pry the shaft forward and lower the back end. Then, slip it out of the extension housing. (DaimlerChrysler)*

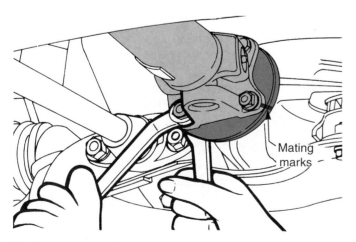

Figure 16-2. *This shows the driveline being detached at the rear axle assembly. Note how the mating flanges are marked before the rear of the driveline is unbolted. (DaimlerChrysler)*

Figure 16-4. *Always use a line wrench to loosen cooler line fittings. Also, use a wrench on the stationary fitting to keep it from moving. If both fittings turn, the cooler line will be damaged.*

must be turned to gain access to all the attaching bolts. This can be done with a special turning tool or a screwdriver. It can also be done by operating the starter with a remote switch (the battery must be temporarily reconnected) or by turning the crankshaft at the front of the engine, **Figure 16-6.**

12. If needed, remove the front exhaust pipes to gain clearance for lowering the transmission. Removing the exhaust pipes will also allow you to lower the back of the engine, which may be required to reach the remaining converter housing attaching bolts.

13. Disconnect any hardware attached to the transmission cross member, such as the parking brake linkage. Removing the linkage will provide more clearance.

14. If an engine holding fixture is not already in place, support the engine with a jack placed under the rear of the engine.

> ⚠ **Warning: Never let the engine hang by only the front engine mounts. Failure to support the engine before removing the transmission can result in serious injury.**

15. Place a **transmission jack** (or floor jack) under the transmission. Raise the transmission slightly to remove weight from the cross member. See **Figure 16-7.**

16. Remove the fasteners holding the cross member to the vehicle's frame and to the transmission mount, which is located on the extension housing. Remove the cross member.

17. If the starter motor is bolted to the converter housing, remove the starter and position it out of the way.

18. Use the transmission jack to remove any strain from between the engine and transmission.

19. Remove the remaining converter housing attaching bolts, which hold the transmission to the engine. You may need to tilt the engine and transmission back slightly to get at the bolts.

> 🗄 **Note: Brackets, such as those holding the filler tube or wiring harnesses, are sometimes attached to the engine with the converter housing attaching bolts. Take note of their position for reinstallation.**

> ⚠ **Warning: *Never* attempt to remove the transmission without the converter housing. To avoid damage to the torque converter or oil pump, the housing should not be separated from the transmission until the transmission is safely on the bench. Also, an automatic transmission is heavy. Use the transmission jack to support and lower the transmission. *Always* have someone help you remove a transmission.**

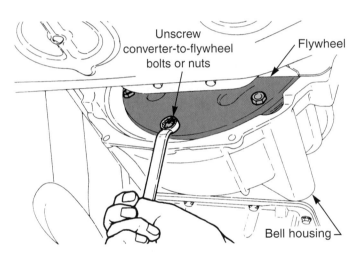

Figure 16-5. *Remove the converter-to-flywheel attaching bolts as part of transmission removal. Note that the inspection shield must be removed to gain access to these bolts. (DaimlerChrysler)*

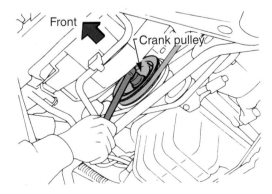

Figure 16-6. *Sometimes it is necessary to turn the engine crankshaft at the front of the engine to gain access to the converter-to-flywheel fasteners. This is easily done with a flex handle and the proper sized socket. (Subaru)*

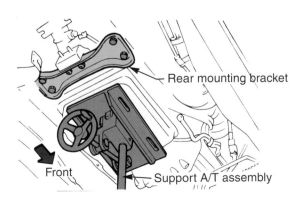

Figure 16-7. *Raise the transmission slightly with a transmission jack before removing the cross member bolts. If the transmission is not supported, it will drop sharply when the cross member is removed. (Subaru)*

20. Slide the transmission straight back and away from the engine, **Figure 16-8.** Attach a small C-clamp to the edge of the converter housing or use a *torque converter holding tool* to secure the converter.

> ⚠ **Warning: Failure to secure the torque converter could result in it sliding out of the unit and dropping to the floor, possibly causing personal injury or damage to the converter.**

21. Lower the transmission. Watch clearances and look for any component you may have forgotten to disconnect. Slowly move the transmission to the workbench with the jack in the lowered position.

Automatic Transmission Installation

Transmission installation is basically the reverse of removal. As with removal, the vehicle should be on a hoist with the engine supported. The following *general* procedures explain how to install an automatic transmission in a vehicle. You should always refer to the manufacturer's service manual for specific instructions.

1. Secure the torque converter with a torque converter holding tool or a C-clamp, **Figure 16-9.**

2. Lubricate the converter pilot hub and crankshaft bore with multipurpose grease. Make sure the converter is fully seated in the transmission housing. Check converter installation with a straightedge and scale if a specification is provided. See **Figure 16-10.**

3. Raise the transmission to the level of the engine.

4. Carefully work the transmission forward until the converter pilot hub enters the crankshaft bore and the converter housing snugs up to the rear of the engine. (Remove the converter holding tool when necessary.) The engine often has alignment dowels to aid in installation. If the torque converter is held by studs and nuts, align the studs with the proper holes in the flywheel.

5. Once the converter housing is flush with the rear of the engine, install at least two converter housing attaching bolts.

6. After the bolts are installed, make sure the torque converter is not jammed. The converter should turn freely, and there should be a slight clearance between it and

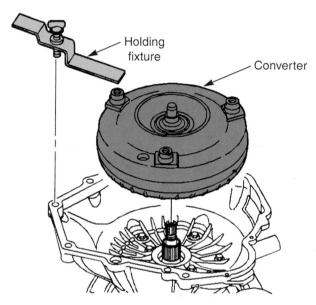

Figure 16-9. *After the converter has been installed in the transmission, install a holding fixture on the converter. This prevents the converter from falling out of the transmission as the unit is moved into position under the vehicle. (General Motors)*

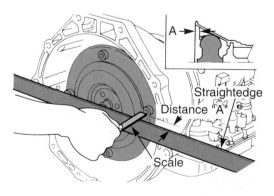

Figure 16-10. *Before installing the transmission, make sure the converter is fully inserted in the transmission. Many manufacturers publish specifications indicating the distance from the front of the converter to the front of the bell housing. It is normal for the converter to be further back than the specifications call for, since it will be pulled forward when the converter-to-flywheel fasteners are installed. If the converter is too far forward, it has not engaged the pump. This must be corrected before installation. (Subaru)*

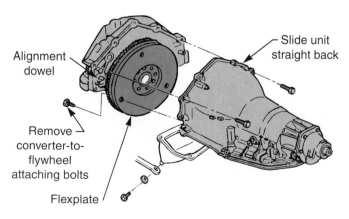

Figure 16-8. *Once the transmission has been disconnected from the engine, slide it backward to remove it. Check for forgotten connections as you move the transmission backward. (General Motors)*

the flywheel. If the converter does not turn freely, it is not properly aligned or engaged with the front pump.

7. If the converter installation is proper, install the remainder of the converter housing attaching bolts. Torque all bolts to specifications.

8. Install and tighten the two lower converter-to-flywheel attaching bolts (or nuts) to specifications. Rotate the torque converter and install the remaining bolts.

9. Raise the rear of the transmission and position the cross member under the extension housing. Then, install the attaching bolts for the cross member to the extension housing and the vehicle's frame. After the cross member is installed, lower the transmission jack and remove it.

10. Install the cooler line fittings by hand, **Figure 16-11.** Starting the cooler lines at this time is easier because the transmission can still be moved slightly.

11. Install miscellaneous items, such as the parking brake linkage, oil cooler lines, modulator vacuum line, electrical connectors, speedometer cable assembly, TV linkage, kickdown linkage, and shift linkage. Install the exhaust pipes, if they were removed.

12. Lubricate the outside of the drive shaft slip yoke and install slip yoke in the output shaft, **Figure 16-12.**

13. Attach the rear U-joint to the rear axle pinion shaft, making sure to line up the marks that were made before removal. Install the center support fasteners, if necessary.

14. Install and/or tighten all remaining fasteners and fittings, as well as the starter.

15. Lower the vehicle but leave the wheels off the ground so the drive wheels are free to turn.

16. Reinstall the negative battery cable.

17. Check the vacuum lines at the engine intake manifold, and make sure the various linkages are properly adjusted.

18. Fill the transmission with the proper ATF. It is advisable to overfill a dry transmission by about 2 quarts (or 2 liters) *as measured on the dipstick* before starting the engine. This will help the pump pick up oil and reduce the chance of damage from a lack of lubrication.

19. Start the engine and immediately recheck the transmission fluid. Add fluid until the proper level is reached. A sucking noise from the oil filler tube is a good indication that more fluid is needed. Be careful not to overfill the transmission.

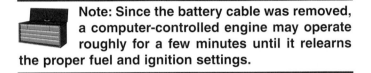 **Note: Since the battery cable was removed, a computer-controlled engine may operate roughly for a few minutes until it relearns the proper fuel and ignition settings.**

20. Place the shift selector lever in reverse and note whether the wheels turn in the reverse direction. Then, place the transmission in drive and allow it to upshift and downshift several times by operating the throttle pedal. This will check for proper shifting and will allow transmission fluid to lubricate the moving parts. Move the selector lever through all quadrant positions.

21. Lower the vehicle and recheck the fluid level. Perform a road test. Check operation in all gears and shift quadrant positions. Note any noises and check for slippage and improper shift points. Look for signs of leakage and recheck the fluid level. Make any final adjustments to the linkage.

Figure 16-11. *Always start the cooler line fittings by hand before tightening them with a wrench. This makes it much less likely that the fittings will be cross-threaded. To make fitting installation easier, start the fittings before the transmission is tightened against the engine.*

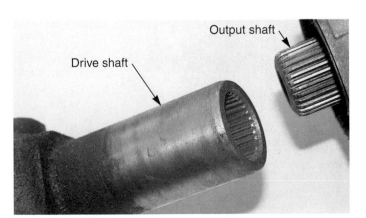

Figure 16-12. *The internal splines of the drive shaft are a match for the external splines on the transmission output shaft. If the splines are lightly lubricated, they should slide together easily. Also, lightly lubricate the outside of the slip yoke to protect the seal.*

Automatic Transaxle Removal

Removal of automatic transaxles is relatively simple if you follow the steps outlined in this section.

Most automatic transaxles can be removed from below the vehicle without removing the engine. A few vehicles are designed so that the transaxle and engine must be removed as a unit. On others, underhood clearances are so close that it is easier to remove them in this way. In this case, you must have a good engine hoist and know how to use it properly.

The following *general* procedures detail how to remove an automatic transaxle from a vehicle. The removal process presented here is for a transaxle that can be removed without removing the engine. Keep in mind that the procedure is much different from that used on rear-wheel drive automatic transmissions.

1. Disconnect the battery negative cable to prevent shorting wires or accidentally operating the starter.

2. While under the hood, also remove any part that can be disconnected before raising the vehicle. See **Figure 16-13.**

3. Install an engine holding fixture to remove the engine and transaxle weight from the mounts. **Figure 16-14** shows a typical engine support fixture.

4. Raise the vehicle to get enough clearance to remove the transaxle. If you do not have access to a hydraulic lift, always use a good hydraulic floor jack and approved jack stands.

 Warning: Never support the vehicle with wood or cement blocks. Be sure the vehicle is secure before working underneath it.

Figure 16-13. *Before raising the vehicle, remove all transaxle connections that can be reached from under the hood. The electrical connector, modulator vacuum line, and ground wire shown here are examples of components that can be disconnected before removing the transaxle.*

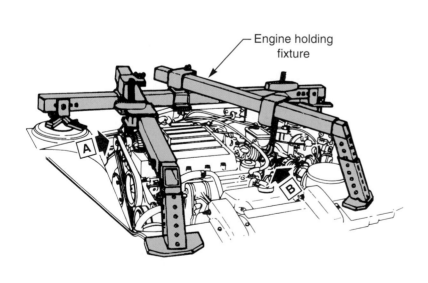

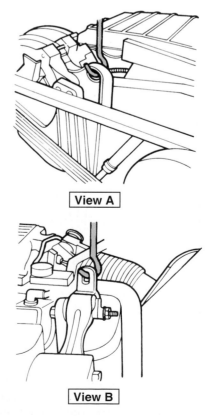

Figure 16-14. *Install the engine holding fixture before raising the vehicle. The holding fixture will slightly raise the engine, removing tension from the motor mounts. This allows the mounts to be removed. (General Motors)*

5. Check the engine mounts to ensure that they are not broken. Broken mounts could be the cause of shifting problems or noises.

6. Drain the transaxle fluid. While it is not necessary to drain the fluid at this point, doing so will prevent fluid spills later. **Figure 16-15** shows draining of transmission fluid and differential lubricant from one make of transaxle.

7. Remove the front wheels.

8. Remove the tie rods at the wheel spindles by removing the cotter pins and attaching nuts and then breaking the tie rod stud loose from the spindle arm with a suitable puller. One type of puller is shown in **Figure 16-16.**

9. Remove the central spindle hub nut on the end of each CV axle using a large socket. The axle can be kept from turning by inserting a screwdriver in the rotor or by having an assistant press on the brake pedal. See **Figure 16-17.** In some cases, the spindle nut is retained by a cotter pin and cage assembly, which must be removed before the nut can be loosened. A few vehicles use a left-hand thread on the spindle nut. Check the manufacturer's specifications before attempting to remove the nut.

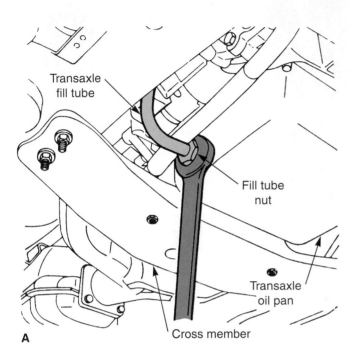

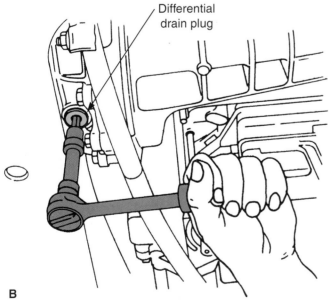

Figure 16-15. *Transaxle transmission and differential are not integral in this design, and different lubricants are required for each. On most transaxles, the transmission and differential are lubricated by the same fluid. A—Draining transmission fluid by disconnecting the fill tube at the transaxle oil pan. B—Draining the differential oil by removing the drain plug. (DaimlerChrysler)*

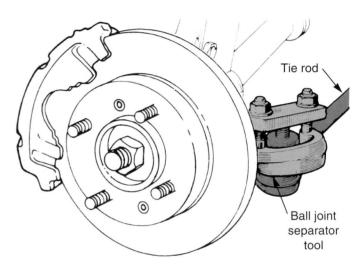

Figure 16-16. *The tie rod end should be removed from the steering knuckle with a special tool. Never pound on the tie rod shaft. Pounding may damage the threads. (DaimlerChrysler)*

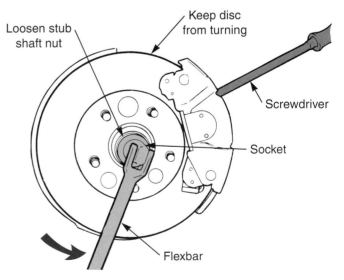

Figure 16-17. *The spindle nut, or hub nut, can be removed with a large breaker bar and a socket. Leave the tire on the floor or have an assistant step on the brake to prevent the rotor from turning. If necessary, a screwdriver can be used to hold the rotor in place. (General Motors)*

10. Remove each front brake caliper as shown in **Figure 16-18.** It is not necessary to disconnect the brake hose from the caliper to perform this operation. Pry the brake pads away from the rotor and remove the caliper fasteners. Calipers may be held in place by large Allen screws or by conventional bolts. A few calipers are two-piece assemblies and the technician must be sure to remove the proper fasteners. Wire the brake calipers out of the way to prevent damage to the brake hoses.

11. Loosen the CV axle shafts at the front hubs. Many CV axles and hubs are a slip fit and will slide apart. If the CV axle shafts are pressed on the front hub, use a pulling tool to remove the axles from the hubs. See **Figure 16-19.** *Do not* hammer on the CV axle to loosen it, as this will damage the CV axle threads. Do not remove the CV axle shafts from the hubs at this time.

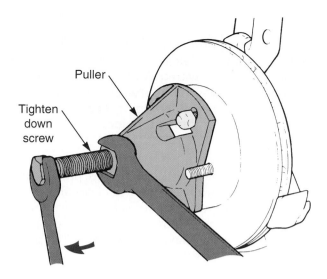

Figure 16-19. *Some front-wheel drive systems use a pressed-on wheel hub and bearing assembly. A special puller tool, such as the one shown here, can be used to remove the spindle from the hub. (General Motors)*

 Caution: Be sure to remove the CV axles in the proper manner, or the axles and other vehicle parts can be damaged.

12. Remove the lower ball joints, **Figure 16-20.** In most cases, the nut holding the lower ball joint can be removed and the ball joint removed with a special puller. Some ball joints are pressed or screwed in, and the service manual should be consulted concerning removal.

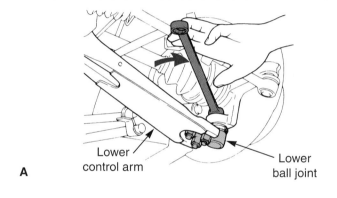

A

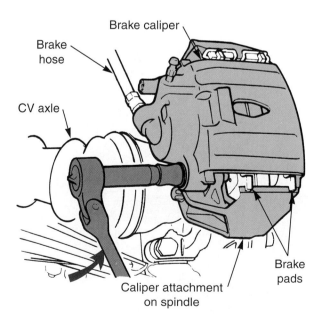

Figure 16-18. *Remove the brake caliper before removing the rotor and the wheel hub and bearing assembly. The caliper can be removed by prying the pads away from the rotor and removing the attaching bolts. The hydraulic system does not have to be opened when removing the caliper. (DaimlerChrysler)*

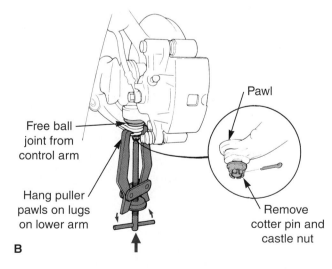

B

Figure 16-20. *This illustration shows the removal process for a lower ball joint. Always support the lower control arm before removing the attaching nut. A—Removing the ball joint attaching nut with a hand wrench. B—Using a special puller to remove the ball joint from the steering knuckle. (Subaru, Honda)*

Warning: Some lower control arms are under spring tension. If this spring tension is suddenly released, the lower control arm will move downward with great force, possibly causing damage or personal injury. Do not remove the lower ball joint before determining whether the lower control arm is under spring tension.

13. Once the ball joint has been removed, push the lower control arm downward until it clears the spindle. Repeat this process for both sides of the vehicle. It may be necessary to loosen the sway bar or strut rod connections to gain sufficient clearance. The spindles can now be removed from the MacPherson struts as explained in the next step.

 Note: On some vehicles, the following step is not necessary.

14. Before removing the spindle from the MacPherson strut, mark the position of the spindle and strut housings. This will help to obtain the original front end alignment during reassembly. Then loosen and remove the attaching bolts as shown in **Figure 16-21**. Some bolts are designed with eccentric (off-center) washers that are used for front end alignment. Mark the position of these washers before removal.

15. Pull the spindles from the MacPherson strut assemblies and pull each CV axle out of its respective hub. (The axles and hubs were loosened in Step 11.) This procedure is shown in **Figure 16-22**. Wire the spindles out of the way to prevent damage.

 Warning: Do not allow the CV axles to hang.

16. Remove the CV axles from the transaxle. Most CV axles are held in place by a retaining ring and can be pulled from the transaxle. If the axle is tight, it can be removed using a special puller attached to a slide hammer, **Figure 16-23**. A few CV axles are held in place by a pin that must be knocked out with a punch, **Figure 16-24**. Other transaxles are bolted to a flange that extends from the transaxle output shafts. These CV axles can be removed by loosening and removing the fasteners. See **Figure 16-25**. A final CV axle retention method makes use of a snap ring or C-lock that holds the axle to the differential assembly. The snap rings can be accessed by removing the differential cover and pulling out the rings, **Figure 16-26**.

 Note: Some vehicles have a two-piece CV axle on the side with the greatest distance between the transaxle output shaft and the drive wheel. To remove these CV axles, the center support must be removed from the vehicle. Most center supports are fastened to the frame. Once the center support has been unfastened, the CV axle and center support can be removed as a unit.

17. Once the CV axles have been removed, place them in a location where they cannot be damaged. Be careful not to damage the CV boots.

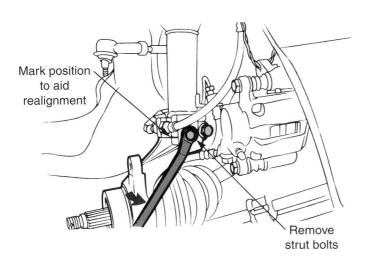

Mark position to aid realignment

Remove strut bolts

Figure 16-21. *Removal of the lower bolts on a MacPherson strut suspension. Always mark the relative position of the parts to aid in front end realignment. (DaimlerChrysler)*

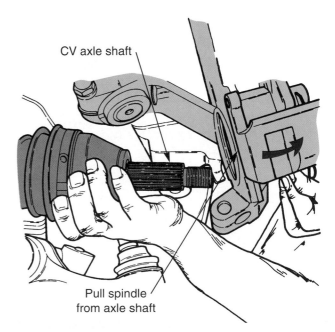

CV axle shaft

Pull spindle from axle shaft

Figure 16-22. *Once the other attachments are removed from the spindle, it can be pulled from the CV axle shaft. (Ford)*

18. Remove the exhaust pipes if necessary to gain clearance for transaxle removal.

> **Note: On some vehicles, it may be necessary to unbolt and swing out the front frame extension or engine cradle on one side to remove the transaxle. On other vehicles, some suspension parts must be removed to gain enough clearance to lower the transaxle. Consult the specific service manual to be sure you perform these procedures properly.**

19. Remove the fill tube and disconnect the throttle valve, kickdown, and shift linkages at the transaxle.

20. Disconnect the transaxle oil cooler lines at the transaxle. Use two wrenches when loosening or tightening fittings to avoid twisting lines. Some cooler lines snap together and require the use of a special removal tool. Position a drip pan under the lines, as they will usually leak oil.

21. Remove the torque converter dust cover.

22. Remove the converter-to-flywheel attaching bolts. The flywheel will have to be turned to reach all the attaching bolts. It may be necessary to remove a cover in the inner fender to gain access to the crankshaft bolt to turn the crankshaft, as shown in **Figure 16-27.** You can also use the starting motor after reconnecting the negative battery cable.

23. Push the converter back away from the flywheel. Remove the starter if it is bolted to the transaxle converter housing.

24. Loosen the transaxle converter housing attaching bolts.

> ⚠ **Warning: Always leave at least two bolts holding the transaxle to the engine until you have a transmission jack securely installed under the transaxle. Although most transaxle cases are aluminum, they are still heavy. You should always use a transmission jack to support and lower the transaxle.**

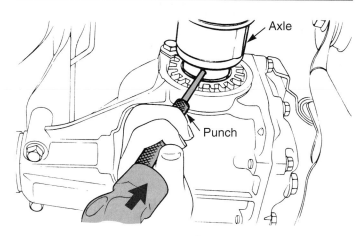

Figure 16-24. *Some CV axles are held in place by a pin that is pressed through the transaxle output shaft and the CV axle shaft. After the pin is driven out, the shaft can be pried out of the transaxle.*

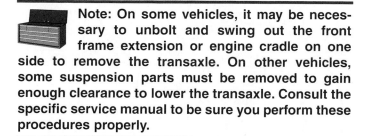
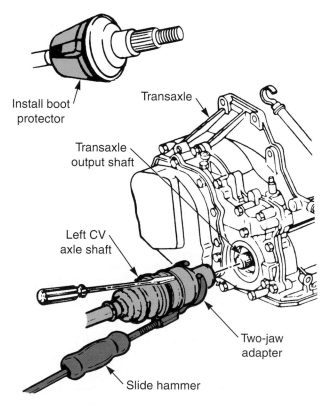

Figure 16-23. *This CV axle is being removed from the transaxle with a special two-jaw adapter that is attached to a slide hammer. (General Motors)*

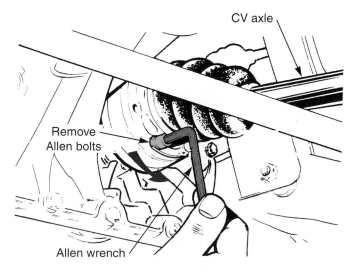

Figure 16-25. *CV axles that are bolted to stub shafts at the transaxle can be unbolted and removed. Always mark the stub shafts and axle shafts before removal to ensure that they are reinstalled in their original position. (Ford)*

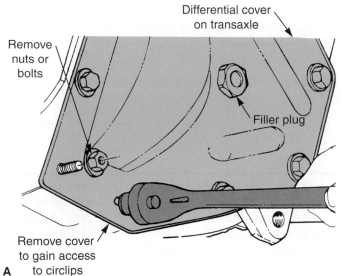

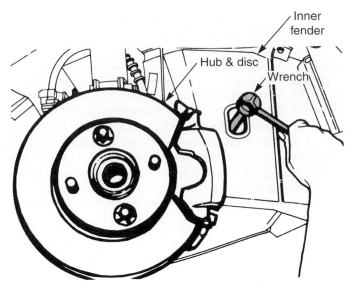

Figure 16-27. *The crankshaft must be turned to gain access to the converter cover attaching bolts. To do so, it may be necessary to remove an access cover in the inner fender. With the cover removed, a wrench and extension can be used to turn the crankshaft. (DaimlerChrysler)*

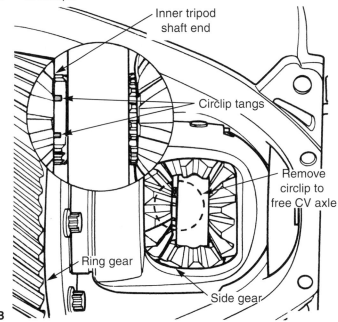

Figure 16-26. *Some CV axles are held in place by a clip that is located inside the transaxle differential assembly. A—Removing the differential cover. B—Removing the circlip. (DaimlerChrysler)*

26. Remove the last transaxle converter housing attaching bolts holding the transaxle to the engine.

27. Slide the transaxle away from the engine, until the converter housing clears the engine flywheel. Attach a small C-clamp to edge of the converter housing or use a torque converter holding tool to secure the converter.

28. Lower the transaxle, watching clearances which are usually close on front-wheel drive vehicles. Also, watch for any component that you may have forgotten to disconnect, especially any electrical connectors, ground wires, or other wiring connected to the transaxle.

29. As soon as the transaxle is clear of the vehicle, lower it to the floor and transport it to the workbench in the lowered position. If the transaxle is very dirty, you may want to remove some of the outside dirt before beginning disassembly.

Automatic Transaxle Installation

Installing the transaxle is basically the reverse of removal. As with removal, the vehicle should be on a hoist with the engine supported. The following procedures detail the procedure for installing an automatic transaxle in a vehicle. Always refer to the manufacturer's service manual for specific procedures.

25. Place a transmission jack under the transaxle so that the assembly is safely supported. Raise the jack slightly to remove the weight of the transaxle from the lower transaxle mounts. Then, remove the lower transaxle mounts and any other parts holding the transaxle to the vehicle body.

 Note: The transaxle may be attached to the mounts through various brackets. Study the transaxle to determine whether the brackets should be left with the transaxle or removed before transaxle removal.

1. Secure the torque converter with a torque converter holding tool or a C-clamp after making sure the converter is fully seated over the input, stator, and pump shafts.

 Caution: Under no circumstances should you install the converter on the flexplate while it is disassembled from the rest of the transaxle. The splines of the transaxle input and stator shafts and the converter can be damaged.

2. Raise the transaxle under the vehicle with a transmission jack or floor jack until it is at the proper height to engage the engine. Then, slide the transaxle converter housing into engagement with the rear of the engine. Make sure that the converter housing fits tightly against the rear of the engine. (Remove the converter holding tool when necessary.)

3. Once the converter housing is flush with the rear of the engine, install at least two of the transaxle converter housing attaching bolts.

4. After the bolts are installed, make sure the torque converter still turns easily and that it is not jammed. If it is, it is not properly fitted over the transaxle shafts. The transaxle must be removed and the converter installed properly.

5. If the converter installation is proper, install the remainder of the transaxle converter housing attaching bolts. Torque all bolts to specifications.

6. Install and tighten the converter-to-flywheel bolts (or nuts) to specifications. Install the starter, if it was removed. Then, install the dust cover.

7. Install and tighten the engine mounts and other fasteners holding the transaxle to the vehicle frame.

8. Begin installation of the CV axles by attaching the inboard end of the CV axle to the transaxle, **Figure 16-28.**

9. Install the outboard end of the CV axle into the spindle and hub assembly, **Figure 16-29.** If the CV axle uses a center support, fasten it to the frame at this time.

10. Reinstall the spindle on the MacPherson strut. Make sure the spindle and strut are reattached as close to their original position as possible. This will aid front end realignment when installation is complete.

11. Pull the lower control arm upward until the ball joint stud can be reinstalled onto the spindle. Install and tighten the clamp bolt, **Figure 16-30.**

12. Reinstall and tighten the sway bar or strut rod connections to the lower control arm.

13. Install and tighten the spindle nut. If the spindle nut uses a washer, be sure to reinstall it. Also, install a new cotter pin if necessary. If the nut should be staked in place, do this now using a hammer and small punch.

14. Reattach the tie rod end to the spindle. Tighten the nut and install a new cotter pin. At this time, you can tighten any other front suspension and steering parts that were removed.

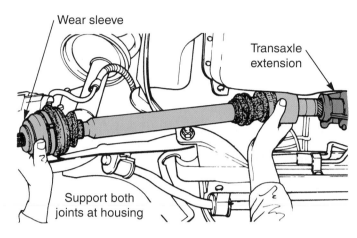

Figure 16-28. *Install the CV axle in the transaxle by sliding the yoke into the housing. Once the yoke is fully seated, the other end of the CV axle can be installed on the spindle. (DaimlerChrysler)*

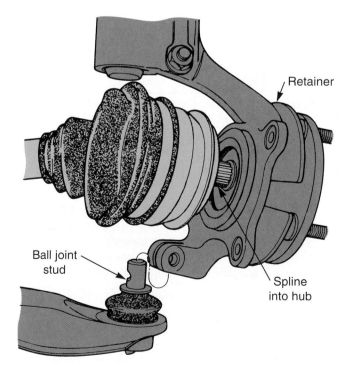

Figure 16-29. *Insert the CV axle in the spindle by lining up the internal splines of the hub with the external splines of the axle. Once the CV axle has been inserted in the spindle, reinstall the ball joint. Note the notch in the ball joint stud that allows the clamp bolt to pass through. (DaimlerChrysler)*

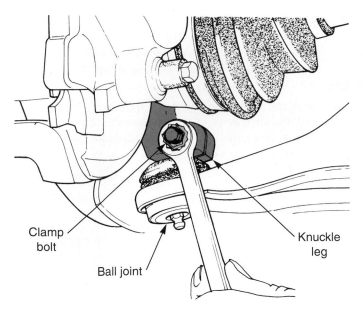

Figure 16-30. *Install the spindle on the ball joint; then install and tighten the clamp bolt. Once the CV axle has been inserted in the spindle, the ball joint can be reinstalled. (DaimlerChrysler)*

15. Install the oil cooler lines; the modulator vacuum line; the electrical connectors; the TV, kickdown, and shift linkages; and other miscellaneous items. Install the exhaust pipes if they were removed. Install the speedometer cable and dipstick tube if they must be installed from under the vehicle.

16. Install the starter.

17. Reinstall any other frame and body parts that were removed for clearance.

18. Lower the vehicle. Install the negative battery cable and other underhood components. Remove the engine support fixture. Be sure to install all electrical components and ground wires. Check the vacuum lines at the engine. Check the various linkages for proper adjustment and movement.

19. Fill the transaxle transmission with the proper ATF. It is advisable to overfill a dry transaxle by about 2 quarts (or 2 liters) *as measured on the dipstick* before starting the engine. If the unit has a separate reservoir for differential oil, refill it also.

20. Start the engine and immediately recheck the transmission fluid. Add fluid until the proper level is reached. Be careful not to overfill the transaxle transmission. Most transaxles hold about 6 quarts of fluid, but check specified capacity.

21. Raise the vehicle by the front lower control arms so the drive wheels are off the ground but are near their normal position. Do not allow the wheels to hang without support. This will damage the CV joints.

22. Operate the transaxle in all gears, allowing it to upshift and downshift several times in drive gear. This will allow you to confirm that the transaxle hydraulic system is operating properly, and it will allow you to check for leaks.

23. Lower the vehicle and perform a road test. If the transaxle shift points are incorrect, check the TV and throttle linkage adjustment or the vacuum modulator connections, as necessary. Recheck the fluid level after the vehicle has been driven a few miles.

Aligning Engine and Drive Train Mounts

In many cases, the rubber or fluid-filled mounts that hold the engine and drive train to the vehicle's frame can be out of alignment. This is more of a problem on front-wheel drive vehicles, but can occur on any vehicle. Misaligned mounts on a rear-wheel drive vehicle can cause vibration on acceleration. Sometimes, a broken or badly misaligned mount can cause the drive shaft to strike the body. In extreme cases, the engine can be so far out of alignment that the fan strikes the radiator core or the fan shroud. A broken or sagging (collapsed) mount is more likely on a rear-wheel drive vehicle than a front-wheel drive vehicle. Always check for broken and sagging mounts before attempting to make adjustments.

On a front-wheel drive vehicle, a misaligned mount can cause several problems. The vehicle may vibrate or pull to one side on acceleration. A severely misaligned mount can break a CV joint or allow the CV axle to disengage from the transaxle output shaft. Broken and sagging mounts can also cause these problems. Some front-wheel drive mounts are not adjustable.

Checking for a Broken Mount

To check for a broken mount on an operating vehicle, open the hood of the vehicle, apply the parking brake, and start the engine. Then have an assistant firmly apply the brake pedal, place the transmission in drive, and slightly open the throttle. Observe the engine as the throttle is opened with the brakes applied. If engine torque causes the engine to rise more than 1″ (2.54 mm), a motor mount may be broken or soft. Repeat this process with the transmission in reverse, since some mounts compress under torque in drive. If the engine rises too far, visually inspect the mounts.

Broken mounts are sometimes difficult to spot. Carefully pry on the engine to raise it in the area of the mount. Attempting to stretch the mount in this manner will usually reveal cracked rubber. Fluid-filled mounts should be checked for oil on their exterior, a sign of leakage. As you check the mount, also look for loose fasteners.

Adjusting Rear-Wheel Drive Mounts

Rear-wheel drive mounts are generally not adjustable and defective mounts must be replaced. The only mount that may be adjustable is the rear mount between the transmission tailshaft and the cross member. Loosen the mounting bolts and move the tailshaft until it is centered in the drive shaft tunnel. Then retighten the mounting bolts.

 Note: If the mount has sagged, it may affect U-joint angles and cause vibration. Some rear mounts can be shimmed to correct this problem.

Checking Front-Wheel Drive Engine Cradles and Brackets

Always check the engine cradle if it holds one of the mounts. The cradle may be out of position, causing the mount to be misaligned. Begin by observing the marks made by the cradle mounting bolts and washers. Notice whether the original marks made by the washers are completely covered. If not, the cradle is incorrectly positioned. Correct this condition by loosening the mounting bolts and moving the cradle to its original position. On some vehicles, an incorrectly installed upper bracket can cause misalignment. Compare the original position of the bolts and washers and readjust the bracket as needed.

Adjusting Front-Wheel Drive Mounts

If the cradle and bracket are in the correct position, the front-wheel drive mount can be aligned by the following procedure.

 Note: On most front-wheel drive vehicles, only one mount is adjustable. Use the proper service literature to determine which mount can be adjusted.

Begin adjustment by removing the load on the mount. The easiest way to do this is to raise the engine and transaxle assembly with a floor jack. Raise the assembly only enough to unload the mount. Be careful not to damage the oil pan. Next, loosen the adjustable mount fasteners and reposition the mount as needed. On many vehicles, mount adjustment is checked by measuring the length of the CV axle on the same side as the adjustable mount. If the length of the CV axle is correct, the mount is properly adjusted. On other vehicles, the distance between the mount and a stationary part of the vehicle's frame must be measured. Sometimes the adjustment procedure simply involves loosening the fasteners and allowing the mount to assume its normal unloaded position.

After the mount has been adjusted, tighten the mount fasteners. Then remove the jack from under the engine and transaxle.

Summary

Before a transmission or transaxle can be overhauled, it must be removed from the vehicle. Removal procedures vary from one transmission or transaxle to another, but many general procedures apply. Have an oil pan handy to catch drips. The battery negative cable must be disconnected before any other steps are taken. The battery cable can be temporarily reconnected to use the starter to turn the flywheel to get at the converter-to-flywheel fasteners.

The general series of transmission/transaxle removal begins with removing anything that can be reached from under the hood with the vehicle on the ground and then raising the vehicle. If a transaxle is being removed, install an engine support fixture. Next, remove the converter-to-flywheel fasteners and the converter housing attaching bolts in the sequence recommended for the unit at hand. Place a transmission jack under the transmission or transaxle. Remove all electrical connectors, lines, and fittings. Then remove the drive shafts or CV axles as applicable, and remove the cross member and brackets that hold the unit to the vehicle. Position the jack to take the load off the engine-converter housing mating area and remove the remaining converter housing attaching bolts. Slide the transmission or transaxle away from the engine. As soon as the transmission or transaxle clears the engine flywheel, lower it and move the unit to a bench, being careful not to let the converter fall out.

To begin reinstallation of the transmission or transaxle, make sure that the converter is fully installed and held in place, and then position the unit on the engine. Install the converter housing attaching bolts and the converter-to-flywheel fasteners as specified. Reinstall the cross member and/or brackets as necessary, and then remove the transmission jack. Next, install the drive shaft or CV axles as applicable. Reattach all fittings and electrical connectors. Finish tightening the converter housing attaching bolts and tighten other connections as needed. Once all connections have been made under the vehicle, lower the vehicle and make all under hood connections. Remove the engine support fixture, if necessary. Add transmission fluid to the unit. Start the vehicle and allow it to operate. Add additional transmission fluid as necessary. Road test the vehicle and make any final adjustments.

Review Questions—Chapter 16

Please do not write in this text. Place your answers on a separate sheet of paper.

1. The first step in transmission/transaxle service is to remove the _____ _____.

2. A two-piece drive shaft will always have a _____ _____.

3. When removing a transmission, what can be done to gain access to the converter-to-flywheel attaching bolts?

4. To turn the flywheel with the starter, you will have to temporarily reconnect the _____ _____.

5. Oil cooler fittings should always be started by _____.

6. Before removing a lower ball joint, make sure the lower control arm is not under _____ _____.

7. Most CV axles are held in place by a _____.
   (A) bolt
   (B) retaining ring
   (C) pin
   (D) None of the above.

8. Why should two wrenches be used when loosening or tightening oil cooler line fittings?

9. After sliding a transaxle away from the engine, the converter can be secured with a small _____ or a torque converter holding tool.

10. Why should you never install the converter to the flexplate and then install the transaxle?

ASE-Type Questions—Chapter 16

1. To keep the engine from falling out of the vehicle, the _____ should be checked before the transmission is removed.
   (A) engine mounts
   (B) cross member bolts
   (C) exhaust hangers
   (D) parking brake linkage

2. Each of the following can be used to gain access to the converter-to-flywheel attaching bolts *except:*
   (A) the starting motor.
   (B) a flywheel turner.
   (C) turning both drive wheels in the same direction.
   (D) turning the crankshaft bolt with a wrench.

3. Which of the following transmission installation steps should be taken first?
   (A) Reinstall the exhaust system pipes.
   (B) Install the converter-to-flywheel bolts or nuts.
   (C) Install the cross member.
   (D) Remove transmission jack.

4. All of the following steps should be taken during transaxle removal *except:*
   (A) remove the converter-to-flywheel bolts or nuts.
   (B) remove the CV axle boots.
   (C) remove the wheels and tires.
   (D) remove the spindle hub center nut.

5. When a transaxle is reinstalled on the engine, it is discovered that the torque converter is jammed against the flywheel. Technician A says the engine should be started to allow the converter to be positioned on the pump ears. Technician B says the transaxle should be removed and the converter should be reinstalled properly. Who is right?
   (A) A only.
   (B) B only.
   (C) Both A and B.
   (D) Neither A nor B.

Chapter 17

Rebuilding Automatic Transmissions and Transaxles

After studying this chapter, you will be able to:
- ❏ Disassemble an automatic transmission.
- ❏ Inspect the internal parts of an automatic transmission.
- ❏ Reassemble an automatic transmission.
- ❏ Disassemble an automatic transaxle.
- ❏ Inspect the internal parts of an automatic transaxle.
- ❏ Reassemble an automatic transaxle.
- ❏ Check endplay of an automatic transmission or transaxle.
- ❏ Perform air pressure checks on an automatic transmission or transaxle.

Technical Terms

Endplay

Soft parts

V-blocks

Leopard spotting

Clutch clearance

Introduction

Rebuilding, or overhauling, automatic transmissions and transaxles requires extensive knowledge and careful work habits. This chapter will outline automatic transmission and transaxle overhaul procedures. If you have studied the information in Chapter 3 and in Chapters 8–10, you will be familiar with the components and tools discussed in this chapter.

This chapter presents *general* procedures for rebuilding automatic transmissions and transaxles. Always refer to the manufacturer's service manual for specific information. The manufacturer's service manual will contain accurate instructions for disassembling, cleaning, inspecting, and rebuilding the unit at hand. Note that certain procedures performed as part of the rebuilding operation—valve body service, for example—were covered in Chapter 15.

Note: For transaxle final drive rebuilding information, refer to Chapter 18.

Although a typical automatic transmission or transaxle contains many parts, it can be successfully rebuilt if you follow logical procedures.

Special service tools are sometimes needed when rebuilding an automatic transmission or transaxle. These tools can be purchased from the vehicle dealer or ordered directly from the manufacturer. Refer to Chapter 2 for more information on special service tools used to service automatic transmissions and transaxles.

Automatic Transmission and Transaxle Disassembly

Before beginning disassembly of a transmission or transaxle, clean the outside of the unit thoroughly and mount the unit in a holding fixture, **Figure 17-1.** If a holding fixture is not available, place the assembly on a clean workbench. The workbench should be constructed so that transmission fluid can drain into a catch basin.

As you remove the major parts of the transmission or transaxle, set them aside for further disassembly. Then work on one subassembly (pump, clutch pack, or servo, for example) at a time. This will help you keep things straight.

Removing External Components

The first step in disassembly is to remove the torque converter holding tool and the torque converter from the transmission or transaxle. Then check input and output shaft endplay as shown in **Figure 17-2.** *Endplay* (back-and-forth movement of the shaft) should be checked before the

transmission or transaxle is disassembled (and again after it is reassembled). Record the endplay readings for later reference. Excessive endplay indicates wear of the thrust washers and other parts. This should be corrected during the rebuilding process.

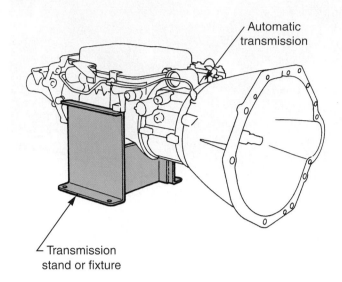

Figure 17-1. *It is much easier to overhaul a transmission or transaxle if it is placed in a special holding fixture. (Nissan)*

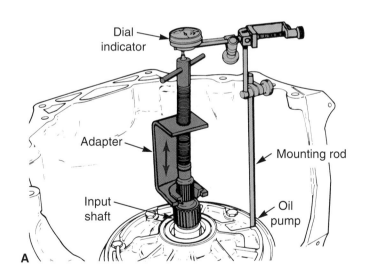

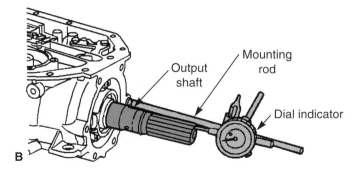

Figure 17-2. *Always check endplay before disassembling the transmission. The service manual will contain the endplay specifications and the procedures for setting up the dial indicators. A—Checking input shaft endplay. B—Checking output shaft endplay. (General Motors)*

After checking endplay, remove the oil pan. Check the pan for metal particles, varnish buildup, and sludge. Excessive sludge indicates that the holding members are burnt and the fluid has been overheated. Metal particles usually indicate that the torque converter, oil pump, planetary gearsets, or other moving parts are badly worn. If this is the first time the oil pan has been removed for service, it is normal to see a few aluminum particles in the pan. These particles were left over from case machining operations at the factory.

Removing the oil pan gives you access to the filter and valve body. The filter may be pressed into a case passageway and held in place with a clip on the valve body, or it may be bolted directly to the valve body. Remove the filter at this time.

After the filter has been removed, the valve body bolts can be unfastened and the valve body can be removed from the case. As you lift the valve body from the case, carefully note the position of all valve body parts, including springs, oil feed tubes, linkage attachments, and check balls. If applicable, you should also remove the vacuum modulator, push rod, and throttle valve from the case at this time. See **Figure 17-3.** Place the assembled valve body and related parts, as well as any modulator parts, together on the workbench.

If you are disassembling a transaxle, remove any side or top covers; then remove the side or top valve body. Note the location of all tubes and check balls. If the transaxle uses a transfer plate, remove it also.

Remove the bolts holding the extension housing to the case. Extension housings are sometimes called tailshaft housings. The extension housing should pull off over the output shaft after the bolts are removed. A few extension housings contain an output shaft bearing, which supports the output shaft. In this design, the bearing snap ring must be expanded before the extension housing can be slid from the shaft.

After removing the extension housing, you can remove the governor. This step applies only to governors mounted on the output shaft, not those mounted in the case. If you are disassembling a transaxle with a drive chain, check the chain for wear as shown in **Figure 17-4.** Then remove the chain and sprocket assembly. If the transaxle uses transfer gears, remove the gears and carefully check the teeth for wear. See **Figure 17-5.**

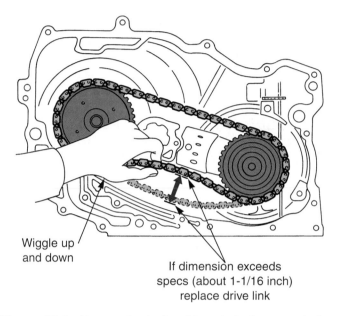

Figure 17-4. *Always check the drive chain for wear before disassembly. To disassemble, remove any snap rings that may be holding the sprockets to their shafts. Position the chain tensioner away from the chain. Then remove the chain and sprockets as an assembly by carefully sliding both sprockets off the transaxle shafts. Save any thrust washers placed between the cover and the sprockets. (DaimlerChrysler)*

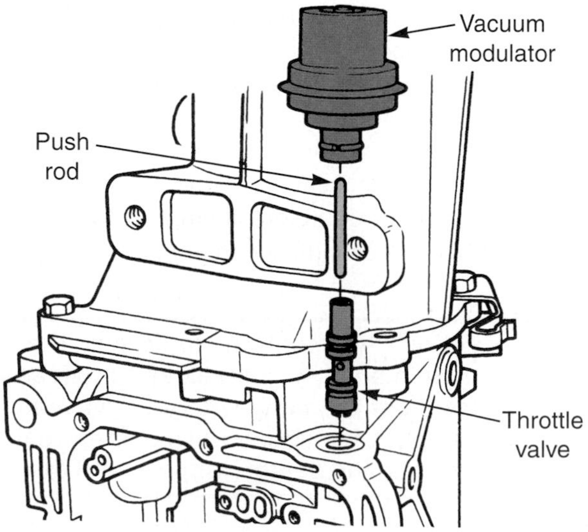

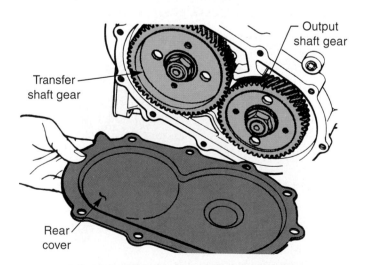

Figure 17-3. *The vacuum modulator and throttle valve can be removed now to prevent damage. Do not lose the pushrod, if the modulator uses one. (Ford)*

Figure 17-5. *Drive gears are installed behind a sheet metal cover. They should be checked for wear or damage. (General Motors)*

Remove the front oil pump seal as shown in **Figure 17-6.** It is generally easier to remove the seal before the oil pump is removed because the extra weight of the transmission/transaxle assembly allows you to use the slide hammer more efficiently. The front pump seal should always be changed when the pump is serviced.

The oil pump can be removed next. Manufacturers sometimes recommend removing the input shaft before removing the oil pump. However, some transmissions and transaxles are constructed so that the input shaft is pressed into the front clutch drum. In these designs, the shaft cannot be removed before the pump.

The pump usually fits tightly in the case, and some pressure must be applied to remove it. Some pumps can be removed by prying them out with a screwdriver or pry bar (once the valve body has been removed). Others must be removed with slide hammers or special pullers, **Figure 17-7.**

When removing the oil pump, always look for thrust washers, which may fall out of place. A few oil pumps have a clutch apply piston installed in their inner face. When

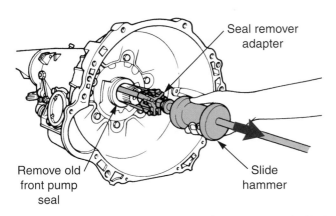

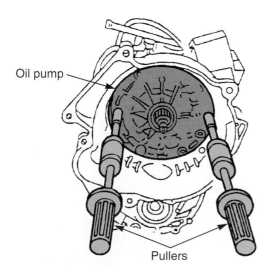

Figure 17-6. *It is easier to remove the front pump seal with the transmission or transaxle assembled. The weight of the unit allows the slide hammer to exert more force on the seal. (Ford)*

Figure 17-7. *Some pumps can be removed by pushing them out from the rear. On many transmissions and transaxles, however, the pump cannot be reached from the rear and must be removed with a slide hammer or special puller. (Ford)*

this is the case, a set of friction discs and clutch plates is installed directly behind the pump. These parts should be removed and placed with the pump for later service.

Removing Internal Components

With all the external parts out of the way, the internal components, such as the input shaft, bands, clutch packs, planetary gearsets, and output shaft, can be removed. Removal steps are similar for all manufacturers. The major factor affecting disassembly is whether or not the transmission or transaxle uses a center support.

On units without center supports, removing internal components is usually a simple matter once the oil front pump is removed. A typical procedure might involve loosening the front band adjuster, sliding the band out of the case, and then removing the forward clutch packs, front planetary gearset, and input shell as an assembly. Be sure to save all thrust washers, as they come out with these parts.

Once the front components have been removed, the rear planetary gearsets and holding members, as well as the output shaft, can be removed. Rear gearsets are removed through the front of the case. The output shaft will usually come out through the rear of the case.

Note that some clutch packs are held in place by a snap ring located on the sun gear, inside the input shell. This snap ring must be removed before the clutch pack and other internal parts can be removed. In addition, some rear planetary gearsets are held in place by a snap ring on the front of the output shaft. This snap ring must be removed in order to completely disassemble these components. Make sure the shaft does not fall out of the case after the snap ring is removed. Also, do not distort or stretch snap rings when removing them. Use snap ring pliers or another suitable tool.

As mentioned, some automatic transmissions have a center support. The center support in an automatic transmission may be retained by one or two case bolts located under the valve body and, sometimes, by bolts on the outside of the case. Other center supports are held in place by a retaining ring in the case, **Figure 17-8.** The center support keeps the rear clutches and gearsets in place. Always make sure the center support bolts or retaining rings have been removed before attempting to remove the center support and mechanical components.

> **Caution:** Attempting to drive the rear clutch, planetary gearset, and shaft assemblies from the case without first removing the center support fasteners will severely damage these components, the center support, and the case.

Once the center support has been removed, the rear planetary gearsets, bands, clutches, and output shaft can

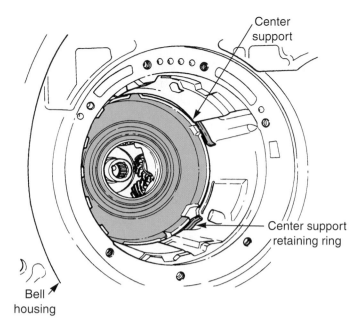

Figure 17-8. *Many modern transmissions and transaxles have a center support. The support shown here is held in place with a retaining ring, which must be removed before removing the rear planetary gearset and holding members, as well as the output shaft. (Ford)*

be removed. The gearsets, holding members, and output shaft can usually be removed through the front of the case. It may be necessary to remove the parking gear before removing the output shaft.

Once the mechanical components mentioned are removed, any internally mounted accessories can be removed. Examples of these parts include some servos, accumulators, clutch apply pistons (usually used at rear of case), and governors. If these parts are not mounted inside the case—for example, if they mount inside the oil pan—they will be removed with the other external components.

Note that many servos and accumulators are under strong spring pressure. Always consult the service manual before removing any retaining rings or bolts. It may be easier to leave the servos and accumulators installed in the case until it is time to inspect them and replace their seals. If the transmission is filled with sludge and debris, however, these parts should be removed at this time for a thorough cleaning.

Automatic Transmission Parts Cleaning

All automatic transmission or transaxle parts must be cleaned thoroughly. This is very important because small dirt particles or varnish formation can cause a passageway to plug up, a valve to stick, or a seal to leak.

Carefully scrape all old gasket material from the case, oil pan, and other parts. Pay attention to the gaskets on hydraulic control assemblies, such as the front pump, the valve body, and the spacer plates.

Clean all metal parts, including the case, with a safe solvent. It may be necessary to soak parts overnight. Do *not* clean clutch friction linings or seals with solvent. To avoid having a confusing pile of parts, always remember to work on one subassembly at a time while cleaning.

 Warning: *Never* clean parts with gasoline or other flammable solvents. A fire could result!

Dry the cleaned parts with compressed air (if possible) or allow them to air dry. Do *not* use rags or shop towels to dry the parts, because they will leave lint on the parts. The lint left behind can cause valves to stick. It is better to leave slight deposits of solvent on the parts than to dry them with rags or shop towels. Once cleaned, cover the disassembled parts with a clean lint-free cloth if they are to sit out overnight.

Drain the torque converter. If there is a drain plug, remove it and set the converter upright with the drain at the lowest point. On converters without a drain plug, lay the center opening face down. This will allow most of the old transmission fluid to drain from the converter.

If the internal parts were very dirty or heavily coated with varnish, the torque converter will be full of dirt and varnish. A torque converter that is filled with dirt or metal particles can be cleaned with a converter flusher. Converter flushers were introduced in Chapter 3. A heavily varnished converter, however, is difficult to clean and is often replaced.

If there is evidence of fluid contamination, the oil cooler and cooler lines must be flushed out. One way to do this is to use an oil cooler and line flusher and follow up with a blast of compressed air directed into one of the cooling lines. Have a drain pan available to catch solvent flowing from the other line. Do *not* apply full air pressure to the cooling lines, as this could rupture the oil cooler. A badly contaminated oil cooler cannot be completely cleaned. Therefore, the radiator should be replaced or the cooler should be bypassed by installing an external oil cooler when a badly contaminated unit is encountered.

Parts Inspection and Repair

Once the transmission or transaxle is disassembled and cleaned, the parts can be inspected, replacement parts can be ordered, and subassemblies can be rebuilt. In this section, parts inspection and repair will be covered when describing the service of each subassembly. During an actual rebuild, internal parts will be inspected during and immediately after the unit is taken apart and cleaned. You should inspect transmission or transaxle parts one subassembly at a time. In this way, you can order replacement parts right away, eliminating time lost waiting for parts.

Note: During inspection, it is not absolutely necessary to check the *soft parts,* such as gaskets, lip seals, O-rings, seal rings, and friction discs. These parts are replaced during a rebuild. Checking the soft parts, however, will often help you to determine what caused the transmission or transaxle to fail.

Shaft Service

Automatic transmission and transaxle shafts, including input, output, and stator shafts, require little service. Check seal ring grooves and bushing surfaces for wear, scoring, and overheating. Inspect shaft splines for wear or damage, **Figure 17-9.** Any shaft with worn or damaged splines should be replaced. If you suspect that a shaft is bent or warped, it can be checked with **V-blocks** and a dial indicator as shown in **Figure 17-10.**

Figure 17-9. *After the transmission or transaxle has been disassembled, check shaft splines, bearing journals, and ring grooves for wear.*

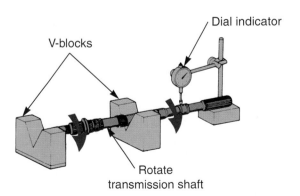

Figure 17-10. *Input and output shafts can be checked for bends or warpage using V-blocks and a dial indicator. Never attempt to straighten a bent or warped shaft. (Toyota)*

Planetary Gearset Service

Check the teeth of the planetary gears for wear, nicks, and chipping. **Figure 17-11** shows a gearset that has been destroyed. **Figure 17-12** shows a sun gear being inspected for tooth damage. Slightly nicked gears can cause noises and vibration, and eventually, they will fail completely. Splines found on the gearset members should also be checked for wear and damage. Check the clearance between the planet gears and the planet carrier to deter-

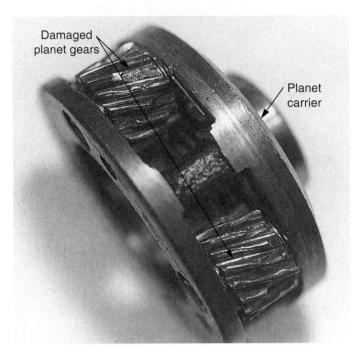

Figure 17-11. *Worn or damaged planet gears can result in a noisy transmission or transaxle. Severely damaged gears will strip off or jam, resulting in a no-drive condition. (BBU, Inc.)*

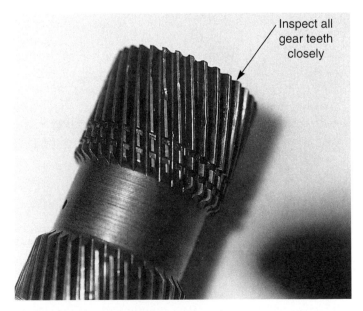

Figure 17-12. *Gear teeth on all of the planetary gears should be closely inspected for chips, cracks, scoring, and wear. Also, check for worn snap ring grooves or bushings. Replace any damaged gears. (DaimlerChrysler)*

mine the amount of endplay, **Figure 17-13.** Planetary gearsets often contain bushings or needle bearings that wear out. These should be carefully checked, as well. Any parts that are worn or damaged should be replaced.

Visually inspect needle bearings and Torrington (flat) bearings and try to rotate them by hand. If you feel any roughness when turning a clean bearing or a gear that rides on a bearing, replace the bearing. If you suspect that a bearing inside a gearset is worn, take the time to disassemble the gearset and check the bearing.

Bushings can be checked visually. If a bushing shows signs of wear or scoring, replace it. A bushing that looks good can be excessively worn. Check bushings for wear by inserting the mating shaft or race into the bushing and trying to rock the shaft or race back and forth. If the shaft or race rocks excessively, the bushing is loose and should be replaced.

Clutch Pack Service

One of the most important parts of automatic transmission overhaul is the proper checking and rebuilding of the clutch packs. These operations are discussed in the following sections.

Clutch Pack Disassembly and Inspection

To begin disassembly, use a screwdriver to remove the snap ring that holds the plates and friction discs in the clutch drum. See **Figure 17-14.** Then, remove the plates and discs. Unless they are excessively worn, the discs and plates will fall out when the drum is turned over. Keep all the steel plates together, noting if there are any reaction plates (thicker steel plates) at the top and bottom of the stack, **Figure 17-15.** Never interchange one set of clutch pack parts with another.

Inspect the steel plates. Look for discoloration, which is caused by overheating and is often a sign of a badly slipping clutch pack. A common sign of overheating is *leopard spotting,* **Figure 17-16.** Look for worn or scored

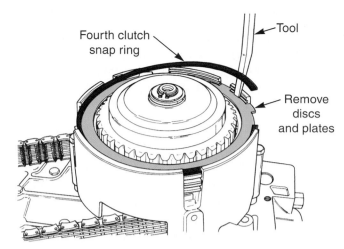

Figure 17-14. *The clutch packs should be disassembled to check for burned friction discs and other problems. Most clutch packs are held together with a large snap ring, as shown here. (General Motors)*

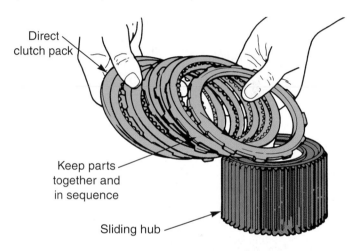

Figure 17-15. *Inspect the clutch plates for wear, burning, scoring, and other damage. Also, check the drum, bushings, hub splines, and thrust washers. Replace parts that show signs of wear or damage. (DaimlerChrysler)*

Figure 17-16. *Steel plates showing leopard spots are often found when the transmission or transaxle is disassembled. Leopard spots are named for their resemblance to a leopard's coat and are a sign of severe clutch overheating. The dark spots are areas where the metal has become so overheated that it has lost its strength and resistance to wear.*

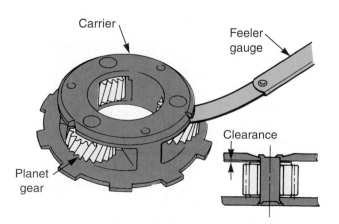

Figure 17-13. *The planet gears and planet carrier can be checked for wear with a feeler gauge, as shown here. Disassembly is not required to make this check. (Nissan)*

plates, a sign of prolonged slippage. See **Figure 17-17.** Also, check the tangs for wear or damage. **Figure 17-18** shows a used steel plate. This plate is shiny, which is normal for a used plate, and could be reused if necessary. If there is any doubt about the condition of the plates, they should be replaced. Steel plates are sometimes included in a transmission overhaul kit.

Friction discs are normally replaced as part of a transmission overhaul and are usually included in a transmission overhaul kit. Worn or damaged discs may be charred, glazed, or heavily pitted. The friction lining may scrape off easily with your fingernail. **See Figure 17-19.** In some instances, all the friction material will be missing from the friction discs. **Figure 17-20** shows a friction disc that has lost almost all its friction material.

Check the condition of the channels on the inside surface of the clutch drum and the splines on the clutch hub. Severe disc damage may cause the hub splines to be stripped. Worn or damaged parts should be replaced.

Removing the Apply Piston

Rebuilding a clutch pack requires the use of a special spring compressor, **Figure 17-21.** Install the spring compressor and compress the return spring. Remove the retaining snap ring with snap ring pliers or a screwdriver. Release the spring compressor; then remove the spring retainer and the return spring or springs.

Remove the clutch apply piston from the drum, **Figure 17-22.** If the piston sticks in the drum, it can sometimes be removed by slamming the drum downward on a block of wood or a wooden workbench. This will

Figure 17-19. *If friction disc material can be scraped off with a fingernail, the friction material is burned. A disc in this condition cannot be reused and should be discarded.*

Figure 17-17. *This plate was scored by the scraping action of foreign material between the plate and the friction disc. The foreign material may be from another part of the transmission or transaxle, or it may be the result of severe damage to the friction discs.*

Figure 17-18. *A normally worn steel plate will be shiny as shown here. This steel plate could be reused if necessary. Before deciding that the plate is good, however, check it for warping and damage to the teeth.*

Figure 17-20. *The friction material of a friction disc will sometimes wear away completely, leaving the steel backing exposed. This disc should be replaced. Water in the transmission will also cause the friction material to separate from the steel backing.*

often jar the piston loose. Compressed air can also be used to blow the piston from the drum. Place the drum downward on the workbench and insert an air nozzle in the oil feed hole in the center of the drum. A short burst of air should push out the piston.

 Warning: When using compressed air to remove a piston, point the open end of the drum away from yourself and others. Direct the piston into a pile of rags or another soft surface so that it does not fly out and become damaged or cause injury. Also, use a regulated amount of air pressure.

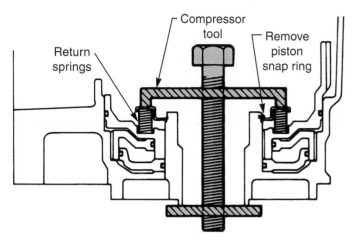

Figure 17-21. *A special compressor must be used to remove the clutch apply piston in order to get at the seals of the clutch assembly. Universal tools are available that fit all types of assemblies. (DaimlerChrysler)*

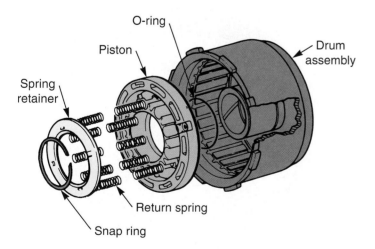

Figure 17-22. *Exploded view of a clutch apply piston and related parts shows parts that must be removed during rebuilding. (Nissan)*

With the piston out of the drum, remove the piston seals. Hard or cracked seals indicate that the clutch has been overheated. If there is also a seal around the interior (hub) of the clutch drum (inner piston seal), remove it, **Figure 17-23.** Then, thoroughly clean the piston and piston bore.

Often, clutch apply pistons and clutch assemblies are installed in the rear of the case. Disassembly and assembly procedures for these assemblies are similar to those for other clutch packs.

Note: If the check ball is missing, do not attempt to install a replacement check ball. Replace the apply piston.

Inspect the piston for cracks. Also check that the air bleed check ball is free.

Obtain new piston seals. Compare the new seals to the old ones to ensure that you are installing the proper seals on the piston. In addition, check the fit of the new seals by placing them in the piston bore. They should fit snugly. A bulge, or buckle, in an outer piston seal means the seal's diameter is too large. A gap between an outer piston seal and the piston bore means the diameter is too small.

Before installing the new piston seals, lubricate them with transmission fluid. Petroleum jelly can also be used, but other types of oil or grease will damage the seals. Carefully install the new piston seals on the clutch drum hub and clutch piston, as applicable. The new seals should fit snugly, but *not too* snugly. Lip seals must be installed on a piston so the sealing lip will be directed toward the hydraulic pressure (toward the back of the drum).

Install the clutch piston in its bore. The type of piston seal determines the installation method. If O-ring seals are

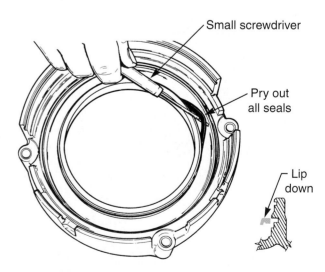

Figure 17-23. *The clutch pack seals should be removed for inspection. Take a close look at them, even though they will be replaced. Hard or cracked seals can often cause upshift and slippage problems. Noting the condition of the original seals will often confirm your initial diagnosis. (General Motors)*

used, the piston can be pressed or tapped into the bore after being thoroughly lubricated.

A piston using a lip seal must be worked into place with a feeler gauge or a special seal installation tool. A satisfactory tool can be made from a stiff piece of wire and a small length of copper tubing. Use the tool to push in the seal as you press on the piston. See **Figure 17-24.** Once the piston is in place, check its installation by trying to turn it. If you cannot turn the piston, the seal is not properly positioned.

Once you are sure the piston is properly installed, put the piston return spring(s) and spring retainer back in position on the piston. Then, use the spring compressor tool to move the return spring(s) and retainer past the groove for the retaining snap ring. Install the snap ring, and make sure it is snug in its groove. See **Figure 17-25.**

If the clutch pack uses a reaction plate next to the apply piston, install it now. Then install the friction discs and steel plates, alternating them and using new parts as required. Always use new friction discs. Some technicians reuse the steel plates after buffing the plates to remove the shiny surface. Some friction disc manufacturers recommend that if reused, the old steel plates be installed without buffing. These manufacturers state that buffing the steel plates causes rapid wear of the new friction discs. It is often quicker, and therefore cheaper, to install new steel plates.

If the clutch pack uses a reaction plate at the top, install it before installing the outer snap ring.

Caution: A dry friction disc will burn severely the first time the clutch is applied. To prevent this, soak new friction discs in clean transmission fluid before installing them. Allow them to soak for about 15 minutes or until they stop bubbling. Always use the correct type of automatic transmission fluid. The vehicle owner's manual should list the proper ATF. The fluid type may also be stamped on the transmission dipstick.

With the clutch pack now assembled, measure the *clutch clearance,* or the distance between the pressure plate and outer snap ring. Clutch clearance is checked with a feeler gauge, **Figure 17-26.** The exact placement of the feeler gauge will vary between makes. Most

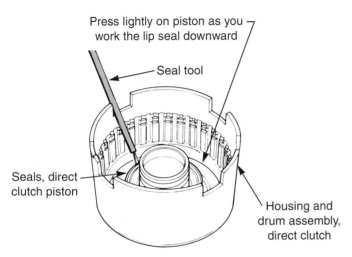

Figure 17-24. *This special tool can be fabricated from copper tubing and wire to make lip seal installation easier. This tool will be much less likely than a feeler gauge to cut the new seal. (General Motors)*

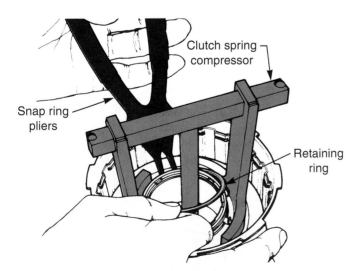

Figure 17-25. *To reassemble the apply piston and return spring assembly, depress the return springs below the snap ring groove and install the snap ring. Sometimes it is easier to install the snap ring by hand or with a small screwdriver than it is to install it using the snap ring pliers. (Ford)*

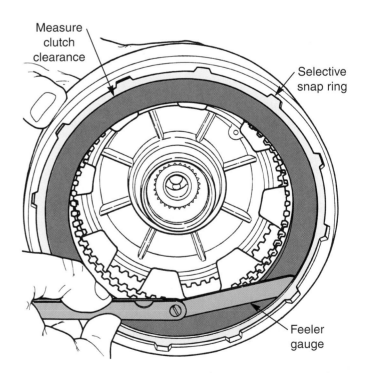

Figure 17-26. *One of the most critical parts of transmission rebuilding is checking the clutch pack clearance. Incorrect clearances must be fixed by replacing the pressure plate or snap ring with one of the proper thickness. (DaimlerChrysler)*

manufacturers have clearance adjustment kits, consisting of snap rings or pressure plates of varying thicknesses. Replacing a snap ring or pressure plate with one of a different thickness will change the clearance.

Clutch clearance is an indicator of how tightly the plates and discs are packed. This is important because it affects transmission operation. If the clearance is too large, the clutch will not be applied tightly and may slip. If the clearance is too small, the shift may be rough or the clutch may be lightly applied when it should be released. Proper clearance helps ensure good shifts and long clutch life.

Clutch pack operation can often be checked with compressed air, **Figure 17-27.** Air is directed to the fluid inlet port of the clutch, applying the piston. Piston action can be observed to verify that the clutch will operate when installed in the transmission.

Caution: *Never* apply air pressure to the clutch piston unless the clutch plates and friction discs are installed. If the plates and discs are not installed, air pressure may push the piston far enough to jam it in its bore.

Repeat the clutch pack rebuilding sequence for every clutch pack in the transmission.

Band Service

Check band friction linings for overall wear. Look for burn marks, glazing, and nonuniform wear. See **Figure 17-28.** Check to see that grooves are still visible and check for flaking. If any material can be scraped off, the band should be replaced. Look for cracks or embedded metal particles. Also, check bands for cracked ends, broken ears, or distortion. See **Figure 17-29.** Bands showing signs of wear or damage must be replaced.

The drum surface that the band rides on should also be inspected. If the surface shows signs of severe scoring,

overheating, or wear, the drum should be replaced; slightly worn drums can be turned down on a lathe. If the surface must be turned down more than a few thousandths of an inch (hundredths of a millimeter) to remove scoring or wear, the clutch drum or gearset member that the band rides on should be replaced. A surface that is shiny must be sanded with emery cloth to remove the shine. Some drums should be sanded *around* the surface perimeter, while others should be sanded in an up-and-down direction. Check the appropriate service literature to determine the direction in which the drum should be sanded. A new band should always be used when the band's mating surface has been sanded.

Caution: Soak new bands in clean ATF for about 15 minutes or until they stop bubbling. Always use the correct fluid.

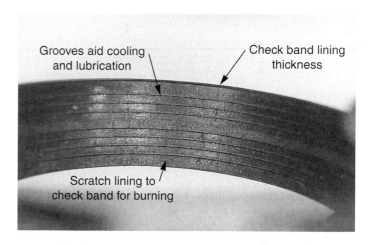

Figure 17-28. *The lining of the transmission bands can become worn or glazed. Worn bands can sometimes be adjusted, but a band that is burned and slipping should be replaced. (BBU, Inc.)*

Figure 17-27. *Clutch pack operation should be checked with air pressure before transmission assembly. This will ensure that the clutch pack operates properly when it is installed. The clutch plates and friction discs should lock together when air is applied to the proper passage. (Subaru)*

Figure 17-29. *When servicing bands, be sure to check the drum surface that the band rides on for signs of scoring, overheating, and wear. Note that the surface of this drum has been sanded to remove the shine.*

Servo and Accumulator Service

Band servos and accumulators should always be disassembled to replace seals, inspect for worn hard parts, and remove accumulated sludge. The inspection and repair of servos and accumulators was covered in Chapter 15.

Overrunning Clutch Service

All overrunning clutches in the transmission or transaxle should be checked to ensure that they turn freely in the direction they should and lock up in the other direction. See **Figure 17-30.** Any overrunning clutch that turns in both directions must be replaced. Also, any clutch showing signs of wear should be replaced. When replacing an overrunning clutch, make sure the new clutch turns in the proper direction when installed.

CVT Belt and Pulley Service

CVT belt and pulley arrangements are similar, but each one has specific repair processes. Always consult the proper service information before beginning disassembly. Belt removal on most CVTs requires a special tool that is a combination of a holding fixture and a clamping tool. The CVT assembly is placed in the holding fixture. A U-shaped clamping ring is positioned around the sliding portion of the drive pulley and locked to levers on the tool. When the levers are depressed, force is applied to the clamping ring, driving the bottom portion of the pulley downward to the fully open position. With both pulleys in the fully open position, the belt can be easily removed.

Once the belt has been removed, check the belt and pulley sheaves for wear and damage. Replace any defective parts. To reinstall the belt, place it in position in the pulleys and lubricate all parts with the correct type of fluid. Then slowly release the levers on the special tool. Finally, inspect the belt to ensure that it is properly reinstalled.

Oil Pump Service

The front oil pump must be disassembled to check it properly. Most pumps are held together by bolts. Some pumps contain the pressure regulator valve and other valves. A few pumps contain a clutch apply piston assembly.

Once the pump has been taken apart, check it for wear in areas where the gears, rotors, or vanes ride against the stationary pump body. Badly worn, scored, or pitted pumps can usually be detected visually. **Figure 17-31** shows a badly scored pump. Scratching the suspect area with your fingernail will usually confirm the presence of wear or scoring. In addition, the condition of the pump can be assessed with the help of feeler gauges, as shown in **Figure 17-32.**

Since the oil pump is designed to provide much more pressure than needed, a slightly worn pump can be reused. If you are reusing an old pump, make sure the wear is slight and the parts are reinstalled in their original positions. The pump's internal elements usually have factory marks to help ensure proper installation.

Always replace the front pump seal, no matter how good it appears. Also, replace the pump bushing, if needed. The pump bushing is critical since a worn bushing can cause the new seal to leak. The proper removal and installation tools should always be used. See **Figure 17-33.** The bushing should be staked in place to prevent it from moving. Some transmissions and transaxles have high pump bushing failure rates. Aftermarket kits, such as the one in **Figure 17-34,** have been developed to solve this problem.

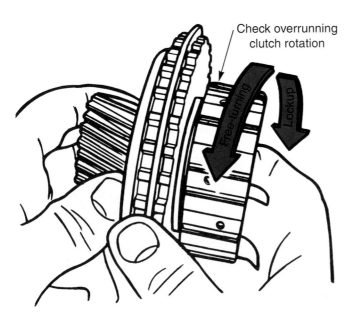

Figure 17-30. *Check every one-way clutch for wear and proper action. One-way clutches should turn in one direction only, as specified. (General Motors)*

Figure 17-31. *This pump body is severely scored at the pump gear-riding surface. A pump body in this condition should not be reused. When the pump body shows this much wear, check the pump cover for similar wear.*

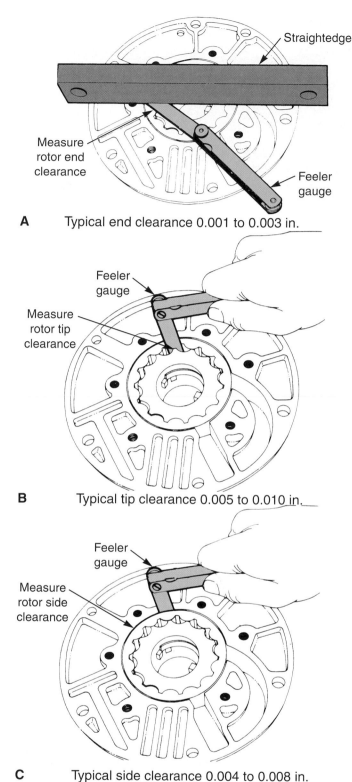

A   Typical end clearance 0.001 to 0.003 in.

B   Typical tip clearance 0.005 to 0.010 in.

C   Typical side clearance 0.004 to 0.008 in.

Figure 17-32. *These three checks will determine whether the pump can be reused. These checks apply to gear and rotor pumps. Always refer to the service manual for wear specifications. A—Checking the clearance between rotor or gear faces and a straightedge laid across pump body (end clearance). B—Checking the clearance between the tips of the inner and outer rotor lobes or gear teeth (tip clearance). C—Checking the clearance between the outer rotor or gear and the pump body (side clearance). (DaimlerChrysler)*

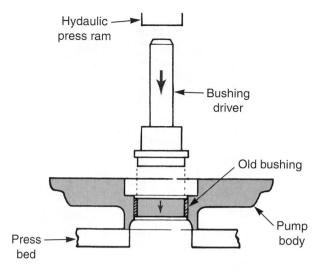

Figure 17-33. *Worn pump bushings can be replaced with new bushings. The old bushing is pressed out, as shown here. Always check the service manual for exact specifications and use the proper bushing removal and installation tools. (Ford)*

If the pump body contains the pressure regulator or other valves, remove them and make sure they are free in their bores. If the valves are sticking, they can be lightly polished to free them. If the valves do not move freely after polishing, replace the pump. **Figure 17-35** shows a typical valve and sleeve removed from the pump body. If the valves in the pump have seals, replace the seals before reinstalling the valves.

Note: There are many variations in pump design and placement. Three major types of pumps are used. The pump may be placed immediately behind the converter and driven by converter lugs, or it may be placed in the valve body and driven by a shaft. Due to these variations, only a general outline of pump rebuilding can be given in this chapter. Always consult the proper service manual before rebuilding any pump.

To reassemble a vane pump, begin by installing the outer slide and the bottom vane ring. Then insert the vanes in the vane rotor, **Figure 17-36**. The vanes must face in the proper direction. Next, install the upper vane ring to hold the vanes in position. Install the pressure regulator components and then install the cover. Finally, torque the fasteners to specifications.

To reassemble a gear or rotor pump, install the gears in the proper order in the pump body. Be sure to install the inner and outer gears so that any marks face upward. Then place the pump cover over the pump body and install and tighten the bolts in the order specified by the manufacturer. Always use a torque wrench to tighten the bolts. If the pump incorporates a clutch apply piston, replace the seals as explained in the clutch pack rebuilding section and then reassemble the pump as explained above.

Figure 17-34. *The kit shown here is used to repair a chronic bushing failure problem on a common truck transmission. It also corrects torque converter slippage and resultant computer control system problems. The replacement bushing is an improved type that resists wear. Instructions furnished with the kit explain how to drill and enlarge lube holes to further increase bushing life, and how to install an improved converter drainback valve. (Superior)*

Figure 17-35. *Exploded view of an oil pump–mounted valve and its aluminum sleeve. The valve and sleeve should be carefully checked for sticking, scoring, or other damage. As with a valve body, be sure that all pump-mounted valve components are reinstalled in their original positions. (Sonnax)*

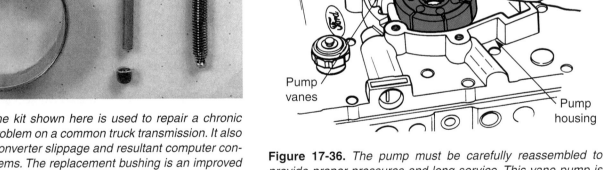

Figure 17-36. *The pump must be carefully reassembled to provide proper pressures and long service. This vane pump is mounted in the side valve body of a common transaxle. (Ford)*

Note: Some pumps installed in the case require that a special tool called an aligning band be placed around the two halves of the pump to ensure that they are properly aligned before the bolts are tightened.

Seal Ring Service

Seal rings are critical to proper transmission and transaxle operation, since they seal in hydraulic fluid going from the valve body to the bands and clutch packs. Seal rings can be made of metal or Teflon. They should always be replaced during a rebuild.

After removing the old seal rings, check the ring grooves, **Figure 17-37**. They must not be worn. Look for nicks or ridges on the groove walls that could prevent the new ring from seating properly, allowing leaks. Groove width can usually be checked with the new ring and a feeler gauge, **Figure 17-38**.

Check the bore surfaces that the seal rings ride against, as well. The bore surfaces must be round and smooth, with no ridges or signs of wear. Place the replacement ring squarely in the bore to see that it conforms to the

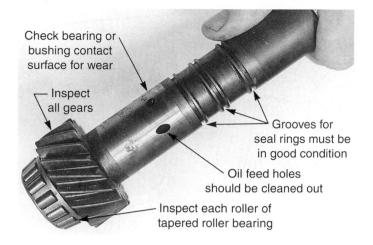

Check bearing or bushing contact surface for wear

Inspect all gears

Grooves for seal rings must be in good condition

Oil feed holes should be cleaned out

Inspect each roller of tapered roller bearing

Figure 17-37. *If any of the shafts contain seal ring grooves, the grooves should be closely inspected for wear.*

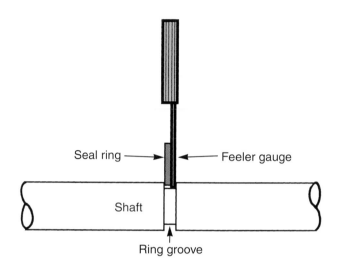

Seal ring — Feeler gauge

Shaft

Ring groove

Figure 17-38. *Seal rings and grooves are vital to proper operation of the transmission or transaxle hydraulic system. They should be carefully checked to ensure that they can do a good job of sealing. Using a feeler gauge as shown here, the maximum end clearance of a new seal ring in its groove should be about 0.003″–0.006″ (0.07mm–0.14mm).*

bore and that the bore is round. The ring should touch the bore all the way around.

Parts having worn or damaged seal ring grooves or bores should be replaced. Always install new seal rings carefully, using plenty of transmission fluid as a lubricant.

Bearing and Thrust Washer Service

All transmission bearings and thrust washers should be checked for wear and scoring. A badly worn thrust washer is shown in **Figure 17-39**. This washer should be replaced. General procedures for checking bushings and washers were discussed in Chapter 4. **Figure 17-40A** shows the procedure for checking bushing wear in a sun gear. **Figures 17-40B** and **17-40C** show the procedure used to replace a bushing. Note the use of special tools to remove the old bushing and install the new one.

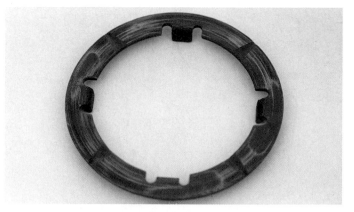

Figure 17-39. *This thrust washer is badly worn and should be replaced. If you encounter a thrust washer showing this degree of wear, remember to inspect the part the washer rides against. Chances are good that the riding surface of the part is also worn out.*

There are many thrust washers in modern automatic transmissions and transaxles. Some of these thrust washers are made in selective thicknesses to control endplay. All thrust washers should be checked with a micrometer to ensure proper endplay. Use thicker or thinner washers as needed. For instance, if the endplay is 0.030″ (0.762mm) too large, a 0.030″ thicker thrust washer should be obtained. If endplay is insufficient, a thinner selective thickness washer should be used. Always recheck endplay to confirm that the proper washer was selected.

Miscellaneous Case Parts Service

Many small parts are installed in the case, including electrical switches and sensors, electrical case connectors, small screens in the pump output passages, seals, and check balls. These parts are as important as the major transmission components. They must be cleaned and carefully checked for wear and damage. Electrical parts should be checked according to the manufacturer's instructions, using the proper electrical test equipment. Seals, such as those used on the T.V. cable or the manual lever and shaft assembly, should always be replaced.

Reassembly

Once all automatic transmission or transaxle subassemblies have been rebuilt, they can be installed in the transmission case. Always refer to the manufacturer's service manual for the exact assembly procedures. It will give procedures and specifications for the type of transmission or transaxle being serviced. Exploded views are very useful during the reassembly process.

 Note: Remember to soak the friction discs, bands, and all rubber parts in fresh transmission fluid before installation.

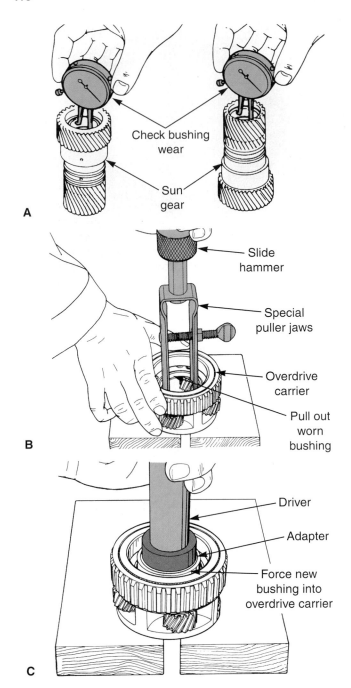

A—Check bushing wear — Sun gear

B—Slide hammer — Special puller jaws — Overdrive carrier — Pull out worn bushing

C—Driver — Adapter — Force new bushing into overdrive carrier

Figure 17-40. *Check all bushings for wear and replace those that are worn. A—Checking a sun gear bushing. B—Removing a planet carrier bushing. C—Installing a new planet carrier bushing. (DaimlerChrysler, General Motors)*

Begin reassembly with the rear of the case. If the case contains a low-and-reverse clutch that has not been serviced prior to this point, disassemble and clean it. Install new seals on the clutch apply piston. Install the piston, as well as the friction discs and steel plates. Make sure the clutch clearance is correct and install the retaining snap ring. If the unit uses a rear band, install the band in the case but do not install the band apply linkage. See **Figure 17-41.**

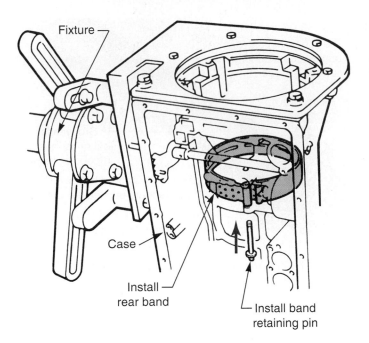

Fixture — Case — Install rear band — Install band retaining pin

Figure 17-41. *The location of a band in the case on one type of transmission is shown. In most instances, the band should be installed before installing the drum or gearset member on which it will ride. (DaimlerChrysler)*

Install the rear planetary gearsets and the output shaft. Be careful to align all parts before installing any snap rings or bolts. After the snap rings or bolts are in place, make sure the output shaft will turn. Also, make sure that all the thrust washers are installed correctly.

If a center support is used, install it now. Install the center support snap ring and/or bolts and make sure the output shaft will still turn.

In many designs, the clutches must be assembled before they are installed in the case. The height of the assembly is being checked to ensure that it meets specifications. With some transmissions and transaxles, the clutches and front gearsets will be assembled prior to installation. Install any such assembly at this point. See **Figure 17-42.** Also, if the unit has a front band, install it now.

If the transmission or transaxle has a set of clutch plates and friction discs directly behind the front pump, install it

 Note: This step does not apply to transaxles with a valve body–mounted pump.

now, making sure to alternate discs and plates. Next, place the proper gasket on the case and install the front pump.

As you slowly tighten the pump fasteners, check that the input and output shafts can be turned by hand. If either shaft begins to bind, find the cause and correct it before proceeding. If a transaxle is being worked on, install the chain and sprockets now. Also, install the transaxle transfer gears, if applicable. See **Figure 17-43.**

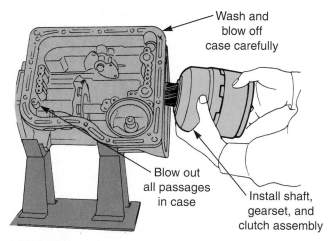

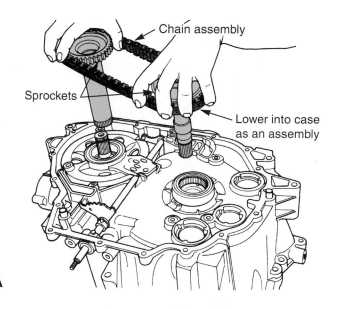

Figure 17-42. *A clutch and gearset assembly is shown being installed in the transmission case. Make sure that parts are properly seated. (Nissan)*

A

 Caution: *Never* **install a transmission on a vehicle if the input and output shafts are binding. The transmission will be badly damaged if it is operated with binding shafts.**

Check the output shaft endplay using a dial indicator, as shown in **Figure 17-44.** If the endplay is not within specifications, the original selective fit thrust washer must be replaced. Most units have only one selective-fit thrust washer. Do not try to adjust endplay by changing other washers.

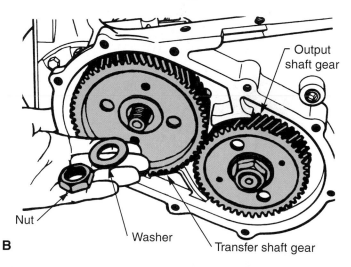

B

Figure 17-43. *Carefully install the drive chain or gears. A—The drive train and sprockets are usually lowered into the case as an assembly. B—The drive gears can be installed separately. Be sure to install any washers and fasteners that were removed. (Ford, DaimlerChrysler)*

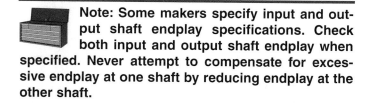 **Note: Some makers specify input and output shaft endplay specifications. Check both input and output shaft endplay when specified. Never attempt to compensate for excessive endplay at one shaft by reducing endplay at the other shaft.**

If not done already, install the servos in the transmission case. Then, install the band apply linkage and adjust the bands to specifications. In many cases, the proper band adjustment is obtained by changing the length of the band apply pin. Proper pin length is usually obtained using a special tool and a torque wrench, as illustrated in **Figure 17-45.** The special tool is installed along with the existing band apply pin. Then the pin is applied against the band using the torque wrench. The correct torque must be applied for the reading to be accurate. If the pin is the correct length, a white line will appear in the tool's window. If the pin is incorrect, a new pin must be obtained.

Other bands are adjusted using a special tool and a dial indicator. The tool is installed with the servo piston. Then, the dial indicator is installed and set to zero. Finally, the piston is pushed downward to tighten the band against the drum and the dial indicator is read. See **Figure 17-46.**

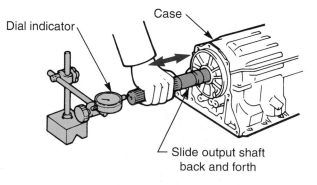

Figure 17-44. *After assembling all the drive train parts, recheck endplay to ensure that you have installed the proper selective thickness thrust washer. (DaimlerChrysler)*

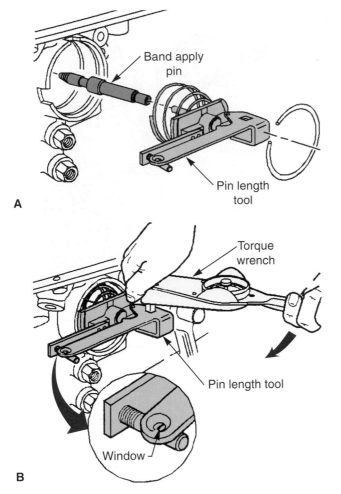

A

B

Figure 17-45. *A special tool must be used to adjust the band on many modern transmissions and transaxles. A—The tool is installed with the existing band apply pin. B—The band is applied with the torque wrench. If the pin is the correct length, a white line will appear in the tool's window. (General Motors)*

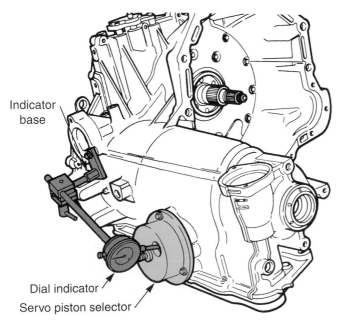

Figure 17-46. *Some bands are adjusted using a dial indicator and a special tool known as a servo piston selector. After installing the tool and the dial indicator, the piston is pushed downward and the indicator is read. (Ford)*

If the dial indicator reading is smaller than the manufacturer's specification, a shorter pin must be substituted. If the dial indicator reading is larger than the manufacturer's specification, a longer pin must be substituted.

Install the governor in the case or on the output shaft, as applicable, and install the extension housing. Then, install a new seal on the extension housing, **Figure 17-47.**

Before installing the valve body, perform an air pressure test on the transmission to make sure the clutch apply pistons and servos are working properly. Air pressure testing was discussed in Chapter 13. **Figure 17-48** shows air pressure being used to check piston operation on one type of transmission.

After air pressure tests are complete, install the valve body-to-case spacer plate, the spacer plate gaskets, and any check balls in the transmission case. Carefully match the holes in the spacer plate and the new spacer plate gaskets to ensure proper fluid flow.

Install the valve body and a new oil filter. Then, install the transmission oil pan using a new gasket. On a transaxle, install the side valve body and oil pan, if neces-

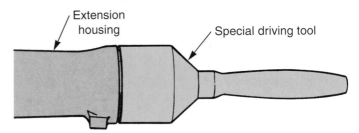

Figure 17-47. *After installing the extension housing, install a new rear seal using the proper driver. Worn extension bushings should also be replaced. (DaimlerChrysler)*

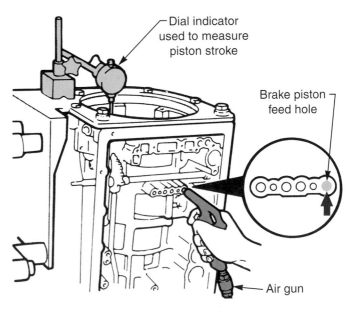

Figure 17-48. *Always perform an air pressure test on the transmission before installing the valve body. Band movement or a sharp thunk from the clutch packs indicates that the air pressure is operating the holding members and that all seals and passages should channel fluid properly. (DaimlerChrysler)*

sary. At this point, you are ready to install the torque converter.

Prior to installation, the converter can be checked for proper endplay as shown in **Figure 17-49.** The tool shown has an expanding pilot that can be tightened firmly into the splined hole for the turbine shaft. Make sure the tool is fully seated on the hub, and lock the tool in place by turning the threaded inner post. Mount a dial indicator on the tool. A typical procedure calls for positioning the plunger so it contacts the converter shell. With the dial indicator in place, zero the indicator. Then, pull up on the tool's handles. The reading obtained is the total endplay of the turbine and stator. If endplay is not within specifications, the torque converter must be replaced.

A rough check of endplay can sometimes be made by inserting a pair of snap ring pliers into the splined turbine hub and pulling upward. Observing the plier movement will indicate how much endplay there is.

The one-way clutch can also be checked at this time. Place snap ring pliers down through the pump drive hub and expand them into the internal splines on the one-way clutch. (These splines mate with external splines on the stator shaft.) Applying torque to the pliers should turn the stator easily in one direction (clutch unlocked) and not so easily in the other direction (clutch locked and turning the stator). A special tool that engages the one-way clutch inner race can also be used to check clutch operation. See **Figure 17-50.**

If the torque converter passes these tests, check the converter for wear at the hub. If there are no signs of wear, add at least 1 quart of fresh transmission fluid to the converter. See **Figure 17-51.**

Note: Some manufacturers recommend lowering the converter onto the transmission shafts, Figure 17-52. In this case, add only about 1 pint of fluid and spin the converter to distribute it to the outside of the converter internals.

Place the torque converter over the input shaft and push it into the converter housing. The internal splines of the converter will move into contact with the splines of the stator shaft and input shaft. It may be necessary to wiggle the converter while pushing on it to get it to engage with the transmission shafts. The converter may also need to be rotated to line up the lugs on the pump drive hub with the slots on the front pump. The lugs should fully engage the front pump. To verify that the torque converter is properly installed, some manufacturers recommend checking it with a measuring rule and a straightedge, **Figure 17-53.** The reading is then compared to factory specifications.

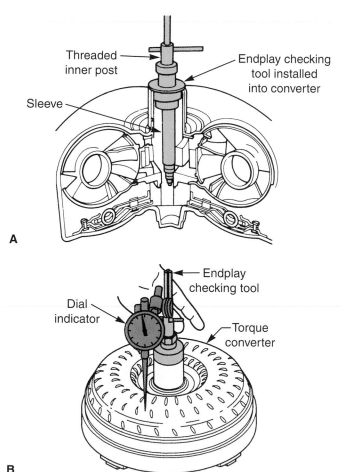

Figure 17-49. *Converter endplay can be checked with the special tool shown here. A—The endplay-checking tool must be fully seated on the converter hub and locked in place. B—After mounting the dial indicator on the tool, zero the indicator and pull up on the tool's handles. The reading obtained is the total endplay of the turbine and stator. (Ford)*

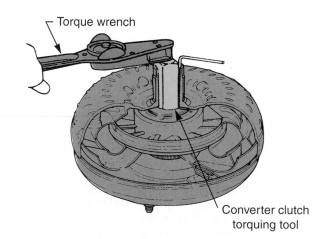

Figure 17-50. *Checking the converter stator one-way (over-running) clutch with a special tool and a torque wrench. If the wrench turns easily in one direction and locks up in the other direction, the one-way clutch is working properly. (Ford)*

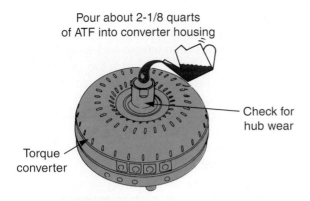

Pour about 2-1/8 quarts of ATF into converter housing

Check for hub wear

Torque converter

Figure 17-51. *Once the torque converter has been cleaned, check it out before reinstalling it. The pump drive hub can be checked for wear, and the internal operation of the converter can be checked. Add a measured amount of transmission fluid to the torque converter. (Nissan)*

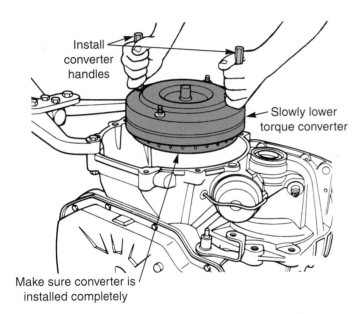

Install converter handles

Slowly lower torque converter

Make sure converter is installed completely

Figure 17-52. *To install the torque converter, turn the transmission or transaxle until the turbine shaft faces up and lower the converter into place. Make sure the converter is completely installed over the shaft splines or it will bind and be damaged when the transmission or transaxle is reinstalled in the vehicle. (Ford)*

Caution: Under no circumstances should you install the torque converter on the flywheel or flexplate before installing the transaxle or transmission on the engine. The splines of the transmission shafts, pump, and converter can be damaged.

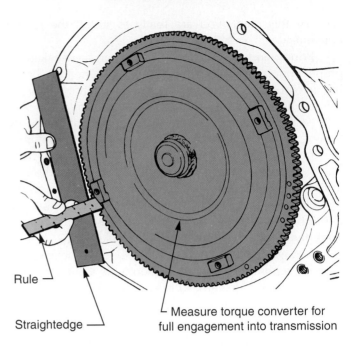

Rule

Straightedge

Measure torque converter for full engagement into transmission

Figure 17-53. *After installing the torque converter, make sure it is completely seated on the input shaft and stator shaft. Also, make sure the lugs on the pump drive hub are fully engaged in the front pump. Measuring the installed depth of the torque converter with a rule and straightedge as shown helps verify that installation is correct. (DaimlerChrysler)*

Summary

The major steps in transmission/transaxle rebuilding are disassembly, cleaning, inspection, and reassembly. To begin rebuilding a transmission or transaxle, remove the torque converter and install the unit in a holding fixture.

Before beginning the disassembly procedure, make sure you have the proper tools and a service manual that covers the transmission model to be rebuilt. Clean the outside of the transmission/transaxle thoroughly and check shaft endplay. Then remove the oil pan, filter, and valve body. On a transaxle, remove the drive chain or transfer gears as necessary. Inspect and save all small parts. After the valve body has been taken off, the front pump (when used) and internal parts can be removed. Removal of the parts will vary according to the internal construction of the transmission or transaxle and whether or not it has a center support.

Carefully note the condition of all parts as they are removed from the case. Then remove any other accessory parts, such as servos, accumulators, and modulators, from the case.

After disassembly, thoroughly clean all internal parts and make further inspections for wear or damage. The converter and oil cooler lines should also be cleaned. Determine what replacement parts will be needed and order them.

The clutch packs are among the most critical parts of the transmission. They should be carefully inspected and serviced. Disassemble the clutch packs and inspect the plates for wear. The friction discs are usually replaced as part of an overhaul. The steel plates are often replaced.

Check the bands for wear, cracking, or burning. Defective bands should be replaced. Band servos should be disassembled, cleaned, and inspected. After the installation of new seals, the servos are ready to be reinstalled. Accumulators are rebuilt in the same manner as servos.

The one-way clutches should be checked to ensure that they turn in one direction only. Planetary gearsets should be checked for obvious defects, such as chipped teeth. Also, check the bushings, bearings, and washers at all locations in the transmission.

Check the transmission/transaxle shafts. Inspect the input and output shafts for wear, bending, or warpage. Also check all the case parts, such as screens, valves, and seals, for wear and damage. Seals should always be replaced.

To reassemble the transmission or transaxle, start by soaking all new friction discs and bands in transmission fluid. Then, install the output shaft, rear planetary gearsets, clutches, bands, and other components at the rear of the case. Remember to reinstall all snap rings in their original positions. Then, install the center support, if the transmission is equipped with one.

Install the front planetary gearsets and clutches as complete assemblies, as required. Measure the assembled height to ensure proper assembly. Install any bands or case-mounted clutches. Then, install the oil pump. Check the transmission endplay and make sure that all shafts turn.

Install the governor on the output shaft or in the case, as applicable. Then, install the extension housing and other output shaft parts. Install the drive chain or transfer gears as necessary.

Install the servos and adjust the bands. Also install any case-mounted accumulators, electrical switches, and other case parts. Then air pressure test the transmission. Install parts that are installed inside the oil pan, such as the valve body, spacer plates, check balls, filter, etc. Then, install the oil pan.

Check the converter for wear or internal damage, and then install it. Make sure the converter is installed properly on the transmission shafts and the pump drive hub (when used) is fully engaged in the front pump.

Review Questions—Chapter 17

Please do not write in this text. Place your answers on a separate sheet of paper.

1. When rebuilding an automatic transmission, you should work on one _____ at a time, so that parts are not mixed up.

2. Before disassembly, clean the _____ of the transmission or transaxle.

3. What can be learned from inspection of the oil pan and its contents?

4. Does the presence of a few aluminum particles in the oil pan mean that the transmission or transaxle is severely damaged? Why?

5. It is usually easier to remove the front pump seal with the _____ still installed.

6. A _____ _____ _____ may be installed on the back of some pumps. In this case, a set of _____ and _____ will be found directly behind the pump.

7. Many clutch packs and planetary gears are held in place by _____ _____.

8. During the rebuilding process, gaskets, seals, and friction discs should be replaced:
   (A) in all cases.
   (B) only if obviously damaged.
   (C) only if removed.
   (D) None of the above.

9. Leopard spotting is a sign of excessive _____.

10. What visual signs indicate that a clutch pack has been slipping?

11. Some technicians _____ steel plates before reinstalling them. This removes the _____ surface. However, it is often cheaper to install _____ plates.

12. If clutch clearance is too _____, the clutch pack may not apply tightly and may slip.

13. To remove a CVT belt, use a special tool to move the drive pulley to the completely _____ position.

14. Badly scored pumps can often be spotted by a _____ inspection. Pump wear can sometimes be determined by the use of _____ gauges. A slightly worn pump can be _____.

15. Fasteners such as _____ or _____ are used to hold the center support to the case.

16. Check the torque converter endplay with a _____ indicator.

Match the transmission component with the measuring device used to check it.

17. Clutch pack clearance _____
18. Band servo operation _____
19. Shaft straightness _____
20. Shaft bushings _____
21. Overheated steel plates _____
22. Thrust washer thickness _____

(A) Dial indicator and V-blocks
(B) Micrometer
(C) Air pressure
(D) Vacuum pump and gauge
(E) Feeler gauge
(F) Visual inspection
(G) Inserting and rocking the shaft

ASE-Type Questions—Chapter 17

1. Technician A says that endplay should be checked before the transmission is disassembled. Technician B says that a few aluminum particles in the oil pan do not indicate a severe transmission problem. Who is right?
   (A) A only.
   (B) B only.
   (C) Both A and B.
   (D) Neither A nor B.

2. All of the following tools can be used to remove the front pump from the transmission case *except:*
   (A) a special puller.
   (B) a slide hammer.
   (C) a hammer and punch.
   (D) a pry bar.

3. If the air bleed check ball is missing from the clutch piston, what should the technician do to correct the problem?
   (A) Replace the clutch apply piston.
   (B) Install an old valve body check ball in the bleed hole.
   (C) Block off the bleed hole with a small screw.
   (D) Block off the bleed hole with an oversize check ball.

4. Technician A says that new friction discs should be soaked in clean transmission fluid before being installed. Technician B says that compressed air is used to check clutch pack clearance. Who is right?
   (A) A only.
   (B) B only.
   (C) Both A and B.
   (D) Neither A nor B.

5. A drum with a slightly worn band apply surface should be:
   (A) turned on a lathe.
   (B) sanded front to back.
   (C) sanded around its diameter.
   (D) replaced.

6. The usual reason for servicing band servos is to replace:
   (A) springs.
   (B) pistons.
   (C) seals.
   (D) snap rings.

7. Technician A says a one-way clutch that turns in both directions can be reused. Technician B says a one-way clutch showing any signs of wear should be replaced. Who is right?
   (A) A only.
   (B) B only.
   (C) Both A and B.
   (D) Neither A nor B.

8. Transmission endplay may be corrected by replacing the:
   (A) case gaskets.
   (B) needle bearings.
   (C) thrust washers.
   (D) snap rings.

9. A transmission has been rebuilt. The output shaft turns easily, but the input shaft does not turn. Technician A says that this is a normal condition when any bushings have been replaced. Technician B says that both shafts should turn freely. Who is right?
   (A) A only.
   (B) B only.
   (C) Both A and B.
   (D) Neither A nor B.

10. Operation on the stator one-way clutch:
    (A) can be checked by applying air pressure to the torque converter assembly.
    (B) can be checked by turning the clutch hub with snap ring pliers.
    (C) can be checked by shaking the torque converter while listening for rattles.
    (D) cannot be checked.

Chapter 18

Automatic Transaxle Final Drive Service

After studying this chapter, you will be able to:
- ❏ Disassemble a planetary gear final drive.
- ❏ Inspect the parts of a planetary gear final drive.
- ❏ Assemble a planetary gear final drive.
- ❏ Disassemble a helical gear final drive.
- ❏ Inspect the parts of a helical gear final drive.
- ❏ Assemble a helical gear final drive.
- ❏ Disassemble a hypoid gear final drive.
- ❏ Inspect the parts of a hypoid gear final drive.
- ❏ Assemble a hypoid gear final drive.

Technical Terms

Cross contamination

Backlash

Tooth contact pattern

Prussian blue

Heel-and-toe patterns

Introduction

A major difference between a transaxle and a conventional rear-wheel drive transmission is that the transaxle incorporates the final drive unit into its case. The final drive consists of the differential unit and the permanently meshed gears that set the final ratio. The final drive axle ratio corresponds to what is commonly called the rear axle ratio of a rear-wheel drive vehicle.

There are three types of final drive units. The planetary gear final drive unit is used on many common domestic automobiles and minivans. The helical gear drive unit is common in many domestic and imported cars and vans. The hypoid gear final drive was more common in the past, but it is still used on one popular line of American cars.

This chapter explains the procedures for disassembling, inspecting, and reassembling the three common final drive types. Most final drive service can be performed with the transaxle installed in the vehicle. The procedures discussed here can also be performed with the transaxle removed from the vehicle. Common final drive adjustments are also covered. Studying this chapter will enable you to repair the final drive unit as part of transaxle service.

Planetary Gear Final Drive Service

Planetary gear final drives are used on many General Motors and Ford transaxles. The planetary final drive on some transaxles cannot be reached without removing the internal transaxle components. Other transaxles are arranged so the final drive can be serviced without removing and disassembling the transaxle.

If the transaxle must be removed and disassembled to access the final drive, refer to Chapters 16 and 17. Many later transaxles have a removable extension housing that allows the final drive to be removed without removing and disassembling the transaxle. Final drive removal procedures for a transaxle with a separate extension housing are shown in **Figure 18-1**. Start by removing the CV axle according to procedures outlined in Chapter 16. Only the CV axle on the final drive side must be removed. Next, remove the speedometer gear, governor, or speed sensor, as necessary.

Remove the bolts holding the extension housing to the transaxle case. Place a drip pan under the extension housing and remove the housing, **Figure 18-1A.** Then remove the clip holding the final drive to the transaxle output shaft, **Figure 18-1B.** After the clip is removed, pull the final drive assembly from the transaxle, **Figure 18-1C.**

Once the final drive assembly has been removed from the transaxle, place it on a clean workbench and begin disassembly. Start by removing the differential side and spider gears. **Figure 18-2A** illustrates one method of

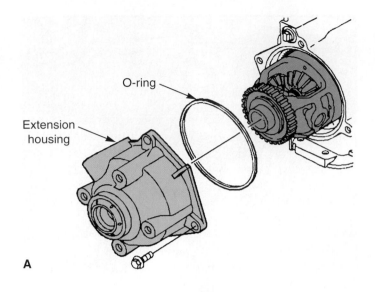

A

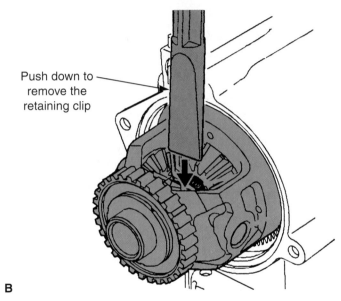

B

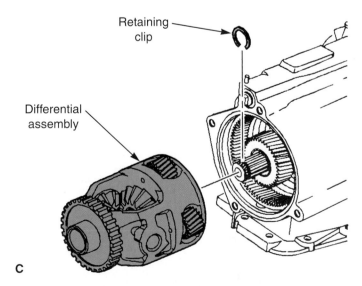

C

Figure 18-1. *Planetary gear final drive removal steps. A—Remove the extension housing from the transaxle. B—Remove the output shaft-to-final drive retaining clip. C—Slide the final drive from the output shaft. (General Motors)*

removing the spider gears. Drive out the roll pin that holds the spider gear shaft to the final drive carrier and slide the pin out of the carrier. Then remove the spider and side gears from the carrier, **Figure 18-2B.**

Figure 18-3 shows the parts of a spider and side gear assembly. Clean all parts and inspect them for wear and damage. To reassemble the differential assembly, install the side gears, making sure that the washers are in position. Then roll the spider gears and their related washers into position and reinstall the spider gear shaft. Install a new shaft-to-case roll pin and check that the differential assembly operates smoothly.

Check the clearance between the planet gears and the final drive carrier, **Figure 18-4.** If the clearance is excessive, the planetary gearset will be noisy and may fail. Excessive clearance can sometimes be corrected by replacing the thrust washers between the gears and the carrier. Thicker service washers may be available for reducing clearance.

Check the final drive planet gears for wear and damage. Defective parts can be replaced to repair the planet carriers. If the planet gears show signs of damage, be sure to check the sun gear for damage. The most commonly replaced parts of a planet carrier are the needle bearings and thrust washers. **Figure 18-5A** shows these parts. The planet gears are usually removed by removing the pin that locks the pinion gear shaft to the final drive carrier. See **Figure 18-5B.** If defective parts are found, replace them. Lubricate all parts as you reassemble them. The needle bearings can be held in place with assembly lube or petroleum jelly. After the defective parts are replaced, recheck the clearance between the planet gears and the final drive carrier. Also, check the pinion gears for smooth operation.

If the speedometer drive gear is installed on the final drive carrier, check it for damage. This is especially important if the speedometer driven gear shows signs of damage.

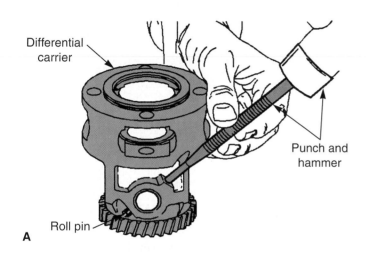

A

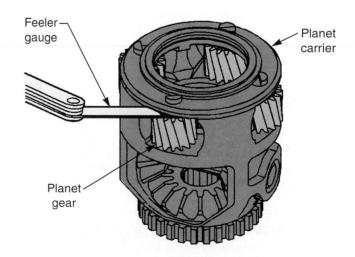

Figure 18-3. *Differential gears and related parts. Check all parts for wear. (DaimlerChrysler)*

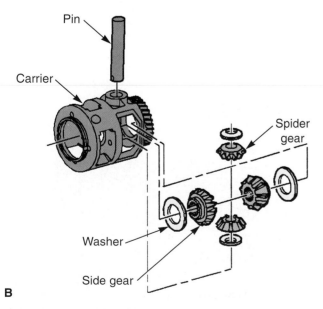

B

Figure 18-2. *Removing the differential gears. A—Punch out the roll pin. B—Slide the shaft from the housing and remove the gears. (General Motors)*

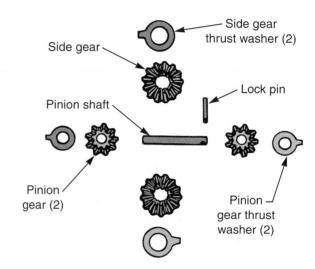

Figure 18-4. *Checking for excessive clearance between the planet gears and the planet carrier. Excessive clearance can sometimes be corrected with thicker thrust washers. (General Motors)*

Some planetary-type drives have a removable speedometer drive gear. If this gear becomes damaged, it can be removed with a puller, **Figure 18-6.** The replacement gear can be installed by heating it in an oven or immersing it in hot water and then lightly tapping it into place.

If the transaxle does not have a removable extension housing, reinstall the final drive along with the other internal transaxle components as explained in Chapter 17. To reinstall a final drive on a transaxle with a removable extension housing, begin by removing all metal and debris from the interior of the unit. If a lot of debris is present, you may want to remove the oil pan and replace the filter.

Then install the final drive assembly over the output shaft. Install the retaining clip to secure the final drive to the shaft. Next, reinstall the extension housing using a new O-ring or gasket. Reinstall the speedometer gear, governor, or speed sensor as applicable, using new seals. If necessary, install a new CV axle seal, and then reinstall the CV axle. After all parts have been installed, check the transaxle fluid level and add fluid as necessary.

Helical Gear Final Drive Service

Helical gear final drives are used on many vehicles. Actual disassembly procedures may vary from those given here, and the proper service manual should always be consulted before beginning disassembly. This procedure can usually be performed without removing the transaxle from the vehicle. To remove a helical ring gear and differential assembly, remove the CV axles as explained in Chapter 17.

Place a drip pan under the final drive cover, sometimes called the differential cover, **Figure 18-7.** Then loosen the cover fasteners and lightly pry the cover away from the transaxle housing. Allow oil to drain from the final drive housing.

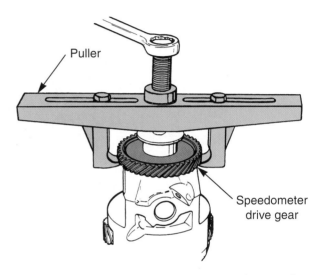

Figure 18-6. *Using a special puller to remove the speedometer drive gear from the final drive housing. Some drive gears must be heated before installing it on the housing. (General Motors)*

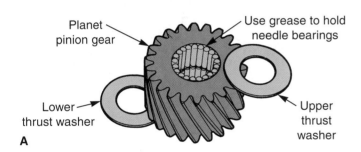

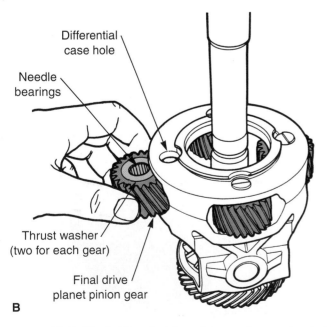

Figure 18-5. *Final drive planet carriers can be overhauled by replacing bearings and washers. A—Typical planet-pinion gear, showing the needle bearings and thrust washers. B—Installing the planet-pinion gear in the carrier. (Ford)*

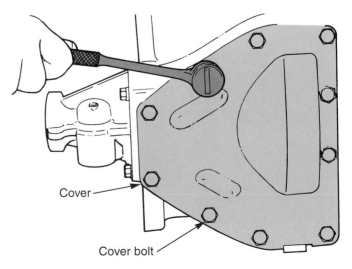

Figure 18-7. *Remove the fasteners to remove the cover over the final drive assembly. Place a pan under the cover to catch spilled oil. (DaimlerChrysler)*

Note: On some transaxles, the cover may have already been removed to reach the CV axle shaft retainers.

Next, remove the differential bearing retainer bolt, **Figure 18-8,** and remove the bearing retainer. Finally, remove the extension housing and remove the final drive and differential assembly as shown in **Figure 18-9.**

To remove the final drive pinion gear, it may be necessary to remove the transfer gear that is attached to the pinion gear shaft. Transfer gear removal is illustrated in **Figure 18-10.** Some transfer shaft gears have selective thickness shims that should be saved for reuse. **Figure 18-11** shows one of these shims. Once the transfer gear has been removed, the pinion gear can be removed from the transaxle case, **Figure 18-12.**

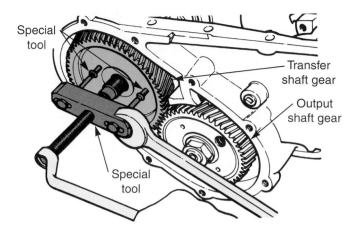

Figure 18-10. *Removing the transfer gear to gain access to the pinion gear. Be sure to remove the correct gear. (DaimlerChrysler)*

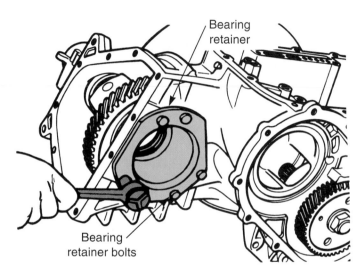

Figure 18-8. *Removing the bearing retainer will allow the final drive assembly to be removed. (DaimlerChrysler)*

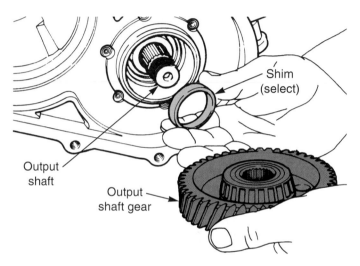

Figure 18-11. *If both transfer gears are removed, save any shims between the transfer gear and the housing. Do not mix the shims between transfer gears. (DaimlerChrysler)*

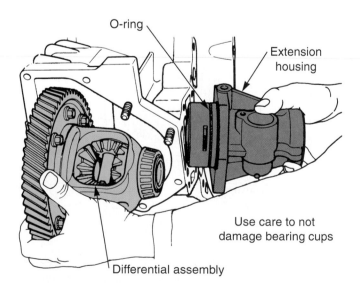

Figure 18-9. *Removing the extension housing and final drive assembly. Carefully observe the position of all parts for reassembly. (DaimlerChrysler)*

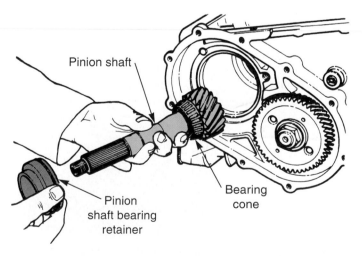

Figure 18-12. *Removing the pinion shaft from the transaxle. Be careful not to damage the tapered roller bearings or the bearing cones during removal. (DaimlerChrysler)*

Once all parts have been removed, clean them thoroughly and check them for damage. Also, clean the interior of the final drive housing. It is very important that all metal and sludge be removed from the interior of the housing, as the movement of new lubricant will throw contaminants onto the rebuilt drive components.

The differential assembly of a helical gear final drive has the same components as a planetary final drive. Service procedures are similar. Remove the pin holding the spider gear shaft in place and remove the shaft, the spider gears, and the side gears. Inspect all parts for wear and damage. Reassemble the differential by first installing the side gears and then rolling the spider gears into position and reinstalling the spider gear shaft. Make sure all spider and side gear washers are in position. If a shaft-to-case roll pin is used, install a new pin. A threaded retaining pin can be reused. Once the differential is reassembled, check that it operates smoothly.

If the ring gear must be replaced, remove the bolts holding the gear to the differential carrier. The ring gear may slide from the carrier, but most require pressing or light hammering to remove. If you remove the ring gear with a hammer, hammer evenly around the outside of the ring gear to avoid distorting the carrier. To install a new ring gear, heat it with a heat gun or by immersing it in boiling water. Then lay the ring gear on a flat surface and quickly place the differential carrier over it, being sure to align the differential carrier and the ring gear bolt holes. **Figure 18-13** shows the ring gear reinstallation process. Most manufacturers call for tightening the bolts in an alternating pattern to avoid distorting the ring gear.

 Note: Helical gear final drive service often involves replacing the tapered roller bearings. Refer to Chapter 4 for bearing replacement procedures.

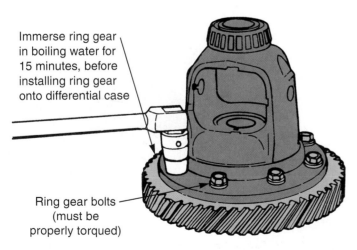

Immerse ring gear in boiling water for 15 minutes, before installing ring gear onto differential case

Ring gear bolts (must be properly torqued)

Figure 18-13. *The new ring gear must be heated before installation. Be sure to line up all bolt holes before the ring gear begins to cool. (DaimlerChrysler)*

Installation of the helical gear assembly is the reverse of removal. Be sure to properly preload the tapered roller bearings. Follow manufacturer's bearing preloading procedures. Replace all seals and gaskets, or replace the RTV sealer bead where indicated.

Caution: If the final drive uses a lubricant other than transmission fluid, the seals separating the transmission and final drive *must* be replaced to prevent *cross contamination* (transfer of either type of fluid between the transmission and final drive).

After the final drive is reinstalled in the case, check endplay as shown in **Figure 18-14.** Change the shims or adjust the threaded bearing retainers as necessary to obtain the proper endplay reading. Once endplay is correct, install the CV axles.

Scrape the old gasket material from the differential and final drive assembly cover. Then, reinstall the cover using a new gasket. Fill the final drive with the proper type of fluid. Some helical final drives are lubricated with the same transmission fluid that lubricates the transmission's hydraulic system. Other final drives are lubricated with manual transmission fluid or hypoid axle gear oil. Always check to determine what type of oil should be used.

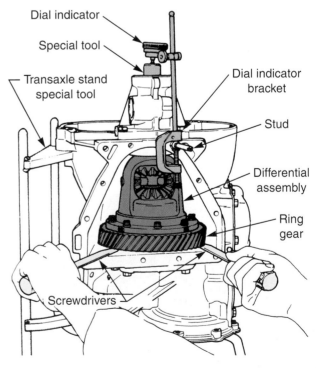

Dial indicator

Special tool

Transaxle stand special tool

Dial indicator bracket

Stud

Differential assembly

Ring gear

Screwdrivers

Figure 18-14. *Checking final drive endplay before finishing the reinstallation. This illustration shows the endplay being checked with the transaxle removed from the vehicle. This check can be made with the transaxle in the vehicle. (DaimlerChrysler)*

Hypoid Gear Final Drive Service

Hypoid gear final drives are used on transaxles installed on vehicles having front-facing (longitudinal) engines. Some older versions were basically rear axle assemblies that were turned around and attached to the end of the transaxle to drive the front wheels. Newer designs are part of the transaxle assembly.

To remove an older type of hypoid gear final drive, remove the CV axles and unbolt the final drive assembly from the transaxle. Then slide the final drive assembly from the transaxle output shaft. To remove a newer type of hypoid gear assembly, remove the CV axles and locate the side cover, **Figure 18-15.** Place a drip pan under the side cover. Remove the differential oil fill plug and loosen the cover-to-transaxle fasteners. Slightly pry the cover away from the transaxle and allow oil to drain from the final drive housing. Finish removing the fasteners and remove the cover from the transaxle.

With the cover removed, pull the ring gear and differential assembly from the transaxle case. Removal of the ring gear and differential assembly is illustrated in **Figure 18-16.** Place the assembly on a clean bench.

Service procedures for the differential assembly of a hypoid gear final drive are similar to those of other final drives. After removing the locking pin, remove the spider gear shaft, the spider gears, and the side gears. Inspect all parts for wear and damage. Replace damaged parts, thoroughly clean and lubricate all parts that are to be reused, and reassemble the side gears, the spider gears, and the shaft. Make sure that all washers are in position, and then install the locking pin. Check for smooth operation once the differential assembly is back together.

A ring gear displaying the damage shown in **Figure 18-17** must be replaced. To replace a ring gear, remove the bolts holding the gear to the differential carrier. Then either press the ring gear from the carrier or gently hammer it loose, alternating hammer blows from side-to-side

to prevent damage to the carrier. Heat the new ring gear with a heat gun or by immersing it in boiling water. Then lay the ring gear on a flat surface and quickly place the differential carrier over it, being sure to align the differential carrier and ring gear bolt holes. Install the bolts and torque them to specifications. Most manufacturers call for tightening the bolts in a specific pattern to avoid distorting the ring gear.

Hypoid ring and pinion gears should always be replaced as a set. To remove the pinion gear, remove the drive chain or transfer gear. Then remove the pinion gear from the case, **Figure 18-18.** Replacement is the reverse of removal. As with helical gear final drives, hypoid drives often require tapered roller bearing replacement. Refer to Chapter 4.

After all repairs and part replacements have been made, adjust the hypoid ring and pinion gears. Hypoid gear adjustment is much more critical than planetary or helical gear adjustment. Improper hypoid gear contact can cause whining noises that will bring the vehicle's driver back immediately. Improper adjustments will also cause

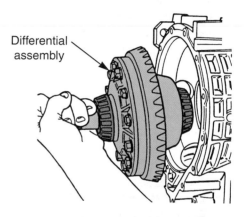

Figure 18-16. *Remove the hypoid ring gear and differential carrier from the transaxle case. All hypoid ring gear carriers use tapered roller bearings. (DaimlerChrysler)*

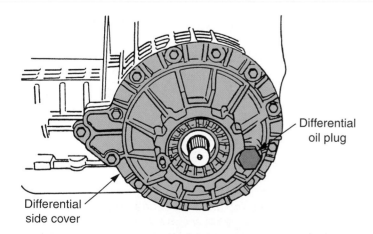

Figure 18-15. *Drain the oil and remove the side cover to begin hypoid gear removal. (DaimlerChrysler)*

Figure 18-17. *Excessive shock loads damaged this ring gear. When severe tooth breakage such as this is evident, suspect that the vehicle has seen severe operating conditions. When a ring gear is replaced, the pinion gear must also be replaced.*

the ring and pinion gear to wear excessively, causing quick failure.

Two adjustments must be made to the hypoid ring and pinion gear: *backlash,* or clearance, and *tooth contact pattern.* To adjust backlash, make all preload adjustments to the tapered roller bearings and then make the preliminary backlash adjustments. This may involve moving threaded adjusters or changing shims and spacers. Then check the backlash using a dial indicator. Attach the dial indicator so the pointer touches the ring gear. While holding the pinion gear shaft, turn the ring gear in either direction as far as it will move against the pinion gear. Zero the dial indicator. Then turn the ring gear in the other direction until it can no longer move. Read the dial indicator. This is the backlash. If the backlash is incorrect, make adjustments as necessary and recheck. **Figure 18-19** illustrates a typical backlash checking procedure.

Once the backlash has been set, check the tooth contact pattern. Begin by coating the teeth of the ring and pinion with a suitable compound. Some technicians use white grease, while others use *Prussian blue* or a similar compound. Turn the pinion gear to rotate the ring gear through two complete revolutions in both directions. Then observe the tooth contact pattern on both sides of the ring gear. **Figure 18-20** shows *heel-and-toe patterns* caused by

typical ring and pinion misadjustments, as well as the pattern shown by a properly adjusted ring and pinion.

Also shown in **Figure 18-20** are the appropriate ring and pinion movements needed to correct tooth contact problems. Tooth contact problems can be corrected by several methods, depending on the design of the unit. Some problems are corrected by turning threaded adjusters or changing shims located at the side of the differential carrier. Some problems can only be corrected by replacing the shim between the pinion gear and bearing with a thicker or thinner one. **Figure 18-21** shows typical ring and pinion adjustment locations. Always consult the proper manufacturer's service manual for specific hypoid ring and pinion adjustment procedures.

Once all adjustments have been made, wash the grease or other tooth pattern-checking compound from the ring and pinion, unless the manufacturer specifically states that it can be left on the gears. If necessary, install the final drive cover using a new gasket. If the final drive assembly is the older self-contained type, slide it onto the transaxle output shaft and install the attaching bolts. Install the CV axles according to instructions in Chapter 16.

Fill the final drive with the proper fluid. All hypoid gear final drives are lubricated with hypoid axle gear oil. Be sure to use the proper weight oil. Do not reuse the old oil.

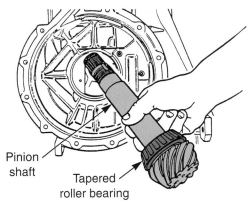

Figure 18-18. *Removing the pinion shaft from the case. Most transaxle final drive pinion shafts will resemble this one. Note that a tapered roller bearing is installed on the shaft at the pinion. An adjustment shim is usually installed under the bearing. (DaimlerChrysler)*

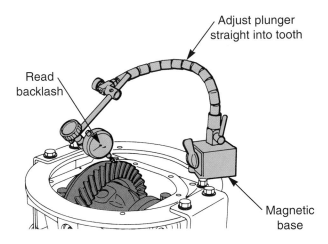

Figure 18-19. *A typical setup for checking backlash. If the backlash is incorrect, the relative positions of the ring and pinion gears must be changed. (DaimlerChrysler)*

Checking item	Contact pattern	Corrective action
Correct tooth contact Tooth contact pattern slightly shifted toward tow under no-load rotation. (When loaded, contact pattern moves toward heel.)		
Face contact Backlash is too large.	 This may cause noise and chipping at tooth ends.	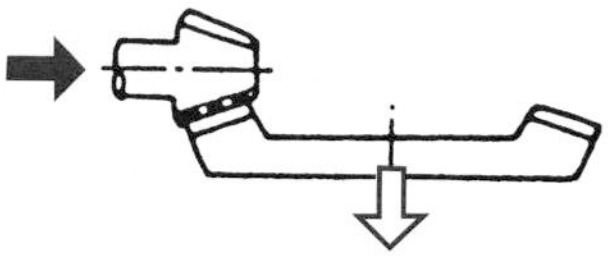 Increase thickness of drive pinion height adjusting shim in order to bring drive pinion close to crown gear.
Flank contact Backlash is too small.	 This may cause noise and stepped wear on surfaces.	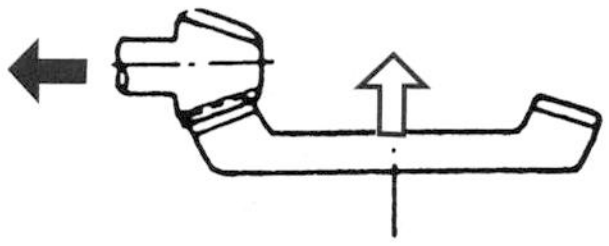 Reduce thickness of drive pinion height adjusting shim in order to move drive pinion away from crown gear.
Toe contact (Inside end contact)	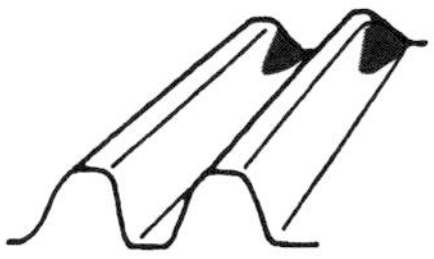 Contact area is small. This may cause chipping at toe ends.	Adjust as for flank contact. 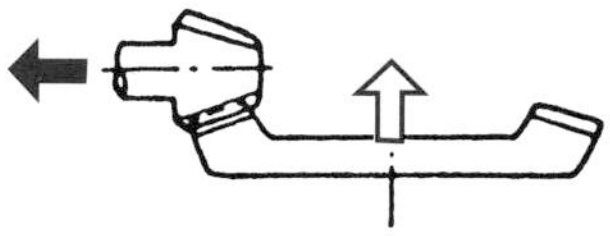
Heel contact (Outside end contact)	Contact area is small. This may cause chipping at heel ends. 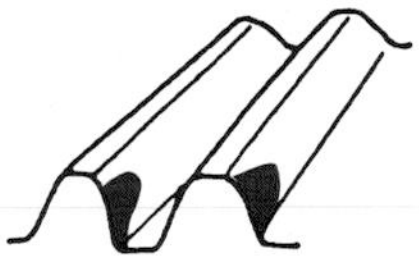	Adjust as for face contact. 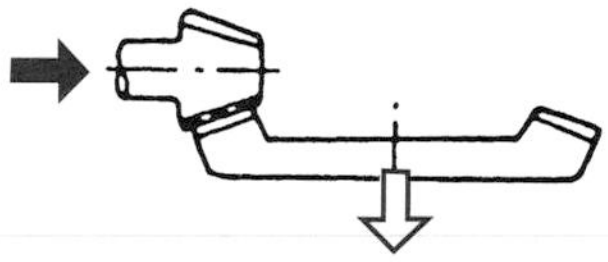

Figure 18-20. *Typical tooth contact patterns. Contrast the correct pattern at the top with the incorrect patterns below it. The far right column shows how to correct the pattern by moving the ring or pinion gear, or both. (Subaru)*

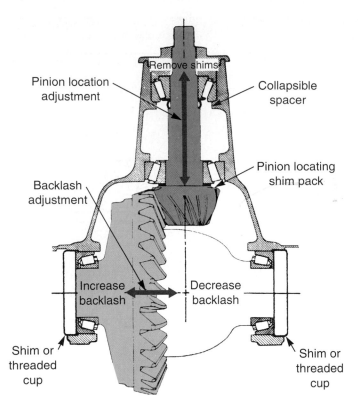

Figure 18-21. *This figure is a general representation showing locations of adjustment devices or adjusting shims on a hypoid gearset. The adjusters and shims are used to correct backlash and tooth contact pattern problems. Actual adjustment devices will vary from one manufacturer to another. (Ford)*

Summary

The three types of final drive units are the planetary, the helical gear, and the hypoid gear. All are currently being used in production front-wheel drive vehicles.

Older planetary gear final drives cannot be accessed without removing and disassembling the transaxle. Newer planetary units can be removed with the transaxle intact. After the CV axle is removed on the final drive side, the extension housing is removed and the final drive is separated from the output shaft. Once the final drive is out of the unit, the differential components and planet gears can be checked, and, if necessary, repaired. Reinstallation is the reverse of removal.

To remove a helical gear drive, remove both CV axles, the final drive cover, the bearing retainer, and the extension housing. Then remove the differential carrier and ring gear from the case. The pinion gear can be slid from the case once the transfer gear is removed. The final drive components can then be cleaned, inspected, and repaired. To reinstall the final drive, reverse the removal steps.

To remove a hypoid gear drive, remove the CV axles, the side cover, the bearing retainer, and the extension housing. Then pull the differential carrier and ring gear out of the transaxle case. The pinion gear can be slid from the case once the drive chain or gear is removed. Clean,

inspect, and if necessary, repair the final drive components. After reinstalling the components, adjust backlash and tooth contact pattern as necessary.

Review Questions—Chapter 18

Please do not write in this text. Place your answers on a separate sheet of paper.

1. The final drive gears are in _____ mesh.

2. The spider gear shaft is usually held to the final drive carrier by a _____ _____.

3. The most commonly replaced parts of a planet carrier are the _____ _____ and _____ _____.

4. What can happen if the interior of the final drive housing is not cleaned out before reinstalling the components?

5. To install a replacement gear on a differential carrier, the gear must be _____ before installation.

Match the final drive part with the service operation it requires.

6. Tapered roller bearing _____

7. Ring and pinion _____

8. Planet carrier _____

9. Extension housing _____

10. Threaded adjuster _____

(A) Replace needle bearings and thrust washers.

(B) Remove metal from the interior.

(C) Turn to make adjustment.

(D) Replace if damaged.

(E) Adjust tooth contact pattern.

ASE Type Questions —Chapter 18

1. Technician A says that some planetary gear final drives can be serviced without disassembling the transaxle. Technician B says that all planetary gear final drives contain spider and side gears. Who is right?

(A) A only.

(B) B only.

(C) Both A and B.

(D) Neither A nor B.

2. To remove a helical gear final drive, you must remove all the following transaxle components *except:*

(A) final drive cover.

(B) bearing retainer.

(C) extension housing.

(D) drive chain.

3. Which of the following procedures is *not* necessary when servicing a helical gear final drive?

(A) Adjusting backlash.

(B) Replacing seals and gaskets.

(C) Checking bearing preload.

(D) Removing transfer gear.

4. Technician A says hypoid gears require that bearing preload be set before backlash is set. Technician B says hypoid gears require that the tooth contact pattern be set before backlash is set. Who is right?

(A) A only.

(B) B only.

(C) Both A and B.

(D) Neither A nor B.

5. Which of the following final drives could be lubricated with gear oil *only*?

(A) Planetary gear without a removable extension housing.

(B) Planetary gear with a removable extension housing.

(C) Helical gear.

(D) Hypoid gear.

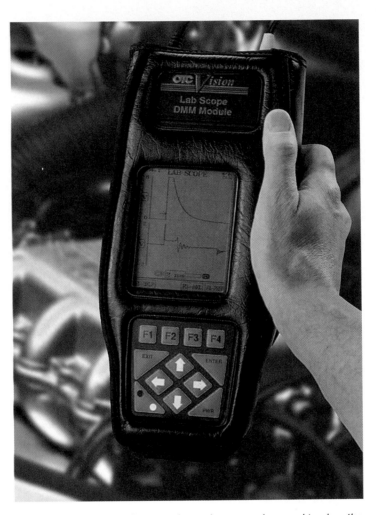

A lab scope, such as the one shown here, can be used to view the waveforms produced by some electronic and electrical devices. (OTC)

Chapter 19

Electronic Control System Service

After studying this chapter, you will be able to:
- ❑ Replace defective components in the transmission or transaxle.
- ❑ Replace engine or vehicle components that affect electronic transmission or transaxle operation.
- ❑ Replace the PROM in a transmission/transaxle computer.
- ❑ Reprogram a transmission/transaxle computer.
- ❑ Check transmission/transaxle operation after service.

Technical Terms

Reluctor

Solenoid pack

EPROM

FEPROM

Flash programming

Direct programming

Indirect programming

Remote programming

Introduction

After a problem with an electronically controlled automatic transmission or transaxle has been located, the technician must make the needed repairs. Some repairs are simple, such as replacing a barometric sensor mounted on the inner fender well. Other repairs, such as reprogramming the vehicle's computer, are more complex.

The technician must make all repairs carefully so that transmission or transaxle performance is maintained. This chapter discusses the procedures for replacing the electrical and electronic components related to the transmission and transaxle.

 Note: Replacement of internal transmission and transaxle parts not directly related to the electronic control system was covered in earlier chapters. These procedures are similar on both electronic and nonelectronic units and will not be covered again.

Occasionally a replacement electronic component will be defective. If a replacement component is to be installed in a hard-to-reach location, such as under a side oil pan, check it with an ohmmeter before installation to ensure that it is good. If ohmmeter specifications are not available, carefully inspect the component for obvious damage.

 Caution: Removing or installing an electrical connector can damage electronic components by causing a voltage surge. The technician should always remove the negative battery cable to prevent electrical damage. Also, after removing the negative cable, touch it to the positive battery post to discharge any remaining electrical charge.

Caution: Loose or corroded electrical connectors cause many electronic control system problems. Always check the condition of the electrical connector before replacing any component.

Replacing Components on the Transmission or Transaxle

Electronic control system components mounted inside the transmission or transaxle oil pan include pressure and temperature sensors, some manual valve position sensors, and output solenoids. To replace most of these components, you must remove the oil pan.

Some transaxles have side-mounted valve bodies located under a sheet metal side pan. It is sometimes difficult to remove this pan. Often, other vehicle components must be removed to gain access to the pan and valve body. **Figure 19-1** shows the location of several solenoids located under the side pan on one particular transaxle.

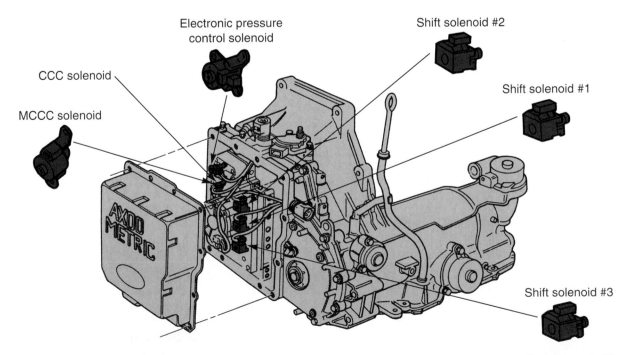

Figure 19-1. *This illustration shows the relative positions of the solenoids used on an electronically controlled transaxle. These solenoids are installed inside the oil pan. (Ford)*

A few components, such as speed sensors and a few pressure switches, are installed on the transmission's case. The oil pan does not have to be removed to service these components.

 Note: Always check the service manual to determine component location before removing the oil pan.

Replacing Pressure and Temperature Sensors

The first step in replacing pressure or temperature sensors is to remove the transmission or transaxle oil pan, if necessary. This was covered in earlier chapters. After removing the oil pan, locate the faulty sensor on the valve body.

 Caution: Many transmission sensors are similar in appearance. Have the service manual handy to ensure that you replace the proper sensor.

Pressure Sensors

Many pressure sensors are separate units threaded into the valve body or the transmission case. However, some pressure sensors are part of an assembly. If one sensor in the assembly fails, the entire assembly must be replaced. To replace a single sensor, remove the electrical connector. Then loosen and remove the sensor with an appropriate socket, **Figure 19-2.** Some pressure sensors require the use of an oil sender removal socket. Install the new sensor and tighten it securely. Finally, reinstall the electrical connector.

To replace a pressure sensor assembly, remove the electrical connector and remove the bolts holding the assembly to the valve body. Then remove the pressure sensor assembly from the valve body, **Figure 19-3.** Place new O-rings on the assembly, install the assembly on the valve body, and tighten the bolts. Since the sensor assembly bolts help secure the valve body, make sure they are tightened to the specified torque. Finally, replace the electrical connector.

Once the new pressure sensor (or sensor assembly) is installed, replace the oil pan and refill the transmission or transaxle with the proper type of transmission fluid.

Temperature Sensors

A typical temperature sensor is shown in **Figure 19-4.** To replace a temperature sensor, remove the electrical connector and unscrew the sensor from the valve body.

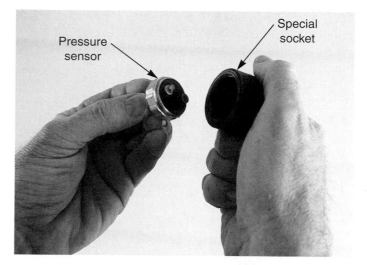

Figure 19-2. *Most pressure switches can be removed with a large socket. A few require a special oil pressure switch socket, such as the one shown here. (General Motors)*

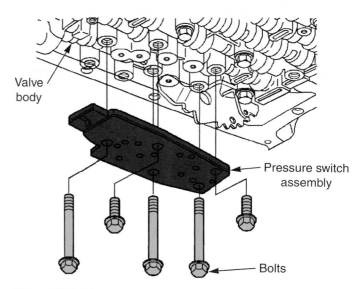

Figure 19-3. *Pressure switch assemblies are usually attached to the valve body with bolts. Remove the bolts to remove the assembly. (General Motors)*

Then install the new sensor and reattach the electrical connector. It is usually not necessary to put sealer on the sensor threads when the sensor is installed on the valve body. However, external (case-mounted) sensors may require thread sealer. If the temperature sensor is part of a pressure sensor assembly, replace the assembly as explained in the previous section.

After installing the new sensor, replace the oil pan and refill the transmission with transmission fluid, if necessary. Road test the vehicle to make sure that replacing the sensor solved the problem.

Output Speed Sensors

Most output speed sensors are located on the transmission/transaxle case or in the extension housing.

See **Figures 19-5 and19-6.** Therefore, it is not necessary to remove the oil pan when servicing these sensors. A few output speed sensors are installed under the side oil pan, and the side pan must be removed to access the sensor.

To replace a case-mounted output speed sensor, remove the sensor's electrical connector. Then remove the bolt holding the sensor to the case and remove the sensor. Compare the old and new sensors to ensure that the new sensor is correct. The new sensor must not be longer than the old unit, or it will contact the moving **reluctor** (toothed

wheel) when installed. Install the new sensor in the case and reinstall the hold-down bolt and electrical connector.

To replace a gear-driven output speed sensor, remove the electrical connector and disconnect the speedometer cable, if it is attached to the sensor. Then remove the driven gear housing retainer and remove the driven gear and housing as a unit, **Figure 19-7.** Check and replace parts as needed. Then reinstall the unit, being sure that the gear engages properly on the output shaft gear. Finally, reinstall the electrical connector and speedometer cable.

If the defective speed sensor is mounted on a wheel, the sensor can usually be replaced by disconnecting the

Fluid temperature sensor

Figure 19-4. *This fluid temperature sensor is installed on the valve body, under a side-mounted pan. Installing the temperature sensor on the bottom of the valve body ensures that it will be submerged in transmission fluid. To remove the sensor, the fluid must be drained and the side pan removed. (Ford)*

Figure 19-6. *Most externally mounted speed sensors resemble this one. The sensor is held to the case by a bracket. Once the speed sensor has been located, it can be easily removed. When reinstalling the sensor, make sure it does not contact the rotating wheel assembly.*

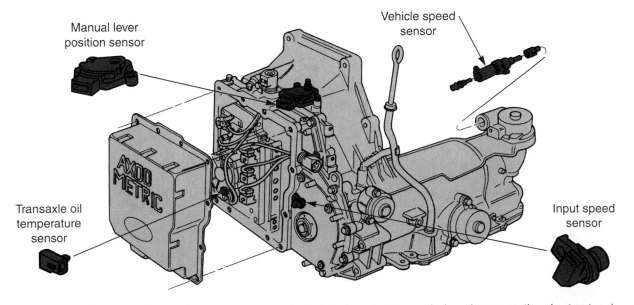

Manual lever position sensor

Vehicle speed sensor

Transaxle oil temperature sensor

Input speed sensor

Figure 19-5. *Electronic transaxle speed sensors are usually installed on the transaxle housing or on the short extension on one side of the transaxle. These can be replaced easily. (Ford)*

electrical connector and removing it from its bracket. The new sensor is then installed in the bracket. A few wheel-speed sensors require that clearance be checked and, if necessary, adjusted after installation. Clearance is checked with a brass or plastic feeler gauge, since a steel gauge would stick on the sensor magnet and give a false reading. See **Figure 19-8.**

Input Speed Sensors

A few transmissions and transaxles have a sensor that measures the speed of the input shaft between the converter and the planet gears. This sensor is called an input speed sensor. It measures the speed of the torque converter turbine. Input speed sensors are installed on the front of the transmission or transaxle case. Input speed sensors can be removed in the same way as output speed sensors. The clearance of some input speed sensors is adjustable. Therefore, the technician should always refer to the appropriate service manual when replacing this type of sensor.

Replacing Pressure Control Solenoids

Most pressure control solenoids are installed inside the transmission or transaxle oil pan. After removing the oil pan, locate the defective solenoid on the valve body.

 Caution: Most pressure control solenoids are similar in appearance. Consult the appropriate service manual to make sure you replace the proper solenoid.

After locating the defective solenoid, remove the electrical connector. Then remove the pin, clip, or bolt holding the solenoid in place and remove the solenoid,

Figure 19-7. *Remove a gear-driven speed sensor as a unit, as shown here. Once it is on the bench, it can be tested and defective parts can be replaced.*

Figure 19-9. Most solenoids have a small filter to reduce the chances of solenoid sticking. Solenoid filters installed in the valve body should be removed and cleaned. Filters that are part of the solenoids should be cleaned if the solenoid is being reused. See **Figure 19-10.**

Before installing the new solenoid, make sure it is fitted with the correct O-ring or gasket. Place the solenoid on the valve body and install the fasteners. If the solenoid is held in place with a bolt, tighten the bolt to specifications. Reconnect the electrical lead, making sure that it makes good contact with the solenoid terminal. Finally, reinstall the oil pan and refill the transmission with the proper transmission fluid.

Some transmission and transaxle solenoids are installed in a solenoid assembly, or ***solenoid pack,*** which contains all the solenoids in a single unit. If one solenoid in the pack goes out, the entire solenoid pack must be replaced. To remove a solenoid pack, remove the electrical connector from the transmission case, **Figure 19-11.** Then remove the oil pan and the bolts or screws holding the solenoid pack to the valve body and transmission case. Finally, gently remove the solenoid pack from the transmission, **Figure 19-12.** In some designs, the solenoid pack is mounted on the outside of the transmission or transaxle case. The oil pan does not have to be removed to access this type of solenoid packs.

Replacing Engine or Vehicle Components

Many of the electronic components that effect transmission or transaxle operation are not located on the transmission or transaxle. Instead, they are located on the engine or elsewhere in the vehicle. Replacement procedures for these components are given in the following sections.

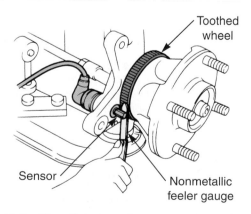

Figure 19-8. *The clearance between some wheel speed sensors and the toothed wheel can be adjusted. Always consult the manufacturer's service manual for exact procedures. (Nissan)*

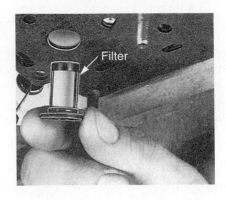

A

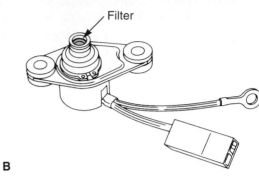

B

Figure 19-10. *It is very important to clean the solenoid filters when servicing a solenoid. A—Some filters are separate and can be removed from the case for cleaning. B—Many filters are part of the solenoid assembly and are replaced with the solenoid. They can sometimes be adjusted. (Ford, Nissan)*

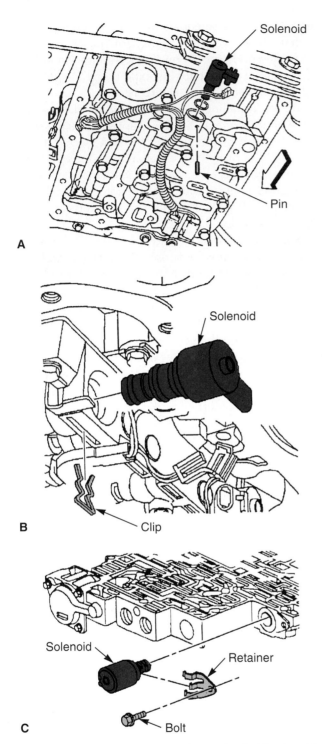

Figure 19-9. *A variety of fasteners is used to hold solenoids in place. A—This solenoid is held in place with a pin. B—This solenoid is held in place by a retaining clip. C—A small bolt and a retainer are used to hold this solenoid in place. (General Motors)*

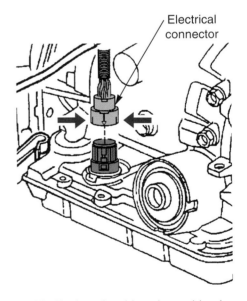

Figure 19-11. *Begin solenoid pack or wiring harness removal by disconnecting the electrical connector. (General Motors)*

Throttle Position Sensors

To replace a throttle position sensor, locate the sensor on the throttle body. Make sure the ignition switch is in the *off* position and remove the sensor's electrical connector. Then remove the attaching screws and slide the sensor

from the throttle shaft. See **Figure 19-13.** To replace the sensor, slide the new unit carefully over the throttle shaft, making sure that any slots or flat spots on the shaft and sensor are aligned. Finally, install and tighten the fasteners and replace the electrical connector.

On some older vehicles, the throttle position sensor is located inside the carburetor. To replace this type of

Figure 19-12. *After removing the bolts holding the solenoid pack to the transmission case, carefully lift the unit from the case. (Ford)*

sensor, the carburetor must be disassembled. Many technicians prefer to overhaul or replace the carburetor instead of merely replacing the sensor. Consult the proper service manual for procedures.

Throttle Position Sensor Adjustment

Some throttle position sensors must be adjusted after replacement. Adjustable sensors generally have slotted mounting holes. To make a throttle position sensor adjustment, obtain a multimeter and set it on the proper scale. If the sensor is a resistance-type unit, it must be adjusted using the ohmmeter. If the sensor is a transducer-type unit, the voltmeter will be used during adjustment procedures.

After determining which multimeter setting to use, remove the electrical connector and connect the multimeter leads to the sensor.

> **Note: If the throttle position sensor is being checked for proper voltage, it will be necessary to use a special harness that allows reference voltage to enter and exit the sensor.**

Make sure the throttle plate is on the hot idle stop. If the vehicle has a carburetor, make sure the choke is completely open. Loosen the connector screws and move the sensor until the proper value is obtained, **Figure 19-14.** Then press the accelerator pedal to the floor (engine off) and make sure the sensor reading for wide open throttle (WOT) is correct. If the WOT reading is *not* correct, check the throttle linkage to ensure that the throttle is being completely opened. Then readjust the sensor position as necessary.

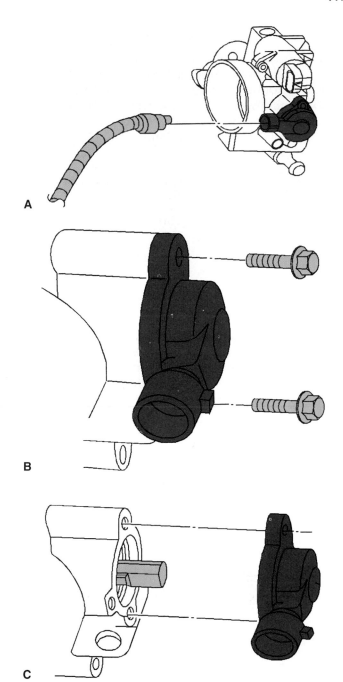

Figure 19-13. *Most throttle position sensors can be removed in a few simple steps. A—Remove the electrical connector. B—Loosen and remove the attaching bolts. C—Remove the sensor from the shaft. Note the position of the throttle shaft for reinstallation.*

Temperature Sensors

Removal and replacement procedures for temperature sensors are explained in the following sections. Coolant temperature sensors are always installed on the engine so that they contact the circulating coolant inside of the engine. Air temperature sensors can be installed on the intake manifold, in the air cleaner, or in the air intake ducting.

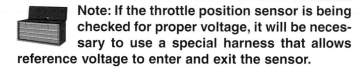

Coolant Temperature Sensors

 Warning: Remove all pressure from the cooling system before attempting to remove any cooling system part or fitting. If pressure is not removed before any part is loosened, hot coolant can spray violently from the opening. Coolant can reach temperatures above 250°F (115°C) and can cause severe burns or eye damage.

Begin coolant temperature sensor replacement by locating the sensor on the engine. Then depressurize the cooling system and drain it to a point below the sensor. Inspect the coolant to determine whether it is in good enough condition to be reused. Many technicians prefer to install new coolant whenever the cooling system is drained.

Note: Some late-model cooling systems use special long-life antifreeze. Always add the proper kind of antifreeze.

Remove the electrical connector from the sensor and use the proper socket to remove the sensor from the engine. See **Figure 19-15.** Some coolant temperature sensors can only be removed with a special socket. Check the end of the sensor and the coolant passages for excessive corrosion and deposits. If corrosion or deposits are evident, the cooling system should be flushed and new coolant should be installed.

To install the new sensor, coat it with thread sealer and install it on the engine. Torque the sensor to specifications. Overtightening may distort the sensor's internal resistor, causing incorrect readings. Finally, refill the cooling system and check for leaks.

 Caution: Some late-model cooling systems require special bleeding procedures to purge air from the engine and radiator. Consult the proper service manual for bleeding procedures.

Air Temperature Sensors

To replace an air temperature sensor, locate it in the engine compartment. The air temperature sensor on most modern vehicles is located in the air duct leading to the throttle valve or elsewhere on the air intake ductwork, **Figure 19-16.** Some older sensors are threaded into the intake manifold. Disconnect the electrical connector and remove the sensor. Compare the old sensor to its replacement. Then install the new sensor and reconnect the electrical connector.

Engine Speed Sensors

Engine speed sensors are usually located on the engine. Some speed sensors are part of the distributor. Others are installed on the engine block near the crankshaft or camshaft. A few engine speed sensors are installed on the transmission case. See **Figure 19-17.**

To replace an engine speed sensor, locate it on the engine and remove the sensor's electrical connector. With the connector removed, loosen and remove the fasteners securing the sensor. Then remove the sensor from the distributor or engine block, as applicable. Compare the old and new sensors to ensure the new sensor is the correct type. Install the new sensor, making sure it will not contact any moving parts. If necessary, align the sensor using

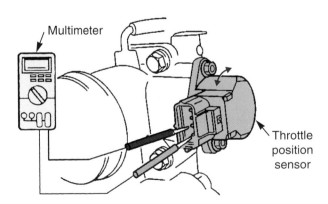

Figure 19-14. *Adjusting a throttle position sensor requires either a voltmeter or an ohmmeter. Loosen the attaching screws and carefully turn the sensor until the specified electrical value is reached. Most modern throttle position sensors are not adjustable. (General Motors)*

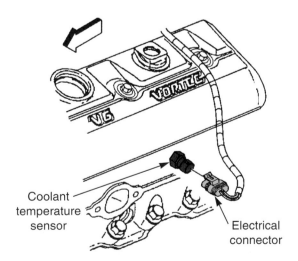

Figure 19-15. *Temperature sensors can be removed after the cooling system has been depressurized and drained. Locate the sensor on the engine and remove the electrical connector. Then loosen and remove the temperature sensor. (General Motors)*

needed special tools. Replace the electrical connector, start the engine, and check system operation.

Exhaust Temperature Sensor

To replace an exhaust temperature sensor, remove the electrical connector and remove the old sensor using the proper size wrench. Compare the old and new sensors to

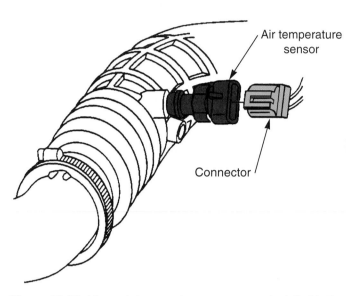

Figure 19-16. *Many air temperature sensors are installed in the duct leading to the throttle valve. (General Motors)*

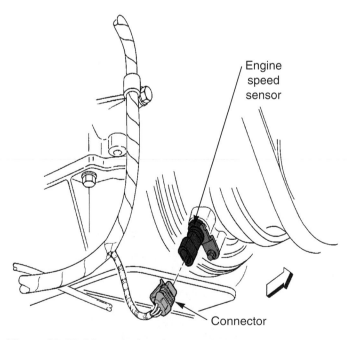

Figure 19-17. *Many engines have more than one engine speed sensor. Be sure to locate the proper sensor using the service manual. This speed sensor is located on the engine's front cover. To remove the sensor, disconnect the electrical connector and remove the bolt holding the sensor to the engine. (General Motors)*

ensure that the new sensor is correct. Coat the threads of the new sensor with high temperature sealer if the vehicle manufacturer requires sealer. Then install the new sensor and reconnect the electrical connector.

Mass Airflow Sensors

Most mass airflow (MAF) sensors are attached to the air intake duct at or near the throttle plate. See **Figure 19-18.** They are always located ahead of the throttle plate in the airstream. Some sensors form a part of the air intake ducting.

Disconnecting MAF sensors from the air intake is usually simple. Begin by making sure the ignition key is in the *off* position and removing the electrical connector from the sensor. Then remove the attaching bolts or clamps and pull the sensor from the duct, **Figure 19-19.** Install the new MAF sensor in reverse order of removal.

Figure 19-18. *This mass airflow (MAF) sensor is installed in the air intake ductwork.*

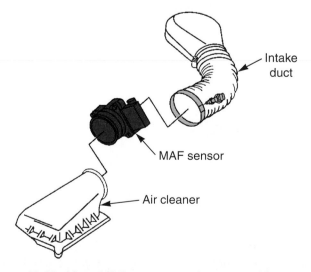

Figure 19-19. *Many MAF sensors can be removed by detaching the entire assembly from the intake ductwork. These sensors are often held to the ductwork with large hose clamps. (General Motors)*

Caution: It is extremely important that there be no air leaks between the MAF sensor and the throttle plate. After installation, recheck all duct clamps and fittings to ensure that no air can enter.

 Note: If the system contains more than one oxygen sensor, refer to an appropriate service manual to ensure that you are changing the correct sensor.

Manifold Absolute Pressure and Barometric Pressure Sensors

Manifold absolute pressure (MAP) sensors and barometric pressure (BARO) sensors can be located just about anywhere under the hood. Some are installed directly to the manifold. Most, however, are connected to manifold vacuum through a hose, **Figure 19-20.** Replacing MAP and BARO sensors is relatively simple. Remove the vacuum line when applicable and disconnect the electrical connector. Then loosen the fasteners holding the sensor to the vehicle and remove the sensor. Finally, install the new sensor and reconnect the vacuum line (if necessary) and the electrical connector.

Oxygen Sensors

Oxygen sensors, or O$_2$ sensors, are installed in the exhaust system, usually on the exhaust manifold or the exhaust pipe. Late-model vehicles have at least two oxygen sensors. On vehicles equipped with OBD II systems, one oxygen sensor is installed in the manifold ahead of the catalytic converter and one oxygen sensor is installed in the exhaust pipe behind the catalytic converter.

A special socket may be needed to remove an oxygen sensor, **Figure 19-21.** Begin by removing the electrical connector from the sensor. Install the socket over the sensor and turn the socket counterclockwise to loosen the sensor. Once the sensor has been unthreaded, pull it out of the manifold or exhaust pipe, **Figure 19-22.** Coat the threads of the new sensor with sealant or anti-seize compound only if recommended by the manufacturer. Then install the new sensor and torque it to specifications. Reattach the electrical connector and make sure the connector wires do not contact the exhaust manifold.

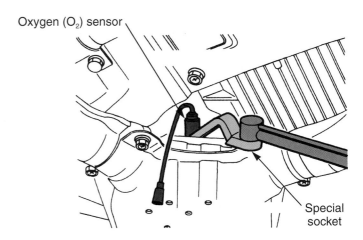

Figure 19-21. *Many oxygen sensors can only be removed by using a special tool to grasp and loosen the sensor body. (DaimlerChrysler)*

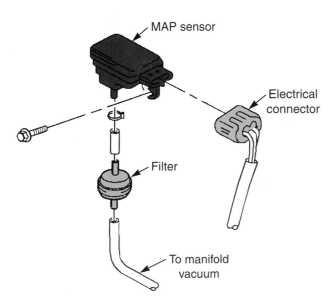

Figure 19-20. *To remove a MAP sensor, remove the electrical connector and vacuum hose and unbolt the sensor from its retaining bracket. Some MAP sensors are installed directly on the manifold and do not use a separate vacuum hose. (General Motors)*

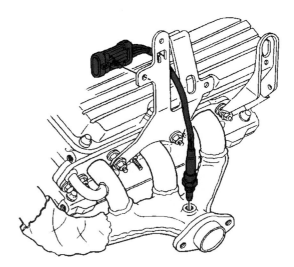

Figure 19-22. *Once the oxygen sensor has been loosened, unscrew it and remove it from the exhaust system part. (General Motors)*

ECM Service

Before replacing an ECM, make sure the ignition switch is turned to the *off* position. Disconnect the negative battery cable, if recommended by the manufacturer. After locating the ECM, disconnect the wiring connector(s) and remove the attaching fasteners. Most ECMs are attached with bolts or screws, but a few are held in place with clips or plastic rivets.

To replace the ECM, set the unit in place and reinstall the fasteners, being sure to reattach any ground straps. Then reconnect the electrical connectors. On some vehicles, it may be necessary to connect the electrical connector before placing the ECM in position on the vehicle.

Figure 19-23. *The PROM is located under a cover on the ECM. Once the cover has been removed, the PROM can be carefully removed from the ECM. A special tool must be used to remove some PROMs.*

 Caution: To prevent static discharge from damaging the new ECM, follow the manufacturer's instructions concerning static discharge and unit grounding.

Many ECMs can be restored to proper performance by providing them with updated information. On some older vehicles, the PROM chip can be replaced. On many newer vehicles, the ECM can be reprogrammed from an outside source.

Replacing a PROM

To replace a PROM, make sure the ignition key is in the *off* position. Unplug any electrical connectors to the ECM and remove the access cover over the PROM. See **Figure 19-23.** Carefully remove the old PROM from the ECM. Then install the new PROM, carefully pushing it into position. Special tools are available to help remove and install the PROM safely. Finally, reinstall the access cover and reinstall the ECM on the vehicle.

Replacing a Knock Sensor Module

In some vehicles, the ECM contains a knock sensor (KS) module. This module can be removed from the ECM when replacement is necessary. The KS module is installed in the ECM, ensuring that the proper timing retard is maintained for engine knock protection.

Flash Programming a Computer

As mentioned, the ECMs in late-model vehicles can be updated by erasing old information from the computer's memory and reprogramming the unit with new information. This new information often cures engine drivability and transmission shifting problems. The computer memory section that can be updated is generally called the **EPROM** (electronically erasable programmable read-only memory)

or **FEPROM** (flash erasable programmable read-only memory). Providing the existing memory with updated information is usually referred to as **flash programming**.

Actual reprogramming details vary between manufacturers, but the basic procedure begins with placing the ECM in the programming mode. One of three methods can be used to reprogram the ECM:
- ❑ Direct programming.
- ❑ Indirect programming.
- ❑ Remote programming.

Direct Programming

Direct programming is the fastest and simplest method of reprogramming an ECM. The new information is downloaded by attaching a shop recalibration device (usually a computerized analyzer) directly to the data link connector. The erasure and reprogramming is done by accessing the proper menu and following the instructions as prompted by the recalibration device. Then, the information (often contained in the device's memory, on a CD-ROM disc, or through a connection to the manufacturer's database) is entered into the ECM through the data link connector. The shop recalibration device is not a scan tool, and a scan tool is not needed for this procedure.

Indirect Programming

To perform **indirect programming,** the proper scan tool must be used to transfer information from a separate *programming computer* to the ECM. The scan tool can also be used to reset some computer-controlled vehicle systems after programming is complete. The programming computer may resemble a personal computer used in the home, or it may be a computerized analyzer that is similar to the one used for direct programming.

In indirect programming, the scan tool is connected to the programming computer and programming information is downloaded from the computer to the scan tool. Most scan tools use a high-capacity memory cartridge to store the programming information. Some newer scan tools have enough fixed memory to hold the programming information and do not use a separate memory cartridge. Once the programming information has been downloaded, the scan tool can be disconnected from the programming computer, taken to the vehicle, and connected to the data link connector. Programming information is then downloaded from the scan tool to the ECM through the data link connector.

Remote Programming

Remote programming is done with the ECM removed from the vehicle. This procedure is used when changes must be made through a direct connection to a manufacturer's database. Remote programming can also be done in cases where normal direct or indirect programming is not practical or possible. Special connectors and tools are required for this type of programming. In most cases, this procedure is done only at vehicle dealerships.

To perform remote programming, remove the ECM as described earlier in this chapter. Once the ECM has been removed from the vehicle, take it to the programming device. The programming device is generally a computer located in the shop. This device may contain the new ECM information, or it may be used with a modem to connect to a remote database. Attach the programming device's electrical connectors to the ECM. Access the device's programming menu and follow instructions given in the menu to program the ECM. Programming normally takes only a few minutes. When programming is complete (as indicated on the menu), remove the programming device's electrical connectors from the ECM and reinstall the ECM in the vehicle.

Rechecking System Operation

After replacing parts or reprogramming the ECM, recheck system operation to ensure that the repairs have been successful. Drive the vehicle long enough for the transmission or transaxle to shift through all the gears. Check part throttle downshifts and detent (passing gear) operation, if possible. While making the test drive, note whether the MIL light comes on. When the road test is complete, use the scan tool to recheck the system for trouble codes.

ECM Relearn Procedures

The electronic control system may require a **relearn procedure,** which is a period of vehicle operation that allows the computer system to adapt to new components

and updated reprogramming. The relearn procedure can often be accomplished by driving the vehicle at various speeds for about ten minutes. Some vehicles require a specific relearn procedure, which may include idling in drive for a specified amount of time or until the engine reaches its normal operating temperature. Always check the manufacturer's service literature for specific relearn procedures. Ignore any unusual engine and transmission conditions until the relearn procedure is complete.

Summary

Replacing components in the transmission or transaxle is relatively easy, but some precautions must be taken. Electronic control system components mounted inside the transmission or transaxle oil pan include pressure and temperature sensors and output solenoids. To replace most of these components, the transmission or transaxle oil pan must be removed. Some pressure and temperature switches are part of a pressure switch assembly. After changing the electronic components, replace the oil pan and refill the transmission or transaxle with transmission fluid.

Most speed sensors are located outside the case. To replace a case-mounted speed sensor, remove the electrical connector, remove the bolt holding the sensor to the case and withdraw the sensor. Compare the old and new sensors to ensure that the new sensor is correct. Then install the new sensor in the case and install the hold down bolt and electrical connector. To install a gear-driven output speed sensor, remove the electrical connector and disconnect the speedometer cable if used. Then remove the driven gear housing retainer and remove the driven gear and housing as a unit. Install the new unit, being sure that the gear engages properly on the output shaft gear. Finally, reinstall the electrical connector and speedometer cable.

Pressure control solenoids can be easily replaced after removing the oil pan. Locate the solenoid on the valve body; then remove its electrical connector and attaching bolts. Install the new solenoid in reverse order of removal. The pan can then be reinstalled.

Many electronic components that effect transmission or transaxle operation are located on the engine or elsewhere in the vehicle. Examples include the throttle position sensor, temperature sensors, engine speed sensor, mass airflow sensor, manifold absolute pressure sensor, barometric pressure sensor, and oxygen sensor. These components should be replaced if faulty.

To replace a throttle position sensor, locate the sensor on the throttle body. Make sure the ignition switch is in the *off* position and remove the electrical connector from the sensor. Remove the old sensor and install the new sensor. Tighten the fasteners and replace the electrical connector. Some throttle position sensors must be adjusted after replacement.

Always depressurize the cooling system before loosening the coolant temperature sensor. Remove the electrical connector and use an appropriate socket to remove the sensor. Coat the new sensor with thread sealer and install it on the engine.

Most air temperature sensors are threaded into the intake manifold or clipped to the air cleaner or ducting. Replace the sensor after disconnecting the electrical connector.

Some engine speed sensors are located in the distributor. Many speed sensors are mounted on the engine block. To replace an engine speed sensor, locate the sensor and remove the electrical connector. Then remove the old sensor, install the new sensor, and replace the electrical connector.

Mass airflow sensors can be installed on the throttle body assembly or on the air intake ducting. After the electrical connector is removed, replacement is simple.

MAP and BARO sensors are easy to replace. Simply remove the vacuum line and electrical connector, and then remove the fasteners holding the sensor to the vehicle. Reverse removal procedures to reinstall.

A special socket may be needed to remove the oxygen sensor. Begin by removing the electrical connector from the sensor and then remove the sensor. Install the new sensor and replace the electrical connector.

ECM removal is simple, but the ignition switch must be in the *off* position to prevent damage. Some makers call for removing the battery negative cable before removing the ECM. ECMs can sometimes be restored to proper operation by replacing the PROM or by flash programming the EPROM from an outside source.

Always recheck system operation after replacing any components or reprogramming the ECM. If necessary, perform the specified relearn procedure before checking the system.

Review Questions—Chapter 19

Please do not write in this text. Place your answers on a separate sheet of paper.

1. Electronic control system components mounted inside the transmission or transaxle oil pan include _____ and _____ sensors.

2. To replace output solenoids, you must remove the _____ _____.

3. A speedometer cable may be part of a(n) _____ _____ sensor assembly.

4. Throttle position sensors are located on the _____ _____ or carburetor.

5. Some throttle position sensors must be _____ after installation.

6. Before attempting to remove any cooling system part, what should the technician do?

7. MAF sensors are always located ahead of the _____ _____.

8. Modern vehicles always have more than one _____ sensor.
   (A) MAF
   (B) oxygen
   (C) coolant temperature
   (D) air temperature

9. OBD II systems always have an oxygen sensor located _____ the catalytic converter.

10. When replacing an ECM, the ignition switch should be in the _____ position.

ASE-Type Questions—Chapter 19

1. Each of the following is an example of an electronic device that could be installed on a transmission or transaxle *except:*
   (A) pressure switch.
   (B) temperature sensor.
   (C) MAP sensor.
   (D) output shaft speed sensor.

2. Technician A says that some transmission temperature sensors are part of an assembly. Technician B says that some transmission pressure switches are part of an assembly. Who is right?
   (A) A only.
   (B) B only.
   (C) Both A and B.
   (D) Neither A nor B.

3. If the mounting holes of a throttle position sensor are slotted, the sensor is:
   (A) open.
   (B) shorted.
   (C) cracked.
   (D) adjustable.

4. Technician A says that a special socket may be needed to replace the coolant temperature sensor. Technician B says that a special socket may be needed to replace the oxygen sensor. Who is right?
   (A) A only.
   (B) B only.
   (C) Both A and B.
   (D) Neither A nor B.

5. Technician A says that MAF sensors are always installed after the throttle plate. Technician B says that air temperature switches can be installed in the air intake ductwork. Who is right?

   (A) A only.
   (B) B only.
   (C) Both A and B.
   (D) Neither A nor B.

6. Which of the following engine sensors is connected to a vacuum line?

   (A) MAP sensor.
   (B) O$_2$ sensor.
   (C) MAF sensor.
   (D) Throttle position sensor.

7. Flash programming can be defined as:

   (A) restoring an ECM to proper performance by providing the EPROM with updated information.
   (B) diagnosing the EPROM with a scan tool to find a problem.
   (C) replacing the EPROM with a new one containing new information.
   (D) replacing the EPROM with a used one containing information from another vehicle.

8. Reprogramming an ECM by using the scan tool to transfer information between an outside data source and the ECM is called _____ reprogramming.

   (A) direct
   (B) indirect
   (C) diagnostic
   (D) remote

9. Always recheck operation by doing all of the following, *except*:

   (A) road testing.
   (B) checking for code reset.
   (C) observing the MIL.
   (D) checking top speed.

10. Technician A says the relearn procedures are the same for all manufacturers. Technician B says that unusual engine and transmission conditions should be ignored until the relearn procedure is completed. Who is right?

    (A) A only.
    (B) B only.
    (C) Both A and B.
    (D) Neither A nor B.

Chapter 20

ASE Certification

After studying this chapter, you will be able to:
- ❏ Explain why ASE certification is beneficial to both the technician and the vehicle owner.
- ❏ Explain the process of registering for ASE tests.
- ❏ Explain how to take the ASE tests.
- ❏ Describe typical ASE test questions.
- ❏ Identify the format of the ASE test results.
- ❏ Explain how the ASE test results are used.

Technical Terms

National Institute for Automotive
    Service Excellence (ASE)

Standardized tests

Certified

Master Technician

ACT

Pass/fail letter

Test score report

Certificate in evidence of competence

ASE Preparation Guide

Introduction

In this chapter, the purpose and organization of the National Institute for Automotive Service Excellence (ASE) is explained. ASE certification and the advantages of being ASE certified are covered. Instructions for applying for and taking the ASE tests are also given.

Reasons for ASE Tests

The concept of setting standards of excellence for skilled jobs is not new. In ancient times, metalworkers, weavers, potters, and other artisans were expected to conform to set standards of product quality. In many cases, the need for standards resulted in the establishment of associations of skilled workers, which set standards and enforced rules of conduct. Ancient civilizations had such associations, and many medieval industries were regulated by guilds (associations of craftsmen). Many modern labor unions descended from early associations of skilled workers.

Certification processes for aircraft mechanics, aerospace workers, and electronics technicians have existed since the beginning of these industries. However, this has not been the case in the automotive industry. Automobile manufacturing and repair began as a fragmented industry made up of many small vehicle manufacturers and thousands of small repair shops. Although the number of vehicle manufacturers decreased, the number and variety of repair facilities continued to grow. Due to the industry's fragmented nature, standards for automotive repair were difficult to establish. For over 50 years, there was no unified set of standards of automotive repair knowledge. Anyone could claim to be an automobile technician, regardless of his or her qualifications. This situation resulted in much unneeded or improperly done repair work. As a result, a large segment of the public came to regard mechanics as unintelligent, dishonest, or both.

This situation began to change in 1972, when the **National Institute for Automotive Service Excellence,** or **ASE,** was established. ASE is a nonprofit organization formed to promote high standards of automotive service and repair through the voluntary testing and certification of technicians. ASE has developed a series of written tests on various subjects in the automotive repair, truck repair, auto body, and engine machinist areas. These tests are called **standardized tests,** which means the same test in a particular subject area is given to everyone throughout the United States.

Any person passing one or more of the ASE tests and meeting certain experience requirements is **certified** (officially recognized as meeting all standards) in the subjects covered by the tests. A technician who passes all the tests in either the automotive or heavy truck areas is certified as a **Master Technician** in that area. Periodic recertification provides an incentive for updating skills and provides guidelines for keeping up with current technology.

The ASE certification program has been extended to Canada. Tests in Canada are similar to those given in the United States. Other countries may become involved with the ASE-certification program in the near future.

Other activities that ASE is involved in are encouraging the development of effective training programs, conducting research on the best methods of instruction, and publicizing the advantages of technician and training program certification. ASE is managed by a 40-member board made up of persons from the automotive and truck service industries, motor vehicle manufacturers, state and federal government agencies, schools and other educational groups, and consumer associations.

The ASE certification test program identifies and rewards skilled technicians. The test program allows potential employers and the driving public to identify good technicians. It also helps good technicians advance their careers. The program is not mandatory, but many repair shops hire only ASE-certified technicians. Close to 500,000 persons are now ASE certified in one or more areas. The advantages the ASE-certification program has brought to the automotive industry include increased respect for and trust of automobile technicians—at least those who are ASE certified. This has resulted in better pay and working conditions for technicians. Thanks to ASE, automotive technicians are taking their place next to other skilled artisans.

Applying for the ASE Tests

You do not have to be employed in the automotive service industry to take the ASE tests. However, you must have at least two year's experience working as an automobile or truck technician to be granted certification. In some cases, training programs or courses, an apprenticeship program, or time spent performing similar work can be substituted for all or part of the work experience.

ASE tests are given twice each year—once in the spring and once in the fall. Tests are usually held on Saturdays and weeknights during a two-week period. The tests are given by **ACT,** a nonprofit organization experienced in administering standardized tests.

The tests are given at designated test centers in over 300 locations in the United States. If necessary, special test centers can be set up in remote locations. However, there must be a minimum number of applicants before a special test center will be set up.

To apply for the ASE tests, obtain an application form like the one shown in **Figure 20-1.** For the most current application form, contact ASE at the following address:

National Institute for Automotive Service Excellence
13505 Dulles Technology Drive
Herndon, VA 22071

Registration Form
ASE Tests

If extra copies are needed, this form may be reproduced.
Important: Read the Booklet first. Then print neatly when you fill out this form.

1. a. Social Security Number:

1. b. Telephone Number: (During the day)

Area Code   Number

1. c. Previous Tests: Have you ever registered for any ASE certification tests before? ❑ Yes ❑ No

2. Last Name: First   Middle Initial

3. Mailing Address:

Number, Street, and Apt. Number

City

State   ZIP Code

4. Date of Birth: 5. Sex: ○ Male ○ Female
Month   Day   Year

6. Race or Ethnic Group: Blacken one circle. (For research purposes. It will not be reported to anyone and you need not answer if you prefer.)

1 ○ American Indian   2 ○ African American   3 ○ Caucasian/White   4 ○ Hispanic/Mexican
5 ○ Oriental/Asian   6 ○ Puerto Rican   7 ○ Other (Specify)

7. Education: Blacken only one circle for the highest grade or year you completed.
Grade School and High School   After High School: Trade or
(including Vocational)   Technical (Vocational) School   College
○7 ○8 ○9 ○10 ○11 ○12   ○1 ○2 ○3 ○4   ○1 ○2 ○3 ○4 ○More

8. Employer: Which of the below best describes your current employer? (Blacken one circle and fill in subcodes from pp. 19-20 as appropriate.)

1 ○ New Car/Truck Dealer/Distributor, Domestic—   Enter code for
2 ○ New Car/Truck Dealer/Distributor, Import—   make of car or truck . . . .

3 ○ Service Station—Enter code . . . . 13 ○ Volume Retailer—Enter code . . . . . .

4 ○ Military—Enter Code . . . . 14 ○ Tire Dealer—Enter code . . . .

5 ○ Indp't Repair Shop— Enter code for
6 ○ Fleet Repair Shop— vehicle type   16 ○ Utility
17 ○ Lift Truck Dealer/Repair Shop

7 ○ Specialty Shop—Enter code . . . . 18 ○ Machinist Facility—Enter code . . . . . .

8 ○ Government/Civil Service

9 ○ Student   19 ○ Leasing & Rental Shop—Enter code

10 ○ Educator/Instructor   15 ○ Other

11 ○ Manufacturer

12 ○ Independent Collision Shop

9. Test Center:
Center Number   City   State

10. Tests: Blacken the circles (regular) or squares (recertification) for the test(s) you plan to take. (Do not register for a recertification test unless you passed that regular test before.) Note: Do not register for more than four regular tests on any single test day.

Regular	Recertification	Regular	Recertification
○ A1 Auto: Engine Repair	❑ A	○ A2 Auto: Automatic Trans/Transaxle	❑ E
○ A8 Auto: Engine Performance	❑ B	○ A3 Auto: Manual Drive Train and Axles	❑ F
○ A4 Auto: Suspension & Steering	❑ C	○ A6 Auto: Electrical/Electronic Systems	❑ G
○ A5 Auto: Brakes	❑ D	○ A7 Auto: Heating & Air Conditioning	❑ H
○ F1 Alt. Fuels: Lt. Vehicle CNG		○ L1 Adv. Level: Auto Adv. Engine Perf. Spec.	
○ S4 School Bus: Brakes		○ T1 Med/Hvy Truck: Gasoline Engines	❑ I
○ S5 School Bus: Suspension/Steering		○ T2 Med/Hvy Truck: Diesel Engines	❑ J
		○ T6 M/Hvy Truck: Elec./Electronic Systems	❑ K
○ T3 Med/Hvy Truck: Drive Train	❑ L	○ S6 School Bus: Elec./Electronic Systems	
○ T4 Med/Hvy Truck: Brakes	❑ M		
○ T5 Med/Hvy Truck: Suspension/Steering	❑ N	○ B2 Body: Painting & Refinish	❑ P
○ T8 Med/Hvy Truck: PMI		○ B3 Body: Nonstructural Analysis	
○ M1 Machinist: Cylinder Head Specialist	❑ Q	○ B4 Body: Structural Analysis	
○ M2 Machinist: Cylinder Block Specialist	❑ R	○ B5 Body: Mechanical & Electrical Components	
○ M3 Machinist: Assembly Specialist	❑ S		

11. Fees: Number of Regular tests (except L1) marked above_____ x $18* = $_____

If you are taking the L1 Advanced Level Test, add $40 here   + $_____

If you marked any of the Recertification tests, add $20 here   + $_____

*Important: School Bus Technicians, see page 8 for a money-saving offer! If you qualify, check here ❑ and adjust your fees (Item 11, first line) accordingly.

Registration Fee   + $ **20**

TOTAL FEE   = $_____

❑ MasterCard   ❑ VISA

Credit Card #

Expiration Date
Month   Year

Signature of Cardholder _____

12. Fee Paid By: 1 ○ Employer   2 ○ Technician

13. Experience: Blacken one circle. (To substitute training or other appropriate experience, see page 3.)

1 ○ I certify that I have two years or more of full-time experience (or equivalent) as an automobile, medium/heavy truck, school bus or collision repair/refinish technician or as an engine machinist. (Fill out #14 on opposite side)
2 ○ I don't yet have the required experience. (Skip 14.)

14. Job History: Required. Provide job history and job details on the opposite side of this form.

15. Authorization: By signing and sending in this form, I accept the terms set forth in the Registration Booklet about the tests and the reporting of results.

Signature of Registrant   Date

Do not send cash. Use credit card, or enclose a check or money order for the total fee, made payable to ASE/ACT and mail with this form to:
ASE/ACT, P.O. Box 4007, Iowa City, IA 52243

Figure 20-1. An ASE certification test registration form. Be sure to fill in all required information and include payment for all test fees. (ASE)

ASE will send the proper form, along with an information bulletin that explains how to complete the form. When you get the form, fill it out carefully, recording all needed information. You may apply for all the tests being given, but you can take fewer tests if desired. Work experience, or any substitutes for work experience, should also be included with the application. If there is any doubt about what should be placed in a particular space, consult the information bulletin. Be sure to choose the test center you wish to attend and record its number in the appropriate space. Most test centers are located at local colleges and schools.

When you send in the application, include a check or money order to cover all necessary fees. A fee is charged to register for the test series, and a separate fee is charged for each test taken. See the latest information bulletin for the current fee structure. In some cases, your employer may pay the registration and test fees. Check with your employer before submitting your application.

It has recently become possible to register over the Internet for ASE tests. ASE's World Wide Web site also contains information on ASE, the certification process,

study materials, etc. See **Figure 20-2.** If you need to take the ASE tests in a language other than English, indicate this

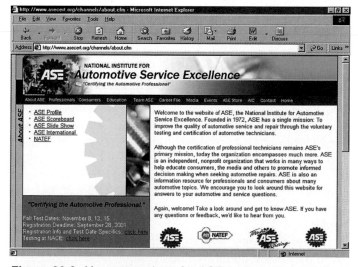

Figure 20-2. You can register for ASE tests or find out more about ASE at ASE's official web site. (ASE)

on the application form. In addition to English, test booklets are available in Spanish and French.

To be accepted for either the spring or fall ASE tests, your application and payment must arrive at ASE headquarters at least one month before the test date(s). To ensure that you can take the test at the test center of your choice, send in your application as early as possible. Two weeks after sending the application and fees, you will receive an admission ticket to the test center. See **Figure 20-3.** If your admission ticket has not arrived and there is less than two weeks until the test date(s), contact ASE using the phone number given in the latest information bulletin. If the desired test center is filled when ASE receives your application, you will be directed to report to the nearest center that has an opening. If it is not possible to go to the alternate test center, contact ACT immediately using the phone number given in the information bulletin.

Preparing for the ASE Tests

ASE tests are designed to measure your knowledge of three things:
- ❏ The operation of automotive systems and components.
- ❏ Diagnosis and testing of systems and components.
- ❏ Repairing automotive systems and components.

Therefore, you should study the basic principles of automotive systems operation, as well as the latest information about diagnostic and repair procedures. Good sources of this material include your textbook, service manuals and factory training material, trade publications such as Motor or Chilton magazines, and service bulletins. Other technical training aids are available. These allow you to study at your own pace. This is an excellent way to secure the knowledge needed to pass the tests.

Remember that the ASE tests are designed to test your knowledge of correct diagnostic and repair procedures. Never assume that the way you have always done something is the correct way.

Taking the ASE Tests

Be sure to bring your admission ticket with you when reporting to the test center. When you arrive at the test center, you will be asked to produce the admission ticket and a driver's license or other photographic identification. In addition to these items, bring some extra number 2 pencils. Although pencils will be available at the test center, extra pencils may save time if your original pencil breaks.

After you enter the test center, follow all instructions given by the test administrators. During the test, read each question carefully before deciding on the proper answer.

ASE tests consist of multiple-choice questions with four possible answers. The specific types of ASE test questions are covered at the end of this chapter.

After completing all the questions in a particular test, recheck your answers to ensure that you did not overlook anything that would change your answer, or that you did not make a careless error on the answer sheet. In most cases, rechecking your answers more than once is unnecessary and may lead you to change correct answers to incorrect ones. Each test session lasts about four hours. However, you may leave after completing your last test and handing in all test material.

Types of ASE Test Questions

Each ASE test contains between 40 and 80 questions, depending on the subject. All test questions are multiple-choice, with four possible answers. These multiple-choice questions are similar to the ASE-type questions at the end of each chapter in this textbook. The following section discusses the types of ASE questions you are likely to encounter.

One-Part Questions

One-part questions require you to answer a single question. For example:

1. Which of the following parts would *not* be found on an electronically controlled transaxle?
   (A) Drive chain.
   (B) Oil pump shaft.
   (C) Governor valve.
   (D) Transfer gears.

When answering any ASE question, always choose the *best answer* out of the four choices. In the previous question, you must remember that the unit being discussed is an electronically controlled transaxle. While some electronically controlled transaxles do not use drive chains, oil pump shafts, or transfer gears, *none* of them uses a governor. Therefore, the correct answer is (C) Governor valve.

Keep in mind that some ASE questions may include irrelevant information. For instance, what if the above question was worded this way:

1. Which of the following parts would *not* be found on an electronically controlled transaxle with a lockup torque converter?
   (A) Drive chain.
   (B) Oil pump shaft.
   (C) Governor valve.
   (D) Transfer gears.

National Institute for Automotive Service Excellence

ACT, P.O. Box 4007, Iowa City, Iowa 52243, Phone: (319) 337-1433   017910042 T

Admission Ticket

Test Center to which you are assigned:

A

John Smith
123 Main Street
Edens, Il. 60000

REGULAR TESTS (Late arrivals may not be admitted.)		
DATE	REPORTING TIME	TEST(S)
11/14	7:00 PM	A1, A2, A8

RECERTIFICATION TESTS (Late arrivals may not be admitted.)

TEST CODE KEY

A1 Auto: Engine Repair
A2 Auto: Automatic Trans/Transaxle
A3 Auto: Manual Drive Train & Axles
A4 Auto: Suspension & Steering
A5 Auto: Brakes
A6 Auto: Electrical/Electronic Systems
A7 Auto: Heating & Air Conditioning
A8 Auto: Engine Performance

M1 Machinist: Cylinder Head Specialist
M2 Machinist: Cylinder Block Specialist
M3 Machinist: Assembly Specialist
T1 Med/Hvy Truck: Gasoline Engines
T2 Med/Hvy Truck: Diesel Engines
T3 Med/Hvy Truck: Drive Train
T4 Med/Hvy Truck: Brakes
T5 Med/Hvy Truck: Suspension & Steering

T6 Med/Hvy Truck: Elec./Electronic Systems
T8 Med/Hvy Truck: Preventive Main. Inspec.
B2 Coll.: Painting & Refinishing
B3 Coll.: Non-structural Analysis
B4 Coll.: Structural Analysis
B5 Coll.: Mechanical & Elec. Components
B6 Coll.: Damage Analysis & Estimating
P1 Parts: Med/Hvy Truck Parts Specialist

P2 Parts: Automobile Parts Specialist
F1 Alt. Fuels: Lt. Veh. Comprsd. Nat. Gas
L1 Adv. Level: Adv. Engine Perf. Spec.
S1 School Bus: Body Sys. & Spec. Equip.
S4 School Bus: Brakes
S5 School Bus: Suspension & Steering
S6 School Bus: Elec./Electronic Systems

See Notes and Ticketing Rules on reverse side. An asterisk (•) indicates your certification in these areas is expiring.

SPECIAL MESSAGES

```
-REVIEW ALL INFORMATION ON THIS TICKET.  CALL IMMEDIATELY TO REPORT AN
 ERROR OR IF YOU HAVE QUESTIONS.
-IF YOU MISS ANY EXAMS, FOLLOW THE REFUND INSTRUCTIONS ON THE BACK OF THIS SHEET.
 THE REFUND DEADLINE IS
-YOU HAVE BEEN ASSIGNED TO AN ALTERNATE TEST CENTER.  THE CENTER
 ORIGINALLY REQUESTED IS FULL.
8010-IL/LOCAL 150 IS LOCATED ON JOLIET AVE, THREE DOORS W. OF LAGRANGE RD ON
SOUTH SIDE OF THE ST.  ENTER THROUGH BACK DOOR.  NO ALCOHOL ON PREMISES.
```

MATCHING INFORMATION: The information printed in blocks B and C at the right was obtained from your registration form. It will be used to match your registration information and your test information. Therefore, the information at the right must be copied EXACTLY (even if it is in error) onto your answer booklet on the day of the test. If the information is not copied exactly as shown, it may cause a delay in reporting your test results to you.

IF THERE ARE ERRORS: If there are any errors or if any information is missing in block A above or in blocks B and C at the right, you must contact ACT immediately. DO NOT SEND THIS ADMISSION TICKET TO ACT TO MAKE SUCH CORRECTIONS.

Check your tests and test center to be sure they are what you requested. If either is incorrect, call 319/337-1433 immediately. Tests cannot be changed at the test center. **ON THE DAY OF THE TEST**, be sure to bring this admission ticket, positive identification, several sharpened No. 2 pencils, and a watch if you wish to pace yourself.

B FIRST FIVE LETTERS OF LAST NAME

S M I T H

C SOCIAL SECURITY NUMBER OR ACT IDENTIFICATION NUMBER

1 2 3   4 5   6 7 8 9

SIDE 1

Figure 20-3. *You will receive your admission ticket approximately two weeks after sending in your registration form.*

The fact that the transaxle has a lockup torque converter is irrelevant. It has no effect on the answer, since an electronic transaxle—with or without a lockup torque converter—would not have a governor. However, if you don't realize that the converter information has no bearing on the question, you could get sidetracked looking for hidden meanings in the other choices. You might end up choosing the wrong answer or you could simply end up wasting time that could be spent on other questions.

Two-Part Questions

Two-part questions used in the ASE tests require you to analyze the statements made by two technicians, Technician A and Technician B. You are asked to determine whether each of the statements is true. For example:

1. Technician A says that Dexron® automatic transmission fluid can be used in any automatic transmission or transaxle. Technician B says that transmission fluid should be changed more often if a vehicle is used for towing. Who is right?

    (A) A only.

    (B) B only.

    (C) Both A and B.

    (D) Neither A nor B.

Notice that both statements can be true, both can be false, or one can be true and the other false. You know from earlier chapters that Technician A's statement is obviously false. Many vehicles use a fluid other than Dexron®. Since (A) is not a correct answer, you can also eliminate (C) since both technicians cannot be right if (A) is wrong. Since fluid should be changed more often under extreme conditions, Technician B is right and the correct answer is (B).

If you can determine whether answer (A) or (B) is correct, you can save time by eliminating either (C) or (D). If either Technician A or Technician B has made a clearly correct statement, (D) cannot be the answer. If either A or B has made a clearly incorrect statement, (C) cannot be the answer.

A variation of the two-part question describes a vehicle problem or condition. Technicians A and B again make statements and you must determine whether both statements are true, both statements are false, or one statement is true and one is false.

1. A transaxle with a hypoid gear final drive makes a sucking noise from the dipstick tube with the engine running. The vehicle will not move. Technician A says that a low transmission fluid level is the most likely cause. Technician B says that a slipping front clutch is the most likely cause. Who is right?

    (A) A only.

    (B) B only.

    (C) Both A and B.

    (D) Neither A nor B.

To correctly answer this type of question, you must read carefully to determine the exact nature of the problem. While a defective front clutch could cause the transaxle to fail, it would not cause a sucking sound, so (A) is not the correct answer. Therefore, (C) is also incorrect. A low oil level could cause a sucking sound and would keep the vehicle from moving. Therefore (B) is correct. Note that (D) cannot be the answer, because both A and B would have to be wrong, and in this instance, B is right.

Notice also that this question contains an irrelevant piece of information. The fact that the transaxle has a hypoid gear final drive has no bearing on the problem and should have been disregarded as you worked your way to the correct answer.

Negative Questions

Negative questions ask you to identify the incorrect statement from four possible choices. The word "except" is often used in negative questions. For example:

1. In neutral, a transmission makes whining noise that varies with engine speed. This could be caused by all of the following defects *except:*

    (A) tight front pump clearances.

    (B) a plugged filter.

    (C) a damaged planetary sun gear.

    (D) internal converter parts rubbing.

If you study this question, you will note that the way to find the correct answer in this case is to determine which answer qualifies as the exception. In other words, determine what three of the above parts have in common that the fourth does not. Since all the above parts except the planetary gears are turning in neutral, the correct answer has to be (C).

Another thing to remember is that the ASE tests are geared to knowledge that applies to current vehicles. If, for instance, you once worked on an old 1960s Powerglide transmission that *did* whine in neutral because the planet gears were defective, you might be tempted to pick something other than (C) as the answer. Try to keep in mind, however, that ASE will not ask a question about a transmission that has been out of production since 1972. ASE is interested in the answer that applies to late-model vehicles. When reading ASE questions that involve problem diagnosis, focus on vehicles that are no more than 10–12 years old.

A variation of the negative question will use the word *least.*

1. A rear-wheel drive transmission has a very hard upshift from second to third. All other shifts and gear engagements are good. Which of the following defects is the *least* likely cause?

    (A) Band adjustment too tight.

    (B) Insufficient clutch pack clearance.

    (C) Modulator vacuum line leaking.

    (D) Stuck 2-3 accumulator piston.

All of these problems could cause a hard 2-3 upshift. What you must do is eliminate the most common causes, leaving the one that is least common. A stuck 2-3 accumulator piston is a very likely cause, leaving (A), (B) and (C). Tight band clearance or insufficient clutch pack clearance could cause a clutch-band "fight" on the upshift. Therefore (A) and (B) are also likely causes. This leaves only (C). A leaking line to the modulator could cause a hard 2-3 upshift. A modulator line leak, however, almost always causes other shifts and gear engagements to be harsh. Therefore, the least likely cause of a harsh 2-3 upshift and no other problem is (C).

When answering this kind of question, remember that all the possible answers may be correct. You have to decide which of them is *not* a likely cause. This relates back to the fact that ASE test questions call for the best answer. This also pertains to ASE questions that use the word *most*. All the answers may be correct, but you must choose the answer that is the *most* correct.

Incomplete Sentence Questions

Some test questions are incomplete sentences, with one of the four possible answers correctly completing the sentence. For example:

1. Air pressure can be used to check:
   (A) free valve movement.
   (B) clutch and band operation.
   (C) solenoid operation.
   (D) pump smoothness.

Once again, you must choose the best answer. Start by eliminating obviously wrong answers. Air pressure cannot be used to check the operation of an electrical device, so (C) is obviously a wrong answer. To choose between (A), (B), and (D), consider what could be checked using air pressure. The pump (D) could be turned by air pressure, but doing so would not accomplish anything. Some valves (A) could also be moved with air pressure applied to various parts of the valve body, but there are better ways to check valve operation. Therefore, (B) is the correct answer.

Test Results

ACT takes about six to eight weeks to process the tests from the various centers and to mail the results. Initially, you will receive a *pass/fail letter.* This letter will tell you only whether or not you have passed each test. A typical pass/fail letter is shown in **Figure 20-4.** Two weeks after receiving the pass/fail letter, you will receive a *test score report.* The test score report is a confidential report of your performance on the tests. The report will list the number of questions that must be answered correctly to pass the test and the number of questions you have answered correctly.

The test questions are also divided into general areas to help you determine which areas require more study. For example, the suspension and steering systems test questions may be divided into the following subsections: steering systems diagnosis and repair; suspension systems diagnosis and repair; wheel alignment diagnosis, adjustment, and repair; and wheel and tire diagnosis and repair. A typical test score report is shown in **Figure 20-5.**

Included with the test score report is a *certificate in evidence of competence.* This certificate lists the areas in which you have been certified. In addition, a pocket card and a wallet card are provided. Like the certificate, they list all the areas in which you are certified. Also included is an order form for shoulder patches, wall plates, and other ASE promotional material.

Note: If you did not indicate that you have two years of automotive experience on the test application, you will not receive a certificate in evidence of competence. After you have met the experience requirement, you must provide ASE with the necessary information to receive your certificate.

All ASE test results are confidential and are provided only to the person who took the test. Test results will be mailed to your home address and will not be provided to anyone else. This is done to protect your privacy. The only test information ASE will release is to confirm to an employer that you are certified in a particular area. This is true even if your employer has paid the test fees. If you wish your employer to know exactly how you performed on the tests, you must provide him or her with a copy of your test results.

If you fail a certification test, you can retake it as many times as you like. However, you (or your employer) must pay all applicable registration and test fees again. You should study all available information in the areas in which you did poorly. The *ASE Preparation Guide* may help you sharpen your skills in these areas. The ASE Preparation Guide is free and can be obtained by filling out the coupon at the back of the information bulletin.

Recertification Tests

Once you have been certified in any area, you must take a recertification test every five years to ensure that your certification remains current and that you have kept up with current technology. The process of applying for the recertification tests is similar to that for the original certification tests. Use the same form and enclose the proper recertification test fees. If you allow your certification to lapse, you must take the regular certification test(s) to regain your certification.

National Institute for
AUTOMOTIVE SERVICE EXCELLENCE

December 20, XXXX   032527

John Smith
123 Main Street
Edens, Il 60000

Dear ASE Test Taker:

Listed below are the results of your November XXXX ASE Tests. You
will soon be receiving a more detailed report.

<u>If your test result is "Pass"</u>, and if you have fulfilled the two-year
"hands-on" experience requirement, you will receive a certificate and
credential cards for the tests you passed.

<u>If your test result is "More Preparation Needed"</u>, you did not attain a
passing score. Check your detailed score report when it arrives. This
information may help you prepare for your next attempt.

If you do not receive your detailed report within the next three weeks,
please call.

Thank you for participating in the ASE program.

A1	ENGINE REPAIR	PASS
A2	AUTOMATIC TRANSMISSION/TRANSAXLE	PASS
A8	ENGINE PERFORMANCE	PASS

23-45-6789

13505 Dulles Technology Drive • Herndon, Virginia 22071-3415 • (703) 713-3800

Figure 20-4. *A pass/fail letter will be sent to you shortly after taking the ASE tests. Note that this letter indicates that all three tests taken were passed. (ASE)*

Your Score is 38. (Passed)
The total score needed to pass A2 is 34 out of 50.

Test A2 Automatic Transmission/Transaxle Content area	Number of questions answered correctly	Total number of questions
General transmission/transaxle diagnosis	19	25
Transmission/transaxle maintenance and adjustment	4	5
In-vehicle transmission/transaxle repair	7	9
Off-vehicle transmission/transaxle repair	8	11
Total test	38	50

Figure 20-5. *You will receive a test score report approximately two weeks after receiving the pass/fail letter. The test score report shows how you performed on each section of individual tests. (ASE)*

Summary

Due to the industry's fragmented nature, unified standards for automotive repair were difficult to establish. This caused a lack of professionalism in the automobile industry, often leading to poor or unneeded repairs, as well as decreased status and pay for automobile mechanics. The National Institute for Automotive Service Excellence, or ASE, was formed in 1972 to improve status of the automotive service industry. ASE tests and certifies automotive technicians in major areas of automotive repair. This has increased the skill level of technicians, resulting in better service for the consumer and improved benefits for technicians.

ASE tests are given twice each year, once in the spring and again in the fall. Anyone can register to take the tests by filling out the proper registration form and paying the registration and test fees.

To be considered for certification, the registrant must have two year's of hands-on experience as an automotive technician. Proof of this experience should also be included with the registration form. About three weeks after applying for the test, the technician will receive an entry ticket, which must be brought to the test center.

The ASE questions will test your knowledge of general system operation, problem diagnosis, and repair techniques. All the questions are multiple choice, with four possible answers. The questions must be read carefully.

The entire test should be gone over one time only to catch careless mistakes.

Test results will arrive 6–8 weeks after the test session. A letter will be sent telling the technician whether he or she has passed the tests. This will be followed by a detailed report showing the scores for all the tests taken. Results are confidential and will be sent only to the home address of the person who took the test. If a test is passed and the experience requirement has been met, the technician will be certified for five years. Anyone who fails a test can take it again during the next session. Tests can be retaken as many times as necessary. Recertification tests are taken at the end of the five-year certification period.

Review Questions—Chapter 20

Please do not write in this text. Write your answers on a separate sheet of paper.

1. Since the automobile industry was so fragmented, it has been hard to come up with a unified set of _____.

2. In what year was ASE founded?

3. If a technician can pass all of the tests in the automotive or heavy truck areas, he or she is certified as a _____ _____.

4. How many times are ASE tests given each year?

5. What three major areas of automotive knowledge are the ASE tests designed to measure?

6. How far in advance of the test date should your application and payment arrive at ASE headquarters?

7. Negative ASE questions generally contain the words _____ or _____.

8. To whom does ASE provide test results?

9. ASE provides test results to the technician's employer. True or False?

10. Name two sources of study material for the ASE tests.

ASE-Type Questions—Chapter 20

1. Technician A says ASE encourages high standards of automotive service and repair by providing a series of standardized tests. Technician B says that ASE encourages high standards of automotive service and repair by establishing a series of instructional courses. Who is right?
   (A) A only.
   (B) B only.
   (C) A and B.
   (D) Neither A nor B.

2. All the following statements about the ASE tests are true except:
   (A) These tests are called standardized tests.
   (B) The number of possible answers varies with the type of test.
   (C) The same test is given to everybody in the United States.
   (D) The tests are always given at official test centers.

3. Which of the following is an advantage that ASE certification has helped to bring to automotive technicians?
   (A) Increased respect.
   (B) Longer working hours.
   (C) Increased use of the commission pay system.
   (D) On-board computer diagnostics.

4. Which of the following should you *not* bring to the ASE test center?
   (A) One or two # 2 pencils.
   (B) Your admission ticket.
   (C) Any needed study materials.
   (D) Photographic identification.

5. Technician A says that a technician can retake any certification test as many times as necessary. Technician B says that certified technicians must take a recertification test every five years. Who is right?
   (A) A only.
   (B) B only.
   (C) Both A and B.
   (D) Neither A nor B.

6. ASE test questions are always _____ types.
   (A) true-false
   (B) completion
   (C) multiple choice
   (D) None of the above. ASE tests have more than one type of test question.

7. Which of the following types of questions are *not* used on ASE tests?
   (A) Incomplete sentence.
   (B) True-False.
   (C) Negative.
   (D) Two-part.

8. Question 3 in the previous column is an example of a(n) _____ test question.
   (A) negative
   (B) one-part
   (C) two-part
   (D) incomplete sentence

9. Question 5 in the previous column is an example of a(n) _____ test question.
   (A) negative
   (B) one-part
   (C) two-part
   (D) incomplete sentence

10. When the word *least* is used in a question, what should you look for among the possible answers?
    (A) The answer that is probably the best choice.
    (B) The answer that is probably the worst choice.
    (C) The answer that is somewhere between the best and worst choices.
    (D) There is no way to tell without actually reading the question.

Chapter 21

Career Preparation

After studying this chapter, you will be able to:
- ❏ Identify three classifications of automotive technicians.
- ❏ Identify the major sources of employment in the automotive industry.
- ❏ Identify advancement possibilities for automotive technicians.
- ❏ Explain how to fill out a job application.
- ❏ Explain how to conduct oneself during a job interview.

Technical Terms

Helper	Government agencies
Installer	Shop foreman
Certified technicians	Service manager
New-vehicle dealerships	Salesperson
Department store service centers	Service advisor
Specialty shops	Parts persons
Independent auto repair shops	Entrepreneur

Introduction

This chapter provides an overview of the career opportunities in the automotive service industry. It discusses the various types of automotive technicians and the type of work each performs. This chapter also includes information on the types of repair outlets, the type of work, working conditions, and pay scales a beginning technician can expect. Also included are tips to help you find employment in the auto repair industry and information on the types of automotive-related jobs into which you can advance. Studying this chapter will help you find a job in the automotive service industry.

Automotive Servicing

The business of servicing and repairing cars and trucks has provided employment for millions of people over the last 100 years. It will continue to provide ample employment opportunities for years to come. Like any career, automotive service and repair has its drawbacks, but it also has its rewards.

Most people in the auto service business work long hours. Diagnosis and repair procedures can be mentally taxing and physically difficult. Working conditions are often hot and dirty. Automotive service has never been a prestigious career, although this is changing as vehicles become more complex and technicians become better trained. The technician often has to deal with difficult, condescending, and sometimes dishonest vehicle owners.

The advantages of the auto service business include the opportunity to work with your hands and the satisfaction of fixing something that is broken. Auto repair offers salaries that are competitive with those for similar jobs and is a secure profession where the good technician can always find work. To ensure that you stay employable, always seek to learn new things and become ASE certified in as many areas as possible.

Levels of Automotive Service Positions

Although the public tends to classify all automotive technicians as "mechanics," there are many types of auto service professionals. The types of auto service professionals include the helper, who changes oil or performs other simple tasks; the installer, who removes and installs components; and the certified technician who is capable of diagnosing difficult problems and repairing complex automotive systems. Although these classifications are unofficial, they tend to hold true throughout the automotive repair industry.

Helpers

The *helper* performs basic service and maintenance tasks, such as installing and balancing tires, cleaning parts, changing engine oil and filters, and installing batteries. See **Figure 21-1**. The skills required of the helper are low, and the pay will be less than that of the other levels. However, working as a helper is a good way for many people to start in the automotive service area. In fact, many technicians started out doing this type of automotive service when they were in their teens.

Installers

The *installer* is an automotive service person who removes and installs parts. See **Figure 21-2**. Installers seldom do complicated repair work, and they generally do not diagnose vehicle problems. Installers are paid more than helpers but less than certified technicians. Many installers take the opportunity to improve their knowledge and skills, and they eventually become certified technicians.

Figure 21-1. *Apprentices generally do basic work. This apprentice is cleaning the outside of a transmission case in preparation for disassembly.*

Figure 21-2. *This installer is preparing to reinstall a transaxle.*

Certified Technicians

Certified technicians are at the top level and have the skills to prove it, **Figure 21-3**. Most technicians are ASE certified in at least one automotive area, and many are certified in all car or truck areas. The certified technician can successfully diagnose and repair every area that he or she is certified in, and can perform many other service jobs. The certified technician is paid more than the helper or the installer.

Types of Auto Service Facilities

There are many types of auto service facilities where the technician can work. The traditional place to get started in automotive repair—the corner service station—is gone, replaced by self-serve stations. However, many opportunities to repair vehicles still exist. Even the smallest community has several automotive repair facilities. Some of these are discussed in the following section.

New-Vehicle Dealerships

All *new-vehicle dealerships* must have large, well-equipped service departments to meet the warranty service requirements of the vehicle manufacturer. These service departments are usually well equipped, with all the special testers, tools, and service literature needed to service a specific make of vehicle. Dealership service departments are also equipped with lifts, parts cleaners, hydraulic presses, electronic test equipment, and other equipment for efficiently servicing vehicles. However, the technician must generally provide his or her own hand and air tools. Dealers stock all the most common parts and are usually tied into a factory parts network, which allows them to quickly obtain any part. See **Figure 21-4**.

Pay scales at most dealerships are based on flat rate hours. The rate per hour is competitive between dealerships in a given area and is usually higher than local industry in general. The number of hours the technician is paid for depends on the work that comes in and how fast he or she can complete it. If you can work fast and enough work comes in, the pay can be excellent. Most modern dealers offer some sort of benefits package (insurance, vacation, etc.).

Dealership working conditions are relatively good, and most of the vehicles requiring service are new or well cared for older models. Since the dealership must be able to fix any part of the vehicle, the technician can perform a large variety of work. Although many dealer service departments have technicians who work only in specific areas, the trend is toward training all technicians to handle any type of work. Most repairs will be on the same make of vehicle, although many large dealerships now handle more than one make. Most modern dealerships are tied into manufacturer hotlines. These hotlines are used to access factory diagnosis and repair information, which makes it easier to troubleshoot and correct problems.

The disadvantages of dealership employment are the lack of salary guarantees, low pay rates for warranty repairs, and fast-paced, often hectic, working conditions. If you welcome the challenge of being paid by the job and do not mind working under deadlines, a dealership may be the ideal employer. Also check out the local large-truck dealerships. Although the work is much heavier, the pay is usually somewhat higher and working conditions are not as fast paced.

Chain and Department Store Auto Service Centers

Many national chain and department stores have auto service centers that perform various types of automotive repairs. These *department store service centers* often hire

Figure 21-3. *This certified technician is rebuilding an automatic transaxle.*

Figure 21-4. *New car dealerships offer excellent working conditions and benefits. However, you have to deal with all types of vehicle problems. There is not a very good opportunity for specialization. (Land Rover)*

technicians for entry-level jobs and may offer opportunities for advancement. Technicians in most of these situations receive a base salary plus a commission for work performed. Pay scales for the various work classifications are competitive, and most national companies offer generous benefit packages. One advantage of working for large companies is the chance of advancement into other areas, such as sales or management.

One disadvantage of working at the average auto service center is the lack of variety. Most auto centers concentrate on a few types of repairs, such as brake work and alignment, and turn down most other repairs. The work can become monotonous due to the lack of variety. Although the job pressures are usually less than those at dealerships, customers still expect their vehicles to be repaired in a reasonable period of time. However, if you would enjoy working on only one or two areas of automotive repair, this type of job may be ideal for you.

Specialty Shops

Specialty shops can offer good working conditions and good pay. Most specialty shops concentrate on one area of service, such as automatic transmission and transaxle repair, and are fully equipped to handle all aspects of their particular specialty. These shops may occasionally take in other minor repair work when business in their specialty is slow. Specialty shops can be ideal places to work if you want to concentrate on one area of repair.

Specialty shops usually offer a base salary, plus commission for work performed. Pay scales are generally competitive. A disadvantage of specialty shops is the lack of variety. Since specialty shops concentrate on a few types of repairs, working at these facilities can become monotonous.

Many specialty shops are franchise operations, and the demands of the franchise can create problems. If the prime purpose of the shop is to sell tires, for instance, the technician who was hired to do steering and suspension repairs may be forced to spend time installing tires. This can be annoying to technicians who want to be doing the job they were hired to perform. However, if this does not bother you, a franchise operation could be a good work situation.

Independent Shops

There are millions of *independent auto repair shops*. Independent repair shops have no connection with vehicle manufacturers, chain stores, or franchise operations. As places to work, they range from excellent to terrible. Many shops are run by competent, fair-minded managers and have first-rate equipment and good working conditions. Other independent shops have almost no equipment, low pay rates, and extremely poor, even dangerous, working conditions. The prospective employee should carefully

check all aspects of the shop environment before agreeing to work there. Technicians at independent shops are usually paid on a salary plus commission basis.

There are two major classifications of independent repair shops: the general repair shop and the specialty shop. The general repair shop takes in most types of work and offers a variety of jobs. General repair shops may avoid some jobs that require special equipment, such as automatic transmission repair, alignment, or air conditioning service. However, they will usually take a variety of other repair work on different makes and types of vehicles. See **Figure 21-5**. A general repair shop can be a good place to work if you like to be involved in many different types of diagnosis and repair.

Specialty shops confine their repair work to one area of repair, such as automatic transmission and transaxle repair, tune-ups, or brake repair. These shops may occasionally take in minor repair work when business in their specialty is slow. Specialty shops can be ideal places to work if you want to concentrate on one area of repair.

Government Agencies

Many local, state, and federal *government agencies* maintain their own vehicles. Government-operated repair shops can be good places to work. Employees are generally paid on an hourly or weekly basis. Pay is usually set by law, with no commission. Although civil service pay scales are lower than those in private industry, the benefits are usually excellent. Pay raises, while relatively small, are regular. Most government shops work a 35–40 hour week and have the same holidays as other government agencies.

The working conditions in most government-operated shops are good, without the stress of deadlines or having to deal with customers. Hiring procedures are more involved than they are with other auto repair shops. Prospective employees must take civil service examinations, which often have little to do with automotive subjects. Some

Figure 21-5. *A general repair shop will commonly perform many types of repairs.*

government agencies require a certain level of education, thorough background checks, lists of former employers or other references, and may require that the employee be a registered voter.

If you think you would be interested in working for a government-owned repair shop, contact your state employment agency for the addresses of local, state, and federal employment offices in your area.

Other Opportunities in the Auto Service Industry

Many other opportunities are available to the automotive technician. These jobs still involve the servicing of vehicles, without some of the physical work. If you like cars and trucks, but are unsure you want to make a career of repairing them, one of these jobs may be right for you.

Shop Foreman or Service Manager

The automotive promotion that you will most likely be offered is **shop foreman** or **service manager**. Many repair facilities are large enough to require one or more foremen or service managers. If you move into management from the shop floor, your salary will increase, and you will be in a cleaner, less physically demanding, position. Many technicians enjoy the management position because it lets them in on the fun part of service—troubleshooting—without the drudgery of actually making the repairs.

The shop foremen and service manager perform duties similar to those performed by managers in many other professions. They perform administrative duties, such as setting work schedules, ordering supplies, preparing bills, determining pay scales, and dealing with employee problems. Some managers also perform financial record keeping, prepare employee paychecks, and pay suppliers. A few service managers place advertising and perform other marketing and public relations duties.

Service managers and shop foremen closely supervise shop operations and may be called in by technicians when a problem arises. Shop managers usually do the hiring. Your first job interview will likely be with a service manager or shop foreman. In addition to their administrative duties, most service managers and shop foremen are directly involved with customer relations, such as tending to billing or warranty disputes.

The disadvantage of a move to management is that you will no longer be dealing with the logical principles of troubleshooting and repair. Instead, you will deal with illogical and arbitrary personalities of people. Both the customers and technicians will have problems and attitudes that you will have to contend with. Unlike a vehicle problem, resolving these problems may require considerable personality and tact. Sometimes the manager has to compromise, which can be hard for a person who is used to being right.

The paperwork load is large for any manager, and it may not be something that a former technician can get used to. Record keeping requires a good bit of desk, and usually, computer time. Automotive record keeping is like balancing a checkbook and writing a term paper every few days. If you do not care to deal with people or keep records, a career in management may not be for you.

Salesperson or Service Advisor

Many people enjoy the challenge of selling. The **sales person** or **service advisor** performs a vital service, since repairs will not be performed unless the owner is sold on the necessity of having them done. Salespersons are not necessary in many small independent shops, but are often an important part of dealership service departments, department store service centers, and specialty repair shops.

The salesperson may enjoy a large income, and be directly responsible for a large amount of business in the shop. However, selling is a people-oriented job, and takes a lot of persuasive ability and diplomacy. If you are not interested in dealing with the public, you would probably not be happy in a sales job.

Parts Person

One often-overlooked area of the automotive service business is the process of supplying parts to those performing repairs. It is as vital as any other area. There are many types of parts outlets, including dealership parts departments, independent parts stores, parts departments in retail stores, and combination parts and service outlets. All these parts outlets meet the needs of technicians and shops, as well as those of the do-it-yourselfer. Due to the variety of vehicles available, as well as the complexity of the modern vehicle, a large number of parts must be kept in stock and located quickly when needed.

Parts persons are trained in the methods of keeping the parts flowing through the system until they reach the ultimate endpoint, the vehicle. Parts must be carefully checked into the parts department and stored so they can be found again. When a specific part is needed, it must be located and brought to the person requesting it. If more of the same parts are needed, they must be ordered. If the part is not in stock, it must be special ordered. This can be a challenging job.

The job of the parts person appeals to many people. This job does not pay as much as some other areas of automotive service, but rates are comparable with other jobs with the same skill level. If this type of work appeals to you, it may be a good job choice. There is an ASE test for parts specialists, and you may want to consider taking it to enhance your employability.

Self-Employment

Many persons dream of going into business for themselves. This can be a good and profitable option for the good technician. However, in addition to mechanical and diagnostic ability, the person with his or her own business must have a certain type of personality to be successful. This type of person must be able to shoulder responsibilities, handle problems, and look for practical ways to increase business and make a profit. This type of person must maintain a clear idea of what plans, both long and short term, need to be made. A person like this is often called an **entrepreneur**.

When you own a business, all the responsibility for repairs, parts ordering, bookkeeping, debt collection, and a million other problems is yours. Starting your own shop requires a large investment in tools, equipment, and workspace. If the money must be borrowed, you will be responsible for paying it back. However, many people enjoy the feeling of independence. If you have the personality to deal with the problems, you may enjoy being your own boss.

Another possible method of self-employment is to obtain a franchise from a national chain. A franchise operation removes some of the headaches of being in business for yourself. Many muffler, tire, transmission, tune-up, and other nationally recognized businesses have local owners. They enjoy the advantages of the franchise affiliation, including national advertising, reliable parts supplies, and employee benefit programs. Disadvantages of a franchise include high franchise fees and start-up costs, lack of local advertising, and some loss of control of shop operations to the national headquarters.

Getting a Job

There are many automotive jobs available at all times. The problem is connecting with the job when it is available. There are essentially two hurdles to getting a job: finding a job opening and successfully applying for the job. Hints for overcoming these hurdles are given in the following section.

Finding Job Openings

Before applying for a job, you must know about a suitable opening. A good place to start your search for job openings is your school. Instructors often have contacts in the local automobile industry, and they may be able to recommend you to a local company. Another good place to begin your search is the classified section of your local newspaper.

Also visit your local state employment agency or job service, **Figure 21-6.** Most of these agencies keep records of job openings, usually throughout the state, and may be able to connect to a data bank of nationwide job openings.

Some private employment agencies specialize in automotive placement. If there is an agency of this type in your area, arrange for an interview with one of their recruiters.

Visit local repair shops that you are interested in working for. Sometimes these shops have an opening that they have not advertised. If you are interested in working for a chain or department store, most of these stores have a personnel department where you can fill out an application. Even if no jobs are available, your application will be placed on file in the event that a job becomes available.

Applying for the Job

Regardless of your qualifications, you will not get a job if you make a poor first impression on the interviewer. If you do not impress your potential employer as competent and dedicated, you will not get the job.

Start creating a good impression when you fill out the employment application. Type or neatly print when filling out the application. Complete all blanks, and completely explain any lapses in your education or previous employment history. List all your educational qualifications, including those that may not apply directly to the automotive industry. **Figure 21-7** shows a typical employment application.

When you are called for an interview, try to arrange for a morning interview since your potential employer is most likely to be in a positive mood in the morning and less likely to be overwhelmed by last minute problems in the shop. Dress neatly and arrive on time or a little early. When introduced to the interviewer, make an effort to remember and repeat his or her name. Speak clearly when answering the interviewer's questions. Do not smoke or chew gum during the interview. State your qualifications for the job without bragging or belittling your accomplishments. At the conclusion of the interview, thank the interviewer for his or her time. If you do not hear from the interviewer in a few days, it is permissible to make a brief and polite follow-up call.

Figure 21-6. *State employment agencies have many job listings that are not normally advertised.*

Employment Application

Date: _____ Social Security Number: _____

Name: _____ Age: _____ Sex: _____

Address: _____

Phone: _____ United States citizen? _____ Can you furnish proof? _____

Employment Desired

Position: _____ Date you can start: _____ Expected salary: _____

Are you currently employed? _____ May we inquire of your present employer? _____

Education

Circle the number for the highest level completed:

High School	Trade/Technical School	Community/Junior College	University
1 2 3 4	1 2	1 2	1 2 3 4

Other: _____

Specialized Training or Certifications: _____
_____

Employment Record

Current/Last employer: _____ From: _____ To: _____

Address: _____ Phone: _____

Salary: _____ Job description: _____

Reason for leaving: _____

Previous employer: _____ From: _____ To: _____

Address: _____ Phone: _____

Salary: _____ Job description: _____

Reason for leaving: _____

Previous employer: _____ From: _____ To: _____

Address: _____ Phone: _____

Salary: _____ Job description: _____

Reason for leaving: _____

Figure 21-7. *Typical employment application. Be sure to fill out the application completely.*

Summary

The automotive service industry provides employment for many people and will continue to do so. Automotive service has some disadvantages, such as long hours, hard work, lack of status, and dealing with difficult customers. Advantages include interesting work, job security, and the satisfaction of diagnosing and correcting problems. Always stay employable by learning new things and becoming certified.

The three general classes of technicians are the helper, the installer, and the certified technician. The helper does simple tasks, such as changing tires and oil. Many helpers move up into the other classes after a short time. The installer installs new parts, such as shock absorbers and strut assemblies, and sometimes moves into brake repair and wheel alignment. The certified technician performs the most complex diagnosis and repair jobs on vehicles, and makes the most money.

There are many places to work as an automotive technician. Among the most popular are new car and truck dealers, auto centers affiliated with department or chain stores, specialty shops, independent repair shops, and government agencies. Some people prefer to have their own businesses, either as independent owners or as part of a franchise system.

Other opportunities in the automotive service field include moving into management or into sales. Another often-overlooked employment possibility is in the automotive parts business.

To obtain a job in the automotive business, first locate possible job openings. Try the local newspaper and state job service. Also visit repair shops in your area. Most department or other large stores with attached auto service centers have personnel departments where you can fill out a job application. To get a job, you must make a good impression. Fill out all job applications carefully and neatly, listing your qualifications honestly. When invited to a job interview, dress neatly, arrive on time, and be courteous. Answer all questions without over- or understating your abilities and experience.

Review Questions—Chapter 21

Please do not write in this text. Write your answers on a separate sheet of paper.

1. What are the advantages of working as an automotive technician?

2. What are the disadvantages of working as an automotive technician?

3. The _____ often performs basic service and maintenance tasks.

4. Installers are often called upon to diagnose suspension and steering problems. True or False?

5. The _____ _____ will be called on to do the most complex jobs.

6. All new vehicle dealers must have well-equipped service departments to meet the _____ service requirements of the _____ _____.

7. Pay scales at most dealerships are based on _____ _____ hours.

8. Pay scales at government garages are based on a weekly or hourly _____.

9. One advantage of a franchise operation is _____ advertising.

10. A good place to start looking for a job is at your local state _____ _____.

ASE-Type Questions—Chapter 21

1. Technician A says that one advantage of working in the automotive service field is the opportunity to work with your hands. Technician B says that one advantage of working in the automotive service field is the high prestige of being an automotive technician. Who is right?
   (A) A only.
   (B) B only.
   (C) Both A and B.
   (D) Neither A nor B.

2. All of the following automotive service jobs will most likely be done by a helper except:
   (A) battery installation.
   (B) tire changing.
   (C) tire repair.
   (D) suspension noise diagnosis.

3. In which of the following businesses can the automotive technician get a job doing only a few kinds of repairs?
   (A) Chain stores.
   (B) Corner service stations.
   (C) Government garages.
   (D) New vehicle dealers.

4. Each of the following statements about working for a specialty shop is true *except:*

   (A) the variety of repairs is limited.
   (B) the opportunity exists to become very skilled in one area of service.
   (C) the technician may be paid by one of several methods.
   (D) new vehicle warranty repairs are commonly performed.

5. One feature of working for a local, state, or federal government operated repair shop, is the lack of _____.

   (A) benefits
   (B) security
   (C) deadlines
   (D) Both A and B.

6. Which of the following is a shop manager *least likely* to do?

   (A) Hire technicians.
   (B) Perform wheel alignments.
   (C) Pay bills.
   (D) Deal with customer complaints.

7. All of the following are job duties of the parts specialist *except:*

   (A) obtaining parts for technicians.
   (B) telling the technicians how to install parts.
   (C) selling parts to the general public.
   (D) special ordering parts.

8. Technician A says that job openings in the automotive service business are plentiful. Technician B says that automotive service shops often advertise in local newspapers. Who is right?

   (A) A only.
   (B) B only.
   (C) Both A and B.
   (D) Neither A nor B.

9. Technician A says that you should fill out the employment application in a sloppy manner to let the owner know that you do not like paper work. Technician B says that filling out the employment application is a good time to start making a good impression. Who is right?

   (A) A only.
   (B) B only.
   (C) Both A and B.
   (D) Neither A nor B.

10. All the following help to make a good impression during the job interview *except:*

   (A) dressing in dirty work clothes.
   (B) arriving a little early.
   (C) not smoking.
   (D) making an effort to remember the interviewer's name.

This vehicle has an automatic transmission with a manual shift program. Note the steering wheel-mounted buttons, which allow the driver to control gear shifting. (Lexus)

Useful Tables

CONVERSION CHART

METRIC/U.S. CUSTOMARY UNIT EQUIVALENTS

Multiply:	by:	to get:	Multiply:	by:	to get:

ACCELERATION

feet/sec^2	x 0.3048	= meters/sec^2 (m/s^2)	x 3.281	= feet/sec^2
inches/sec^2	x 0.0254	= meters/sec^2 (m/s^2)	x 39.37	= inches/sec^2

ENERGY OR WORK (watt–second = joule = newton–meter)

foot–pounds	x 1.3558	= joules (J)	x 0.7376	= foot–pounds
calories	x 4.187	= joules (J)	x 0.2388	= calories
Btu	x 1055	= joules (J)	x 0.000948	= Btu
watt–hours	x 3600	= joules (J)	x 0.0002778	= watt–hours
kilowatt–hrs.	x 3.600	= megajoules (MJ)	x 0.2778	= kilowatt–hrs

FUEL ECONOMY AND FUEL CONSUMPTION

miles/gal	x 0.42514	= kilometers/liter (km/L)	x 2.3522	= miles/gal

Note:
235.2/(mi/gal) = liters/100km
235.2/(liters/100 km) = mi/gal

LIGHT

footcandles	x 10.76	= lumens/meter^2 (lm/m^2)	x 0.0929	= footcandles

PRESSURE OR STRESS (newton/sq meter = pascal)

inches Hg(60 °F)	x 3.377	= kilopascals (kPa)	x 0.2961	= inches Hg
pounds/sq in	x 6.895	= kilopascals (kPa)	x 0.145	= pounds/sq In
inches H$_2$O(60 °F)	x 0.2488	= kilopascals (kPa)	x 4.0193	= inches H$_2$O
bars	x 100	= kilopascals (kPa)	x 0.01	= bars
pounds/sq ft	x 47.88	= pascals (Pa)	x 0.02088	= pounds/sq ft

POWER

horsepower	x 0.746	= kilowatts (kW)	x 1.34	= horsepower
ft–lbf/min	x 0.0226	= watts (W)	x 44.25	= ft–lbf/min

TORQUE

pounds–inches	x 0.11298	= newton–meters (N-m)	x 8.851	= pound–inches
pound–feet	x 1.3558	= newton–meters (N-m)	x 0.7376	= pound–feet

VELOCITY

miles/hour	x 1.6093	= kilometers/hour (km/h)	x 0.6214	= miles/hour
feet/sec	x 0.3048	= meters/sec (m/s)	x 3.281	= feet/sec
kilometers/hr	x 0.27778	= meters/sec (m/s)	x 3.600	= kilometers/hr
miles/hour	x 0.4470	= meters/sec (m/s)	x 2.237	= miles/hour

COMMON METRIC PREFIXES

mega	(M)	= 1 000 000	or 10^6	centi	(c)	= 0.01	or 10^{-2}
kilo	(k)	= 1 000	or 10^3	milli	(m)	= 0.001	or 10^{-3}
hecto	(h)	= 100	or 10^2	micro	(μ)	= 0.000 001	or 10^{-6}

METRIC/U.S. CUSTOMARY UNIT EQUIVALENTS

Multiply:	by:	to get:	Multiply:	by:	to get:

LINEAR

inches	x 25.4	= millimeters (mm)	x 0.03937	= inches
feet	x 0.3048	= meters (m)	x 3.281	= feet
yards	x 0.9144	= meters (m)	x 1.0936	= yards
miles	x 1.6093	= kilometers (km)	x 0.6214	= miles
inches	x 2.54	= centimeters (cm)	x 0.3937	= inches
microinches	x 0.0254	= micrometers (μm)	x 39.37	= microinches

AREA

inches^2	x 645.16	= millimeters^2(mm^2)	x 0.00155	= inches^2
inches^2	x 6.452	= centimeters^2(cm^2)	x 0.155	= inches^2
feet^2	x 0.0929	= meters^2(m^2)	x 10.764	= feet^2
yards^2	x 0.8361	= meters^2(m^2)	x 1.196	= yards^2
acres^2	x 0.4047	= hectares (10^{4}m^2)	ha x 2.471	= acres
miles^2	x 2.590	= kilometers^2 (km^2)	x 0.3861	= miles^2

VOLUME

inches^3	x 16387	= millimeters^3 (mm^3)	x 0.000061	= inches^3
inches^3	x 16.387	= centimeters^3 (cm^3)	x 0.06102	= inches^3
inches^3	x 0.01639	= liters (L)	x 61.024	= inches^3
quarts	x 0.94635	= liters (L)	x 1.0567	= quarts
gallons	x 3.7854	= liters (L)	x 0.2642	= gallons
feet^3	x 28.317	= liters (L)	x 0.03531	= feet^3
feet^3	x 0.02832	= meters^3 (m^3)	x 35.315	= feet^3
fluid oz	x 29.57	= milliliters (mL)	x 0.03381	= fluid oz
yards^3	x 0.7646	= meters^3 (m^3)	x 1.3080	= yards^3
teaspoons	x 4.929	= milliliters (mL)	x 0.2029	= teaspoons
cups	x 0.2366	= liters (L)	x 4.227	= cups

MASS

ounces (av)	x 28.35	= grams (g)	x 0.03527	= ounces (av)
pounds (av)	x 0.4536	= kilograms (kg)	x 2.2046	= pounds (av)
tons (2000 lb)	x 907.18	= kilograms (kg)	x 0.001102	= tons (2000 lb)
tons (2000 lb)	x 0.90718	= metric tons (t)	x 1.1023	= tons (2000 lb)

FORCE

ounces—f (av)	x 0.278	= newtons (N)	x 3.597	= ounces—f (av)
pounds—f (av)	x 4.448	= newtons (N)	x 0.2248	= pounds—f (av)
kilograms—f	x 9.807	= newtons (N)	x 0.10197	= kilograms—f

TEMPERATURE

°F -40   0   32   40   80 98.6 120 160 200 212 240 280 320 °F
°C -40 -20   0   20   40   60   80 100 120 140 160 °C

°Celsius = 0.556 (°F – 32)   °F = (1.8 °C) + 32

TAP/DRILL CHART

	COARSE STANDARD THREAD (N.C.) Formerly U.S. Standard Thread				FINE STANDARD THREAD (N.F.) Formerly S.A.E. Thread			
Sizes	Threads Per Inch	Outside Diameter at Screw	Tap Drill Sizes	Decimal Equivalent of Drill	Sizes	Threads Per Inch	Outside Diameter at Screw	Tap Drill Sizes	Decimal Equivalent of Drill
1	64	.073	53	0.0595	0	80	.060	$^3/_{64}$	0.0469
2	56	.086	50	0.0700	1	72	.073	53	0.0595
3	48	.099	47	0.0785	2	64	.086	50	0.0700
4	40	.112	43	0.0890	3	56	.099	45	0.0820
5	40	.125	38	0.1015	4	48	.112	42	0.0935
6	32	.138	36	0.1065	5	44	.125	37	0.1040
8	32	.164	29	0.1360	6	40	.138	33	0.1130
10	24	.190	25	0.1495	8	36	.164	29	0.1360
12	24	.216	16	0.1770	10	32	.190	21	0.1590
$^1/_4$	20	.250	7	0.2010	12	28	.216	14	0.1820
$^5/_{16}$	18	.3125	F	0.2570	$^1/_4$	28	.250	3	0.2130
$^3/_8$	16	.375	$^5/_{16}$	0.3125	$^5/_{16}$	24	.3125	I	0.2720
$^7/_{16}$	14	.4375	U	0.3680	$^3/_8$	24	.375	0	0.3320
$^1/_2$	13	.500	$^{27}/_{64}$	0.4219	$^7/_{16}$	20	.4375	$^{25}/_{64}$	0.3906
$^9/_{16}$	12	.5625	$^{31}/_{64}$	0.4843	$^1/_2$	20	.500	$^{29}/_{64}$	0.4531
$^5/_8$	11	.625	$^{17}/_{32}$	0.5312	$^9/_{16}$	18	.5625	0.5062	0.5062
$^3/_4$	10	.750	$^{21}/_{32}$	0.6562	$^5/_8$	18	.625	0.5687	0.5687
$^7/_8$	9	.875	$^{49}/_{64}$	0.7656	$^3/_4$	16	.750	$^{11}/_{16}$	0.6875
1	8	1.000	$^7/_8$	0.875	$^7/_8$	14	.875	0.8020	0.8020
$1^1/_8$	7	1.125	$^{63}/_{64}$	0.9843	1	14	1.000	0.9274	0.9274
$1^1/_4$	7	1.250	$1^7/_{64}$	1.1093	$1^1/_8$	12	1.125	$1^3/_{64}$	1.0468
					$1^1/_4$	12	1.250	$1^{11}/_{64}$	1.1718

BOLT TORQUING CHART

METRIC STANDARD						SAE STANDARD/FOOT POUNDS						
Grade of Bolt	5D	.8G	10K	12K		Grade of Bolt	SAE 1 & 2	SAE 5	SAE 6	SAE 8		
Min. Tensile Strength	71,160 P.S.I.	113,800 P.S.I.	142,200 P.S.I.	170,679 P.S.I.		Min. Tensile Strength	64,000 P.S.I.	105,000 P.S.I.	133,000 P.S.I.	150,000 P.S.I.		
Grade Markings on Head	5D	8G	10K	12K	Size of Socket or Wrench Opening	Markings on Head	⬢	⬢	⬢	✳	Size of Socket or Wrench Opening	
Metric		Foot Pounds			Metric	U.S. Standard	Foot Pounds				U.S. Regular	
Bolt Dia.	U.S. Dec Equiv.				Bolt Head	Bolt Dia.					Bolt Head	Nut
6mm	.2362	5	6	8	10	10mm	1/4	5	7	10	10.5	3/8	7/16
8mm	.3150	10	16	22	27	14mm	5/16	9	14	19	22	1/2	9/16
10mm	.3937	19	31	40	49	17mm	3/8	15	25	34	37	9/16	5/8
12mm	.4720	34	54	70	86	19mm	7/16	24	40	55	60	5/8	3/4
14mm	.5512	55	89	117	137	22mm	1/2	37	60	85	92	3/4	13/16
16mm	.6299	83	132	175	208	24mm	9/16	53	88	120	132	7/8	7/8
18mm	.709	111	182	236	283	27mm	5/8	74	120	167	180	15/16	1.
22mm	.8661	182	284	394	464	32mm	3/4	120	200	280	296	1-1/8	1-1/8

DECIMAL CONVERSION CHART

FRACTION	INCHES	M/M	FRACTION	INCHES	M/M
1/64	.01563	.397	33/64	.51563	13.097
1/32	.03125	.794	17/32	.53125	13.494
3/64	.04688	1.191	35/64	.54688	13.891
1/16	.6250	1.588	9/16	.56250	14.288
5/64	.07813	1.984	37/64	.57813	14.684
3/32	.09375	2.381	19/32	.59375	15.081
7/64	.10938	2.778	39/64	.60938	15.478
1/8	.12500	3.175	5/8	.62500	15.875
9/64	.14063	3.572	41/64	.64063	16.272
5/32	.15625	3.969	21/32	.65625	16.669
11/64	.17188	4.366	43/64	.67188	17.066
3/16	.18750	4.763	11/16	.68750	17.463
13/64	.20313	5.159	45/64	.70313	17.859
7/32	.21875	5.556	23/32	.71875	18.256
15/64	.23438	5.953	47/64	.73438	18.653
1/4	.25000	6.350	3/4	.75000	19.050
17/64	.26563	6.747	49/64	.76563	19.447
9/32	.28125	7.144	25/32	.78125	19.844
19/64	.29688	7.541	51/64	.79688	20.241
5/16	.31250	7.938	13/16	.81250	20.638
21/64	.32813	8.334	53/64	.82813	21.034
11/32	.34375	8.731	27/32	.84375	21.431
23/64	.35938	9.128	55/64	.85938	21.828
3/8	.37500	9.525	7/8	.87500	22.225
25/64	.39063	9.922	57/64	.89063	22.622
13/32	.40625	10.319	29/32	.90625	23.019
27/64	.42188	10.716	59/64	.92188	23.416
7/16	.43750	11.113	15/16	.93750	23.813
29/64	.45313	11.509	61/64	.95313	24.209
15/32	.46875	11.906	31/32	.96875	24.606
31/64	.48438	12.303	63/64	.98438	25.003
1/2	.50000	12.700	1	1.00000	25.400

SOME COMMON ABBREVIATIONS

U.S CUSTOMARY		METRIC	
UNIT	ABBREVIATION	UNIT	ABBREVIATION
inch	in.	kilometer	km
feet	ft.	hectometer	hm
yard	yd.	dekameter	dam
mile	mi.	meter	m
grain	gr.	decimeter	dm
ounce	oz.	centimeter	cm
pound	lb.	millimeter	mm
teaspoon	tsp.	cubic centimeter	cm^3
tablespoon	tbsp.	kilogram	kg
fluid ounce	fl. oz.	hectogram	hg
cup	c.	dekagram	dag
pint	pt.	gram	g
quart	qt.	decigram	dg
gallon	gal.	centigram	cg
cubic inch	in^3	milligram	mg
cubic foot	ft^3	kiloliter	kl
cubic yard	yd^3	hectoliter	hl
square inch	in^2	dekaliter	dl
square foot	ft^2	liter	L
square yard	yd^2	centiliter	cl
square mile	mi^2	milliliter	ml
Fahrenheit	F°	square kilometer	km^2
barrel	bbl.	hectare	ha
fluid dram	fl. dr.	are	a
board foot	bd. ft.	centare	ca
rod	rd.	tonne	t
dram	dr.	Celsius	C°
bushel	bu.		

Acknowledgments

The production of a textbook of this type would not be possible without assistance from the automotive industry. The authors would like to thank the following companies for their assistance in the preparation of **Automatic Transmissions and Transaxles.**

American Honda Motor Co.
BBU, Inc.
Caterpillar Inc.
DaimlerChrysler
Deere & Co.
Dorman Products
FAG Bearings
Federal Mogul Corp.
Fel-Pro Inc.
Fluke Corp.
Ford Motor Co.
General Motors Corp.[†]
Hayden
Hunter Engineering Co.
L.S. Starrett Co.
Lexus
Lincoln Automotive
Lisle Corp.
Mercedes-Benz of North America, Inc.
Nissan Motor Corp.
Owattona Tool Co., Div. of SPX Corp.
Pennzoil Co.
RTI Technologies, Inc.
Saab-Scandia of America, Inc.
Sachs
Snap-on Tool Corp.
Sonnax
Subaru of America, Inc.
Superior
TIF Instruments, Inc.
Toyota Motor Sales, USA, Inc.
Vaco Products Co.
Volkswagen
Volvo of America
ZF Transmission Group

The authors would also like to thank the following persons and organizations that provided vehicles, parts, and test equipment for the photographs, as well as other items used throughout this text.

Ross Fossbender, B&B Electronics, Ottawa, IL
Don Russell, Ferret Instruments, Cheboygan, MI
Henrik Jacobs, Timothy Jacobs, Jason Powers, Mike Baskins, Ryan Shannon, and Robert Rivera,
    Greenville Transmission Clinic, Greenville, SC
Joe Gallagher, Sonnax Industries, Bellows Falls, VT
Vincent J. McKenna, Spectronics Corporation, Westbury, NY
Paul Erickson, Superior Transmission Parts, Tallahassee, FL
Karin Markenstein, ZF Transmission Group, Saarbrucken, Germany

[†]Portions of materials contained herein have been reprinted with permission of General Motors Corporation, Service Technology Group.

Glossary

A

Accumulator: A hydraulic device used in the apply circuit of a band or clutch to cushion initial application.

ACT: A nonprofit organization experienced in administering standardized tests.

Actuators: Components that drive hydraulic system output devices. These may be pistons actuated by hydraulic pressure, or they may be electric solenoids.

Adhesives: Compounds used to hold parts together. Adhesives are sometimes used to hold gaskets in place or to hold small parts together during assembly.

Air pressure tests: Tests that involve applying compressed air to the hydraulic pressure ports that supply hydraulic fluid to the holding members, and sometimes, to the governor valve. When air pressure is applied, operation of these components can be seen or heard to determine if they are working correctly.

Air temperature sensor: See *Intake air temperature sensor.*

Alignment pin: Pin used to adjust a neutral start switch.

Alternating current (ac): An electrical current that moves in one direction and then the other.

Ammeter: Instrument used to measure electrical current in a given circuit.

Ampere (amp): Unit of electrical current. The higher the amperes, the more electrons are moving in the circuit.

Anaerobic sealer: Sealer that cures to a plastic-like substance in the absence of air. It will remain fluid as long as it is exposed to the atmosphere.

Annulus gear: See *Ring gear.*

Antifriction bearing: Bearing that uses an assembly of balls or rollers contained within a housing, where they are aligned and free to roll.

Apply piston: See *Clutch piston.*

ASE Preparation Guide: Free booklet available from ASE to help you prepare for the ASE certification tests.

Atmospheric pressure: The pressure exerted by the earth's atmosphere on all objects. Measured with reference to the pressure at sea level, which is about 14.7 psi (101 kPa).

Automatic shift diagram: Chart that shows a vehicle's approximate shift points. The shift points in the chart can be compared with actual shift points to determine whether the transmission or transaxle is operating properly.

Automatic transmission fluid (ATF): A special type of lubricating oil, or fluid, designed for use in automatic transmissions and transaxles.

Automatic transmission shafts: Shafts that transmit power from the torque converter to the driveline. One transmission shaft provides a stationary support for the torque converter stator.

Auxiliary oil cooler: Air-cooled heat exchanger that connects to the existing oil cooling system. It may be installed to completely replace the main oil cooler in the radiator, or it may be helper units that further cool fluid that has passed through the main oil cooler. Auxiliary coolers are usually mounted ahead of the radiator.

Auxiliary valve bodies: Secondary valve bodies that are often bolted to the main valve body or to the transmission case. Auxiliary valve bodies are separated from the main valve body because of design considerations, such as clearance restrictions.

Axial load: Load from a rotating shaft that is parallel to the axis of the shaft.

B

Backlash: Amount of "play" between two parts. In the case of gears, backlash refers to how much one gear can be moved back and forth without moving the gear into which it is meshed.

Ball bearings: Bearings that provide a rolling contact for reduced friction.

Band and clutch application chart: Chart that shows which holding members are applied in a particular gear. This type of chart can be very useful if trying to isolate a defective holding member when the transmission or transaxle is slipping in a particular gear.

Bands: Flexible metal friction devices designed to hold planetary gearset members stationary.

Barometric pressure sensor (BARO sensor): Sensor that measures the pressure of the surrounding air and converts the pressure reading to an electrical signal.

Bearings: Devices used to reduce friction between rotating and stationary parts. Further, they guide and support rotating parts, preventing damage from misalignment or excessive clearance.

Bending tool: Tool used to bend tubing in a relatively sharp angle without kinking the inside of the bend.

Blown gasket: Condition in which pressure from within a component will force out a piece of a gasket. Blown gaskets are usually caused by loose fasteners or warped mating parts that do not tightly hold the gasket.

Bolt grade: Indication of the amount of pulling, or stretching, force a fastener can withstand before it stretches or, in some cases, breaks. Bolt grade, then, is an indication of tensile strength.

Bounce-back effect: Condition in which return flow from the torque converter turbine hits the forward (oncoming) faces of the impeller vanes with a great deal of force, opposing impeller rotation. This action, which would occur without a stator, reduces the effectiveness of the impeller.

Bushings: One-piece bearings that provide a sliding contact with a moving part. They are installed where a rotating part passes through a stationary part or two rotating parts are in contact with each other.

C

Calipers: Tools used to take external or internal measurements. They essentially consist of a pair of movable legs. In measuring, the legs are adjusted to fit the dimension in question. The caliper can then be laid over a rule, and the span can be measured.

Capacitor: Electrical device used to damp out voltage fluctuations or control electronic frequencies. The capacitor serves as a trap for voltage surges (sometimes called spikes) before they can affect electrical or electronic circuits.

Case: An aluminum casting that is machined to serve as a mounting and aligning surface for the moving parts of the transmission or transaxle. Most other transmission or transaxle parts are housed in the case. Passageways in the case allow hydraulic fluid to travel between hydraulic components.

Central processing unit (CPU): The section of the ECM that receives the input sensor information, compares this information with information stored in memory, performs calculations, and makes output decisions.

Certified technician: Automotive service person who is ASE certified in at least one area and has the required work experience.

Chain drive: A power-transfer system using a chain and two or more sprockets. It is used in some automatic transaxles to transmit motion in a small space.

Chain tensioner: Device used to take up the slack in a chain. Most chain tensioners are spring-loaded devices that maintain constant pressure on the return, or nondrive, side of the chain.

Check ball: See *Check valve*.

Check valve: One-way valves that allow fluid to flow in one direction only.

Chip capacitors: Very small capacitors used on late-model vehicles.

Circuit: Source of electricity, resistance unit, and wires that form a path for the flow of electricity from the source, through the resistance unit, and back to the source.

Circuit breaker: A circuit protection device that consists of a contact point set attached to a bimetallic strip. The bimetallic strip will bend as it heats up. When the strip becomes hot enough, it bends enough to open the point set, breaking the circuit. When the strip cools off, it straightens out and allows the point set to close. The advantage of the circuit breaker is that it can reset itself.

Circuit protection devices: Devices designed to protect circuits from damage due to excessive current. These devices include fuses, fusible links, and circuit breakers.

Clearance: Space between the root of one gear tooth and the top of the mating tooth.

Clutch apply piston: An aluminum or steel disc that fits into a bore in the lower portion of the clutch drum in a multiple-disc clutch. The piston is moved by hydraulic pressure. Movement of the apply piston clamps the drive discs and clutch plates together.

Clutch clearance: The distance between the pressure plate and the outer snap ring in a multiple-disc clutch.

Clutch drum: The housing for all the other multiple-disc clutch components. The outer surface of the clutch drum may be a holding surface for a band. The inner surface contains large splines that mate with teeth on the clutch plates or drive discs.

Clutch hub: A part within a multiple-disc clutch or another component, such as a planetary ring gear, that fits into the inside diameter of the set of drive discs and clutch plates. External splines on the hub engage the internal teeth of the drive discs. The clutch hub may also be splined to the transmission input shaft or to a part of a gearset.

Clutch pack: See *Multiple-disc clutch*.

Clutch piston. Hydraulically operated piston that applies clutches in a multiple- disc clutch.

Clutch plates: Metal plates with external teeth that lock into channels on the inside of the clutch drums. The drive discs and clutch plates form a sandwich of alternating layers. Their purpose is to lock together the drum and hub, as well as the parts connected to each when the clutch is engaged.

Clutch pressure plate: Serves as a stop for the set of drive discs and clutch plates when the clutch apply piston is applied. The pressure plate is always installed at the end of the clutch pack opposite the apply piston.

Clutch return springs: Springs that push the apply piston away from the drive discs and clutch plates after hydraulic pressure is released from the piston.

Clutches: A series of flat, ribbed plates.

Collars: Steel rings used to hold gears or bearings on shafts. Collars are usually used where space is too limited to allow for a different kind of retainer.

Color-coding: Process of assigning a specific color to the insulation of every vehicle wire to facilitate circuit tracing.

Complete circuit: A circuit that forms a complete path for electricity.

Composition plates: See *Drive discs.*

Compound planetary gearset: Two or more planetary gearsets that are linked together. This arrangement can provide more forward gear ratios than a basic planetary gearset.

Compressed-air system: Pneumatic system in which the pressure inside the system is greater than the pressure of the outside air.

Conductors: Materials that consist of atoms that easily give up or receive electrons. Examples are copper and aluminum.

Continuity: Condition in which a circuit or component forms a complete path for the flow of electricity.

Continuously variable transmission (CVT): Automatic transaxle that uses a steel belt and two variable-diameter pulleys to transfer engine power to the drive wheels.

Control loop: A continuous circle of causes and effects used to operate part of the vehicle. When the control loop is operating, the input sensors furnish information to the computer, which makes output decisions and sends commands to the output devices.

Control valves: Hydraulic valves used to regulate the operation of the system and of other hydraulic components.

Converter clutch torque converter: See *Lockup torque converter.*

Converter cover: Enclosure that fits over the turbine, attaches to the impeller, and indirectly attaches to the engine crankshaft. In some designs, a ring gear is welded to the perimeter of the cover.

Converter flusher: Cleaning tool that moves a pulsating flow of solvent in and out of a torque converter to remove debris from inside the converter. In addition, the converter flusher rotates the converter to further agitate the solvent inside the unit.

Converter housing: A stationary enclosure that surrounds and protects the torque converter.

Converter housing attaching bolts: Bolts securing the transmission to the engine.

Converter pilot hub: A cylindrical metal extension located at the center of the converter cover.

Converter-to-flywheel attaching bolts: Bolts securing the torque converter to the flywheel.

Coolant temperature sensor: Sensor that is threaded into an engine coolant passage and sends a varying electrical signal based on coolant temperature changes to the ECM.

Core value: Return value of a rebuildable part.

Countershaft gear assembly: Assembly consisting of a shaft and several gears that is located between input and output shafts.

Coupling phase: Condition when torque converter turbine speed approaches about 90% of impeller speed.

Cross contamination: Transfer of fluid between the transmission and final drive in transaxles that use different fluids in each section.

Current: The movement of free electrons through a conductor.

Cut back valve: Control valve that uses governor pressure to reduce throttle pressure. This allows the 2-3 shift valve to upshift when the engine speed becomes too high in passing gear.

D

Data link connector (DLC): Connector used to retrieve information about electronic control system problems. It allows the technician to directly access the information stored in the ECM's memory. All OBD II systems use standardized 16-pin connectors. The connectors used in OBD I systems may vary from one manufacturer to another.

Dedicated solenoid: Solenoid that operates only its associated valve. In transmissions or transaxles using dedicated solenoids, there is one solenoid for each shift valve. For example, a four-speed transmission with three shift valves has three dedicated solenoids.

Depth micrometer: Instrument used to measure the depth of openings in machined surfaces.

Detent solenoid: Solenoid is energized by closing the kickdown switch, which is located near the accelerator.

Detent valve: Valve used on some transmissions to aid downshifting.

Diagnostic charts: Charts prepared by transmission/transaxle or vehicle manufacturers help pinpoint the exact cause of a problem. Each chart applies to one specific type of transmission or transaxle and is designed to match a transmission or transaxle problem to a specific part malfunction.

Dial indicator: Gauge used to measure small amounts of part movement. Frequently used to check gear backlash, shaft endplay, cam lobe lift, etc.

Diodes: Semiconductor devices that act as one-way check valves for electricity. They allow current to flow in only one direction.

Direct current (dc): Current that flows in only one direction, from negative to positive.

Direct drive: Gear ratio in which the driven gear makes one revolution for each revolution of the drive gear. Direct drive has a gear ratio of 1:1. With this gear ratio, there is no change in either speed or torque between the engine output and the transmission output. In other words, the input and output shafts turn at the same speed.

Direct programming: Method of reprogramming an ECM in which new information is downloaded by attaching a shop recalibration device directly to the data link connector.

Direct-air cooler: Add-on oil cooler installed ahead of the radiator. Air passing through the cooler removes heat by direct contact.

Directional control valves: Valves used to direct the path of flow in a hydraulic circuit. These valves are used in automatic transmissions and transaxles to control the application of different hydraulic output devices.

Drain pans: Pans used to catch oil that drips from assemblies as they are drained or disassembled.

Drifts: Removal and installation tools designed to be struck with a hammer. They are used to remove pins, plugs, and other pressed-in parts from bores.

Drive discs: Metal plates covered with a friction lining. The drive discs have internal teeth that engage splines on the clutch hub. Some discs have external splines. These splines engage clutch splines inside the clutch drum.

Drive shell: Bell-shaped part that is commonly used to transfer power to the planetary sun gear.

Drivers: Special ECM circuits that consist of power transistors and related electronic devices. The drivers operate the output solenoids. See also *Power transistors.*

Dust shield: Device used to protect seals that are exposed to large amounts of dirt and water.

Duty cycle: A ratio of on to off.

Dynamic seals: Seals that flex slightly to allow a small amount of fluid leakage. Slight fluid leakage keeps the seals lubricated.

E

Electromagnet: A magnet produced by placing a coil of wire around a steel or iron bar (core). When current flows through the wire, the magnetic fields of each wire loop combine to create a very strong magnetic field. The core helps increase magnetic field strength.

Electronic control module (ECM): Computer that can precisely control transmission operation through solenoids, replacing the hydraulic and mechanical components previously used.

Electronic control system: System that contains an electronic control module (ECM), input sensors, and output devices to operate the hydraulic components of a transmission or transaxle. The ECM processes information from the sensors. It then uses this information to operate solenoids and other output devices installed in the transmission or transaxle to control pressure flow through the hydraulic system.

Electronically controlled pressure regulator: Pressure regulator operated by an ECM-controlled solenoid. The solenoid is pulsed to produce a duty cycle. The duty cycle creates a precisely controlled pressure leak at the main pressure regulator. This controlled leak accurately modifies pressure.

Electronically erasable programmable read-only memory (EPROM):. Type of computer memory used in late-model vehicle. Information in EPROM can be updated by reprogramming.

End clearance: The distance between the two parts.

End thrust: A thrusting action that occurs whenever power is being transmitted through a helical gear. This action is due to the helix angle (angle of the helical gear teeth).

Endplay: Amount of axial (lengthwise) movement of a component. Back-and-forth movement of a shaft.

Engine braking: Using the engine (via engine resistance) to slow down the vehicle, rather than using the brakes.

Engine holding fixture: Device that supports the weight of the engine once the rear engine mount has been removed.

Engine speed sensor: Sensor mounted on the camshaft or crankshaft and provides the computer with the engine speed signal.

Environmental Protection Agency (EPA): Agency of the United States government charged with enforcing laws against environmental destruction.

Epoxies: Adhesives that are two-part mixtures applied in a one-step process. Epoxies are mixed immediately before application.

Exhaust temperature sensor: Computer sensor used to measure the temperature of the engine exhaust gases.

Extension housing: Enclosure that supports and encloses the tail end of the transmission output shaft.

External gaskets: Gaskets that keep fluid from leaking out of a component. They also keep dirt or moisture from entering.

F

Face: The contact surface of a gear tooth.

Face width: The width of a gear tooth measured parallel to the gear axis.

Feeler gauge: Measuring instrument consisting of a thin strip of hardened steel ground to an exact thickness and used to check clearances between parts.

Final drive unit: Portion of the transaxle that corresponds to the rear axle gears and differential assembly of a rear-wheel drive vehicle.

Flash erasable programmable read-only memory (FEPROM): Type of computer memory that can be updated by reprogramming.

Flash programming: Procedure that involves providing existing computer memory (EPROM or FEPROM) with updated information by reprogramming.

Fluid coupling: A hydrodynamic device designed to transmit power through a fluid. This device essentially consists of a drive member (impeller) and a driven member (turbine).

Flywheel: A lightweight disc with a ring gear that is used to engage the starter motor.

Flywheel holder: Tool used to keep the flywheel from turning during transmission or transaxle service.

Flywheel turners: Tool used to move the flywheel by engaging two or more teeth on the flywheel ring gear.

Form-in-place gaskets: Gaskets made from a silicone sealer.

Freeze frame: See *Snapshot*.

Frequency: The speed, or rate, of the change in current flow.

Friction bearing: Bearing that is pressed into place and does not move. A rotating shaft slides on the friction bearing's surface.

Friction discs: See *Drive discs*.

Fuse: Circuit protection device made of a soft metal that melts when excess current flows through it. The metal melts before the current can damage other components or circuit wiring.

Fuse block: Housing for fuses and other circuit protection devices. Generally located under the dashboard or in the glove compartment.

Fusible link: A length of wire made of soft metal that melts when excess current flows through it. Fusible links are usually installed in the wiring leading from the battery or starter solenoid to the main electrical circuits.

G

Gaskets: Type of static seal used on nonmoving parts. Used to keep fluid from leaking out of the transmission or transaxle and to keep pressure from leaking internally. Gaskets are used where major components are joined together.

Gauge pressure scale: Scale on which normal atmospheric pressure (14.7 psi [101 kPa] at sea level) is chosen as a zero reference pressure. Gauges using this scale measure pressure relative to that of the surrounding atmosphere.

Gear: A toothed wheel that engages other gears or mechanical parts for the purpose of transmitting power.

Gear backlash: The amount of movement between mating gear teeth when one gear is held and other is moved to the limit of travel, first one way and then the other.

Gear drive: A system of gears that engage other gears or mechanical parts for the purpose of transmitting power.

Gear pump: Pump commonly used in automatic transmissions. The simplest form consists of two meshed gears inside a housing.

Gear ratio: The speed relationship between two gears determined by the difference in the number of teeth on the gears. The gear ratio relates the speed of the drive gear to the speed of the driven gear.

Gear reduction: A gear ratio in which the drive gear makes more than one revolution to turn the driven gear through one complete revolution. A 3:1 gear reduction means that it takes three revolutions of the drive gear to turn the driven gear through one complete revolution.

Gear skipping: Condition in which a transmission or transaxle skips gears when shifting (shifts from first to third, for example).

Governor: See *Governor valve*.

Governor pressure: Governor valve output pressure, which is proportional to road speed.

Governor valve: Unit that senses output shaft speed to help control shifting. It works with the throttle valve to determine shift points.

Grade markings: Bolt head markings used to identify the tensile strength of a bolt.

Ground: Return circuit to the battery formed by the vehicle's frame or body.

H

Hardening sealers: Sealers used on parts that will remain assembled for long periods, possibly the life of the vehicle.

Harness: A series of wires bound together to form a compact, protected unit. Harnesses used inside the transmission or transaxle are usually not wrapped.

Helical gears: Gears used to connect parallel shafts. The teeth of helical gears are cut at an angle, called a helix angle, across the gear surface.

Helper: Automotive service person who performs basic service and maintenance tasks, such as installing and balancing tires, cleaning parts, changing engine oil and filters, and installing batteries.

Holding fixtures: Stands used to hold transmissions and transaxles during service procedures.

Holding members: Units that hold or drive the various parts of the planetary gear assembly.

Hole gauge: Instrument used to measure small-diameter holes or openings.

Hot tank: See *Immersion cleaner*.

Housings: Units that surround the torque converter and the output shaft and are generally cast as an integral part of the case. When used, separate bell housings (housing around the torque converter) and output shaft housings are bolted to the case.

Hub and shaft assembly: Input shaft between the driven sprocket and the other transmission parts on a transaxle that uses a chain drive.

Hydraulic circuit diagrams: Diagrams used to trace circuits in the hydraulic system.

Hydraulic control system: A set of hydraulic parts and passages that applies and releases the holding members to obtain the needed gear ratios at any vehicle speed and throttle position; controls system pressures for proper shift feel and long holding member life as loading and acceleration vary.

Hydraulic pressure test: Diagnostic test that involves connecting one or more pressure gauges to the transmission or transaxle pressure fittings and observing the hydraulic fluid pressures. Actual pressures are then compared to published values to decide if there is a hydraulic system problem.

Hydraulic pump: Pump that produces the fluid pressure to operate other hydraulic components and lubricate the moving parts of the transmission or transaxle.

Hydraulic valves: Devices used in a hydraulic circuit to control pressure, as well as the direction and rate of fluid flow.

Hydraulics: The science and technology of liquids at rest and in motion.

Hypoid gear oil: Lubricant used in many final drives. This is the same type of oil used in the rear axles of rear-wheel drive vehicles.

Hypoid gears: Gears used in applications in which power flow must be diverted by some angle.

I

Idler gear: See *Reverse idler gear.*

Immersion cleaner: A powerful parts washer that is usually filled with strong, corrosive chemicals. Cleaning is done automatically.

Impeller: A set of curved vanes welded to the inside of a shell that forms the rear half of the torque converter. The impeller is attached to the converter cover. Since the converter cover is attached to the flywheel, the impeller turns whenever the engine turns. The impeller is the torque converter drive member.

Indirect programming: Method of reprogramming an ECM in which the proper scan tool must be used to transfer information from a separate programming computer to the ECM.

Induction: Process of moving a wire through a magnetic field or moving a magnetic field through a wire to create current flow.

Input sensor: Any sensor that provides information to a computer.

Input shaft: Heat-treated steel shaft that transfers power from the torque converter turbine to the front planetary gearset of the transmission or transaxle.

Input shell: See *Drive shell.*

Input speed sensor: Sensor that tells the ECM how fast the transmission input shaft is turning.

Inside micrometer: Instrument used for measuring inside diameters or other part openings. The micrometer reading is the size of the opening.

Installer: An automotive service person who removes and installs parts.

Insulators: Materials made of atoms that resist giving up or accepting atoms. Glass and plastic are good insulators.

Intake air temperature sensor: Computer sensor used to measure the temperature of the air entering the intake manifold.

Integrated circuit (IC). A circuit made by combining many diodes, transistors, chip capacitors, resistors, and other parts into a complex electronic circuit by etching the circuitry on small pieces of semiconductor material.

Interference fit: Method of holding one part in a mating component. The part is slightly larger than the opening into which it is fit. A shaft that is fit into a smaller hole is an example. The parts fit together tightly, and they cannot be removed easily.

Internal gaskets: Gaskets that prevent loss of pressure or crossover of fluids within a component.

J

Jumper wire: A short piece of insulated wire with an alligator clip on each end that is used to temporarily bypass components and to apply voltage to a component or section of a circuit.

K

Karmann vortex sensor: A MAF sensor that operates from turbulence created by a calibrated restriction in the air intake passage.

Key: Fastener inserted in a keyway (slot) to keep a shaft and machine part, such as a gear or pulley, from rotating in relation to each other.

Keyways: Slots cut in mating parts to accept a key.

Kickdown valve: See *Detent valve.*

L

Leopard spotting: Condition of clutch discs that is a sign of severe clutch overheating. Named for its resemblance to a leopard's coat.

Lepelletier gear train: Gear train that creates six forward speeds with less complexity and approximately half the parts of typical four- and five-speed automatic transmissions.

Limp-in mode: Mode that the ECM may shift into when there is an electronic system malfunction. In limp-in mode, the ECM ignores most input sensor readings and operates the engine and drive train output devices based on internal settings.

Line bias valve: Valve that allows the throttle pressure to the main pressure regulator to increase quickly when the throttle valve is first moved, but slows the rate of increase after the first opening. This allows the line pressure increase to more closely match increases in engine torque.

Line pressure: Pressure produced by the hydraulic pump and regulated by the pressure regulator valve. The pressure of the overall transmission hydraulic system.

Lip seals: Seals used on rotating shafts. The casing is a rigid support for the other seal components. The sealing element is designed to contact the rotating shaft.

Lockup solenoid: Solenoid used to engage the torque converter lockup clutch.

Lockup torque converter: Torque converter with an internal friction-clutch mechanism that mechanically locks the turbine to the impeller and converter cover when the transmission or transaxle is in high gear. It is designed to eliminate the 10% slippage that takes place between the impeller and turbine during the coupling phase of operation.

Lubricant leakage: Unwanted loss of fluid from inside the transmission or transaxle.

M

Magnetic powder clutch: A power transfer device using a metallic powder contained between two hubs. Magnetizing the powder causes it to lock the two hubs together.

Main oil cooler: A heat exchanger built into a side or bottom tank in the engine cooling-system radiator used to remove heat from the transmission fluid. Cooler lines connect the transmission to the oil cooler.

Maintenance indicator light (MIL): A dashboard-mounted warning light that illuminates when a problem is detected by the computer control system.

Manifold absolute pressure sensor (MAP sensor): Sensor that measures the intake manifold vacuum and converts it to an electrical signal, which is sent to the ECM. The MAP sensor is attached to intake manifold vacuum.

Manual linkage: A cable or a series of rods and levers that connects the shift lever to the manual valve inside the transmission or transaxle.

Manual shift program: A method of manually changing gear ratios on electronically controlled automatic transmissions and transaxles. The manual shift program allows the driver to change between gears by moving the shift lever.

Manual valve: Valve that is located in the valve body and operated by the driver through the shift linkage. This valve allows the driver to select *Park, Neutral, Reverse,* or different *Drive* ranges. When the shift selector lever is moved, the shift linkage moves the manual valve. As a result, the valve routes hydraulic fluid to the correct transmission components.

Manual valve position sensor: Sensor that is a set of on-off switches. Each gear shift lever position selected by the driver closes a particular switch in the manual valve position sensor, telling the ECM which transmission operating position has been selected.

Manufacturers' service manuals: Books that contain the information needed to repair transmissions and transaxles or other vehicle components. They contain information specific to one vehicle make or model.

Mass airflow sensor (MAF sensor): Computer sensor that is installed in the air intake system and measures the amount of air entering the engine.

Master Technician: Technician who has passed all the ASE certification tests in either the automotive or heavy truck areas and has the required work experience.

Material Safety Data Sheet (MSDS): Informational sheets that list all the known dangers of a chemical or material, as well as the first aid procedures to follow in the event of skin, eye, or respiratory system contact.

Micrometer: Precision measuring tool that will give readings accurate to within a fraction of one thousandth of an inch.

Modulator TV valves: Valves that modify the pressure supplied by the throttle valve for more precise shifting.

Multimeter: Meter that can read all major electrical values (voltage, resistance, and amperage) and may be able to read voltage waveforms and provide other information.

Multiple-disc clutch: Component used to hold gearset members in place and to lock two elements together for power transfer.

Multi-shift solenoids: Solenoids used in combination with pressure passages in the valve body. The pressure passages are arranged so that multi-shift solenoids affect some valves directly and others indirectly. In this arrangement, a four-speed transmission with three shift valves might have only two shift solenoids.

N

National Institute for Automotive Service Excellence (ASE): A nonprofit organization formed to promote high standards of automotive service and repair through the voluntary testing and certification of technicians.

Needle bearings: Bearings with tiny rollers that resemble needles.

Neutral safety switch: See *Neutral start switch.*

Neutral start switch: Switch that prevents engine starting when the selector lever is in a position other than Park or Neutral.

No-drive condition: Condition in which the transmission or transaxle fails to transmit power to the drive wheels.

Nonhardening sealers: Sealers used on parts that will be removed occasionally, such as transmission or transaxle pans.

Nonpositive displacement pumps: Pumps with output that varies with pump speed and system internal resistance. These pumps are used for lower-pressure, high-volume applications—primarily, for transporting fluids from one location to another.

Nonpowered test light: Testing device used to check a circuit for power. This type of test light essentially consists of a probe with a light and a lead with an alligator clip. The light is powered by the circuit.

Normally closed switches: Switches that are closed until hydraulic pressure is present.

Normally open switches: Switches that are open until hydraulic pressure is present.

O

Ohm: Unit of measurement for resistance to the flow of electric current in a given unit or circuit.

Ohmmeter: An electrical instrument used to measure the amount of resistance in a given unit or circuit.

Ohm's law: Formula for computing unknown voltage, resistance, or current in a circuit by using two known factors to find the unknown value.

Oil clearance: The clearance between a shaft and bearing that allows the bearing lubricant to enter and circulate properly.

Oil cooler: A device that cools the transmission or transaxle fluid. On in-radiator oil coolers, excess heat in the transmission fluid is transferred to the engine coolant through the walls of the cooler passages. On external oil coolers, the temperature of the transmission fluid is lowered by air passing across the cooler passages.

Oil cooler and line flusher: Cleaning tool that uses solvent to remove contaminants from the oil cooler and cooler lines.

Oil diverters: Devices that keep excess oil away from parts that could be overlubricated. Diverters are also used to keep excess oil from sealing areas or vents that could leak lubricant.

Oil filter: Device used to remove particles from the automatic transmission fluid before the fluid is circulated by the pump.

Oil pan: Sheet metal pan installed on the bottom of the transmission case. It fits over the valve body and serves as a reservoir for transmission fluid.

Oil pressure diagrams: See *Hydraulic circuit diagrams.*

Oil pump shaft: Shaft that drives the oil pump on most chain-drive transaxles.

Oil scoops: Mechanisms used to pick up oil being thrown from moving parts and deliver it to other portions of a transaxle, which would otherwise be starved for lubrication.

Oil transfer tubes: Tubes used to connect valve body pressure chambers when it would be impractical to cast the needed passages in the valve body or the case.

On-board diagnostics, generation 1 (OBD I): Classification given to diagnostic systems found on vehicles produced before 1996.

On-board diagnostics, generation 2 (OBD II): Enhanced diagnostic system required on 1996 and newer vehicles. Protocol called for standardization of codes, data link connectors, and terminology.

One-way clutch: See *Overrunning clutch.*

Open circuit: A circuit that is not complete. Current cannot flow in an open circuit.

O-rings: Seals that have a small diameter through the cross section, as compared to the annulus, or ring. O-rings are used with both stationary and moving parts.

Output shaft: Shaft that transfers power from the planetary gearsets to the driveline components.

Outside micrometer: A micrometer designed to measure the outside dimensions of objects.

Overdrive: Gear ratio in which the driven gear makes more than one revolution for each complete revolution of the drive gear. Most overdrive gearsets have a ratio of around 1:1.5 (sometimes written as 0.66:1). With this type of ratio, there is a speed increase through the gearset.

Overrun clutch: Hydraulically applied clutch that is separate from the mechanical overrunning clutches.

Overrunning: Release of the one-way clutch when the vehicle is coasting.

Overrunning clutch: Mechanical clutch mechanism that will drive in one direction only. If torque is removed or reversed, the clutch slips.

Oxygen sensor (O_2 sensor): An oxygen-sensitive sensor placed in engine exhaust stream. The sensor produces electrical signals, which vary with oxygen content of exhaust. The ECM reads this signal as a rich or lean mixture.

P

Parallel circuit: Circuit in which current flow is split so that each electrical component has its own current path. Different amounts of current will flow in different parts of the circuit, depending on the resistance of each part.

Parallel shaft transaxles: Automatic transaxles that have sliding gears that move in and out of mesh.

Park lock: A lever that engages a toothed wheel (parking gear) on the output shaft. When the park lock is engaged, the vehicle cannot roll.

Parking gear: A toothed wheel splined to the output shaft. The linkage extends to the rear of the transmission and locks the output shaft to the case.

Parking pawl: Component that moves in and out of engagement with the parking gear.

Parts washer: Device used to clean various automotive parts. Most parts washers have a pump to recirculate solvent.

Pascal's law: Law of hydraulics. States that a liquid that has been pressurized transmits the same amount of pressure to every part of its container.

Passing gear: Term used to describe condition in which the accelerator is pushed all the way to the floor and the detent valve provides a forced downshift, increasing torque to the drive wheels.

Pattern failure: A problem that is common to a certain type of vehicle.

Peening: Indentations in the gear faces. Peening is usually caused by severe loads on the gear teeth or by large particles of foreign matter in the lubricant.

Piezoelectric crystal: Pressure-sensitive material that produces an electrical current proportional to the pressure exerted upon it.

Pitting: Small holes, or pits, in the gear teeth that reduce the gear contact area. Pitting is usually caused by abrasive particles, corrosion, or a lack or lubricant.

Planet carrier: Component that holds the planet gears in a planetary gearset so they are free to rotate.

Planet gears: Gears that surround and mesh with the sun gear in a planetary gearset.

Planet pinions: See Planet gears.

Planetary gearsets: Constant-mesh gears arranged in such a manner that they resemble the solar system. They create different forward gear ratios or provide reverse operation when certain members are held stationary and others are driven.

Planetary holding members: Components that function by either holding a planetary member stationary or applying drive power to it.

Planet-pinion carrier: See Planet carrier.

Plug-in connector: Wiring connector that has male and female ends that are plugged into each other.

Pneumatics: The study of the mechanical properties (reactions to applied forces) and physical properties of air and other gases.

Positive displacement pumps: Output from this type of pump is the same during every rotation, no matter what the pump speed or internal resistance. This type of pump can generate very high pressure.

Power transistors: Large transistors designed to carry heavy current loads.

Press: A piece of equipment consisting essentially of a frame and a ram. It is used for applying pressure on an assembly.

Press fit: Type of interference fit in which considerable pressure is applied to the mating pieces to force them together.

Pressure gauges: Instruments that measure pressure of a fluid.

Pressure regulator valve: Control valve installed in the outlet line from the pump that controls overall transmission pressures.

Pressure relief valve: Check valve that opens to exhaust excess pressure. It keeps pressure from exceeding a preset maximum value. Its purpose is to prevent excess pressure from damaging the hydraulic system.

Pressure sensors: Simple on-off switches connected to pressure passages in the transmission or transaxle that feed certain clutches and bands. When the clutch piston or band servo is pressurized, the pressure also operates the switch. Whether the pressure switches are on (pressure at the switch port) or off (no pressure at the switch port) tells the ECM what gear is engaged.

Programmable read only memory (PROM): Computer memory section that contains information similar to that in ROM. However, the information in PROM can be changed. On older vehicles, PROM is a removable chip, which can be replaced if necessary. On newer vehicles, a variation of the PROM called EPROM (erasable programmable read only memory) or a flash EPROM is used. Some of the information in EPROM can be replaced by erasing and reprogramming.

Prussian blue: A deep blue pigment (dye) mixed with a grease-like carrier. By spreading a thin film on one part, placing the other part firmly in position, and then removing it, it is possible to check contact surfaces.

Pullers: Tools used to remove parts that are pressed together.

Pulsed: Turned on and off rapidly.

Pump drive hub: Hollow rear shaft that fits over the torque converter stator and turbine shafts and into the oil pump at the front of the transmission case.

Punches: Removal and installation tools designed to be struck with a hammer. They are used to remove pins, plugs, and other pressed-in parts from bores.

R

Radial load: Load produced by a rotating shaft that is perpendicular to the axis of rotation.

Random access memory (RAM): Computer memory section that is a temporary storage place for data from the input sensors.

Ratio control motor: A small dc motor used on CVTs to move the variable ratio control valve through linkage.

Ravigneaux gear train: One of the first compound planetary gear designs. There are several versions of the Ravigneaux gear train, but all have two sun gears, one ring gear, and one planet carrier.

Reaction shaft: See Stator shaft.

Read only memory (ROM): Computer memory that contains operating instructions and specifications for the system. ROM is a permanent memory that will be retained if the battery is disconnected or the system fuse is removed.

Reamers: Cutting tools used to salvage damaged transmission parts by enlarging a hole cut in the metal of the part.

Reduction gear: A set of at least two gears that cause the output shaft speed to be much lower than the input shaft speed.

Relay: Switching device that uses a magnetic field to close one or more sets of electrical contacts, allowing electrical flow in a circuit.

Relearn procedure: Period of vehicle operation that allows the computer system to adapt to new components and updated reprogramming.

Release springs: See *Clutch return springs.*

Remote programming: Method of reprogramming an ECM that is performed after the ECM has been removed from the vehicle. This procedure is used when changes must be made through a direct connection to a manufacturer's database.

Resistance: The opposition of the atoms in a conductor to allow the flow of electrons. In other words, it is the opposition to current flow in a conductor or component.

Resistors: Components placed in a circuit to reduce current flow.

Retainer plates: Plates used to hold gears and bearings in position. They are attached to a stationary housing with bolts, machine screws, or snap rings.

Reverse idler gear: Gear used to turn the output shaft in the same direction as the countershaft gear assembly.

Ring gear: Gear with internal teeth that surrounds and is meshed with the planet gears in a planetary gearset.

Ring seals: Seals used to seal stationary parts or hydraulic pistons that slide in their bores. The ring seal fits into a shallow groove cut into one or both parts. Ring seals can be round, half-round, or square in cross section.

Road test: Process of driving a vehicle to observe transmission or transaxle operation under actual road conditions.

Roller bearings: See *Ball bearings.*

Roller chain: Chain that employs rollers that rotate on drive pins. The rollers contact the sprocket teeth. The spaces between the rollers match the size of the teeth. The rollers provide a rolling contact, reducing friction between the chain and the sprocket.

Root cause: Actual cause of a problem.

Rotor pump: Pump consisting of inner and outer rotors. The inner rotor is usually driven by the torque converter drive lugs. It causes the outer rotor to turn. As the rotors move apart, fluid is drawn in at the inlet port. The fluid is carried to the outlet port and is discharged when the rotors move together.

RTV (room temperature vulcanizing) sealer: Silicone sealer that cures to a rubberlike consistency to form a gasket.

Running fit: Condition in which the oil clearance between a shaft and bearing is sufficient to enable parts to turn freely and receive proper lubrication.

S

Scan tool: Hand-held testing device used to obtain trouble codes and other information from the vehicles' on-board computer.

Schematic: A graphic representation of a hydraulic, pneumatic, or electrical system. Tracing the flow of power along the path shown in the schematic allows the technician to determine which components should be energized under what conditions.

Scoring: Scratches on gear teeth faces caused by improper machining or lack of lubrication. If the scoring is not corrected, it will lead to spalling or welding.

Screw extractors: Tools used to remove broken screws or other broken fasteners from parts.

Seal protector: A metal or plastic sleeve used to prevent seal damage when a seal is installed over a shaft with a sharp area, such as a keyway.

Seal rings: Metal or Teflon® rings that seal oil pressure directed to or through passageways in rotating shafts or drums (not to be confused with ring seals). Seal rings are designed to allow slight leakage for lubrication of moving parts.

Sealants: See *Sealers.*

Sealers: Compounds that can improve the sealing ability of a gasket and help hold a gasket in place during component reassembly. In some cases, sealers can be used in place of gaskets.

Seals: Used to keep fluid from leaking out of the transmission or transaxle and to keep pressure from leaking internally. Seals are used at moving parts, such as the torque converter, drive shaft, and various internal rotating parts. They are also used as sliding pressure seals at the band servos and clutch apply pistons.

Selector lever: Lever mounted on the floor or on the steering column. Movement of the selector lever causes the manual valve inside the transmission to move.

Self-diagnostics: The ability of a computer to continuously monitor the operation of a specific system and send warning signals (dashboard light and trouble codes) when an abnormal condition is detected.

Self-powered test light: Testing device used to check for circuit continuity. This device resembles the nonpowered test light, but it has an internal battery.

Semiconductors: Electronic components made of silicon or germanium. Small amounts of impurities are added to these materials to cause them to act as either conductors or insulators, depending on how voltage is directed into them.

Semisynthetic transmission fluid: Transmission fluids that are blends of synthetic fluid and conventionally refined fluids.

Series circuit: Circuit with two or more resistance units wired in such a way that current must pass through one unit before reaching the other. This is the simplest type of automotive circuit.

Series-parallel circuit: Circuit that has some components that are wired in series and others that are wired in parallel. All the current flows through some parts of the circuit, while the current path is split in other parts.

Servo: Hydraulic device that consists of a piston and cylinder and is used to apply and release bands.

Servo-cover depressor: Tool used to push the servo cover in when removing the snap ring.

Shellac: A gummy and sticky substance that remains pliable. It is frequently used on fibrous gaskets, both for sealing and for holding the gaskets in place during assembly.

Shift kits: Kits containing upgraded springs, valves, seals, and other parts, most of which are installed in the valve body. These parts modify internal pressures and fluid flow to improve shift feel and transmission/transaxle durability.

Shift linkage: Mechanical linkage that operates the manual valve. This forms a direct mechanical connection between the driver and the transmission. The shift linkage also operates the parking gear mechanism.

Shift points: Speeds at which a particular transmission or transaxle should shift.

Shift problems: Problems that result in improper gear changes. Examples include late or early upshifts, no upshifts or no downshifts, harsh or soft shifts, and no passing gear.

Shift valves: Valves located in the valve body that control transmission or transaxle upshifts and downshifts.

Shift-brake interlock: Device designed to keep the driver from moving the gear selector out of Park without pressing on the brake pedal. It keeps the vehicle from being placed in gear accidentally or at high engine speeds.

Short circuit: Occurs when a wire's insulation fails or is removed and the bare wire contacts the frame, body, or other grounded part of the vehicle.

Silent chain: Chain made up of a series of flat metal links that provide quiet operation. These links have teeth that engage matching teeth in the sprockets.

Simpson gear train: Compound planetary gearset consisting of two separate ring gears, two separate planet carrier sets, and one common sun gear.

Size gauges: Linear measurement devices used for measuring part dimensions and clearances.

Sliding-vane pump: Pump using two or more spring-loaded vanes operating in slots in a rotor body. As the rotor spins, the vane ends rotate and rub against pump walls.

Slippage: Condition in which the transmission or transaxle does not transmit all the engine power to the drive wheels.

Snap ring: A split ring used to hold parts on shafts or inside bores. The snap ring fits tightly in or around a groove that is machined to accept it.

Snap ring pliers: Special service tools used to remove and install snap rings, or retainer rings.

Snapshot: A scan tool feature that records vehicle operating conditions immediately before and after a malfunction occurs.

Soft parts: Parts that are replaced during an automatic transmission or transaxle rebuild. These parts include gaskets, lip seals, O-rings, seal rings, and friction discs.

Solenoid: Switching device that uses a magnetic field to perform a mechanical task, such as moving a transmission valve.

Solenoid pack: Assembly that contains all the transmission or transaxle solenoids.

Spalling: A condition in which sections of the gear teeth flake or chip off. Spalling can be caused by lack of lubricant or severe gear loads.

Spanner wrenches: Tools used to tighten large parts or to hold parts in place while other tools are used.

Specialty shops: Repair shops that concentrate on one area of service, such as automatic transmission and transaxle repair, and are fully equipped to handle all aspects of their particular specialty.

Spline: Longitudinal slot cut into a shaft or other part. A spline may be straight, running parallel to the axis of the shaft or part, or it may be a spiral, running at an angle. Splines can be internal or external.

Splined shaft: Shaft that has many equally spaced, parallel splines.

Spool valve: Hydraulic control valve that is so named because of the spool-like flow-controlling element. This type of valve is located in holes in the automatic transmission valve body.

Spring compressor: Tool used to compress coil springs for removal and installation.

Spur gears: Type of gear used to connect parallel shafts. The teeth on spur gears are always cut straight across, or axially. In other words, they run parallel to the axis of rotation.

Stall speed: The maximum engine speed that can be obtained when the converter turbine is stationary.

Stall test: A method of loading the engine through the transmission or transaxle to isolate a problem in the torque converter or transmission. A stall test can be used to check the holding ability of the clutch packs and bands, as well as the operation of the stator one-way clutch. The test is sometimes useful to diagnose slipping holding members, low hydraulic pressures, or problems with the torque converter stator.

Static seals: Stationary ring seals that fit into nonmoving components and are found in many parts of automatic transmissions and transaxles. They are intended to form a perfect seal.

Stator: Torque converter component that consists of a set of vanes curved in the opposite direction of the turbine vanes. This unit fits between the impeller and turbine and is smaller than either. It redirects fluid coming from the turbine to the impeller's direction of rotation, helping the torque converter multiply power. The stator is the component that makes a torque converter out of a simple fluid coupling.

Stator shaft: Solidly mounted shaft on the front of the transmission oil pump. It provides a stationary support for the torque converter stator and overrunning clutch.

Straight roller bearings: Bearings consisting of inner and outer steel races separated by hardened roller bearings.

Straightedges: Hardened steel bars used to check for warping of part surfaces.

Sump: Bottom or side oil pan.

Sun gear: Center gear in a planetary gearset.

Sun-gear shell: See *Drive shell.*

Switch: A device that uses electrical contacts to energize or de-energize a solenoid or motor, or to send an electrical signal to another electrical device.

Synthetic transmission fluid: Transmission fluids made by chemically rearranging petroleum molecules to produce a uniform molecular structure suited to modern transmission and transaxle operating conditions.

T

Tapered roller bearing: Bearing utilizing a series of tapered, hardened steel rollers operating between outer and inner hardened steel races. Tapered roller bearings are useful for heavy loads. They are used in rear axles, front drive axles, and transaxle final drives.

Telescoping gauge: Instrument used to measure relatively large holes. The gauge is placed in the hole, properly tightened, and locked. It is then withdrawn from the hole and measured with an outside micrometer.

Temperature sensors: Input sensors that measure the temperature of the transmission fluid.

Tensile strength: An indication of how much stretching force a material can withstand before it breaks.

Thermostatic element: Device used on transaxles with two oil pans to allow the side pan to act as a transmission fluid reservoir when the transaxle is warmed up. This is necessary because the lower oil pan cannot hold enough fluid for cold operation without being overfilled when hot.

Thread sealer: Compound used to seal threaded fasteners that extend into the fluid passages of a component.

Thread-locking compound: Type of anaerobic sealer used to prevent threaded fasteners from loosening.

3-2 shift timing valve: Control valve that causes all holding members to briefly disengage during detent (passing gear) downshifts. This allows the engine to speed up for more power in passing gear when the holding members re-engage.

Throttle linkage: Connects the engine's throttle plate to the transmission's throttle valve.

Throttle pressure: Line pressure modified by movement of the throttle valve. The throttle pressure is transmitted to the shift valve (and a few other valves).

Throttle pressure booster: Valve that works with the 2-3 accumulator valve to precisely time and cushion the shift during forced downshifts.

Throttle valve: Control valve that senses how hard the engine is working and delays upshifting accordingly.

Throttle valve (TV) linkage: A mechanical connection between the throttle valve in the transmission and the throttle lever on the vehicle's throttle body or carburetor.

Thrust bearings: Flat, disk-like bearings that are made up of needle rollers. The rollers are arranged radially.

Thrust load: See *Axial load.*

Thrust washers: Washers used between the end of a rotating shaft and a stationary housing or between two rotating shafts to prevent excessive axial shaft movement by limiting the amount of end clearance.

Tightening sequence: Tightening pattern for threaded fasteners that ensures that parts are tightened evenly. Tightening is usually done gradually and follows a crisscross pattern, starting in the middle of the part and working outward.

Titania oxygen sensor: Oxygen sensor that varies its resistance in response to changes in the oxygen content of exhaust gases.

Tooth contact pattern: Refers to shape, size, and location of actual contact area between two mating gear teeth.

Torque converter: A hydrodynamic device that uses transmission fluid to transmit and multiply power. The torque converter is mounted directly behind the engine and is turned by the engine crankshaft.

Torque multiplication: The ability of a component, such as a torque converter or gearset, to increase output torque above input torque.

Torque specifications: Tightening values for threaded fasteners.

Torque wrench: Tool used to apply a measured turning force, or torque, to a threaded fastener.

Torrington bearing: See *Thrust bearings.*

Transducers: A coil of wire wrapped around a moveable iron core.

Transfer gears. Transaxle gear drive consisting of two large helical gears that are in constant mesh.

Transfer shaft: Shaft that transmits power from the transaxle output shaft to the differential assembly.

Transistor: A semiconductor device that can be used as a switch or an amplifier.

Transmission alignment tools: Devices used to align some internal transmission parts during assembly.

Transmission fluid: A combination of petroleum oils and various additives pressurized by the pump and used to operate the hydraulic system.

Transmission fluid filter: Filter that catches and removes impurities as the fluid passes through it.

Transmission oil cooler: See *Main oil cooler.*

Transmission oil pump: Pump that produces the fluid pressure to operate other hydraulic components and lubricate the moving parts of the transmission or transaxle. It is driven by lugs on the torque converter.

Trouble code: Diagnostic output consisting of a numeric or alphanumeric designator that identifies the general location of a problem or defective component.

Troubleshooting charts: Charts that contain the logical steps used to determine the cause of a problem.

Tubing cutter: Tool consisting of a cutting wheel and opposing rollers that guide the tool and provide a surface against which to cut tubing.

Tubing flaring tool: Tool used to flare tubing, which means to make an enlarged lip at the end of the tubing. Flaring tools consist of a ram and a block that contains different-sized holes to match various tubing sizes.

Turbine: Torque converter component consisting of a set of curved vanes welded to a turbine shell. The turbine is driven by the impeller through fluid contained within the converter. The turbine is the torque converter driven member.

Turbine shaft: See *Input shaft*.

Turbine speed sensor: See *Input speed sensor*.

TV limit valve: Valve that limits the amount of pressure that can reach the throttle valve. This prevents the development of excess throttle valve pressures, which could cause excessive line pressures.

2-3 backout valve: Control valve that prevents a rough upshift if the driver lifts up on the accelerator just as the shift occurs.

V

Vacuum: Pressure in an enclosed area that is less than atmospheric pressure.

Vacuum gauge: A pressure gauge used to measure vacuum, which is pressure below normal atmospheric pressure.

Vacuum modulator: Device that moves the throttle valve with changes in engine load, causing throttle pressure to vary.

Valve body: A complex aluminum or iron casting that serves as the control center for the transmission's hydraulic system. It contains many internal passageways and components, including hydraulic valves.

Valve bores: Holes drilled into a valve body to accept the spool valves. These holes are a very close fit with the valves.

Vane pump: See *Sliding vane pump*.

Variable ratio control valve: A valve that controls the CVT drive pulley by varying pressure to the drive pulley piston.

Varnish formation: Problem that results from fluid oxidation and appears as a sticky, burned-on substance that resembles furniture-refinishing varnish.

V-block: A block of steel with a deep "V" groove cut in one or more spots. Can be used in pairs to support a shaft while it is turned to check for runout.

Vehicle identification number (VIN): Individual series of letters and numbers assigned to a vehicle at the factory by the manufacturer.

Vehicle speed sensor: Sensor that is mounted on the output shaft of the transmission or transaxle and tells the ECM how fast the vehicle is traveling.

Vernier caliper: Very accurate sliding caliper used for measuring inside and outside diameters.

Volt: Unit of measurement of electrical pressure that will move a current of one ampere through a resistance of one ohm.

Voltage: The electrical pressure created by the difference in the number of electrons between two terminals. It provides the push that makes electrons flow.

Voltmeter: Instrument used to measure voltage in a given circuit.

W

Wear: Erosion of a surface caused by relative motion between two parts.

Weirs: See *Oil diverters*.

Welding: Occurs when mating gear teeth overheat, melt together, and then pull apart as the gear turns. This will quickly destroy the gear.

Wheel hop: Condition in which one wheel breaks loose from the pavement when the vehicle makes a turn.

Wilson gear train: Compound planetary gearset used on five-speed, rear-wheel drive transmissions installed in some late-model European vehicles. The Wilson gear train is a combination of three simple gearsets that are connected and located at the rear of the transmission case.

Wire gauge: Rating system for wire diameter. The larger the gauge number, the smaller the wire.

Wiring diagram: A drawing showing electrical devices and the wires connecting them. It also shows wire colors and terminal types. Use of the wiring diagram allows the technician to trace defective components in the wiring system (or schematic).

Z

Zirconia oxygen sensor: Oxygen sensor that produces a varying voltage signal with changes in the oxygen content of the exhaust gases.

Index

D

E

F

G

T

Automatic Transmissions and Transaxles

Automatic Transmissions and Transaxles covers the design, construction, operation, diagnosis, service, and repair of automatic transmissions and transaxles. This comprehensive text details both hydraulic and electronic transmission controls, with extensive coverage of electronic control systems. *Automatic Transmissions and Transaxles* is a valuable resource for anyone who needs a thorough understanding of today's automatic transmissions and transaxles, as well as those preparing for ASE Certification Test A2, Automatic Transmission/Transaxle.

Features

- Correlated to Section A2 of the ASE Task List, Automatic Transmission/Transaxle.
- Color-coded illustrations are used throughout this text to enhance understanding.
- Includes a detailed chapter on troubleshooting electronic control systems, as well as a chapter on electronic control system service.
- Emphasizes the use of the latest diagnostic equipment to troubleshoot electronic transmission and transaxle control systems.
- Includes coverage of OBD II as it relates to automatic transmissions and transaxles.
- Contains information on continuously variable transmissions and manual shift systems.

Other Automotive Titles from Goodheart-Willcox

- **Auto Brakes.** Details the theory, operation, diagnosis, and service of modern brake systems.
- **Auto Electricity and Electronics.** Covers the operation, diagnosis, and service of electrical, electronic, and computer-controlled systems found on today's cars and light trucks.
- **Auto Engine Performance and Driveability.** Teaches the skills necessary to diagnose and repair engine performance and driveability problems.
- **Auto Engine Repair.** Covers the construction, operation, diagnosis, and repair of gasoline and diesel engines.
- **Auto Heating and Air Conditioning.** Details the operation, diagnosis, and service of automotive heating, ventilation, and air conditioning systems.
- **Auto Suspension and Steering.** Teaches the skills needed to successfully diagnose, service, and repair all types of automotive suspension and steering systems.
- **Manual Drive Trains and Axles.** Contains the latest information on the operation, diagnosis, service, and repair of manual transmissions and transaxles, as well as clutches, drive axles, drive shafts, half shafts, CV joints, universal joints, and four-wheel drive/all-wheel drive components.

GOODHEART-WILLCOX ENCOURAGES

ASE CERTIFIED®

PROFESSIONALISM THROUGH TECHNICIAN CERTIFICATION

ISBN 1-59070-426-6

9 781590 704264